WILLIAM F. MAAG LIBRARY
YOUNGSTOWN STATE UNIVERSITY

Seafood Toxins

ACS SYMPOSIUM SERIES **262**

Seafood Toxins

Edward P. Ragelis, EDITOR
Food and Drug Administration

Based on a symposium sponsored by
the Division of Agricultural and Food Chemistry
at the 186th Meeting
of the American Chemical Society,
Washington, D.C.,
August 28–September 2, 1983

American Chemical Society, Washington, D.C. 1984

Library of Congress Cataloging in Publication Data

Seafood toxins.
(ACS symposium series, ISSN 0097-6156; 262)

Bibliography: p.
Includes indexes.

1. Seafood—Toxicology—Congresses. 2. Marine toxins—Congresses. 3. American Chemical Society. Division of Agricultural and Food Chemistry.

I. Ragelis, Edward P., 1932- . II. American Chemical Society. Division of Agricultural and Food Chemistry. III. Series.

RA1242.S48S43 1984 615.9'45 84-18551
ISBN 0-8412-0863-8

Copyright © 1984

American Chemical Society

All Rights Reserved. The appearance of the code at the bottom of the first page of each chapter in this volume indicates the copyright owner's consent that reprographic copies of the chapter may be made for personal or internal use or for the personal or internal use of specific clients. This consent is given on the condition, however, that the copier pay the stated per copy fee through the Copyright Clearance Center, Inc., 21 Congress Street, Salem, MA 01970, for copying beyond that permitted by Sections 107 or 108 of the U.S. Copyright Law. This consent does not extend to copying or transmission by any means—graphic or electronic—for any other purpose, such as for general distribution, for advertising or promotional purposes, for creating a new collective work, for resale, or for information storage and retrieval systems. The copying fee for each chapter is indicated in the code at the bottom of the first page of the chapter.

The citation of trade names and/or names of manufacturers in this publication is not to be construed as an endorsement or as approval by ACS of the commercial products or services referenced herein; nor should the mere reference herein to any drawing, specification, chemical process, or other data be regarded as a license or as a conveyance of any right or permission, to the holder, reader, or any other person or corporation, to manufacture, reproduce, use, or sell any patented invention or copyrighted work that may in any way be related thereto. Registered names, trademarks, etc., used in this publication, even without specific indication thereof, are not to be considered unprotected by law.

PRINTED IN THE UNITED STATES OF AMERICA

RA
1242
.S48S43
1984

ACS Symposium Series

M. Joan Comstock, *Series Editor*

Advisory Board

Robert Baker
U.S. Geological Survey

Martin L. Gorbaty
Exxon Research and Engineering Co.

Herbert D. Kaesz
University of California—Los Angeles

Rudolph J. Marcus
Office of Naval Research

Marvin Margoshes
Technicon Instruments Corporation

Donald E. Moreland
USDA, Agricultural Research Service

W. H. Norton
J. T. Baker Chemical Company

Robert Ory
USDA, Southern Regional
Research Center

Geoffrey D. Parfitt
Carnegie-Mellon University

Theodore Provder
Glidden Coatings and Resins

James C. Randall
Phillips Petroleum Company

Charles N. Satterfield
Massachusetts Institute of Technology

Dennis Schuetzle
Ford Motor Company
Research Laboratory

Davis L. Temple, Jr.
Mead Johnson

Charles S. Tuesday
General Motors Research Laboratory

C. Grant Willson
IBM Research Department

WILLIAM F. MAAG LIBRARY
YOUNGSTOWN STATE UNIVERSITY

FOREWORD

The ACS SYMPOSIUM SERIES was founded in 1974 to provide a medium for publishing symposia quickly in book form. The format of the Series parallels that of the continuing ADVANCES IN CHEMISTRY SERIES except that in order to save time the papers are not typeset but are reproduced as they are submitted by the authors in camera-ready form. Papers are reviewed under the supervision of the Editors with the assistance of the Series Advisory Board and are selected to maintain the integrity of the symposia; however, verbatim reproductions of previously published papers are not accepted. Both reviews and reports of research are acceptable since symposia may embrace both types of presentation.

CONTENTS

PREFACE

POISONINGS CAUSED BY THE INGESTION of toxin-containing seafood have been a part of mankind's existence and concern for centuries. While we have slowly progressed toward a better understanding of the nature and cause of seafood toxins, they continue to present serious economic and public health problems.

In recent years, strong evidence of an increase in the magnitude, duration, and geographical distribution of food-borne illnesses caused by seafood toxins has been presented. The effect can be seen in the form of impediments to the development, growth, and stability of the local and commercial seafood industries in various areas of the world—most noticeably throughout the tropics where an abundant, accessible seafood resource resides.

Among the contributing factors is the steady rise, over the past two decades, in the demand for seafood by a rapidly increasing world population. Statistics from the U.S. Department of Commerce show that the world's consumption of fish and shellfish went from 68.7 million tons in 1977 to 74.8 million in 1981, and, from all projections, is expected to increase considerably in the future. The enormous improvements in the technology used in the harvesting, transporting, processing, and marketing of seafood and seafood products are another contributing factor. Furthermore, the growing, harvesting, and marketing of certain algae as food protein supplements have dramatically accelerated, especially in developing countries. Consequently, control measures are needed to ensure that non-toxic strains of algae are collected and processed.

This volume examines the ways in which the safe use of fish and shellfish can be hindered by toxic fresh- and saltwater organisms, and by spoilage. We have attempted to present the entire range of this topic from the impact of these toxins on seafood use to their chemistry, pharmacological actions, responsible organisms, geographical distribution, and the methods used for their detection and determination. The seafood toxins covered include paralytic shellfish poison (PSP), ciguatera toxins, scombroid-related toxins, tetrodotoxin (pufferfish), *Ptychodiscus brevis* toxins, and diarrhetic shellfish poison. We have also included the neurotoxins, endotoxins, and peptidic toxins associated with blooms of certain strains of blue-green algae (cyanobacteria). All of these organisms can affect water environment and supplies and have caused contact irritations and poisonings in animals and humans.

It is fitting that this volume commemorates the 75th anniversary of the Agricultural and Food Division of the American Chemical Society, dedicated to the knowledge and use of world food resources. I hope it will serve to summon immediate attention and bring rapid solutions to the problems posed by seafood toxins. To all those who contributed I express my sincere gratitude.

EDWARD P. RAGELIS
Food and Drug Administration
Washington, D.C.

July 3, 1984

OVERVIEW

1

Development of Marine Resources

THOMAS J. BILLY and BETTY M. HACKLEY

National Oceanic and Atmospheric Administration, National Marine Fisheries Service, Washington, DC, 20235

The Fisheries Conservation and Management Act (FCMA), as amended has afforded an unprecedented opportunity for U.S. commercial and recreational fishermen and processors to take advantage of the vast fishery resources off our coasts. This in turn has resulted in an Administration policy to assist in the development and full utilization of U.S. fishery resources. Many opportunities exist to further develop our fisheries. As consumers become more diet conscious, there should be increased demand for fishery products. Alternative technologies are opening up new resources and the increasing acceptance of fabricated products will provide uses for less utilized species. There are also growing political pressures to use our resources. The U.S. fisheries however, are faced with a number of problems which impede development of the domestic industry and its ability to penetrate foreign markets. Included are the awareness, detection and control of toxins and contaminants which affect product safety and quality of the fish we consume. In the area of marine toxins, past efforts by the National Marine Fisheries Service have included cooperative research with FDA as well as several S-K projects on ciguatera and PSP. Only through continued research can we hope to remove these impediments to the full utilization of U.S. fishery resources.

This is an overview of the U.S. fishing industry, its anticipated growth and development, the role of the National Marine Fisheries Service (NMFS) and its concerns regarding the safety of fishery products for human consumption.

Marine fish and shellfish constitute a valuable renewable resource that, with proper management, can contribute significantly to the nation's economic growth. The nation's marine resources support both recreational and commercial fishing and are managed to reflect both.

This chapter not subject to U.S. copyright.
Published 1984, American Chemical Society

The Magnuson Fishery Conservation and Management Act (MFCMA) of 1976, as amended, is the most significant fisheries legislation in the nation's 200-year history. It provides for the conservation and exclusive management by the U.S. of all fishery resource within the Fishery Conservation Zone or FCZ which extends 200 nautical miles from our shores. The land area equivalent to the U.S. FCZ is approximately 3.6 million square miles. It has afforded an unprecedented opportunity for U.S. commercial and recreational fishermen to reach new levels of production and to expand further their already significant contributions to the nation's food supply and economy.

This Act forms the keystone of our national program for marine fisheries. One of its goals is to further develop and maintain healthy commercial and recreational fishing industries. It should be noted that public health is one of the legitimate concerns to be addressed in the management of our fisheries. In the implementation of a management measure in a fishery management plan, the MFCMA authorizes the DOC Secretary to promote public health and safety, so long as the measure applies to fishing activities and is a recognized fishery management technique. The measure also must be consistent with regulations issued under other statutes.

Primary Federal responsibility for fisheries management and development in coastal waters rests with National Oceanic and Atmospheric Administration (NOAA) in the Department of Commerce. Within NOAA, Federal fisheries programs are administered by NMFS. It should be noted here that Sea Grant is also a part of NOAA and sponsors a significant amount of fisheries research.

The NMFS role in fisheries development is catalytic in nature by encouraging and assisting in the achievement of full utilization of U.S. fishery resources to derive maximum economic and social benefits from those resources under U.S. jurisdiction. In practical terms, this will require greater initiative on the part of the commercial and recreational fishing industries to utilize America's non-traditional fish species, while the Federal Government will work to reduce foreign fishing in our fishery conservation zone and remove or reduce other impediments.

The 1982 U.S. commercial harvest of marine fish and shellfish was 2.9 million metric tons. To illustrate one result of reducing foreign fishing effort, since the enactment of the MFCMA, foreign fishing vessels in the mid-1970's harvested approximately 2.3 million metric tons per year off U.S. coasts compared to 1.6 million metric tons today. On the other hand, domestic commercial landings in 1981 were about 20 percent higher than in the years immediately preceding the MFCMA, with nearly all regions of the country sharing in the increases. The estimated U.S. catch for 1990 is 3.6 million metric tons which reflects further reduced foreign fishing as well as the development of many underutilized species.

As stability and prosperity in traditional fisheries provide the basis for private sector investment in new, underutilized fisheries, fisheries management and development activities will be coordinated to ensure that increased investment in harvesting and processing operations do not result in excess capacity and recurring marketing crises in traditional fisheries. Traditional fisheries are those for which the existing fishing industry has the capability to fully harvest, process, and market the allowable catch available and historically has done so. There is, however, enormous potential in underutilized species.

In supporting the increased use of underutilized fisheries, the NMFS fisheries development program over the next few years will concentrate on the development of those fisheries with greatest potential for growth. Many fisheries can be developed. However, emphasis on the development of specific underutilized fisheries will allow NMFS to use its resources more effectively.

The current fisheries targeted for commercial development are Alaska groundfish, coastal pelagics in the South Atlantic and Gulf of Mexico, and East coast squid. These new fisheries have been identified as providing significant benefits to the nation if developed and as having a high potential for success in the short term. Focusing on these fisheries would be consistent with other national objectives of fully utilizing fishery resources within the FCZ, and reducing foreign fishing activity. Many other species are of interest to industry, including some associated with marine toxins. Fisheries development is a key component of the planned program resulting from the recently passed Caribbean Basin Economic Recovery Act which provides for the economic revitilization and facilitation of economic opportunities for certain countries in this area. The presence of ciguatoxin in certain reef fishes from this area could present a problem. Another species that could be developed in the future is the large clam resource recently identified in the Bering Sea. However, successful development of this resource will require further definition of the PSP problem and a method for selecting harvest areas where the potential for PSP is minimal.

In addition to stressing the development of targeted commercial fisheries, the Agency's fisheries assistance programs are making a more concerted effort to support the expansion of Marine Recreational Fishing (MRF). The Agency has funded an in-depth study to assess the significance of the economic activities associated with MRF and recommend specific growth prospects. As with commercial fisheries, the development strategy will focus on the opportunities showing the best prospects for growth and will stress the Federal Government's catalytic role. Emphasis will be placed on close coordination with the coastal states, the several sports fishing constituencies, and the industries which support MRF.

The NMFS assists industry in identifying and analysing the opportunities for fisheries development and provides services to strengthening existing fisheries. Efforts are made to reduce impediments such as lack of fishing and processing technology, a lack of suitable products of acceptable quality and safety, including problems with seafood toxins, and consumer education.

The NMFS assists in transferring available information about lesser used species to fishermen, sellers, and consumers. The NMFS also provides services to the domestic fishing industry. These services include:

- product standards development,
- product inspection;
- product quality and safety information, and;
- market development and assistance.

The NMFS development programs are financed both from funds appropriated by the Congress and from tariffs on imported fisheries products better known as Saltonstall-Kennedy funds.

Research is an essential element to the successful establishmemt and maintenance of markets for U.S.-produced seafoods both domestically and abroad. Several kinds of research need to be carried out in order to achieve continued successful product utilization.

These include:

- biological assessment of availability;
- basic food technology research;
- product quality and safety research;
- economic research; and
- market research.

The results of these kinds of research collectively are needed in order to place acceptable seafood products in target markets at prices that constitute a favorable economic return to the U.S. industry and are competitive with similar seafoods from non-U.S. sources.

Failure to provide timely answers through substantive research in any of the aforementioned areas will result in unsuccessful attempts to develop and commercialize a particular harvestable resource, and possibly, unnecessary human illness.

Recognizing the importance of research to successful development of any fishery and the limited monies available for research purposes, we are directing both our inhouse and sponsored research toward:

1) achieving a balanced program mix on the limited number of targeted species previously mentioned;

2) maintaining an acceptable level of effort on industry problems which continue to arise in connection with the use of traditional species; and

3) maintaining a "longer range" food research effort which would be applicable to species as yet unscheduled for exploitation for food use.

Current NMFS seafood safety research, both inhouse and through contract studies addresses concerns such as heavy metals, pesticides, petrochemicals, other industrial chemicals, marine biotoxins, and pathogenic organisms. The NMFS is also conducting seafood quality research and developing Federal grade standards and specifications for seafoods.

In our program to provide consumers with safe, high quality products and respond to critical national and international needs to eliminate major impediments to expanding markets, a primary seafood utilization issue is the area of marine toxins that continue to present serious economic problems to the fishing industry and impact consumers. Ciguatera is one of the most treacherous and common forms of seafood poisoning in tropical waters. Also, sporadic outbreaks of PSP in shellfish from most of the major growing areas of North America pose a threat to public health. Both can result in illness to the consumer and economic disaster to the fisheries. In September 1979, NMFS co-sponsored a workshop with FDA to review the problem of toxins in fish and fishery products, the current state of knowledge, and plan further research. This resulted in an exchange of information which has contributed to the progress that has been made since that time. Over the past four years NMFS has supported marine toxin research both inhouse and through cooperative studies with FDA and the S-K funding process. From 1980 to 1983, a total of $27.3 million dollars has been provided to fund fishery development projects. Of this amount, $1.2 million dollars have been used for research on marine toxins.

Clearly, the occurrence of marine toxins is a national problem impacting on commercial and recreational fisheries and involving both traditional species and those scheduled for development. The NMFS looks to the research leaders in this area to help in overcoming these problems so that the species of concern can be dealt with effectively. Although the total funding for research has been reduced the past two years, NMFS will continue to support, to the extent possible, research efforts in this area.

RECEIVED April 28, 1984

WILLIAM F. MAAG LIBRARY
YOUNGSTOWN STATE UNIVERSITY

2

Paralytic Shellfish Poisoning

An Emerging Perspective

CLARICE M. YENTSCH

Bigelow Laboratory for Ocean Sciences, W. Boothbay Harbor, ME 04575

Dinoflagellates are rapidly gaining a reputation for being the "nasties" in the marine environment. The slate now includes organisms causing Paralytic Shellfish Poisoning (PSP), Neurotoxic Poisoning (NSP), Ciguatera Poisoning, and more recently described Diarrhetic Shellfish Poisoning (DSP).

Problems with toxic dinoflagellates are indeed global (1). I will confine my comments to PSP, the well-documented human affliction resulting from the consumption of shellfish containing dinoflagellate-derived toxins, namely saxitoxin and related compounds. Biological oceanographers have traditionally wrestled with sampling strategies. Inappropriate sampling has hindered our understanding of organisms with patchy distribution (e.g. dinoflagellates). High technology developments with regard to flow cytometry/sorting offer promise to increase our understanding of the cell biology, and remote sensing offers promise to increase our understanding of the oceanography (Figure 1). Despite these advancements, immediate gains in an appreciation of the biogeographical distribution will only be possible once a sound simple global monitoring program is effected.

In an attempt to achieve a conceptual balance, let us consider the cell biology, the biogeographical distribution, and the oceanography. They are all interrelated and thus an increase of knowledge in one area strengthens our interpretation of another area.

Cell Biology

Surprisingly, most of the definitive cell biology on toxic dinoflagellates has yet to be observed, understood and described. What was once thought to be the major causative genus, *Gonyaulax*, is now known to contain a wide variety of similar strains/varieties/species (2).

Although the taxonomy of these organisms is still in a state of uncertainty, definite biochemical differences are obvious -- primarily toxin content (3). While initially saxitoxin was considered a single toxin causing shellfish poisoning, it is now

0097-6156/84/0262-0009$06.00/0
© 1984 American Chemical Society

clear that there are suites of toxins, that these suites vary from one geographic region/organism to another, and that these toxin suites vary with life stage of the producing dinoflagellate and are altered in molecular structure within the shellfish which acts as a vector of the toxins (4). In short, the situation once interpreted as a very simple one, is highly complex. Toxigenesis, biochemistry, physiology and cell cycle analysis are basically still unknowns.

Flow cytometers/sorters offer promise as a research tool here. Flow cytometry combines revolutionary laser, electronics and optics technology (Figure 2). One is able to analyze cells at very rapid rates (greater than 2000 cells per second) on a cell-by-cell basis. Feasible measurements include: autofluorescence of the chlorophyll pigments and accessory pigments, as well as induced fluorescence, such as adding hydrogen peroxide, which results in toxin fluorescence (5); and uptake of various dyes specific for DNA (6-7) (therefore direct growth rate estimates are possible), RNA and protein (therefore nutritional concentration estimates are possible). Lipid stains, pH stains, membrane potential stains and immunofluorescent antibody preparations are reagents which provide additional information. Another part of this instrumentation is a charging collar and deflection plates. Basically the cells, in single file, are surrounded by a sheath of saline which can be charged if the cells going through the charging collar are of interest. Charged cells are then sorted into different containers by the deflection plates. Cells are viable after this process, and can be used for establishing cultures or subsequent EM investigation and biochemical experimentation.

A study of interest to the PSP problem is photoadaptation. We know that phytoplankton organisms grow in what is often termed an unbalanced environment. That is, light and nutrients, both basic requirements, do not occur in the same location. Holligan et al. (8) show potential new growth graphically (Figure 3). The cells of toxic dinoflagellates, as well as others, seem to "prefer" a low light environment, just at the interface of the low-light-high-nutrient zone. Thus, there are masses of dinoflagellates at subsurface maxima near frontal boundary layers (Figure 4). To continue to grow, cells must adapt to the low light environment. It is hypothesized that this is accomplished by an increase in pigmentation per cell. In the laboratory, increased pigmentation has been measured in bulk analyses. We are now able to measure such phenomena on an individual cell basis (Figure 5). In these laboratory experiments, we have grown the cells in various light intensities. Chlorophyll indeed does increase on a per cell basis. Upon microscopic observation it is clear that there is organelle photomorphogenesis, that is the chloroplasts *per se* actually change from the spaghetti-shaped to very tightly packed more ovoid-shaped entities (9; Figure 6). Whether or not this occurs in the natural environment has yet to be demonstrated.

Biogeography

On the other side of the balance is the biogeography. Distribution patterns were once thought to be simple, but are now

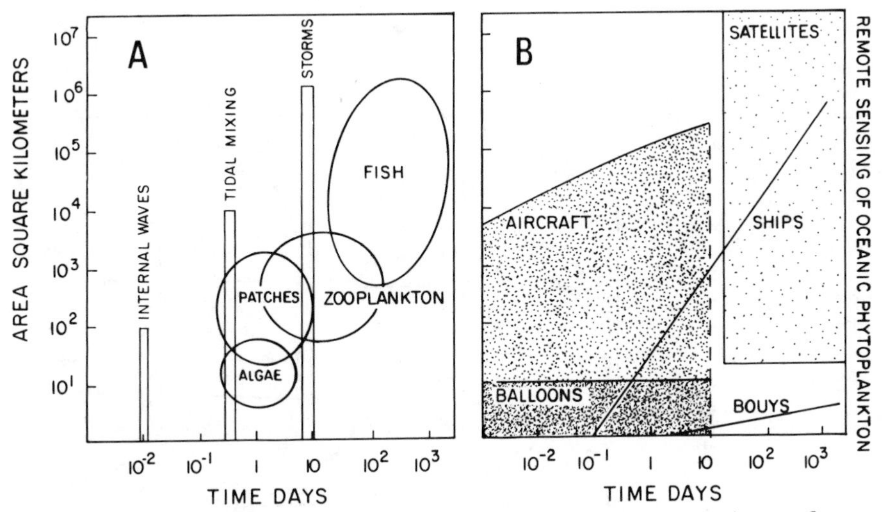

Figure 1. Time and space scales of algae, patches of algae, zooplankton and fish, A, and various means of remote sensing of biological properties in the oceans, B. Note that ships and buoys give no synopticity. Adapted from Esaias (36).

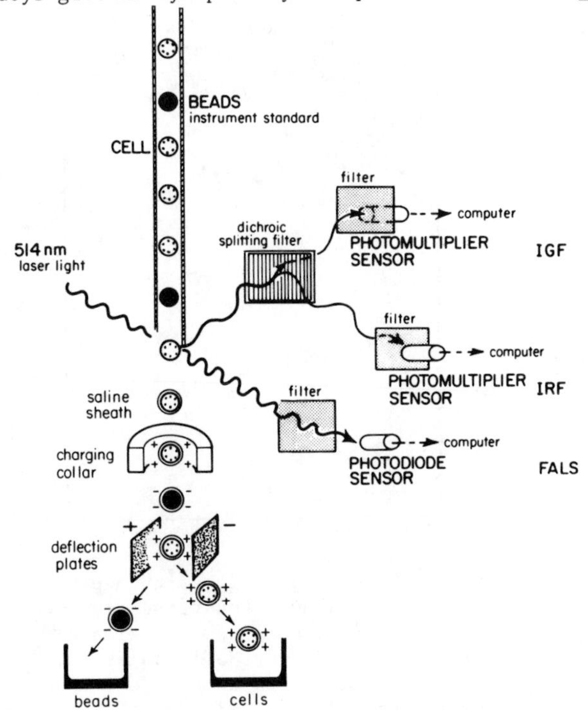

Figure 2. Representation of measurement of multiple parameters of single cells via flow cytometer fluorescence analysis and sorting. IGF= Integrated green fluorescence. IRF= Integrated red fluorescence. FALS= Forward angle light scatter.

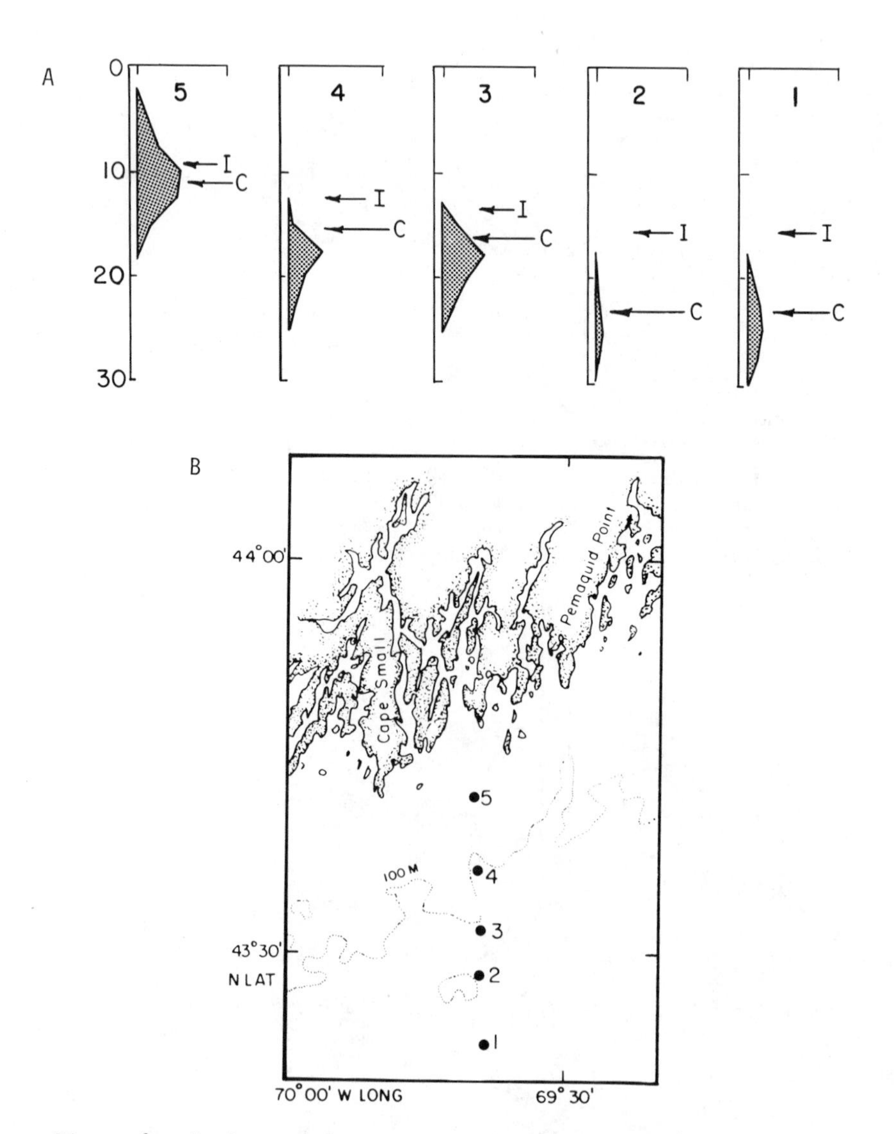

Figure 3. A. Potential new production with depth (meters) at stations off the coast of Maine. Locations are shown in B. Station 5 is nearshore. Station 1 is approximately 25 miles offshore. Summer (July) 1979. After Holligan et al. (submitted) (8). Gonyaulax tamarensis var. excavata comprised a major percent of phytoplankton biomass from 0-60 m. Station 5 = 0%; station 4 = 30%; station 3 = 46%; station 2 = 4%; station 1 = 0%.

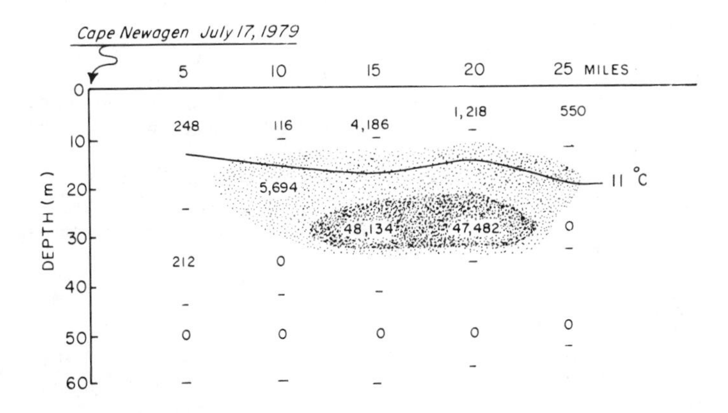

Figure 4. Subsurface dinoflagellate maxima, stations 5, 4, 3, 2, and 1. Data courtesy of P. M. Holligan.

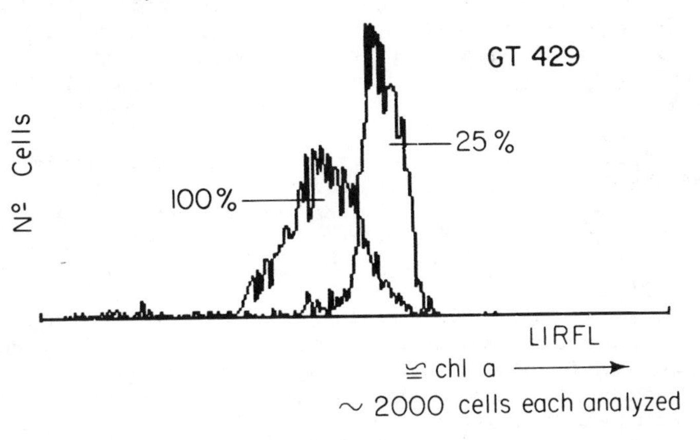

Figure 5. Pigment per cell of Gonyaulax tamerensis var excavata grown under 100% (266 μEin·cm^{-2}·s^{-1}) and 25% (66 μEin·cm^{-2}·s^{-1}) light regimes. X-axis is fluorescence intensity per cell and y-axis is number of events; 2,000 cells were analyzed in each case using the flow cytometer.

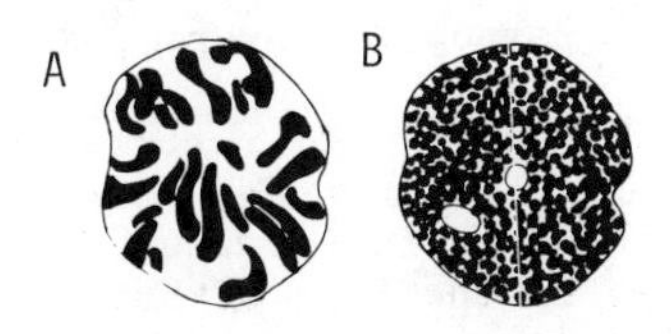

Figure 6. Diagram of chloroplast size and shape observed using epifluorescence microscopy. Gonyaulax tamarensis var. excavata-A. grown at 266 μEin·cm^{-2}·s^{-1} and B. at 50% or 133 μEin·cm^{-2}·s^{-1}.

recognized as complex. While biogeographic shifts are common in all species of plants and animals, and are particularly well documented in dinoflagellates due to the 600 million year fossil record of the cyst and thecal forms, this phasing in and out has been the cause for alarm when a toxic species is involved. It is now well recognized that there are biogeographical shifts in distribution of toxic species. There is the claim that toxic dinoflagellates are increasing in intensity; spreading in distribution, and extending in duration (10). Certainly, in some spot locations this is true. Yet, such observations may be strictly reduced to natural fluctuations.

Oceanography

This part of the situation is becoming simplified. Fortunately, there are many known unifying similarities as far as physiology and biochemistry are concerned, thus dinoflagellates responsible for causing shellfish toxicity and subsequently PSP tend to occupy similar environmental niches. The oceanography, here defined as the water mass characteristics and movement, is following a unifying theme. For each area of the globe that is being seriously studied, some definite patterns emerge and these patterns of dinoflagellate growth and/or accumulation, are similar from one region to another. Ecological parallels are now being recognized (11-21).

Data now indicate that prior to a bloom of toxic dinoflagellates and subsequent shellfish intoxification there are a few prerequisites. Namely: 1) bay, estuary and/or shelf sea where there is 2) seasonal stratification, 3) frontal boundary layers (22), and/or 4) tremendous energy resulting from major tidal dissipation (23).

Our present understanding indicates that accumulations of dinoflagellates develop and persist at water mass discontinuities. While this is an interplay between cell biology and oceanography, the study of frontal regions has lead to a far greater understanding of the subsurface dinoflagellate populations (24-25). While there is to date little documentation as to whether these are regions of dinoflagellate cell growth, or dinoflagellate cell accumulation, it is certain that these frontal areas have both a temperature and color signature and their movement can be detected remotely by small aircraft infra-red radiometry, satellite infrared radiometry, and Coastal Zone Color Scanner (CZCS) imagery resulting in expanded synoptic coverage in both time and space (26; Figure 7).

When a geographic pattern of occurrence has been identified (13) and annual (Figure 8) and year-to-year fluctuations analyzed (Figure 9), cause/effect relationships can be hypothesized and tested. For example, in the Gulf of Maine, red tides appear to arise in one of four ways which are listed in order of impact: 1) during the progressive thermocline development (May) and degradation of stratification (September and October); 2) any major storm event which disrupts the thermocline, even temporarily; 3) when mild but steady meteorological events dictate offshore movement of surface waters and onshore movement of the shelf-break fronts which are rich with subsurface accumulations of dinoflagellates persisting at near thermocline depths (10 to 25 meters); and

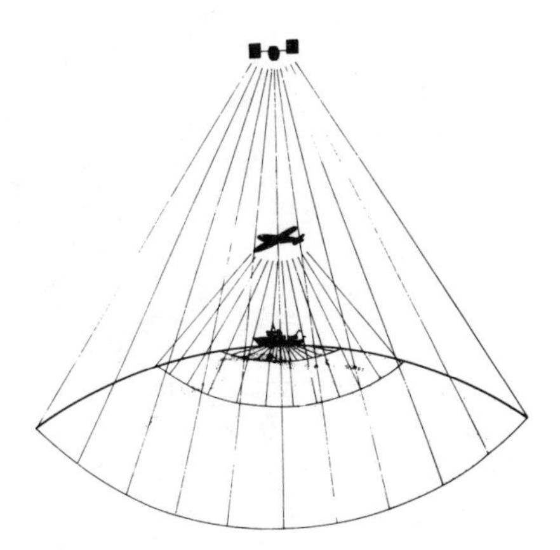

Figure 7. Time and space synopticity gained by remote sensing by aircraft and satellite.

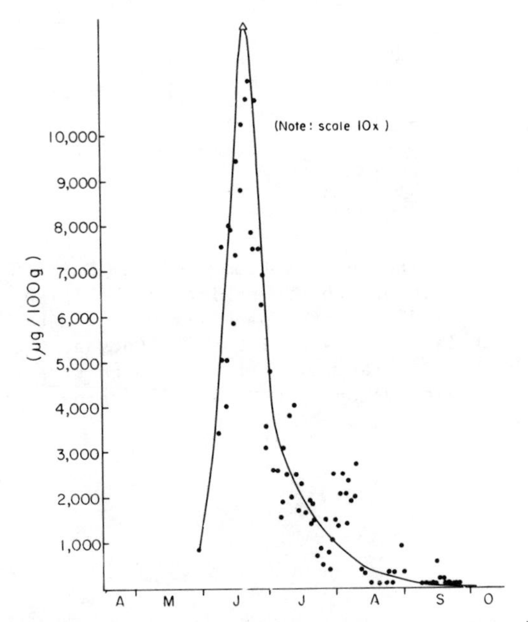

Figure 8. Toxin content in shellfish (Mytilis edulis) recorded daily from Monhegan Island off the coast of Maine in 1978. Toxin expressed as μg/100g in fact is μg saxitoxin-equivalents per 100g tissue of shellfish from April (A) to October (O). Depicts information gained from annual survey.

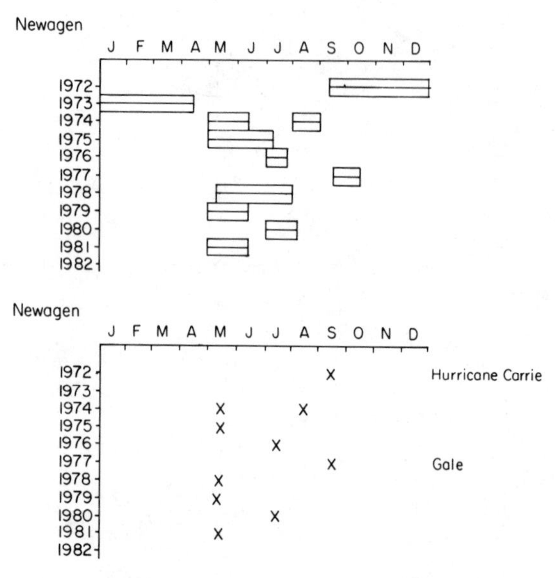

Figure 9. Shellfish toxin data from Newagen, Maine. Upper panel is times of State closure to the taking of shellfish (>80μg/100g) for years 1972 to 1982. Bottom panel is "start dates" indicating when initial levels over quarantine are witnessed. These dates fall in categories 1) transition from mixed to stratified water column (May) and vice versa (Sept.) and 2) during summer meteorological events.

4) creation of a microclimate in shallow ponds, embayments and estuaries, appropriate for encystment and/or dinoflagellate duplication. In our region, the latter is generally associated with the spring rise in toxicity in nearshore areas coincident with excystment of the dinoflagellates (J.W. Hurst, pers. comm.) while the first three are generally associated with summer and autumn toxicities, which can result in very dangerous toxin levels. Currently, shellfish toxin pattern data and/or species enumeration patterns are coalescing with physical oceanographic data hopefully leading to an environmental predictive model which may be applied to broad geographic regions throughout the Gulf of Maine.

Human Impact

The past decade has been characterized by the human concern of science for a better environment. Coincidently, there has been a surge of media presentations reporting or implying spreading of toxic red tides. It is no wonder that the public has been provoked into asking numerous relevant questions. The questions can be reduced to two. Although the questions appear related, they are not identical. Basically 1) Are red tides caused by pollution? and 2) Do human activities serve to aggravate red tide problems?

Perhaps we need reminding that the regions of the world's oceans most severely affected by toxic red tides have waters which are relatively pristine. In the U.S., Alaska and Maine coastal areas are severely affected by organisms leading to Paralytic Shellfish Poisoning, yet each state has populations of approximately one million, and there is little industrial development along coastlines of over 3,000 miles for Maine and 7,500 miles for Alaska. In each case the rocky coast protrudes into the cold waters with finger-like projections and island systems, and each region has a shelf-break several miles offshore which sets up physical conditions which encourage generation of massive subsurface populations of toxic dinoflagellates. These populations persist throughout the time of water stratification, commonly mid-April to mid-October in these latitudes. It is now hypothesized that when a proper sequence of meteorological events, such as persistent offshore winds, drive the surface waters offshore, these subsurface waters are forced inshore over the clam flats and mussel beds. Hence the shellfish feed on the toxic dinoflagellates contained in this water mass and thus the shellfish consumed by humans may be toxic. Red tides are indeed a natural phenomenon and have been known to exist prior to recorded history. There are noted accounts in the Bible as well as Indian legend. Dinoflagellates have flourished for over 600 million years as compared with human 20 million year venture on Earth.

The assessment that red tides are a natural phenomenon, however, does not absolve humans from concern. Some data suggest an apparent spread in geographic distribution, increase in intensity, and extension in duration. Note the use here of the word "apparent." An argument could be posed from our recent data (Table I) against this. While we do not argue that there is an apparent decline, these data serve to demonstrate several fallacies in using scant and inconsistent data. First, we assume data are collected in a regular manner. While Maine can boast an

Table I. Shellfish Toxin Content. Numbers of Marine Shellfish Samples Which Meet Stated Criteria

	total number of samples	500μg/ 100g	% of total	>2000μg/ 100g	% of total	>5000μg/ 100g	% of total
1979	2983	136	4.6	11	0.4	2	0.6
1980	4191	330	7.9	119	2.8	31	0.7
1981	4572	158	3.5	34	0.7	9	0.2
1982	2257	55	2.4	0	-	0	-
1983	2701	63	2.3	15	0.5	0	-

Original data courtesy of J.W. Hurst and P. Hoyt, Maine Dept. Marine Resources.

outstanding monitoring program, sampling frequency and spacing is not constant. Both are increased during times when toxin levels are high, thus biasing the data sets.

There is great value in current toxin data analysis and I would strongly encourage researchers in regions where large data sets exist to see whether or not there has been an increase in high toxin level scores.

From the scant bits of analyzed data, there is no way to ascertain whether there is an actual global increase or whether this impression is generated by increased human awareness and observational powers which have been fine-tuned. In some local regions where long-term phytoplankton studies have been conducted (e.g. Oslofjord; Norway, Seto Island, Japan; and Split, Yugoslavia) investigators have found correspondence in increased pollution and increased dinoflagellate blooms. Correspondence, however, does not necessarily reflect a cause-and-effect relationship. In some cases, the type of pollution (i.e. adding increased organics and nutrients) would be predicted to enrich the environment to encourage increased phytoplankton growth, yet we do not understand why dinoflagellates outcompete other plankton.

A recent research thrust concerns the benthic resting cysts of toxic dinoflagellates (the so-called dormant "over-wintering" stage in the life history common in our sediments from October to April) (17, 27-34). Benthic monitoring of cysts, possibly with corers, could tell the investigator historically whether the species has been in the region and possibly for how long.

Basically, benthic cysts reseed blooms on a seasonal basis. This has serious implications to human activities. The possible effects may be manifested in a wide range of marine activities. In particular, those engaged in projects such as the seeding of shellfish beds, shellfish culture, and marine dredging operations should be altered to possible dangers. Microscopic cysts are easily carried with dredged-up sediment and may be transported far from the source after dumping at sea. Caution should be exercised when shellfish are being transferred from one geographical region to another. Cysts may be buried and lost, or by contrast, numbers of motile cells sufficient for a bloom condition can result from merely one cyst in less than one month (assuming one doubling per day, a rate easily documented in culture and field conditions). Once introduced into a new area, cysts may directly contaminate shellfish and they may establish a more permanent local toxic dinoflagellate populations.

The Future: Global Monitoring

While early work claimed that PSP is confined to the temperate latitudes, reports from Venezuela, Malaysia, the Phillipines and Mexico lead us now to believe that oceanographic conditions, not some feature associated with latitude, are dictating the onset. There is a striking overlap of areas affected by toxic dinoflagellates of the genus Gonyaulax and regions of high tidal dissipation (35; see Figure 10). This hypothesis needs to be tested using rigorous statistical procedures.

A proper impact statement of human activities and resulting informed decisions will need to await national and global information bases with data collected in a controlled uniform manner so that intercomparisons are valid. In my mind, this is the singular most important effort necessary for future understanding and prediction of toxic dinoflagellates. While remote sensing, a tool useful for the study of oceanographic features yielding synoptic data sets, and flow cytometry and sorting, a tool useful for the study of cellular biology, are promising high technology developments, the major understanding will result from far simpler approaches.

We believe that shellfish toxin data can be used as a chemical tracer in the marine environment (13). The signal integrator in the water column can be the ubiquitous and common mussel (Figure 11) and the signal integrator in the benthic layer can be the scallop (Figure 12). Both cases can yield important data on seasonal and year-to-year variation. It is difficult to conceive of a more useful continuous sampling device than the common mussel. It works well in a variety of climatic conditions as well as extreme weather conditions. It is nearly indestructible, and the signal is uniquely integrated over time, relieving the investigator from concern about "filtering" diurnal migration, tidal variation, advection fluctuations and the like from the data.

I believe that we need to be realists. At the time of this writing, there is no clear evidence to indicate whether there is an unnatural spreading in distribution of the causative organisms, increased duration of the seasonal occurrence or increased intensity of the toxification. To respond to any claims of apparent spreading a sound global dinoflagellate monitoring effort should be immediately established. Figure 13 is a reasonable approach.

The equipment is not costly. It involves: a small pump (Little Giant at ~\$100), garden hose (30 meters for ~\$30), a small generator (Honda 6K Watt at ~\$400) and a small grab (~\$400). For less than \$1,000 one can be totally operational from a row boat. Additional gear can be added as affordable to follow temperature, salinity, pigments, nutrients and bioluminescence in the water column (Figure 14).

For a global monitoring program to be effective, I believe that there must be monitoring of the environment, the oceanography, the motile cells in the water, the benthic resting cysts in the sediments (which can be undertaken year-round), and the toxin in the shellfish. The program must be consistently embarked

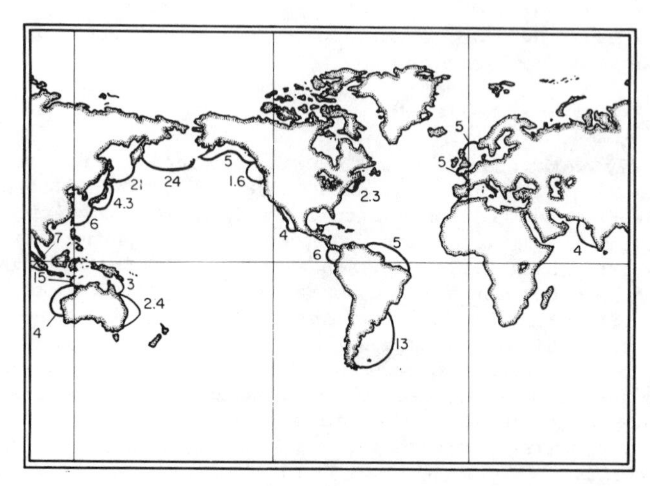

Figure 10. Major tidal dissipation areas of the world, the greater the number, the greater the energy involved. After Miller (23).

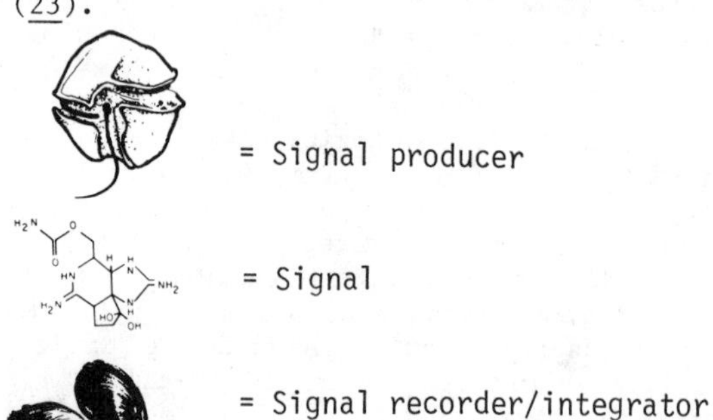

Figure 11. Using toxin molecules as water column tracers of presence of toxin producing dinoflagellates in motile form. The common mussel, found world-wide, is used as the signal recorder/integrator.

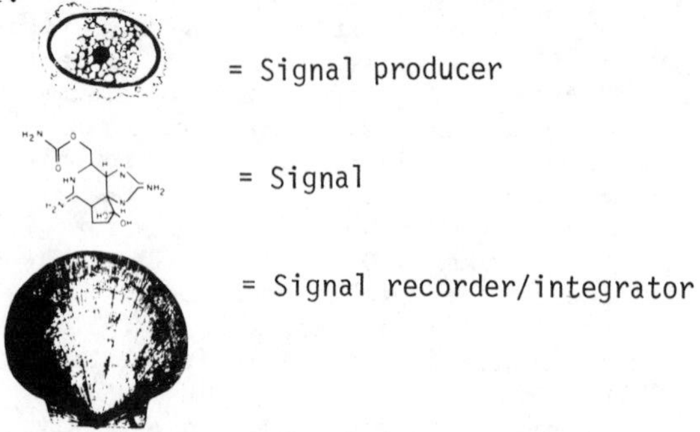

Figure 12. Using toxin molecules as benthic tracers of presence of toxin producing dinoflagellates in cyst form. The scallop is used as the signal recorder/integrator.

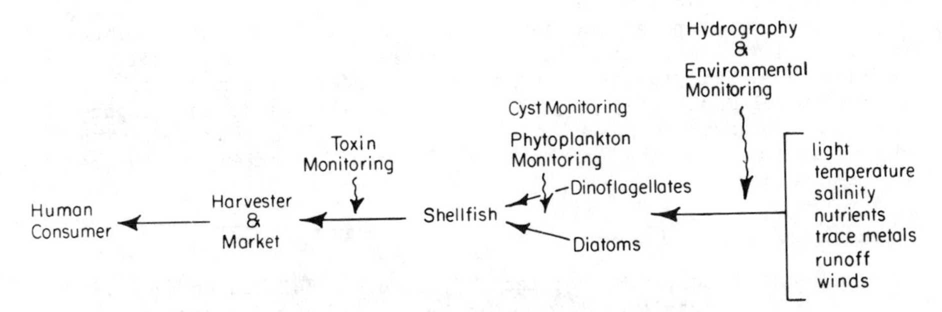

Figure 13. Suggested monitoring program design which includes toxin monitoring in shellfish, phytoplankton monitoring, and hydrography and environmental monitoring.

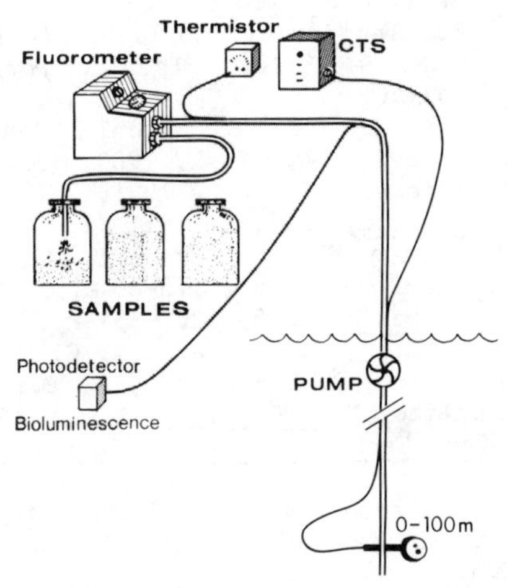

Figure 14. Water column analysis involving pump profiling system. In the system pictured, conductivity (an index of salinity) and temperature are profiled. The fluorometer measures in vivo fluorescence from chlorophyll (from all phytoplankton) and discrete samples are taken for cell counts and nutrient chemistry. Photodetectors can be employed for the measurement of bioluminescence. Light measurements range from Secchi disc readings to more sophisticated transmissometry and spectral radiometry instruments.

on globally, thus most hopefully sponsored by World Health Organization (WHO), Food and Agriculture Organization (FAO), International Union of Pure and Applied Chemists (IUPAC) or similar non-partisan interest groups, and must be long term -- that is, no less than 10 years initially. The program must not fluctuate with funding sentiments: a case in point is the California situation which in 1980 proved to be disasterous (19).

Public awareness does need enlightenment. Assay methods do need significant improvement. Yet, I feel that no thrust will be as important in providing leaps in our understanding as will a sound global monitoring program. I believe that it is the responsibility of government, and collective governments concerned with human welfare and the maintenance of sound fisheries, to pursue such programs.

Acknowledgments

I thank W.M. Balch, P.M. Holligan, J.W. Hurst, F.C. Mague and C.S. Yentsch for rewarding discussions. P. Boisvert prepared the manuscript and J. Rollins prepared the illustrations. Partial funding was provided through FDA 223-77-2314, NIH 5-RO1-ES01329-03, NMFS/NEFC NA-82-FA-C-00043 and the State of Maine. This is Bigelow Laboratory for Ocean Sciences contribution number 84002.

Literature Cited

1. Baden, D. G. Intnl. Rev. Cytol. 1983, 82, 99-150.
2. Taylor, F. J. R. In "Toxic Dinoflagellate Blooms"; Taylor, D. L.; Seliger, H. H., Eds.; Elsevier North Holland: New York, 1979; pp. 47-56.
3. Steidinger, K. A. In "Prog. Phycol. Res."; Round, F.; Chapman, Eds.; Elsevier: New York, 1983; Vol. 2, pp. 147-188.
4. Shimizu, Y.; Yoshioka, M. Science 1981, 212, 547-49.
5. Yentsch, C. M. Toxicon, 1981, 19, 611-21.
6. Yentsch, C. M.; Mague, F. C.; Horan, P. K.; Muirhead, K. J. exp. mar. Biol. Ecol. 1983, 67, 175-83.
7. Karentz, D. J. Protozool. 1983, 30, 581-8.
8. Holligan, P. M.; Balch, W. M.; Yentsch, C. M. submitted.
9. Yentsch, C. M.; Cucci, T. L.; Glover, H. E.; Phinney, D. A.; Selvin, R. submitted, Mar. Biol.
10. White, A. W. Can. Tech. Rept. Fish. Aquat. Sci. 1982, 1064, 12 pp.
11. Carreto, J. I.; Lasta, M. L.; Negri, R.; Benavides, H. Instituto Nacional de Investigacion y Desarroclo Pesquero 1981, Argentina, 52 pp.
12. Davison, P.; Yentsch, C. M. Technical Report 1982, Instituto Nacional de Pesca, Montevideo, Uruguay.
13. Hurst, J. W.; Yentsch, C. M. Can. J. Fish. Aquat. Sci. 1981, 38, 152-6.
14. Lembreye, G. V. Apartado Anales del Instituto de al Patagonia 1981, Magallanes, Chile, 12, 273-89.
15. Okaichi, T. Rept. Fac. Agriculture 1982, Kagawa Univ., Japan, 239 pp.
16. Sampayo, M. A. de; Cobecadas, G. Instituto Nacional de Investigacao dao Pescas 1980, Alges-Praia, Lisbon, Portugal, 25 pp.

17. Prakash, A.; Medcof, J. C.; Tennant, A. D. Bull. Fish. Res. Bd. Can. 1971, 177, 87 pp.
18. Saunders, S.; Sample, T.; Matsuda, R. Metro Rept. Water Pollution Control Dept. Seattle, WA, 31 pp.
19. Sharpe, C. A. 1982, California Dept. Health Services, Sanitary Engineering Section, 75 pp.
20. Tangen, K. In "Toxic Dinoflagellate Blooms"; Taylor, D. L.; Seliger, H. H., Eds.; Elsevier North Holland: New York, 1979; 179-82.
21. White, A. W. Rept. to ICES Advisory Comm. on Marine Pollution 1980, 11 pp.
22. Simpson, J. H.; Bowers, D. Deep-Sea Res. 1981, 28A, 727-38.
23. Miller, G. R. J. Geophys. Res. 1966, 2485-89.
24. Pingree, R. D.; Pugh, P. R.; Holligan, P. M.; Forster, G. R. Nature London 1975, 258, 672-77.
25. Holligan, P. M.; Harbour, D. S. J. Mar. Biol. Assoc. U.K. 1976, 1075-93.
26. Yentsch, C. S. In "Remote Sensing Applications in Marine Science and Technology"; Cracknell, A. P., Ed.; D. Reidel Publ. Co., 263-97.
27. Steidinger, K. A. Int. Conf. Toxic Dinoflagellate Blooms; LoCicero, V., Ed.; Mass. Sci. Tech. Fdn.: Wakefield, MA, 1975, 153-62.
28. Dale, B. In "Toxic Dinoflagellate Blooms"; Taylor, D. L.; Seliger, H. H., Eds.; Elsevier North Holland: New York, 1979; 443-52.
29. Dale, B.; Yentsch, C. M.; Hurst, J. W. Science 1978, 201, 1223-5.
30. Anderson, D.; Wall, D. J. Phycol. 1978, 14, 224-34.
31. Yentsch, C. M.; Lewis, C. M.; Yentsch, C. S. Bioscience 1980, 30, 251-4.
32. White, A. W.; Lewis, C. M. Can. J. Fish. Aquat. Sci. 1982, 39, 1185-94.
33. Thayer, P. E.; Hurst, J. W.; Lewis, C. M.; Selvin, R.; Yentsch, C. M. Can. J. Fish. Aquat. Sci. 1983, 40, 1308-14.
34. Selvin, R. C.; Lewis, C. M.; Yentsch, C. M.; Hurst, J. W. Toxicon, in press.
35. Yentsch, C. M.; Holligan, P. M.; Balch, W. M.; Tvirbutas In "Tidal Mixing and Plankton Dynamics"; Bowman, M.; Yentsch, C. M.; Peterson, R., Eds.; Springer Verlag: New York, in press.
36. Esaias, W. In "Primary Productivity in the Sea"; Falkowski, P. G., Ed.; Plenum Press: New York, 1980; 321-37.

RECEIVED February 6, 1984

3

Ciguatera Seafood Poisoning

Overview

EDWARD P. RAGELIS

Food and Drug Administration, Washington, DC 20204

Ciguatera poisoning is an important and serious cause of morbidity in humans that results from the consumption of a large variety of reef associated fishes throughout the circumtropical regions of the world. It is estimated that well over 50,000 people may be afflicted with the disease yearly and the number may be increasing. Besides affecting the public directly, ciguatera has had a dramatic impact on the development, growth, and stability of the in-shore fisheries in tropical and subtropical areas. Moreover, reports of ciguatera, coupled with adverse publicity have led to a ban on the sale of selected fish species, causing huge economic losses. Ciguatera is separate and distinct from human illnesses (botulism and scombroid poisoning) that result from eating seafood that has spoiled because of improper handling and/or processing. Cooking (e.g., frying, baking, broiling, boiling, steaming), smoking, drying, salting, or freezing does not appear to destroy the toxin in the fish flesh, and one cannot tell from smell or appearance whether or not a fish is ciguatoxic. The victim usually recovers from ciguatera within a few days (death occasionally occurs), but symptoms may last for several weeks, months, or possibly years. Ciguatera is the most commonly reported poisoning associated with eating fish in the United States and its territories.

In spite of the long history of ciguatera, the origin or identity of the toxin(s), the organism(s), and ecological factors responsible for the disease are still not completely known. However, despite an incomplete description of the molecular structure of ciguatera toxin(s), methods are being developed for the detection of ciguatoxic fish.

This chapter not subject to U.S. copyright.
Published 1984 American Chemical Society

History

Ciguatera poisoning, which has been attributed to the agent "ciguatoxin" (1), occurs throughout the regions between the Tropic of Cancer and the Tropic of Capricorn. The malady probably dates to antiquity. References to fish poisoning can be found in Homer's Odyssey (800 B.C.) and were noted during the time of Alexander the Great (356-323 B.C.), when his soldiers were forbidden to eat fish to avoid the accompanying maladies and malaise that could threaten his conquests (2). The term ciguatera is of Spanish origin, dating to the eighteenth century. It was originally used to refer to intoxications caused by the ingestion of a marine snail, *Turbo Livona pica*, known in the Spanish Antilles by the Cuban name "cigua" (2-4). Current usage of ciguatera refers to human intoxication from ingestion of tropical and subtropical finfish (2). While the name has its origin in the West Indies, the phenomenon was observed and recorded in the Indian and Pacific Oceans as early as the sixteenth century (5). The recorded outbreaks of ciguatera, both in the New Hebrides aboard the vessel of the Portuguese explorer Pedro Fernandez de Queiros in 1606 and in 1774 aboard Captain Cook's Resolution, were probably two of the earlier vivid accounts of the illness (5,6).

In general, outbreaks of ciguatera poisoning are sporadic and unpredictable both in geographic distribution and time. Moreover, of the more than 400 species implicated (7), not all of the fish of the same species caught at the same time in the same place are toxic. Only a few miles can separate ciguateric and safe fish of a given species. Within the tropical Western Atlantic, e.g., Florida and the Caribbean, barracuda, grouper, and snapper are the fish most often implicated in the disease. In the Hawaiian Islands region, the kahala (amberjack) and uluau or Papio (Hawaiian names for 25 species in the *Carangidae* family) are the most common offenders. A general list of the species of fish likely to be toxic and their distribution has been documented (7-10).

Medical, Public Health, and Economic Aspects

The symptoms of ciguatera poisoning are complex, involving the digestive, cardiovascular, and neurological systems (see below); the symptoms can occur in various combinations (11).

Symptoms of Ciguatera Poisoning

1. Digestive

   - nausea, vomiting
   - diarrhea, abdominal cramps

2. Cardiovascular

   - slow pulse rate between 40 and 50 beats/min.
   - irregular or accelerated pulse between 100 and 200 beats/min.
   - reduced blood pressure

3. Neurological

- headache
- severe pruritus
- temperature reversal
- paresthesia
- arthralgia
- myalgia
- convulsions, muscular paralysis
- audio and visual hallucinations
- vertigo and loss of equilibrium

Usually the illness begins with gastrointestinal inflammation, which causes severe dehydration and weakness, followed by cardiovascular and neurological syndromes. The distinctive features of the poisoning are severe pruritus, temperature reversal, and paresthesia -- tingling and numbness of the extremities. The neurological symptoms may persist for months or years. Victims can develop sensitization, whereby the neurological symptoms can recur with stress, drinking alcoholic beverages, or eating seemingly non-toxic fish (8). Treatment for ciguatera is symptomatic, with no proven antidote known. However, protamide, used in the treatment of shingles (*Herpes zoster*), has been used successfully in treating ciguatera cases on Grand Bahama Island (12). Protamide is available in 15 countries (e.g., Belgium, Spain, United Kingdom, and Portugal) but is not approved for use in the United States.

Over the last eight years, most of the estimated cases of world-wide poisoning in humans due to the three major kinds of seafood toxins that are found in fresh and unspoiled marine organisms, namely, paralytic shellfish poison (PSP-saxitoxins/gonyautoxins), ciguatoxin(s), and tetrodotoxin, (13-16) were caused by ciguatera (Table I).

Table I. Worlwide Poisoning in Humans

Toxin	Lethal Dose, mg	No. Cases (1976-1983)
PSP (saxitoxin)	1-2	>400 (27 deaths)
Ciguatoxin	< 1	>10,000 yearly (<1% mortality)
Tetrodotoxin	1-2	>100 yearly (~50% mortality)

Other estimates suggest that as many as 50,000 individuals per year worldwide are afflicted with ciguatera poisoning (17). Ciguatera morbidity in the Pacific averages five cases per 1,000 (8). It is a sad commentary that on many islands in the Pacific, where ciguatera can be endemic, fish have to be imported, causing economic loss and hardship both to the island and the commercial and local fishermen whose livelihood depends on harvesting fish (8).

Most startling is the report for 1983 of the morbidity rate of ciguatera poisoning for southern Queensland, Australia, which averaged 18/1,000 (18). Implicated in the poisonings has been the Spanish mackerel (*Scomberomorus commersoni*), a pelagic fish economically important (1,000 tons landed annually) to the mackerel fishing industry in Queensland (19). Pelagic fish are generally considered safe and are not normally associated with ciguatera poisoning (6).

Impact of Ciguatera in the United States. While reliable morbidity statistics have been difficult to obtain because of the lack of a precise diagnostic means to clearly identify clinical cases of ciguatera, information has been gathered which shows that it is a widespread public health problem in the United States and its territories (20-22).

The number of cases of ciguatera may average over 2,000 per year. Data accumulated from a three-year study of southern Florida indicate an average of 1,300 cases per year (23) or for Dade County, Florida, 0.5 cases per 1,000 population (24). Moreover, 1.6-4.4% of the local population on St. Thomas, U.S. Virgin Islands, is annually afflicted with ciguatera, representing a morbidity statistic of 27 cases per 1,000 persons in a population of about 60,000 (25). In Hawaii, over a period of 82 years (1900-1981), 653 people have been afflicted, with only two reported fatalities, indicating a low mortality rate for ciguatera in Hawaii (21). In the Commonwealth of Puerto Rico, where ciguatera was essentially unrecognized before 1976 (26), it has recently become a considerable health and economic concern. Between April and June 1981, 49 persons were stricken with ciguatera. Two fatalities were reported among 22 victims who became ill from eating freshly caught barracuda (27). [Unofficial estimates (28) indicate that there may be over 100 cases per year on the Island that are unreported or the disease is misdiagnosed.] This episode and the threat of ciguatera has led to a ban by the Commonwealth in the buying or selling of barracuda, amberjack (medregal), and the urel negro or blackjack. In addition, the "ciguatera scare" caused the Island's demand for fish to fall drastically in 1981, particularly for red snapper, which dropped 80%. Red snapper and grouper together represent one-third of the annual catch; and ciguatera continues to be an impediment to the growth and stability of the fledgling Puerto Rican fishing industry. Two months before the Puerto Rican incident, an outbreak of ciguatera poisoning occurred at nearby St. Croix, U.S. Virgin Islands. This outbreak involved at least 69 people, most of whom had eaten red snapper purchased from a local vendor (29). An unofficial count of those affected may have been as high as 150 (30).

Although ciguatera is a problem limited to circumtropical regions, with the intensified commercialization of tropical reef fish, ciguatera is increasing in frequency. Two examples are the outbreak of ciguatera in Montgomery County, Maryland (31) in September 1980 involving 12 persons (two seriously ill with symptoms lasting a year) and one case in Boston, Massachusetts (32) in November 1982. Both events involved the consumption of grouper shipped from Florida to the local restaurants where the incidents occurred. A third but conflicting account occurred in October 1980, involving five persons in east Tennessee who became ill (one seriously) after eating a combination of shellfish (hard clams and scallops) and King mackerel. The poisonings were initially attributed to the consumption of the clams (33), but analysis of the King mackerel showed that it was ciguatoxic (34).

The economic loss to the Florida/Caribbean/Hawaiian seafood industry from constraints in harvesting potentially ciguatoxic fish, coupled with the adverse publicity, has not been clearly determined. It is estimated that over $10,000,000 is lost annually (17); this does not include the higher price of liability insurance being paid by the industry or recent court cases of ciguatera poisoning and pending litigation (35).

Origin of Toxin(s)

In spite of the long history of ciguatera, the origin or identity of the toxin(s) is still not completely known. Lewis (8) has referenced a number of theories that have been put forth. These have included the effects of pollution, the presence of heavy metals such as copper, climatic changes, and diseased fish. Also, a food chain theory (36) was proposed, whereby fish become toxic through the food web, beginning with the benthic and detrital herbivorous fish feeding on a toxic organism (either an algae, fungus, protozoan, or a bacterium), and the toxin(s) is transmitted to the higher tropic omnivores and carnivores that prey on the herbivores (Figure 1). We now believe, as with PSP (37-39), that the genesis of ciguatera toxin(s) is a toxic dinoflagellate(s) (Figure 1).

Pacific Organism. In the course of examining the food habits of the detrital feeder *Ctenochaetus striatus*, a surgeonfish that has been implicated in numerous cases of ciguatera in Tahiti (40), it was discovered that the detrital fraction of coral rubble in a ciguatera-endemic area in the Gambier Islands in French Polynesia was toxic (41). Further examination of the detrital material revealed the presence of a benthic dinoflagellate that was shown to bear a direct relationship to the occurrence of ciguateric fish (41,42). This epiphytic dinoflagellate was recognized as a new species, *Gambierdiscus toxicus* (43), which was later found and identified in Hawaii (44,45). The organism appears to spawn and flourish following major natural or human-contrived disturbances and destruction of coral reefs, e.g., by dredging and construction (41,43).

Caribbean Organism. Parallel to this discovery, the Food and Drug Administration (FDA) in 1979 supported work at the University of Southern Illinois, Carbondale, to identify the organisms(s) responsible for ciguatera poisoning in the Caribbean. Over 70 different sites in the British and U.S. Virgin Islands were examined for epiphytic/benthic flora, especially the presence of dinoflagellates. A collection of 65 strains representing 18 species of the most common dinoflagellates was established,

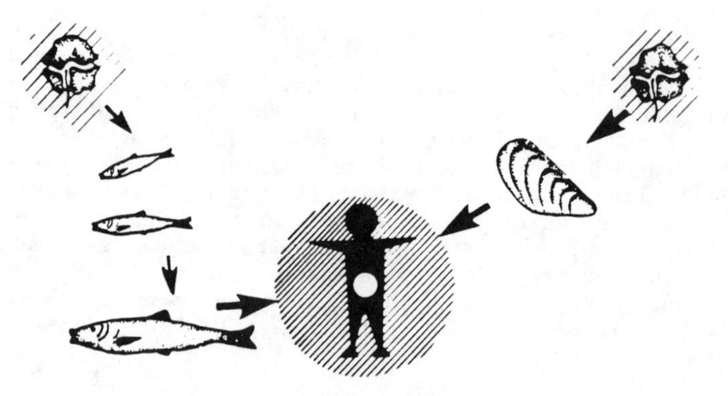

Figure 1. Food chain theory.

including G. toxicus, which is associated in these waters with other planktonic dinoflagellates and literally "glues" itself to macroalgae (46,47). Five of the species, namely, G. toxicus, Prorocentrum concavum, P. rhathymum, Gymnodinium sanquineum, and Gonyaulax polyedra, elaborate one or more toxic isolates that kill mice within 48 hours (46). Extracts from P. concavum were more potent than those from G. toxicus, containing an unknown, very potent, fast-acting toxin. Chromatographic analysis of extracts from G. toxicus indicates the presence of ciguatoxin, one other lipid-soluble toxin, and maitotoxin, findings similar to those with cells of G. toxicus collected in the Pacific (13). Maitotoxin is associated with ciguatoxin and scaritoxin in ciguateric fish (5). These observations suggest that more than one organism may produce a combination of toxins contributing to ciguatera poisoning in the Caribbean and in the Pacific (21,48), which may account for the bizarre and diverse syndromes accompanying the illness.

Chemistry

The principal active agent of ciguatera in carnivores is still thought to be ciguatoxin (49), and attempts to elucidate its molecular structure are currently based on the toxin isolated from the viscera of moray eels, Lycodontis (Gymnothorax) javanicus (1, 49,50). The yield of toxin has been extremely low, on the order of 2×10^{-6}% or 1.3 mg from 62 kg of eel viscera (49,51). Ciguatoxin is a highly oxygenated, white solid lipid (LD_{50} 0.45 ug/kg, ip, in mice) with a molecular weight of 1111.7±0.3 determined by ^{252}Cf plasma desorption mass spectrometry. Scheuer (49) proposes a molecular formula of $C_{53}H_{77}NO_{24}$ or $C_{54}H_{78}O_{24}$. On the basis of extensive ^{1}H-nuclear magnetic resonance spectroscopic studies at 360 and 600 MHz, the toxin contains four olefinic, five hydroxyl, and five methyl groups; the bulk of the oxygen atoms are present as ether linkages (49). These features closely resemble those of brevetoxin C (51), isolated from Florida's red tide organism, Ptychodiscus brevis (39), and okadaic acid, a toxin isolated from marine sponges (52) and the dinoflagellate Prorocentrum lima (53). It still has not been unequivocally established whether the toxin isolated from moray eel viscera is the same as that produced by G. toxicus, but incomplete data seem to indicate a direct similarity (13,45).

Other Toxins. Much less is known concerning the molecular structures of the secondary or associated toxins, maitotoxin and scaritoxin, which coexist with ciguatoxin in toxic fish (5). They have been found together in the grazing Okinawan turban shell, Turbo argyrostoma (54). Scaritoxin, a lipid-soluble toxin isolated from the muscular tissue of parrotfish, Scarus gibbus (55), produces symptoms physiologically similar to ciguatoxin in mice but is chromatographically different (56). It is speculated to be a metabolite of ciguatoxin (5). Maitotoxin, with increased water-solubility (insoluble in acetone), was first isolated from the surgeonfish, C. striatus, and is a major toxic component elaborated by G. toxicus (46,57). It is a potent marine toxin with a minimum lethal dose (ip) in mice of 0.2 ug/kg (58,59). Maitotoxin is considered to be a non-peptidic material with a molecular weight of >10,000 as determined by ultrafiltration (59,60). The infrared spectra of the toxin indicate numerous hydroxyl functions and the presence of an amide group (42). With its large molecular weight, maitotoxin could be a precursor of ciguatoxin (5). Besides these three toxins, other unidentified toxins have been detected in ciguateric fish (50). These could be alteration or degradation products of either maitotoxin, ciguatoxin, or scaritoxin, or entirely different toxins accumulated from a variety of toxic organisms (46).

Detection of Ciguatoxin

Presently, there is no validated quantitative method for the determination of ciguatoxin, because of the scarcity and unknown structure of the toxin. A variety of bioassays including feeding suspect fish to cats and mongooses (61-62), injecting extracts into mice (63-65), using brine shrimp (66) and guinea pig atrium (67), have been used for the detection of ciguatoxic fish. None of these procedures have achieved the specificity, sensitivity, and practicality necessary for quantitative routine testing, although the guinea pig atrium procedure does appear to differentiate between maitotoxin and ciguatoxin. However, a recent bioassay (68), using the smooth muscle of the guinea pig ileum, appears to be able to distinguish three separate lipid-soluble toxins at the nanogram level in isolates from cultures of *G. toxicus*, and has the potential to be a screening method for these toxins.

A more promising method for the detection of ciguatoxin includes a radioimmunoassay (RIA) procedure, which was developed at the University of Hawaii (69). Through support received from the FDA and the National Marine Fisheries Service, this RIA was extensively evaluated for its potential as a positive screening test for ciguatoxic fish. During a two-year study (April 1979-81), the RIA was used to detect and remove potentially toxic kahalas or amberjacks (*Seriola dumerili*) from the Hawaiian market. Fifteen percent of the 5,529 fish examined (approximately 45 tons) was rejected (70). During this period, no incident of ciguatera poisoning due to marketed *S. dumerili* occurred, although poisonings (71) due to other fish species were reported to the Hawaii Department of Health (30 cases involving 88 individuals). Furthermore, the RIA was shown to be applicable and useful for developing and surveying fishing grounds in areas where ciguatera outbreaks occur (72). It also proved to be helpful in identifying the toxic snappers and groupers implicated in the previously cited (29) episodes of ciguatera poisoning in St. Croix and Tennessee (34). However, the RIA was too costly and time consuming for routine testing. Also, the procedure is specific for polyether lipid residues and cross reacts with okadaic acid, brevetoxin, monensin, and other polyether-containing lipid soluble residues (71,73,74). For these reasons, the RIA was converted to a simpler enzyme-immunoassay (EIA), which gives results that compare favorably with the RIA method and the mouse bioassay ($P < 0.01$ and $P < 0.001$, respectively) and can detect approximately 0.01-0.05 ng of ciguatoxin (73). However, the EIA method needs further examination and evaluation before it can be accepted for the routine direct assessment of ciguatoxin in fish tissues, and preparation of monoclonal antibodies to ciguatoxin should increase the specificity of the EIA method (74). In addition to these methods, a counterimmunoelectrophoresis (CIEP) procedure has been reported that appears to discriminate between nontoxic and ciguatoxic Caribbean fish samples involved in episodes of ciguatera poisoning (75). Toxicity was established "on the basis of human toxic reactions and by an *in vivo* mouse bioassay." However, the investigators acknowledge that the procedure needs substantial testing and modification before it can be used as a screening test for toxic fish. It will be interesting to compare the CIEP and EIA methods once they are perfected for their value and ability to detect ciguatoxic fish.

Summary

The strategy and challenges (76) to control the widespread nature of ciguatera would appear to remain:

1. To identify the dinoflagellates responsible for ciguatera.
2. To understand the process by which dinoflagellates introduce the toxin(s) into the ecosystem.
3. To isolate quantities of ciguatera toxin(s) from fish and dinoflagellate cultures.
4. To determine the molecular structure of these toxins and their interconversion.
5. To develop highly sensitive, rapid, and inexpensive methods for the detection of toxin(s) in suspect fish, i.e., the need for a "marketplace" test.
6. To perfect diagnostic means to clearly identify and evaluate clinical cases of ciguatera.
7. To discover the mechanisms of toxicity and to develop effective safe antidotes for the treatment of victims.
8. To develop and utilize an information base on ciguatera for the seafood industry, health personnel, the public, and researchers.
9. To identify the factors contributing to the entry of ciguateric fish into the marketplace and to develop procedures (e.g., surveillance/compliance programs) to minimize the consumption of toxic fish.

With these goals realized, we should be able to alleviate the economic hardship and health hazard associated with ciguatera and be able to harvest a vast, untapped protein-rich seafood resource.

Literature Cited

1. Scheuer, P. J.; Takahashi, W.; Tsutsumi, J.; Yoshida, T. *Science* 1967, 155, 1267-8.
2. Halstead, B. W. In "Poisonous and Venomous Marine Animals of the World"; Darwin: Princeton, N. J., 1965; Vol. I.

3. Parra, D. A. Descipcion de Diferentes Peizas de Historia Natural Las Mas del Ramo Maritimo Representadas en Sententa y Cinco Laminas, 1787, Havana, Cuba.
4. Gudger, E.W. Am. J. Trop. Med. 1930, 10, 43-55.
5. Bagnis, R. Oceanol. Acta 1981, 4, 375-87.
6. Banner, A. H. In "Biology and Geology of Coral Reefs"; Jones, O. A.; Endean, R., Eds.; Academic: New York, 1976; Vol. III, Chap. 6, pp. 177-213.
7. Bagnis, R.; Berglund, F.; Elias, P. S.; Van Esch, G. J.; Halstead, B. W.; Kojima, K. Bull. W.H.O. 1970, 42, 69-88.
8. Lewis, N. Ph.D. Thesis, University of California, Berkeley, 1981.
9. Halstead, B. W. In "Poisonous and Venomous Marine Animals of the World"; Darwin: Princeton, N. J., 1978.
10. Brody, R. W. Proc. Gulf Caribb. Fish. Inst., 1972, 24th Annual Session; pp. 100-16.
11. Bagnis, R.; Kuberski, T.; Langier, S. Am. J. Trop. Med. Hyg. 1979, 28, 1067-73.
12. Bailey, V., personal communication, 1983, Cooper Laboratory, Freeport, Bahama.
13. Withers, N. W. Annu. Rev. Med. 1982, 33, 97-111.
14. Expert Committee Report on Ciguatera, Suva, Figi, South Pacific Commission, Noumea, New Caledonia, February 16, 1981.
15. Verber, J. "Shellfish Borne Disease Outbreak," Food & Drug Administration, Northeast Technical Services Unit, Davisville, Rhode Island, 1984; pp. 18-20.
16. Shimizu, Y., University of Rhode Island, Kingston, RI, personal communication, 1982.
17. Sea Grant Workshop, National Oceanic and Atmospheric Administration, Washington, D. C., 1983.
18. Lewis, N. D. This publication.
19. Lewis, R. J.; Endean, R. Toxicon 1983, 21, 19-24.
20. Tacket, C. "Studies on the Epidemiological and Clinical Aspects of Ciguatera," A Conference on Ciguatera, Caribbean Fisheries Management Council, November 6-7, 1981.
21. Anderson, B. S.; Sims, J. K.; Wiebenga, N. H.; Sugi, M. Hawaii Med. J. 1983, 42, 326-34.
22. Hokama, Y., as cited in Ragelis, E.P. J. Assoc. Off. Anal. Chem. 1982, 65, 327.
23. Poli, M. M.S. Thesis, University of Miami, Coral Gables, 1982.
24. Lawrence, D. N.; Enriquez, M.; Lumish, R. M.; Maceo, A. J. Am. Med. Assoc. 1980, 244, 254-8.
25. McMillan, J. P.; Granade, H. R.; Hoffman, P. J. Coll. Virgin Is. 1980, 6, 84-107.
26. Payne, C., Metropolitan Hospital, Los Lomas, Rio Piedros, Puerto Rico, personal communication, 1981.
27. Ghigliotty, J. The San Juan Star, October 4, 1981.
28. Lynch, H., Food and Drug Admin., San Juan, Puerto Rico, personal communication, 1982.
29. Morbidity and Mortality Weekly Report, 1981, 30, 138.
30. Ramos, E., Food and Drug Admin., San Juan, Puerto Rico, personal communication, 1981.
31. Morbidity and Mortality Weekly Report, 1980, 29, 610.
32. Rotner, H., Union Hospital, Lynn, Massachusetts, personal communication, 1982.

33. Cosby, A. G., Food and Drug Admin., Nashville, TN, personal communication, 1980.
34. Kimura, L. H., "Evaluation of a Radioimmunoassay System for the Detection of the Seafood Toxin(s) Responsible for the Disease Ciguatera," Contract 223-79-2271, Food and Drug Admin., 1981.
35. Lewis, R. 186th ACS National Meeting, Division of Agricultural and Food Chemistry, 1983, Paper No. 97.
36. Randall, J. E. Bull. Mar. Sci. Gulf Caribb. 1958, 8, 236-67.
37. Yentsch, C. M. This publication.
38. Steidinger, K. A. In "Progress in Phycological Research"; Elsevier Science Publishers BV: 1983; Vol. II, p. 147.
39. Baden, D.B. Int. Rev. Cytol. 1983, 82, 99-150.
40. Bagnis, R. Hawaii Med. J. 1968, 28, 25-8.
41. Yasumoto, T.; Nakajima, I.; Bagnis, R.; Adachi, R. Bull. Jpn. Soc. Sci. Fish. 1977, 43, 1021-6.
42. Yasumoto, T.; Inoue, A.; Bagnis, R.; Garcon, M. Bull. Jpn. Soc. Sci. Fish. 1979, 45, 395-9.
43. Adachi, R.; Fukuyo, Y. Bull. Jpn. Soc. Sci. Fish. 1979, 45, 67-71.
44. Taylor, F. J. R. In "Toxic Dinoflagellate Blooms"; Taylor, D. L.; Seliger, H. H., Eds.; Elsevier North Holland: New York, 1979; p. 71.
45. Shimizu, Y.; Shimizu, H.; Scheuer, P. J.; Hokama, Y.; Oyama, M.; Miyahara, J. T. Bull. Jpn. Soc. Sci. Fish. 1982, 48, 811-3.
46. Tindall, D. R.; Dickey, R.W. This publication.
47. Besada, E.G.; Loeblich, L. A.; Loeblich, A. R. Bull. Mar. Sci. Gulf Caribb. 1982, 32, 723-35.
48. Yasumoto, T.; Oshima, Y.; Murakami, Y.; Nakajima, I.; Bagnis, R.; Fukuyo, Y. Bull. Jpn. Soc. Sci. Fish. 1980, 46, 327-31.
49. Scheuer, P. J. Naturwissenschaften 1982, 69, 528-33.
50. Tachibana, K. Ph.D. Thesis, University of Hawaii at Manoa, Honolulu, 1980.
51. Kukina, M.; Tachibana, K.; Koyanagi, L. M.; Scheuer, P. J. 186th ACS National Meeting, Division of Agricultural and Food Chemistry, 1983, Abstract No. 135.
52. Tachibana, K.; Scheuer, P. J.; Tsukitani, Y.; Kikuchi, H.; Van Engen, D.; Clardy, J.; Gopichand, Y.; Schmitz, F. J. J. Am. Chem. Soc. 1981, 103, 2469-71.
53. Murakami, Y.; Oshima, Y.; Yasumoto, Y. Bull. Jpn. Soc. Sci. Fish. 1982, 48, 69-72.
54. Yasumoto, T.; Kanno, K. Bull. Jpn. Soc. Sci. Fish. 1976, 42, 1399-1404.
55. Chungue, E.; Bagnis, R.; Yasumoto, T. Biochimie 1977, 59, 739-41.
56. Hashimoto, Y. In "Marine Toxins and Other Bioactive Marine Metabolites"; Japan Scientific Societies: Tokoyo, 1979; p. 102.
57. Yasumoto, T.; Nakajima, I.; Oshima, Y.; Bagnis, R. In "Toxic Dinoflagellate Blooms"; Taylor, D.; Seliger, H. H., Eds.; Elsevier North Holland: New York, 1979; p. 65.
58. Yasumoto, T. Igaku No Ayumi 1980, 112, 886-92.

59. Takahashi, M.; Ohizumi, Y.; Yasumoto, T. J. Biol. Chem. 1982, 257, 7287-9.
60. Ohizumi, Y.; Yasumoto, T., Mitsubishi-Kasei Institute of Life Sciences, Tokyo, Japan, unpublished data (1982).
61. Banner, A. H.; Sasaki, S.; Helfrich, P.; Alender, C. B.; Scheuer, P. J. Nature 1961, 189, 229-30.
62. Banner, A. H.; Scheuer, P. J.; Sasaki, S.; Helfrich, P.; Alender, C. B. Ann. N. Y. Acad. Sci. 1960, 90, 770-87.
63. Bagnis, R. Bull. W.H.O. 1973, 49, 67-76.
64. Kimura, L. H.; Hokama, Y.; Abad, M. A.; Oyama, M.; Miyahara, J. T. Toxicon 1982, 20, 907-12.
65. Hoffman, P. A.; Granade, H. R.; McMillan, J. P. Toxicon 1983, 21, 363-9.
66. Granade, H. R.; Cheng, P.C.; Doorenbos, N. J. Pharm. Sci. 1976, 65, 1414-5.
67. Miyahara, J. T.; Akau, C. K.; Yasumoto, T. Res. Commun. Chem. Pathol. Pharmacol. 1979, 25, 177-80.
68. Miller, D. M.; Dickey, R.W.; Tindal, D.R. This publication.
69. Hokama, Y; Banner, A. H.; Boyland, D. Toxicon 1977, 15, 317-25.
70. Kimura, L. H.; Abad, M. A.; Hokama, Y. J. Fish. Biol. 1982, 21, 671-81.
71. Hokama, Y., as cited in Ragelis, E. P. J. Assoc. Off. Anal. Chem. 1983, 66, 367.
72. Ito, B. M.; Uchida, R. N. "Proceedings Symposium on the Status of Resource Investigations in the Northwestern Hawaiian Islands"; Grigg, R. W.; Pfund, R.T.; Eds.; University of Hawaii Sea Grant College Program: Honolulu, 1980, pp. 81-6.
73. Hokama, Y.; Abad, M. A.; Kimura, L. H. Toxicon 1983, 21, 817-24.
74. Hokama, Y.; Kimura, L. H.; Abad, M. A.; Yokochi, L.; Yasumoto, T.; Shimizu, Y.; Baden, D. G.; Nukina, M.; Scheuer, P. J. This publication.
75. Emerson, D. L.; Galbraith, R. M.; McMillan, J. P.; Higerd, T. B. Arch. Intern. Med. 1983, 143, 1931-3.
76. Doorenbos, N.J. This publication.

RECEIVED May 25, 1984

4

Miscellaneous Seafood Toxicants

BRUCE W. HALSTEAD

World Life Research Institute, Colton, CA 92324

The expansion of marine fisheries into tropical waters has increased the risks of widespread human intoxications in temperate latitudes. This situation has been facilitated by increased travel, rapid transport of food stuffs, and improved refrigeration facilities. This presentation directs attention to a broad spectrum of toxic protozoans, coelenterates, echinoderms, molluscs, arthropods, fishes, and mammals that may serve as transvectors of these poisons. The need for additional epidemiological, toxicological, and chemical research is noted.

The expansion of marine fisheries into tropical waters, improved rapid transport of fisheries products, and increased travel to remote areas of the world have introduced new and little-known intoxicants into temperate zone populations. With the expansion and intensification of things toxicological, the subject of paralytic shellfish poisoning, ciguatera fish poisoning, and puffer poisoning have gradually made their debut into the consciousness of toxicologists and some clinicians. In reviewing the broad spectrum of marine biotoxicology it becomes obvious that these three biotoxic categories represent only a small facet of the enormous array of toxic molecular structures that inhabit the marine environment. The purpose of this presentation is to direct attention to some of the other areas of toxicological interest involving man's food resources and the enormous phylogenic range of marine biotoxins.

Marine Invertebrate Intoxications

Protozoans (Protistans). The relationship of dinoflagellates to paralytic shellfish poisoning and the role of *Gambierdiscus toxicus* in ciguatera fish poisoning are well documented. However, recent studies have shown the toxic dinoflagellates are also involved in other forms of human biotoxications as a result of ingesting molluscs which serve as their transvectors. The dinoflagellate species involved in these intoxications include *Prorocentrum redfieldii*,

0097–6156/84/0262–0037$06.00/0
© 1984 American Chemical Society

P. micans, and Dinophysis acuminata in the Netherlands (1). Dinophysis fortii has been incriminated in Japan (2), and an unknown toxic species of the same genus has caused intoxications along the coast of Chile (3). The clinical symptomatology is of the gastrointestinal type, consisting of nausea, vomiting, and diarrhea. No neurological disturbances have been reported. It is felt that the problem of diarrhetic shellfish poisoning is probably more widespread than is generally recognized and deserves greater investigational interest. The toxin involved has been designated by Yasumoto et al (2) as dinophysitoxin. The chemical and pharmacological characteristics of the poison have not been fully determined.

Cnidarians (Coelenterates). Jellyfishes are commonly eaten in Japan and elsewhere in the Orient without any known cases of poisoning. The nematocysts of jellyfishes and other coelenterate species contain proteinaceous toxins which are inactivated by heating and gastric juices and therefore are not responsible for dietary intoxications. However, biotoxications have resulted in the Philippines, New Guinea, and Samoa from the ingestion of sea anemones which have been eaten raw. Rhodactis howesii, commonly known as metalelei, and Physobranchia douglasi, known as humane in Samoa, are poisonous when eaten raw, but safe to eat when cooked. Radianthus paumotensis, locally known as matamalu samasama in Samoa, and another unknown species of Radianthus, are considered to be poisonous when eaten either raw or cooked. The symptomatology consists of nausea, vomiting and abdominal pain, pallor, and prostration. The victim may lose consciousness (8-36 hours) and then dies in an apparent state of shock. There is believed to be a central nervous system involvement. Studies by Martin (4, 5) report that the sea anemone possesses a nondialyzable and heat labile toxin containing a polysaccharide moiety and exhibiting UV absorption at 278 nm. Since the poison was inactivated by tryptic digestion, its activity was believed to be due to a protein or polypeptide moiety. No curarelike action was found on rabbit phrenic nerve-muscle preparation. The toxin has an LD_{99} (2-4 hours) of 3 mg/kg (IP in mice). Martin claims that "R. howesii contains a paralytic poison which differs from other known poisons of this category."

Echinoderms. Sea cucumbers are commonly eaten throughout the Orient and in some of the Pacific Islands where they are sold under the vernacular names of trepang (Malaysia), sea slugs, sea cucumbers (English), erico (Japan), and hai shen (China). Gastrointestinal disturbances may result from the ingestion of these echinoderms. It is believed that these intoxications are caused by saponins, which are well-known toxic components of echinoderms. Saponins are complex compounds, a group of glycosides composed of sugars, steroid or triterpenoid moieties. They are widely distributed in plants, but are relatively uncommon in animals. The saponins derived from sea cucumbers are of the triterpenoid type. Several saponins have been isolated from various species of holothurians. Chanley et al (6, 7) isolated a crystalline saponin which was termed holothurin A, having an empirical formula of $C_{50-52}H_{81-85}O_{25-26}SNa$, and IR absorption at 1748 and 1629 cm^{-1}, but no absorption in the UV region. Yasumoto et al (8) isolated a saponin which they designated as holothurin B, having an empirical formula of $C_{45}H_{75}O_{20}SNa$, with a sugar composition and other properties similar to those of holothurin A. In addition, other saponins have been isolated and their structures studied which have been designated as holotoxin, stichoposide, cumarioside, and thelothurin. The chemical characteristics of these compounds have been discussed at length by Scheuer (9, 10) and by Hashimoto (11). Echinoderm saponins have recently been reviewed in depth by Burnell and ApSimov in Marine Natural Products, Volume V (12).

Molluscs. Molluscs have been incriminated in a number of intoxications aside from paralytic shellfish poisoning and diarrhetic shellfish poisoning.

Abalone Poisoning. Abalone viscera is commonly eaten in Japan where it is locally known as tsunowata or tochiri. It is eaten boiled, in the fresh state, or salted. Three abalone species have been incriminated, *Haliotis sieboldi*, *H. japonicus*, and *H. discus hannai*, in human intoxications. A similar type of intoxication has been caused by ingestion of the Japanese turban shell, *Turbo cornutus*. Abalone poisoning is known to cause a skin eruption, swelling, itching, and finally a loss of the skin and hair of the ears of cats. Takenaka, et al (13), Konga (14), and Hashimoto, et al (15) reported outbreaks of abalone poisoning in humans. The symptoms consist of a sudden onset of burning, itching, erythema, subcutaneous infiltration, and sometimes ulceration of the skin. The toxic substance is a pyrophenophorbide *a*, a derivative of algal chlorophyll *a*, which causes photosensitization.

Callistin Shellfish Poisoning. Numerous persons have been poisoned from eating the Japanese Callista, *Callista brevisiphonata* in the vicinity of Mori-machi, Hokkaido. The ovary of this clam contains a high concentration of choline during the spawning season. Intoxications have occurred only during the months of May to September in Japan. The onset of symptoms is rapid, usually less than one hour, and consist of itching, flushing of the face, urticaria, sensation of constriction of the chest, epigastric and abdominal pain, nausea, vomiting, paralysis or numbness of the throat, mouth and tongue, sweating, chills, and fever. The victim usually recovers within a period of two days. No fatalities have been reported. The Japanese government has prohibited the harvesting of these shellfish for a certain period after outbreaks have occurred. Cooking does not destroy the toxic properties of the shellfish and the degree of freshness of the organism is not a factor in the occurrence of the poisoning. Yasumoto and Sano (16) and Yasumoto and Shimizu (17) reported finding betaines present in *Callista* clam ovaries, including the following: glycine betaine, homoserine betaine, valine betaine, -butyrobetaine, - valerobetaine, homarine, and stachydrine, but the choline content was very low.

Cephalopod Poisoning. Intoxications resulting from the ingestion of toxic cephalopods have been caused by squid and octopus taken from certain specific areas in Japan. From 1952 to 1955, there were 779 outbreaks involving 2,974 persons and 10 deaths. Routine bacteriological tests were negative. It was believed that the poisoning was due to a biotoxin of an unknown nature. Symptoms develop within a period of 10-20 hours and consisted of gastrointestinal upset, abdominal pain, headache, weakness, paralysis, and convulsions. Most of the victims recovered within a period of 48 hours. Ordinary cooking procedures apparently do not destroy the poison (18-20). This poisoning has been due to an unknown pollutant since no further outbreaks have been reported.

Ivory Shell Poisoning. Human intoxications have occurred in the Niigata and Shizuoka Prefectures from eating the Japanese ivory shell, Babylonia japonica. This shellfish is widely distributed in sandy mud areas along the coast of Japan and is commonly eaten. The symptoms consist of gastrointestinal upset, nausea, vomiting, and abdominal pain, diarrhea, numbness of the mouth, loss of consciousness, and convulsions. The toxin has been found only in the midgut gland of the shellfish. The flesh and other viscera are said to be nontoxic. Toxin levels have been recorded as high as 6,600 MU/g. The toxin, which has been designated as surugatoxin, is believed to be produced by a toxic gram-negative bacterium, as yet unidentified, which the snail feeds upon. The toxin has an empirical formula of $C_{25}H_{26}BrN_5O_{13}7H_2O$. The structure has been determined by Kosuge, et al (21, 22) by single crystal x-ray diffraction. Hirayama, et al (23, 24) have isolated a toxic crystalline substance from the digestive gland of B. japonica, which they have termed IS-toxin, having an empirical formula of $C_{22}H_{36}BrN_4O_{14}$ (MW 660). IS-toxin is said to be 30 times more potent than surugatoxin. Surugatoxin and IS-toxin have a specific autonomic ganglionic blocking action which is attributable to a specific inhibitory action on the microtinic ganglionic receptors (25). Tetrodotoxin has recently been found to be present in B. japonica (26, 27).

Turban Shell Poisoning. The turban or top shells are members of the gastropod family Turbinidae. An outbreak of poisoning has been reported from Marcus Island, Western Pacific by Hashimoto, et al (28) from eating Turbo argyrostoma. The clinical characteristics were similar to those of ciguatera fish poisoning. The recovery period was about two months. A similar outbreak has resulted from the ingestion of T. setosus in Hao, Tuamotu Islands (29). Several toxic substances have been isolated from the midgut gland of Turbo, both fat- and water-soluble fractions. Three sulfur containing amines were extracted, (3-methylthiopropyl)trimethylammonium chloride; [3-(dimethylsulfonio)propyl] trimethylammonium dichloride, and toxin B; and candicine chloride, (ρ-hydroxyphenethyel) trimethylammonium chloride. These basic substances of low molecular weight exhibit chromatographic behavior and symptoms in mice similar to paralytic shellfish poison. Toxin B and (3-methylthiopropyl)trimethylammonium chloride both have an MDL of 6 mg/kg (IP injection in mice). It is not certain that the isolated toxins were the same as those having caused human intoxications (30-35). Paralytic shellfish poisoning has been found to be present in the turban shells T. marmorata, T. argyrostoma, and in the top shells Tectus pyramis, T. nilotica maxima from Ishigaki Island, Japan (36).

Tridacna Clam Poisoning. Giant clams of the species Tridacna maxima, commonly eaten in French Polynesia, may sometimes give rise to digestive disorders and serious disturbances of the nervous system, such as paresthesia, lack of motor coordination, ataxia, and tremor. Such on outbreak occurred at Bora-Bora in the Society Islands in 1964 (37). About 30 persons were affected and there were two deaths. Numerous domestic animals which had eaten the remnants of the toxic products also died. Toxicological study of the specimens taken at the time did not lead to the isolation of a pure compound. The maximum concentration of the poison is in the mantle and viscera of the clam (38). Banner described the presence of both water- and fat-soluble toxins in the Tridacna, but their properties are unknown.

Whelk Poisoning. This form of poisoning is caused by the ingestion of whole whelks or ivory shells, i.e., molluscs of the genus Neptunea, Buccinum, and Fusitriton. The poison is found in the salivary glands of the mollusc and consists of tetramine ($C_4H_{12}N$). Human intoxications have occurred in Japan. Symptoms consist of intense headaches, dizziness, nausea, vomiting, visual impairment, and dryness of the mouth. No deaths have been reported. Ordinary cooking procedures do not destroy the poison (39-45).

Arthropods. Crab Poisoning. Human intoxications have been reported from eating crabs in the Amami, Palau, Ryukyu, Singapore, Mauritius, and Philippine Islands (46-55). Most of the toxic crab species are members of family Xanthidae and include the genera Demania, Carpilius, Atergatis, Platypodia, Zosimus, Lophozozymus, and Eriphia. Clinical symptoms develop within 15 minutes to several hours after ingestion, involving gastrointestinal upset, numbness of the lips and limbs, atoxic gait, stupor, aphasia, respiratory distress, loss of consciousness, and generally death within 4-6 hours. The case fatality rate is very high. The clinical picture compares favorably with that of paralytic shellfish poisoning. The poison from Zosimus aeneus has been described by Hashimoto, et al (56), Konosu, et al (57), and Noguchi et al (58) and has been found to be pharmacologically and chemically similar to saxitoxin. Yasumoto, et al (27, 36, 59) have reported saxitoxin, neosaxitoxin, and gonyautoxins to be present in Okinawan crab species Neoanthias impressus, Actaeodes tomentosus, Eriphia scabricula, Pilumnus vespertillo, Schizophrys aspera, Thalamita sp., and Percnon planissimum. The food source of the PSP toxins was found to be a calcareous red alga Jania sp. (60). Raj et al (61) reported finding PSP toxins in two species of xanthid crabs from Fiji, Atergatis floridus and Zosimus aeneus.

Coconut Crab Poisoning. The land coconut crab, Bigrus latro, is commonly eaten in tropical areas. However, on occasion it may be poisonous. Outbreaks have been reported in the Ryukyu, Rap, and Tuamotu Islands (62-66). Symptoms include nausea, vomiting, headache, chills, joint aches, muscular weakness, and exhaustion. Deaths have been reported. The nature of the toxin is unknown.

Horseshoe Crab Poisoning. Numerous intoxications have been reported in Thailand from eating Asiatic horseshoe crabs, Carcinoscorpius rotundicauda and Tachypleus gigas, of the family Xiphosuridae, but there has been some question as to the toxicity of the latter species. These intoxications are locally referred to as mimi poisoning. The symptoms consist of nausea, vomiting, abdominal cramps. dizziness, cardiac palpitation, weakness, paresthesias of the lower limbs, aphonea, burning sensation of the mouth and throat, hypersalivation, loss of consciousness, muscular paralysis, and death. The case fatality rate is allegedly high. The precise chemical and pharmacological properties of the poison are unknown. However, Fusetani, et al (67) have investigated the toxin of C. rotundicauda and found the poison to be chemically similar to saxitoxin. The fat-soluble fractions were all nontoxic. The hepatopancreas was most toxic (10 MU/g of wet tissue); the eggs (5 Mu/g) and muscle (2 MU/g) were only slightly toxic. They isolated two toxic fractions, following the procedures of Shimizu, et al (68) and Onoue, et al (69). The major fraction was similar to saxitoxin, but the minor fraction differed from any known marine biotoxin. One specimen had a total of 2600 MU of the major toxin in the eggs, which would be lethal to a human when compared to saxitoxin. Tests on specimens of T. gigas were found to be nontoxic.

Marine Vertebrate Intoxications

Most marine biotoxications are caused by fishes, but turtles and mammals have been incriminated in rare instances. A tentative classification of ichthyotoxic fishes is as follows:

Poisonous Fishes. Defined as fishes which, when ingested, cause a biotoxication in humans due to a toxic substance present in the fish. Fishes that may become accidentally contaminated by bacterial food pathogens are not included.

Ichthyosarcotoxic Fishes. Those fishes that contain a poison within the flesh in the broadest sense, i.e., musculature, viscera, skin or slime (mucus), which when ingested by humans will produce a biotoxication. The toxins are oral poisons believed to be substances of small molecular size that are generally not destroyed by heat or gastric juices. The various kinds of biotoxicity recognized among ichthyosarcotoxic fishes are as follows: 1) Poisonous cyclostomes (lampreys and hagfishes) causing cyclostome poisoning; 2) Poisonous elasmobranchs (sharks and rays) causing elasmobranch poisoning; 3) Ciguatoxic fishes causing ciguatera poisoning; 4) Tetrodotoxic fishes causing pufferfish poisoning; 5) Clupeotoxic fishes causing clupeotoxism; 6) Scombrotoxic fishes causing scombroid fish poisoning; 7) Hallucinogenic fishes causing hallucinatory fish poisoning; and 8) Gempylotoxic fishes causing gempylid fish poisoning.

Ichthyootoxic Fishes. Those fish that produce a poison which is generally restricted to the gonads of the fish. The musculature and other parts of the fish are usually edible. There is a definite relationship between gonadal activity and toxin production. Fishes in this group are mainly freshwater species, but a few marine species have been incriminated.

Ichthyohaemotoxic Fishes. Those fishes having poisonous blood. The poison is usually destroyed by heat and gastric juices.

Ichthyocrinotoxic Fishes. Those fishes that produce a poison by means of glandular structures, independent of a true venom apparatus, i.e., poison glands are present but there is no traumagenic device.

Venomous or Acanthotoxic Fishes. Those fishes that produce poisons by means of glandular structures and are equipped with a traumagenic device to transmit their venoms. The poisons are parenteral toxins, usually large molecules, and are readily destroyed by heat or gastric juices. Venomous fishes are of no direct concern as oral marine intoxicants, the most serious biotoxications of public health importance being produced by ichthyosarcotoxic fishes.

Types of Ichthyosarcotoxism. Chimaera Poisoning. The musculature and viscera of some of the chimaeras or ratfishes have been found to be toxic. Unfortunately very little is known concerning either the clinical characteristics of the biotoxication or the chemical nature of the poison.

Cyclostome Poisoning. The slime and flesh of certain lampreys and hagfishes are reported to produce a gastrointestinal distrubance including nausea, vomiting, and dysenteric diarrhea. The slime and skin are said to contain a poison which is not destroyed by gastric juices or heat. The chemical and pharmacological properties of the poison are not known (70-74), and no control measures exist. There is a possible danger of using these fish in marine protein concentrates.

Elasmobranch Poisoning. The musculature of some sharks, such as the Greenland shark, *Somniosus microcephalus*, is said to be poisonous to eat (75,76) and the livers of several species of tropical sharks may cause severe intoxication (70, 77-81). While the musculature may cause symptoms of a mild gastroenteritis, ingestion of toxic shark livers may have very severe effects, with the onset of symptoms within a period of less than 30 minutes. Nausea, vomiting, diarrhea, abdominal pain, headache, weak pulse, malaise, cold sweats, oral paresthesia, and a burning sensation of the tongue, throat and esophagus may be present. The neurological symptoms develop later and consist of extreme weakness, trismus, muscular cramps, a sensation of heaviness of the limbs, loss of superficial reflexes, ataxia, delerium, incontinence, respiratory distress, visual distrubances, convulsions, and death. The recovery period, if the victim recovers, varies from several days to several weeks. The mortality rate is not known. The severity of the symptoms varies with the amount of shark liver eaten, the species of shark, physical condition of the victim, and other factors which are not clearly understood.

The nature of poisons in sharks is a complex one. Vitamin A may be a causative factor in some shark species. The liver of sharks contains waxes, diacyl glyceryl ether, and hydrocarbons of which squalene and pristane are the most predominant (11). Sakurai and Masukara (82) and Kaneda, et al (83) demonstrated that squalene and wax can cause seborrhea and lethalities in rats. These studies suggest that the shark livers are not suitable for food. There is also clinical evidence that the ingestion of livers from tropical shark species may produce a ciguatera-like syndrome.

Clupeotoxism. This is a form of ichthyosarcotoxism caused by fishes of the order Clupeiformes, which includes the families Clupeidae (herrings), Engraulidae (anchovies), Elopidae (tarpons), Albulidae (bonefishes), Pterothrissidae (deep sea bonefishes), and Alepocephalidae (deep sea slickheads). The families most commonly incriminated in human clupeotoxications are Clupeidae and Engraulidae. Clupeotoxism is a sporadic, unpredictable public health problem of the tropical Pacific Ocean, the Caribbean Sea, and the tropical Atlantic Ocean. Most poisonings have occurred in tropical island areas and were caused by fishes that had been captured close to shore. The viscera are regarded as the most toxic part of the fish. It is believed that the poison is derived from a planktonic source. Tropical clupeiform fishes are most likely to be toxic during the warm summer months. There is no possible way to detect a toxic clupeiform fish by its appearance and the degree of freshness has no bearing on its toxicity. The clinical characteristics of clupeotoxism are distinct and usually violent. The first indication of poisoning is a sharp metallic taste which may be present immediately after ingestion of the fish. This is rapidly followed by a severe gastrointestinal upset, which may be accompanied by a drop in blood pressure, cyanosis, and other evidence of a vascular collapse. Concurrently, or within a short period, a variety of neurological disturbances develop, including nervousness, dilated pupils, violent headaches, numbness, tingling, hypersalivation, muscular cramps, respiratory distress, paralysis, convulsions, coma, and death. Death may occur in less than 15 minutes. There are no accurate statistics on the case fatality rate, but it is reported to be very high. Treatment is symptomatic. There is no information available on the pharmacological or chemical properties of the poison and there are no public health control measures.

Hallucinatory Fish Poisoning (Ichthyoallyeinotoxism). This form of ichthyosarcotoxism is caused by the ingestion of certain types of reef fishes which occur in the tropical Pacific and Indian Oceans. The families incriminated in ichthyoallyeinotoxism include the following: Acanthuridae, Kyphosidae, Mugilidae, Mullidae, Pomacentridae, Serranidae, and Siganidae. Ichthyoallyeinotoxism may result from eating the flesh or the head of the fish where the poison is reputedly concentrated. This biotoxication is sporadic and unpredictable in its occurrence. The poison affects primarily the central nervous system. The symptoms may develop within a few minutes to two hours and persist for 24 hours or longer. Symptoms are dizziness, loss of equilibrium, lack of motor coordination, hallucinations, and mental depression. A common complaint of the victim is that "Someone is sitting on my chest," or there is a sensation of a tight constriction around the chest. The conviction that he is going to die, or some other frightening fantasy, is a characteristic part of the clinical picture. Other complaints consist of itching, burning of the throat, muscular weakness, and abdominal distress. No fatalities have been reported, and in comparison with other forms of ichthyosarcotoxism, hallucinogenic fish poisoning is relatively mild. There is no information available concerning the pharmacological and chemical properties of the poison. There are no public health control measures. Ordinary cooking procedures do not destroy the poison.

Gempylid Fish Poisoning. Gempylotoxism is caused by ingestion of the flesh of fishes of the family Gempylidae, the escolars or pelagic mackerels, which contain an oil with a pronounced purgative effect. The purgative oil is present also in the bones. Gempylid poisoning is usually not a serious matter and many native groups esteem these fishes despite their purgative effects. Ordinary cooking procedures to not destroy the purgative effect of the oil. The flesh of these fish contains a large amount of wax esters (84, 85).

Miscellaneous Types of Fish Poisoning. Barracuda Poisoning. Historically barracudas have most commonly been involved in ciguatera fish poisoning. However, an outbreak of poisoning in Tokyo, Japan, in 1949, suggested that another type of poison might also be involved. The symptoms consisted of numbness of the lips and mouth, ataxia, and hypersalivation. The paradoxical disturbance as found in ciguatera was absent. There were no gastrointestinal disturbances. Hashimoto (86) extracted a toxin from the barracuda, *Sphyraena picuda*. The poison extracted was found to affect cats, but mice were insensitive to it. The poison appeared to be in the fat-soluble fraction. Unfortunately there was not sufficient quantities of the toxin to permit adequate pharmacological and chemical studies.

Filefish Poisoning. Members of the filefish family, Monacanthidae, have produced human intoxications in the western Pacific islands. It was once believed that this was merely a form of ciguatera fish poisoning. However, studies by Hashimoto et al (87) have found that the poison was not a ciguatoxin fraction. It was found to be a water-soluble substance and was initially referred to as aluterin. However, they later found it to be chemically identical to palytoxin, which was first discovered in the anthozoan *Palythoa tuberculosa*. Ingestion of the poison produces vomiting, diarrhea, and joint aches in humans, and rapid death in pigs. The viscera of filefish are said to be extremely toxic. Palytoxin has an LD_{50} (IP, mice) of 0.15 ug/kg (88).

Goby Poisoning. At least one species of the family Gobiidae, namely *Gobius criniger*, has been found to contain tetrodotoxin (89, 90). Fatalities to humans and animals have been reported in Taiwan, and this goby is known to be toxic in the Philippines and Ryukyu Islands.

Parrotfish Poisoning. For many years intoxications resulting from the ingestion of parrotfish, members of the family Scaridae, were considered to be ciguatera. However, Hashimoto et al (91) have isolated a poison from *Scarus gibbus* from the Gambier Islands, which they designated scaritoxin. The poison was found to be largely concentrated in the musculature. The flesh was extracted by the method adopted for extraction of ciguatoxin and the crude toxin obtained was fractionated by column chromatography on silicic acid with chloroform-methanol (9:1). Two toxic factors, however, were separated by DEAE-cellulose column chromatography. Scaritoxin was eluted with chloroform, whereas the toxin identical with ciguatoxin was eluted with chloroform-methanol (1:1). The scaritoxin fraction was further purified by two gel filtrations on Sephadex LJ-20 to give a product with a lethality of 30 ug/kg in mice. The purified toxin gave several spots on a thin layer plate. Scaritoxin and ciguatoxin are readily differentiated by thin layer chromatography. Hashimoto et al (91) believe that scaritoxin is present in a variety of ciguatoxic fish. The symptoms initially resemble ciguatera fish poisoning, but after five to ten days develop into a second phase consisting of a loss of equilibrium, ataxia, tremors, and dysmetria. There is definite evidence of a cerebellar disturbance. Recovery is very slow, requiring a month or more.

Rabbitfish Poisoning. Human intoxications have resulted from the ingestion of rabbitfish, members of the family Siganidae. The symptoms are headache, flushing, redness of the eyes, itching, and bitter taste in the mouth after meals. No gastrointestinal disturbances have been reported. According to Hashimoto et al (63) the poison is non-ciguatoxic, and may be derived from the marine blue-green alga Lyngbya majuscula (Microcoleus lyngbyaceus). However, Banner questions the relationship of Lyngbya to ciguatera because of the difference in symptomatology characteristics (92, 93).

Fish Liver Poisoning. The livers of large fish species such as sharks, tunas, and sea bass may cause intense headaches, vomiting, facial edema, fever, and severe desquamation. Onset of symptoms is from 30 minutes to 12 hours. All of these fishes contain extremely high levels of vitamin A in their livers. As the fish grows, the concentration of vitamin A in the liver increases (94-96).

Ichthyootoxism. Ichthyootoxism is one of the lesser-known forms of fish poisoning. Ichthyootoxic fishes constitute a group of organisms that produce a poison which is generally restricted to the gonads of the fish. There is a definite relationship between gonadal activity and toxin production. The musculature and other parts of the fish are usually edible. Fishes in this group are mainly freshwater, but others are anadromorous, brackish water, or marine species. Many of the fishes involved in ichthyootoxism are phylogenetically unrelated. Although toxic fish roe poisoning has been recognized for many years, very little chemical or pharmacological work has been done except for the recent studies of Japanese scientists. The following data are based largely on their work.

Blenny Roe Poisoning. Human intoxications have resulted from the ingestion of the roe of the northern blenny, Stichaeus grigorjewi, in Japan (97-101). The symptoms consist of nausea, vomiting, malaise, diarrhea, prostration, dizziness, and abdominal pain. The toxin present in S. grigorjewi was first designated as dinogunellin, after the former generic name of the fish, but was later changed to lipostichaerin after the name of the fish was changed. The poison has been found to be a high-density lipoprotein fraction similar to lipovitellin of bird's egg yolk. The purified poison has been found to have an LD_{50} (IP, mice) of 25 mg/kg. Hatano and Hashimoto (102) extracted pure dinogunellin and found it to have a MW of 800, and exhibits strong UV absorption at 259 nm, which is characteristic of adenine. Structural studies have been conducted by Hatano et al (103) in which they isolated two moles of ammonia and one each of adenine, ribose, phosphoric acid, and aspartic acid, which they designated as compound A, and compound B, identified as 2-aminosuccinamide. A closely-related species to the northern blenny, S. nozawai, from Hokkaido, has also been found to contain dinogunellin. No human intoxications have been reported from this species.

Cabezon Poisoning. The roe of the cabezon or marbled sculpin, Scorpaenichthys marmoratus of the Pacific coast of the United States, has edible flesh, but poisonous roe. Fuhrman et al (104, 105) and Hashimoto et al (106) studied the roe of S. marmoratus. The purified poison has an LD_{50} of 200 mg/kg (IP, mice), and is believed to be chemically identical to dinogunellin.

Numerous other species of freshwater fishes are reported to contain toxic roe, and have caused human intoxications, but the poisons have not been studied to any great extent.

Poisonous Marine Turtles

Toxic marine turtles may cause a type of poisoning about which little is known. However, the cases reported bear witness to its serious nature. While most species may be eaten without fear, in the tropical parts of the Pacific, particularly around the islands of Japan, some turtles may become extremely toxic.

Species known to have caused poisoning in man are the green turtle, *Chelonia mydas*, the leatherback turtle, *Dermochelys coriacea*, and, above all, the hawksbill turtle, *Eretmochelys imbricata*, found in the waters around the Philippines, India, New Guinea, Tahiti, and Japan. Symptoms may become manifest in a few hours to several days after consumption of the turtle meat and consist of nausea, vomiting, diarrhea, abdominal cramps, vertigo, a dry, burning sensation in the lips and the tongue, and irritation of the mouth and pharynx. Difficulty in swallowing, excessive salivation and a stomatopharyngitis may occur quite late, but are still the more serious in that they bring about difficulty in breathing. Later still, red papules the size of a pin head appear on the tongue and finally ulcerate. If the poisoning is severe, the patient becomes somnolent and prostrate, which may precede death from damage to the liver and kidneys. The case fatality rate is about 44 percent. The pharmacological and chemical properties are unknown.

Marine Mammals

Several marine mammal species have been blamed for poisonings. The polar bear, *Thalarctos maritimus*, is encountered throughout the Arctic regions and there have been numerous cases of poisoning due to ingestion of the liver. The main symtpoms are intense formication, stabbing frontal headaches, nausea, vomiting, diarrhea, apathy, giddiness, irritability, collapse, photophobia, and convulsions. The condition is rarely fatal and the patient usually recovers in a few days. The toxicity is due to excessive quantities of vitamin A present in the tissues. Certain species of seals, such as the bearded seal, *Erignathus barbatus*, sea lions, such as Peron's sea lion, *Neophoca cinerae*, the sei whale, *Bálaenoptera borealis*, and dolphins have been blamed in various parts of the world for disorders of the same kind (107, 11).

Comments

Numerous other marine organisms, invertebrates, fish and algae are known to contain poisons, but most of them have not caused human intoxication due to ingestion. It is becoming obvious that some of these poisons which were previously believed to originate in certain organisms are known to be distributed in other organisms; e.g., palytoxin from the coelenterate *Palythoa*, more recently found in filefish (Monacanthidae); saxitoxin from dinoflagellates found in turban shells (Turbinidae), crabs (Xanthidae), horseshoe crabs (Ziphosuridae), and red calcareous alga (*Jania*); tetrodotoxin from pufferfish (Tetraodontidae) found in gobyfish (Gobiidae), ivory shells (Buccinidae); and an unknown ciguatoxin-like poison in rabbitfish (Siganidae), possibly derived from a blue-green alga *Lyngbya majuscula* (*Microcoleus lyngbyaceus*).

The number of poisonous marine organisms is rapidly expanding into the food resources of man and there is an ever-increasing economical and public health requirement to intensify scientific investigation into the fascinating and, at times, baffling arena of marine biotoxicology. Greater research efforts need to be expanded in the epidemiology, toxicology, and chemistry of these poisons.

Literature Cited

1. Kat, M. In "Toxic Dinoflagellate Blooms"; Taylor, D. L.; Selinger, H. H., Eds.; Elsevier: New York, 1979; pp. 215-220.
2. Yasumoto, T.; Oshima, Y.; Sagawara, Y.; Fukuyo, Y.; Oguri, H.; Igarashi, T.; Fujita, N. *Bull. Japan Soc. Sci. Fish.* 1980, 46, 11, 1405.
3. Avaria, S. In "Toxic Dinoflagellate Blooms"; Taylor, D. L.; Selinger, H. H., Eds.; Elsevier: New York, 1979; pp. 161-164.
4. Martin, E. J. *Pac. Sci.* 1960, 14, 4, 403.
5. Martin, E. J., unpublished data.
6. Chanley, J. D.; Ledeen, R.; Wax, J.; Nigrelli, R. F.; Sobotka, H. *J. Am. Chem. Soc.* 1959, 81, 5180.
7. Chanley, J. D.; Perlstein, J.; Nigrelli, R. F.; Sobotka, H. *Ann. N.Y. Acad. Sci.* 1960, 90, 902
8. Yasumoto, T.; Nakamura, K; Hashimoto, Y. *Bull. Japan Soc. Sci. Fish.* 1967, 31, 7.
9. Scheuer, P. J., Ed. "Chemistry of Marine Natural Products"; Academic: New York, 1973; 201 p.
10. Scheuer, P. J., Ed. "Marine Natural Products, Chemical and Biological Perspectives"; Academic: New York, 1978; Vol. II, 392 p.
11. Hashimoto, Y. "Marine Toxins and Other Bioactive Marine Metabolites"; Japan Scientific Societies Press: Tokyo, 1979; 369 p.
12. Burnell, D. J.; ApSimov, J. W. In "Marine Natural Products"; Scheuer, P. J., Ed.; Academic: New York, 1983; Vol. V, pp. 287.
13. Takenaka, S.; Sawada, G.; Yoshioka, M. *Tokyo Iji Shinshi* 1899, No. 1114, 1359.
14. Koga, S. *Tokyo Iji Shinshi* 1900, No. 1179.
15. Hashimoto, Y.; Naito, K.; Tsutsumi, J. *Bull. Japan Soc. Sci. Fish.* 1960, 26, 1216.
16. Yasumoto, T.; Sano, F. *Bull. Japan Soc. Sci. Fish.* 1977, 40, 1163.
17. Yasumoto, T.; Shimizu, Y. *Bull. Japan Soc. Sci. Fish.* 1977, 43, 201.
18. Motohiro, T.; Tanikawa, E. *Bull. Fac. Fish. Hokkaido Univ.* 1952, 3, 142.
19. Kawabata, T.; Halstead, B. W.; Judefind, T. F. *Am. J. Trop. Med. Hyg.* 1957, 6, 5, 935.
20. Halstead, B.W. "Poisonous and Venomous Marine Animals of the World"; U.S. Government Printing Office: Washington, D.C., 1965; Vol. 1, 994 p.

21. Kosuge, T.; Zenda, H.; Ochiai, A.; Masaki, N.; Noguchi, M.; Kimura, S.; Narita, H. Tetrahedron Letts. 1972, 2545.
22. Kosuge, T.; Tsuji, K.; Hirai, K.; Yamaguchi, K.; Okamoto, T.; Iitaka, Y. Tetrahedron Letts. 1981, 22, 35, 3417.
23. Hirayama, H.; Sugihara, K.; Tsuyama, S.; Wakigawa, K.; Ohkuma, H.; Gohgi, K. Folia Pharmacol. Japan 1973, 69, 583.
24. Hirayama, H.; Sugihara, K.; Wakigawa, K.; Ohkuma, H.; Gohgi, K. Japan J. Pharmacol. 1974, 24, 559.
25. Hayashi, E.; Yamada, S. Brit. J. Pharmacol. 1975, 53, 207.
26. Noguchi, T.; Maruyama, J.; Ueda, Y.; Hashimoto, K.; Harada, T. Bull, Japan Soc. Sci. Fish. 1981, 47, 7, 909.
27. Yasumoto, T.; Oshima, Y.; Hosaka, M.; Miyakoshi, S. Bull. Japan Soc. Sci. Fish. 1981, 47, 7, 929.
28. Hashimoto Y.; Kamiya, H.; Shibota, M., unpublished data.
29. Bagnis, R. Rev. Corps Sante Armes 1969, 10, 783.
30. Yasumoto, T.; Kanno, K. Bull. Japan Soc. Sci. Fish. 1976, 42, 1399.
31. Yasumoto, T.; Bagnis, R.; Vernoux, J. P. Bull. Japan Soc. Sci. Fish. 1976, 42, 359.
32. Kanno, K.; Yasumoto, T. In "Abstract of Paper, Autumn Session"; Japan Soc. Sci. Fish:Kyoto, 1974, p. 123.
33. Yasumoto, T.; Endo, M. Bull. Japan Soc. Sci. Fish. 1973, 39, 1055.
34. Yasumoto, T.; Endo, M. Bull. Japan Soc. Sci. Fish. 1974, 40, 217.
35. Yasumoto, T. Bull. Japan Soc. Sci. Fish. 1974, 40, 1169.
36. Yasumoto, T.; Oshima, Y.; Kotaki, Y. Toxicon 1983, Suppl. 3, 513.
37. Bagnis, R. Bull. Soc. Pathol. Exotique 1967, 60, 6, 580.
38. Banner, A. H. In "Animals Toxins"; Russell. F. E.; Saunders, P. R., Eds.; Pergamon: Oxford, 1967; p. 157.
39. Asano, M. Bull. Japan Soc. Sci. Fish. 1952, 17, 18-19, 73.
40. Emmelin, N.; Fange, R. Acta Zool.(Stockholm) 1958, 39, 47.
41. Asano, M.; Itoh, M. Tohoku J. Agric. Res. 1959, 10, 2, 209.
42. Asano, M.; Itoh, M. Ann. N.Y. Acad. Sci. 1960, 90, 3, 674.
43. Fange, R. Nature 1957, 180, 196.
44. Fange, R. Acta Zool. (Stockholm) 1958, 39, 39.
45. Fange, R. Ann. N.Y. Acad. Sci. 1960, 90, 3, 689.
46. Mote, G. E.; Halstead, B. W.; Hashimoto, Y. Clin. Toxicol. 1970, 3, 597.
47. Hashimoto, Y.; Konosu, S.; Yasumoto, T.; Inoue, A.; Noguchi, T., unpublished data.
48. Alcala, A. C.; Halstead, B. W. Clin. Toxicol. 1970, 3, 609.
49. Garth, J. S. Micronesica 1971, 7, 179.
50. Teh, Y. F.; Gardiner, J. E. Pharmacol. Res. Commun. 1970, 2, 251.
51. Gonzales, R. B.; Alcala, A. C. Toxicon 1977, 15.
52. Inoui, A.; Noguchi, T.; Konosu, S.; Hashimoto, Y. Toxicon 1968, 6, 119.
53. Carumbana, E. E.; Alcala, A. C.; Ortega, E. P. Silliman J. 1976, 23, 265.
54. Guinot, D., unpublished data.
55. Garth, J. S.; Alcala, A. C. Proc. 3rd Int. Coral Reef Symp., 1977, p. 645.

56. Hashimoto, Y.; Konosu, S.; Yasumoto, T.; Inoue, A.; Noguchi, T.; Toxicon 1967, 5, 85.
57. Konosu, S.; Inoue, A.; Noguchi, T.; Hashimoto, Y. Toxicon 1968, 6, 113.
58. Noguchi, T.; Konosu, S.; Hashimoto, Y. Toxicon 1969, 7, 325.
59. Yasumoto, T.; Oshima, Y.; Tajiri, M.; Kotaki, Y. Bull. Japan Soc. Sci. Fish. 1983, 49, 4, 633.
60. Kotaki, Y.; Tajiri, M.; Oshima, Y.; Yasumoto, T. Bull. Japan Soc. Sci. Fish. 1983, 49, 2, 283.
61. Raj, U.; Haq, H.; Oshima, Y.; Yasumoto, T. Toxicon 1983, 21, 4, 547.
62. Hashimoto, Y.; Konosu, S.; Yasumoto, T.; Kamiya, H., unpublished data.
63. Hashimoto, Y.; Fusetani, N.; Kimura, S. Bull. Japan Soc. Sci. Fish. 1969, 35, 1086.
64. Yonabaru, S., unpublished data.
65. Bagnis, R. Clin. Toxicol. 1970, 3, 4, 585.
66. Hothuis, L. B. Crustaceana 1968, 15, 215.
67. Fusetani, N.; Endo, H.; Hashimoto, K.; Takahashi, K. Toxicon 1982, 20, 3, 662.
68. Shimizu, Y.; Alam, M.; Oshima, Y.; Fallon, W. E. Biochem. Biophys. Res. Commun. 1975, 66, 731.
69. Onoue, Y.; Noguchi, T.; Hashimoto, K. Toxicon 1969, 7, 325.
70. Coutiere, H., unpublished data.
71. Pawlowsky, E. N., unpublished data.
72. Halstead, B.W. Public Health Report 1958, 73, 4, 302.
73. Halstead, B.W. Clin. Pharmacol. Ther. 1964, 5, 615.
74. Russell, F. E. Adv. Marine Biol. 1965, 3, 255.
75. Jensen, A. S. Mindeskr, Japetus Steenstrups Fods. 1914, 30, 12.
76. Jensen, A. S. Spolia Zool. Mus. Haun. 1948, 9, 20.
77. Coutand, H. Thesis, Faculte de Medecine, Universite de Montpellier.
78. Phisalix, M., unpublished data.
79. Fish, C. J.; Cobb, M. C. "U.S. Fisheries and Wildlife Service Research Report, No.36," Washington, D.C., 1954, p. 14.
80. Halstead, B.W. "Dangerous Marine Animals"; Cornell Maritime: Cambridge, Md., 1959; 146 p.
81. Helfrich, P., unpublished data.
82. Sakurai, Y.; Masukara, T. Rep. Food Res. Inst. Japan 1952, 7, 107.
83. Kaneda, T.; Sakai, H.; Ishii, S.; Arai, K. Bull. Tokai Reg. Fish Res. Lab. 1955, 12, 1.
84. Kimura, H. J. Chem. Ind. Japan 1926, 29, 620.
85. Cox, W. M.; Reid, E. E. J. Am. Chem. Soc. 1932, 54, 220.
86. Hashimoto, Y. Bull. Japan Soc. Sci. Fish. 1956, 21, 1153.
87. Hashimoto, Y.; Konosu, S.; Yasumoto, T.; Kamiya, H. Bull. Japan Soc. Sci. Fish. 1969, 35, 316.
88. Moore, R. E.; Scheuer, P. J. Science 1971, 172, 495.
89. Hashimoto, Y.; Noguchi, T. Toxicon 1971, 9, 79.
90. Hashimoto, Y.; Noguchi, T.; Mitsuura, N.; Kao, K., unpublished data.
91. Hashimoto, Y.; Kamiya, H.; Kinjo, K.; Yoshida, C. Bull. Japan Soc. Sci. Fish. 1975, 41, 903.

92. Banner, A. H. Hawaii Med. J. 1959, 19, 35
93. Banner, A. H.; Scheuer, P. J.; Sasaki, S.; Helfrich, P.; Alender, C.B. Ann. N.Y. Acad. Sci. 1960, 90, 770.
94. Miura, M. Tokyo J. Med. Sci. 1889, 3, 7, 366.
95. Abs. O. Ernahrungsforschung 1958, 3, 488.
96. Shimma, Y.; Taguchi, H. J. Food Hyg. Soc. Japan 1961, 2, 28.
97. Takayanagi, F.; Kitamura, T.; Satoh, T. unpublished data.
98. Asano, M.; Itoh, M. Tohoku J. Agric. Res. 1963, 13, 151.
99. Asano, M.;Itoh, M. Tohoku J. Agric. Res. 1966, 16, 299.
100. Sakai, M.; Shinano, H.; Kimura, T.; Ezura, Y.; Ban, M.; Hayashi, I. Food Sanit. Res. 1962, 12, 7, 53.
101. Asano, M. Bull. Japan Soc, Sci. Fish. 1952, 15, 113.
102. Hatano, M.; Hashimoto, Y. Toxicon 1974, 12, 231.
103. Hatano, M.; Marumoto, R.; Hashimoto, Y. In "Animal, Plant and Microbial Toxins"; Ohsaka, A., et al, Eds.; Plenum: New York, 1976; Vol. 2, p. 145.
104. Fuhrman, F. A.; Fuhrman, G. J., Dull, D. L.; Mosher, H. S. J. Agric. Food Chem. 1969, 17, 417.
105. Fuhrman, F. A.; Fuhrman, G. J.; Roseen, J. S. Toxicon 1970, 8, 55.
106. Hashimoto, Y.; Kawasaki, M.; Hatano, M. Toxicon 1976, 14,141.
107. Rodahl, K.; Moore, T. Biochem. J. 1943, 37, 166.

RECEIVED January 31, 1984

Alaska's Shellfish Industry

RICHARD A. NEVÉ[1] and PAUL B. REICHARDT[2]

[1]Alaska Department of Environmental Conservation, Juneau, AK 99811
[2]Department of Chemistry, University of Alaska, Fairbanks, AK 99701

This is a recounting of the history and a discussion of the potential of the Alaskan shellfish industry. Among the issues addressed are the varieties of clams available for commercial cultivation, the current locale and relative productivity of commercial clam banks, the current efforts at extending commercial harvesting and the contingent technological innovations, the deleterious effect upon the industry by paralytic shellfish poisoning (PSP), and the recent methods of monitoring for PSP. The article concludes with a citing of the difficulties this emerging renewable resource industry is facing.

Along Alaska's 33,000 miles of available and relatively unpolluted shoreline - exceeding the coastline of all other states combined - are found many habitats especially suitable to a commercial clam fishery. From the Bering Sea Peninsula in the north, along the southcentral Alaskan coast and throughout the southeast portion of the state are found several species of clams of sufficient size and abundance to merit commercial marketing. It is perhaps the largest, untapped fisheries resource in the United States. Along the Bering Sea Peninsula alone, it has been estimated that an annual harvest of whole clams could yield 19 to 25 million pounds of meat. More conservative estimates place the true sustainable harvest for the entire state in the neighborhood of 50 million pounds per year including shell weight. In either case, this represents considerable underutilization of a renewable resource (1).

Of the over one hundred species of clams found in Alaska's waters, the razor clam, *Siliqua patula*, is generally the most prized and has a very great commercial potential. This clam is found in abundance on many Alaskan beaches, and in the past has been successfully marketed; however, due to the unpredictable presence of PSP, its dual use now is as a bait for dungeness crab, and a relatively small consumable commercial market.

At the present time, there are only three approved beaches from which razor clams may be harvested for commercial purposes. These are Swikshak, Polly Creek, and the Cordova area. It is estimated that these three beaches alone represent a potential annual harvest of one million pounds (shell weight) and a minimum of two million dollars on the wholesale market.

Historically, the razor clam has been Alaska's one real clam fishery. Commercial harvest of razor clams in Cook Inlet began in 1919 with processing taking place at Snug Harbor on the

0097-6156/84/0262-0053$06.00/0
© 1984 American Chemical Society

west side. Very little data is available prior to 1945 other than noting that Polly Creek was the main harvest area. Razor clams are found up to Cape Kasilof on the east side and to the Kustatan River on the west side.

From 1945 to 1949, production was insignificant, averaging less than 10,000 pounds annually. For 1950 and 1951, production increased to 304,000 and 112,000 pounds, respectively (2). There was no production at all from 1952 through 1959. In 1954 Alaska was expelled from the National Shellfish Council, the national regulatory body for shellfish, which meant clams could no longer be shipped out of state for human consumption.

Operations (2) began again in 1960, with commercial harvesting being limited to the Polly Creek area. The fishery continued at Polly Creek through 1963 with the harvest declining from approximately 372,000 pounds in 1960 down to 195,000 pounds in 1962. From 1963 through 1970, no commercial harvest occurred. The annual harvest averaged less than 30,000 pounds from 1971 through 1973, at which time, the State was readmitted to the National Shellfish Sanitation Program. The Cook Inlet razor clam harvest from 1919 to the present came from Polly Creek exclusively. The extreme fluctuations in harvest throughout the history of this fishery are attributed more to fluctuations in effort and market conditions than to the abundance of clams.

The 1981 razor clam fishery in Upper Cook Inlet increased substantially in both harvest and effort. Early that year the Alaska Department of Environmental Conservation (ADEC), coordinating with the National Shellfish Council, obtained certification of the Crescent River bar as an extension to the Polly Creek beach.

This effectively doubled the area available to harvest for human consumption by allowing two processors to operate simultaneously without interfering with each other. One processor operated at Polly Creek with 15 to 20 hand-diggers while the other moved into the Crescent River bar area with about 30 individuals hand-digging. Approximately 70 percent of the 1981 harvest, or 315,000 pounds, came from Crescent River bar (2).

In 1981, the expanded harvest area and increased effort combined to yield the highest recorded commercial harvest of razor clams for Upper Cook Inlet since 1922. Over 440,000 pounds of clams in-the-round were harvested from April through September from the Poly Creek/Crescent River bar area. This equates to roughly one million razor clams. The razor clam fishery in Cook Inlet has garnered increasing commercial interest in recent years with the 1982 season resulting in a noticeable upswing in requests for information concerning harvesting and processing information; 1983 information is not yet available.

Applications for provisional permits to operate hydraulic clam dredges have been received. On-site inspection of these dredges prior to harvesting is prescribed in the permit. One of the more notable operations seeking a provisional permit is a vessel-mounted subtidal razor clam dredge capable of harvesting clams in depths up to ten fathoms. This enterprise initially targeted the bait clam area but will shift to the human consumption area (i.e., Polly Creek) if and when bait contracts are filled.

Presently, the Alaska Department of Fish and Game (ADF&G), (Commercial Fisheries Division, Soldotna office), carries out biweekly sampling for PSP and water quality as set forth in the Interagency Memorandum of Agreement. These sampling trips take place April through September and are the extent of the department's monitoring and data collection activities directed at razor clams in Cook Inlet. A monitoring program directed at gathering information on age structure, density, set frequency and other basic biological parameters is being considered, but budgetary constraints leave the initiation of such a program in doubt.

While it sounds like this is a *Siliqua patula* soliloquy, other species have potential also. There has been a new interest (starting in 1977) in the commercial utilization of geoducks (*Panope generosa*). Geoducks of acceptable commercial size are approximately 30 years of age, with some old-timers running over 100 years (3).

A limited exploratory cruise was conducted in 1980-81 in the Ketchikan area, and samples were taken to determine levels of PSP. The results of that cruise were sufficiently encouraging that additional funds were sought to do a more in-depth study. The goal was to establish commercial harvest techniques and to determine areas where commercial quantities of geoducks could be found. This would allow the new industry to be established, still protect the consuming public from any possible PSP contamination, and delineate areas of commercial concentration. The beds surveyed were chosen as a result of previous knowledge and tips from fishermen, or favorable-looking bottom conditions.

The second part of the study involved test marketing. Samples were market tested after undergoing various levels of processing. Some were shipped live; some were eviscerated; and others were butchered and shipped separately as necks or bodies. Different grades were also shipped to test the marketability of the lower quality geoducks. Market samples were sent to Japan, Los Angeles, New York, Seattle, Santa Barbara, San Francisco, San Diego, and Korea. All products were acceptable to the markets tested. Prices ranged in 1982 from $.70 per pound to $3 per pound, depending on quality and the degree of processing (3). It is anticipated that, in the future, local processors could buy directly from the fishermen and develop wholesale markets from their plants.

Samples were also taken from the geoduck beds for PSP analysis (3). Spot check samples were taken, with results indicating that high levels existed only in the viscera of the clam. Of all samples taken, only one sample of body meat taken near Gravina Island (Ketchikan, 1980) revealed PSP levels above the quarantine limit of 80 mg/100 g for human consumption.

Additional studies were taken to investigate clam abundance and distribution of potential harvest areas and to test for PSP homogeneity in geoduck beds to determine probable methods for product control. (At the same time samples of butter and horse clams were also collected for PSP testing.)

Indications are that numerous commercial beds of geoducks

are available, and that a viable industry could begin throughout the lower half of southeast Alaska.

In fact, this summer ADEC has received from the U.S. Food and Drug Administration (FDA), market clearance for human consumption of geoducks from beds near Noyes Island, west of Ketchikan. Other areas will likely open up next season.

Butter clams, (Saxidomus gigantea), found throughout southcentral and southeast Alaska, were commercially fished until the presence of PSP destroyed the industry. There are no plans for immediate utilization of this clam commercially because of the high PSP content of the meat, and lack of a rapid and suitable depuration procedure.

Although not commercially harvested in the past, the steamer clam, Protothaca staminea, is also available on some beaches in commercial quantities. At the present time, steamers are imported from outside the state as a luxury item.

Results (4) of a 1976-1977 clam survey (by National Marine Fisheries Services, FDA, the University of Alaska, ADF&G, and ADEC) indicate a potential commercial market for the surf clam (Spisula polynema). The beds, found along the Alaskan Peninsula between Port Moller and Port Heiden, can be harvested using an east coast-style hydraulic clam harvester which operated effectively in the Bering Sea substrates. A surf clam resource area extends over 1600 square miles at a depth identified between 13 to 18 fathoms. While detailed production fishing tests were not conducted, catch rates obtained with the three-foot wide harvester indicate commercial potential with a ten-foot wide harvester. The resource biomass has been conservatively calculated at 248,000 to 324,000 metric tons of whole clams, with an annual potential yield of about 19 to 25 million pounds of meat. The average size of Alaskan surf clams was found to be smaller than the East Coast species; however, the meat yield was considerably higher. Tests have indicated that surf clams can be held live aboard ship, either tanked or dry, without excessive dead loss. Bacteriological tests indicate that clams and the environment meet FDA requirements. No PSP toxins were detected in the edible portion of surf clams using the mouse bioassay.

In 1917 the industry produced five million pounds of shellfish products (5). Today Alaska's commercial clam fishery is virtually nonexistent. The destruction of this fledgling clam industry in the 1940s was a direct result of product contamination by PSP. Forty years later we are still seeking a solution; but, we are beginning to see a light at the end of the tunnel.

It is obvious that there is great potential for a successful commercial clam fishery in Alaska - if several problems can be resolved. First and foremost are the problems associated with PSP, caused by the presence of toxic metabolites produced by dinoflagellates of the genus Protogonyaulax (6).

In addition the Department of Environmental Conservation has recently contracted with two independent laboratories for an improved method for monitoring PSP. One is an immuno chemical method. Another method under study to determine the presence of PSP is an on-site monitoring system - a black box, if you will -

whose operating principle is of a proprietary nature. It cannot be revealed until the patent is secure. It is suffice to say that ADEC has invested $60,000 in each study. Additional research is directed toward a method of neutralizing the toxin.

Second, in addition to the PSP problem, the commercialization of Alaska's shellfish industry is hampered by high labor and transportation costs along with a scarcity of readily available capital. However, these problems lend themselves to solutions through innovation, mechanization and some forms of low interest state loans.

A third concern is the industry's ability to compete with clam products harvested from the East Coast, Canada, Japan and Korea. To successfully compete, we must establish a reputation for a safe and wholesome quality product. To this end, Alaska has become a member of the National Shellfish Sanitation Program which allows us to participate in interstate marketing. In addition, the state has instituted a seafood certification program.

Fourth, technological innovations (such as the use of dredges) to increase harvest yields and reduce labor costs will be necessary if a successful commercial industry is to be established.

Finally, if Alaska's clam industry is to grow, it will be necessary to devote more resources to beach certification and monitoring. If clam bait harvesting is transferred to presently noncertified beaches, it will serve to assist in the identification of the most productive new clam beaches.

At a time when the available resources are decreasing due to pollution and overfishing elsewhere in the United States, a consumer demand for shellfish products has created the conditions for an expanding and profitable market. Alaska, with its relatively uninhabited and unpolluted shoreline, with its many species of commercially harvestable shellfish, has a great potential for economic growth, if a renewed shellfish industry can be established.

Here is what we now know about the prevalence of PSP in Alaskan waters (7). We have recently conducted surveys in southeast Alaska and the Bering Sea - the two regions of most obvious economic potential - on the presence of PSP-producing dinoflagellates.

Our method consisted principally of incubating sediment from shellfish beds and observing the growth of *Protogonyaulax*. In many cases we subsequently grew cultures of clonal isolates and demonstrated their toxicities.

In southeast Alaska we extensively surveyed the Icy Straits cross sound areas and - to a lesser extent - the Ketchikan area. Basically we found toxic dinoflagellates everywhere we looked. Although quantitative results were not as clear cut as we may have hoped, it appears that toxic dinoflagellate cysts are widespread and tend to concentrate in the fine sediment found in "holes" or "pockets" in the ocean bottom. Unfortunately, this appears to be the case even in the vicinity of supposedly "safe" beaches where shellfish seem to be protected only by serendipitous hydrology and/or environmental conditions. Thus the prognosis for commercial shellfish development is not good in southeast Alaska.

Things, however, appear to be quite different in the Bering Sea. The first thing we discovered is that this area is an expensive and difficult one to sample. The limited sediment samples collected near shore in the Port Moller/Port Heiden study area consisted mostly of coarse materials which contained few, if any, viable dinoflagellate cysts. Sediments which could be cultured to produce toxic dinoflagellates were, however, found in areas having historical and recent PSP episodes - e.g. Lost Harbor and Dutch Harbor. In addition, a near uniculture of a toxic *Protogonyaulax* species was found in waters over the Middle Shelf in the Spring of 1981 (8).

While PSP-producing organisms are present in this area, they appear to be less widespread than in southeast Alaskan waters, a finding consistent with the commercial shellfish survey findings in the late 70's. Although the potential for PSP problems in these centers clearly exists, it seems that - unlike southeast Alaska - PSP will not preclude commercial exploitation of the Bering Sea shellfish resource.

Literature Cited

1. Feder, H.M.; Paul, A.J. *Alaska Seas and Coasts* 1974, 2, 1-7.
2. Reusch, P.; Browning, J. "Upper Cook Inlet Razor Clam Report"; Alaska Department of Fish and Game, Division of Commercial Fisheries, 1982; pp 1-6.
3. Blankenbeckler, D.; Koeneman, T.; Gunstrom, G. "A Research and Management Approach to an Exploratory Fishery on Geoducks in S.E. Alaska"; Alaska Department of Fish and Game, Division Commercial Fisheries, 1982; pp 1-15.
4. Hughes, S.E. "Abundance, Quality and Production Fishing Studies on the Surf Clam, *Spisula polynyma*, in the S.E. Bering Sea", 1978, NW Alaska Fisheries Center Report.
5. Orth, F.L.; Smeller, C.; Feder, H.M; Williams, J. "Alaska Clam Fishery: A Survey and Analysis of Economic Potential"; Alaska Sea Grant Report 75-5, 1975; pp 1-148.
6. Hall, S.; Reichardt, P.B.; Neve', R.A. *Biochem. Biophys. Res. Commun.* 1980, 97, 649.
7. Hall, S. Ph.D. Thesis, University of Alaska, 1982.
8. Sambrotto, R. Institute of Marine Science, University of Alaska, personal communication, 1981.

RECEIVED May 21, 1984

Risks and Benefits of Seafood

BRENDA C. HIGGINS and ALBERT C. KOLBYE

The Nutrition Foundation, Inc., Washington, DC 20006

Health Benefits

Seafood is a wholesome and healthful food. Today's health conscious consumer would do well to eat more fish for a variety of reasons.

First, most fish products are lower in calories than meat and poultry because (1) they usually contain more moisture and less fat. Shellfish, too, are very low in fat, but some may be high in cholesterol.

Fish protein differs from that of meat because it has less connective tissue and no elastin (2). It is easily digested and recommended for many therapeutic diets because it lacks elastin and the collagen converts to gelatin during cooking. The fat content of fishery products is lower than that of meats, but the protein content of fish is as high because fish protein is of equivalent biological value (3). Knowing the relative value of fish protein, one can compare a serving of meat with a serving of fish: a three-ounce serving of steak contains 330 calories, 27 grams of fat and 20 grams of protein. A serving of halibut, on the other hand, contains only 155 calories, 7 grams of fat and 23 grams of protein. The halibut, therefore, has half the calories and one-fourth of the fat contained in the steak, but provides just as much protein of equivalent biological value (4).

Fish may be divided into two classes: lean fish, such as haddock, whiting and flounder, which contain little fat -- less than two and a half percent -- and fat fish, such as herring, trout, salmon and bluefish, which usually contain 10 to 25 percent fat (3). Shellfish, too, are low in calories, have good quality protein and are particularly excellent sources of trace elements such as copper, iron, zinc and manganese. Oysters are especially rich in iron, zinc and manganese. Salt water fish

0097-6156/84/0262-0059$06.00/0
© 1984 American Chemical Society

are rich sources of iodine, but most fish are poor sources of the minerals calcium and iron (1). Fish are low in sodium, even if they are salt water fish, and can be used in low sodium diets. Some shellfish, however, do not have this same distinction; but, oysters and soft clams are low in sodium and are the exception (5).

Sixty to eighty-five percent of the fatty acids in fishery products are either monounsaturated or polyunsaturated. Polyunsaturated fats have been shown to reduce blood cholesterol levels and the same phenomenon exists for fish oils. Epidemiological studies of the Greenland Eskimos show their diet to be high in marine oils and the population has correspondingly low serum cholesterol values (6)(7).

In addition, the myocardial infarct incidence among these Eskimos is low and their blood plasma and platelets contain high levels of the fatty acid, eicosapentaenoic acid, which is an analog of arachidonic acid (6). The action of this omega-3 fatty acid may be its ability to inhibit the formation of thromboxane A_2, a blood aggregating agent. Indeed, the Greenland Eskimo data show bleeding time to be longer than among a control group of Danes who had less fish in the diet and, consequently, less eicosapentaenoic acid (7). The threshold amount of eicosapentaenoic acid needed to precipitate these changes in blood clotting time and the incidence of myocardial infarct is unknown. The mechanism for these observations is not completely understood.

Most tissues of the body incorporate omega-3 fatty acids from a diet rich in fish. It has been suggested that omega-3 fatty acids are essential for life, but they cannot be synthesized by the human body. Other possible sources of these fatty acids, in addition to fish, are not completely defined (7).

Some fish oils are used to make margarine and shortening. The end products of the hydrogenation of fish oils used in margarines have been examined for their relation to the development of heart disease, a concern which is in direct contrast to fish oil's potentially protective effects.

Upon hydrogenation, fish oils, especially the longer chain fatty acids (C22's), form many new positional and geometric isomers. In some cases, partially hydrogenated fish oils may contain as many as 50% of the docosenoic acids in the trans form as opposed to the naturally occurring cis form. It is these trans isomers which may be responsible for the lipidosis of heart and skeletal muscle found in monkeys. Epidemiological studies of populations with high intakes of docosenoic acids, however, do not yield higher rates of myocardial lesions. Thus, the excessive susceptibility of laboratory animals is not confirmed in populations (6).

Obviously, in order for fishery products to be of a health benefit, they must be consumed. The high perishability of fishery products creates some difficulty for both the government and industry to provide wholesome and appealing products (8). The consumption pattern of fishery products in the United States attests to the difficulty in handling marine products. Per capita consumption of fish is not uniform throughout the country; some data indicate that consumption is much higher in coastal areas (about twice the average) where fish of high quality is readily accessible. It is much lower in inland areas where the quality of fresh fish available to consumers may be relatively poor. Unfortunately, when individuals have a bad experience with a fishery product, they tend to exclude all fishery products from the diet (9).

It seems ironic, but one of the factors which makes fishery products so attractive from a health standpoint also makes it not only undesirable, but actually unpalatable. That factor is the oil. Fish oil oxidizes readily in air, which changes the color to brown or dark red, and has a distinctive smell. Not only is the quality reduced, but the peroxide in the lipid may be toxicologically active (8). This phenomenon of rancidity is a problem, particularly when making salted or dried fish products. Lipid oxidation takes place in fresh and frozen seafood, too, and can be catalyzed by metal ions. This factor becomes a concern throughout the handling of fishery products from the time of harvesting all the way through processing. Another problem is that oxidized unsaturated lipids bind to protein and form insoluble lipid-protein complexes. These complexes account for some of the toughened texture, poor flavor and unappealing odor of poorly stored, chilled or frozen seafood (10). Storage temperatures are critical to the safety and palatability of seafood. Fresh, wet fish should be kept as close to $32^{O}F$ as possible, and frozen fish should be kept as cold as possible -- not higher than $0^{O}F$ (9). While cold storage is an excellent way to preserve fishery products, excessive cold storage causes toughening of the tissues (3)(11).

Another major factor affecting texture, and thus the acceptability of cooked fish, is the *post mortem* pH. The lower the pH, the tougher the flesh (3).

In spite of the effects of cold storage on fish quality, how fish are handled from the time of harvesting dramatically affects palatability. Countries such as Norway and Iceland have strict handling procedures, mandated by law, which has contributed to their reputation for quality groundfish. Dramatic improvements in appearance, yield, shelf life, texture and overall quality have been reported after comparisons have been made between groundfish bled and boxed at sea and groundfish handled and stored in the traditional manner. Advances in maintaining seafood quality are now allowing Iceland and Norway to ship fresh fish directly into the U.S. market and sell it for a premium price. U.S. producers will have to make changes in their methods in order to avoid invasion from foreign markets (12).

Not every fishing system works as well and preserves fish as long as the ones used in Iceland and Norway. Spoilage at sea is very common and contributes substantially to the loss of fishery resources. Several factors affect the keeping quality of fish held on trawlers. One factor is the temperature of the water from which the catch is taken. Fish from high temperature water needs less chilling to inhibit enzyme activity; fish from very low temperature waters need very low temperature chilling to retard enzyme activity (13).

Aside from the complications imposed by the oil in fish and the need to keep the catch cold, there is also a compound in fish called trimethylamine oxide (TMAO), which is converted to trimethylamine (TMA). Trimethylamine oxide is essentially an odorless compound which is converted by bacteria in and on the fish to the fishy tasting and smelling compound, trimethylamine. Retarding this conversion is one of the prime objectives of adequate cold storage of fishery products (3)(8).

Consumption

Japan and Russia are the two most aggressive countries in the field of large-scale fishing; Japan also has the highest per capita consumption of fish. The United States is a relatively small consumer of fishery products, with 50 to 60% of the fish consumed being imported. On a per capita basis, Americans consumed 11.4 pounds of edible weight of fish products in 1970 — less than one-fourth the amount of poultry consumed (11).

In the United States, almost half of the fish consumed is fresh or frozen; the other half is canned. Of the nearly 200 commercial species of fish, most consumers are familiar with fewer than twenty (11).

Tuna is by far the most popular fish consumed in the United States. In recent years, Americans have eaten an average of three pounds or more of canned tuna per capita. Tuna accounts for almost one-fourth of the total of all species of fish and shellfish consumed in the United States (14).

One out of every three fish fillets or steaks eaten in the United States is cod. Other popular fish are flounder and other turbot, which comprise another 20% of consumption. About 15% of total consumption consists of haddock or pollock. Salmon ranks fourth in overall consumption behind tuna, shrimp and cod (14).

On a worldwide basis, a fifth of the total catch of fish and shellfish has been of sea herring and pilchards. The catch has been increasing over the years and in 1979 it exceeded 34 billion pounds -- more than the catch of cods, tunas or any other group of fish (14).

Most of the pilchards and herring-like fish nourish man indirectly. These fish are mainly used as fish meal which is used, in turn, to feed poultry and livestock, especially pigs (15).

World Catch

About 90% of the world catch comes from oceans, seas and tributaries. Rivers, lakes and other bodies of fresh water provide the remainder of the supply (11).

World production of aquatic life, excluding whale, doubled twice in the decades between 1950 and 1970. Growth in world fishery production throughout the seventies, however, was lower than the rate of growth in population. Consequently, on a worldwide basis, the contribution of fish to human nutrition has declined (16).

There is a limit to the amount of fish which can be caught before the supply is seriously compromised. The stabilization and, indeed, decline in some species of the world catch in the past decade is largely due to the diminishing number of stocks of conventional fish. The limit of future production lies between four and ten times the present harvest. Part of this increase will come from increased fishing, but possibly fifty percent will be achieved only through better management of this renewable natural resource (16).

There is no doubt that for much of the world population fish are the main source of animal protein. Twenty years ago, it was thought that increasing the world catch would solve the world's animal protein problem. More realistic projections currently show that the stock may not meet the demand, especially in developing countries. The thought today is that better management, such as through technological advances, is needed to take the strain off the catch and make better use of fish end-products (16).

One such suggestion is to turn fish materials now being used for animal feed and fertilizer into palatable human food, instead. Since fish spoil rapidly, much of the world catch becomes unfit for human consumption, but freezing fish and fish products conserves fish which otherwise would have been wasted and it allows greater distribution. This technology, however, is relatively expensive for most consumers of the world (8).

Aquaculture is another means of meeting the demand for fishery products. This method, however, is rather costly in developed nations, although many countries have practiced some form of aquaculture for many years. China practiced fish farming as far back as 2000 B.C., but it was introduced only recently in Africa where about 700,000 tons are produced annually. Fish farming has maintained its importance through the centuries in Russia, which today produces about 200,000 tons of pond fish (17).

Although aquaculture can be costly, one of the least costly and most productive species is the pond-raised catfish. Most catfish require a living organism for food, but the channel catfish will eat almost anything and can grow as large as twenty pounds (18). As yet, only about 30% of the U.S. consumers have ever tasted catfish, but with increased production and greater consumer awareness, this figure is sure to rise (19).

Catfish are not the only species suitable for aquaculture. Bass, trout, prawns and salmon are successfully being cultivated for human consumption (18).

Risks of Consuming Seafood

Risks to the consumer from seafood can come from either naturally occurring bacteriological contaminants or manmade environmental contamination.

Type E Clostridium botulinum is one microbiological contaminant which is appearing more frequently in fishery products. Most botulism outbreaks have been traced to foods that have been poorly processed and held in an anaerobic environment conducive to Clostridium botulinum growth. It is worthy of note that fresh or frozen fishery products historically never have been implicated in outbreaks of botulism among humans (20).

The smoked fish industry was responsible for several botulism outbreaks in the 1960's, but it has produced millions of pounds of safe products since that time after processing techniques were modified (21).

Some of the other organisms which cause major health problems around the world and which are found in seafoods are diphyllobothriasis, anisakiasis, heterophyiasis and chlonorchiasis. These infectious organisms can be retarded or controlled by precautionary storage and preparation techniques (22).

Vibrio cholerae outbreaks have occurred on occasion in the United States among persons eating seafoods. Although the dangers from this organism are not at critical proportions, there is some concern that the contamination might become more widespread unless preventive measures are taken (17)(20)(23).

A toxicant worthy of note is one which occurs primarily in Caribbean fishes and is known as ciguatera toxin. As many as three hundred types of fishes from the Caribbean and equatorial regions of the Pacific Ocean have been found to contain this toxin (15).

Environmental contamination is another source of concern, not only for the safety of the consumer, but also for the fishing industry because of the potential cost to marine life as well as to human life (20)(24).

Since the late 1960's problems related to contamination of fish and shellfish have been recognized. The problem compounds share one characteristic: biopersistence. Compounds such as various salts of lead or mercury have contaminated shellfish or fish living in either fresh or salt water. Chlorinated compounds such as DDT, PCB and dioxins have been found in fish living in fresh water rivers and lakes and to some extent in the Baltic and Mediterranean Seas (25).

In more recent years, efforts have been stepped up to control environmental pollution, with the result that in many areas residues of such compounds in fish and shellfish are decreasing substantially from previous levels. The chlorinated compounds mentioned earlier have the ability at sustained high doses in rodents to promote the development of liver cancer; thus, they have been called "carcinogenic." When viewed in terms of projected hazards to human health, however, human exposure is usually low level and sporadic. Most of the detectable levels of chlorinated compounds in various fish and shellfish pose no hazard to human health.

In summary, from the standpoint of risks and benefits, the benefits far outweigh the risks. There are many health benefits to be derived from eating fish and shellfish as long as the nutritional quality is assured. Seafood and fresh water fish, as well as shellfish, are excellent foods. They are a renewable natural resource which can feed many who are hungry. Some obstacles to the realization of fishery products' full potential revolve around production, spoilage and distribution problems. There are many individuals and organizations, as well as the federal government, who are cooperating to ensure that the public safety is secure. The greatest threat to the public health is from biological toxins and infections as opposed to chemical toxins, but greater attention should be given to controlling environmental contamination.

Literature Cited

1. Anthony, J.E.,; Hadgis, P.; Milam, R.; Herzfeld, G.; Taper, L.; Ritchey, S. J. Food Sci. 1983, 48, 313-314.
2. Fox, B. "Food Science - A Chemical Approach;" Hodder and Stoughton: London, 1978; p. 210-211.
3. Love, R. In "Advances in Fish Science and Technology;" Connell, J.J., Ed; Fishing News Books Ltd.: Surrey, England, 1979; p. 130-138.
4. "Teacher's Manual. Fishery Products Inspection," U.S. Department of Commerce, National Oceanic and Atmospheric Administration, National Marine Fisheries Service.
5. Stillings, B.; Thompson, M. "Seafoods and Health," Fishery Market Development Series No. 17; U. S. Department of Commerce, National Oceanic and Atmospheric Administration, National Marine Fisheries Service.
6. Carpenter, K. In "Advances in Fish Science and Technology;" Connell, J.J., Ed; Fishing News Books Ltd.: Surrey, England, 1977; p. 124-130.
7. Gunby, P. J. Amer. Med. Assoc. 1982, 247, 729-731.
8. Suzuki, T. "Fish and Krill Protein: Processing Technology;" Applied Science Publishers Ltd.: London, 1981.
9. Ronsivalli, L. J.; Gorga, C.; Kaylor, J.; Carver, J.; Mar. Fish Rev. 1978, 1, 1-4.
10. Khayat, A.; Schwall, D.; Food Tech. 1983, 37, 130-140.
11. Vail, G.; Phillips, J.; Rust, L.; Griswold, R.; Justin, M. "Foods - An Introductory College Course;" Houghton-Mifflin Company: Boston, 1973.
12. Sackton, J. Sea. Bus. Rep. 1982, 80, 57-60.
13. Hansen, P.; Jensen, J. Infofish Mkt. Dig. 1982, 6, 26-28.
14. Slavin, J. "Fish Facts;" Food Marketing Institute: Washington, D.C., 1982.
15. Chichester, C., personal communication.
16. Robinson, M. "Prospects for World Fisheries to 2000;" Food and Agriculture Organization of the United Nations: Rome, 1982, June, 1-16.
17. Kreuzer, R. In "Fishery Products;" Fishing News Ltd: Surrey, England, 1974, 22-47.
18. Hopkins, H; FDA Cons. 1981, 15, 10-15.
19. Food Eng. 1982, 54, 151.

20. Finch, R. In "Fishery Products;" Fishing News Ltd: Surrey, England, 1974, 53-59.
21. Eklund, M.W. Food Tech. 1982, 36, 107-112.
22. Asner, M. FDA Cons. 1982, 16, 4-7.
23. De Paola, A. J. Food Sci. 1981, 46, 66-70.
24. "Interchange of Pollutants between the Atmosphere and the Oceans," Reports and Studies No. 13, World Meteorological Organization, 1981.
25. "The Review of the Health of the Oceans," Reports and Studies No. 15, United Nations Educational, Scientific and Cultural Organization, 1981.

RECEIVED April 23, 1984

7

Ciguatera Toxins: Where Do We Go from Here?

NORMAN J. DOORENBOS

University of Wisconsin-Eau Claire, Eau Claire, WI 54701

Although much useful information on ciguetera toxins has been generated, the presence of these toxins in some seafood remains a serious international health problem. Future research goals should include the development of practical tests for product safety, the development of specific antidotes for the toxins, the introduction of methods to minimize the accidental consumption of toxic seafood, the discovery of the genesis of the toxins and methods to control the problem, and the development and dissemination of an information base on all aspects of this and other seafood toxin problems for the seafood industry, health personnel, researchers, and the public.

It is almost presumptuous for me to present a paper at the quarter point of this Symposium on Seafood Toxins on the subject "Where Do We Go From Here?" This morning, we heard a number of fine presentations on the need for and the results of research on seafood toxins. We will learn even more in the three remaining half-day sessions. Please let me share with you the observations and reflections of a chemist with a long time interest in toxicology, who has also experienced the difficulties and frustrations of investigating seafood toxins.

The points which I would like to make can be illustrated in large measure by an examination of the ciguatera problem. Ciguatera is an international problem as made clear in Halstead's Poisonous and Venomous Marine Animals of the World (1). It is a threat to public health and has curtailed fisheries development. Let's look at some recent reports. Dr. Nabil A. Mansour (2), University of Alexandria, noted that ciguatera is a Red Sea problem and may prevent the planned development of an Egyptian Red Sea fisheries. Hundreds of ciguatera cases are recorded annually in Sri Lanka (3) and Australia (4). The annual totals are thought to be in the thousands. New Guinea fisheries researchers (5) believe that thousands are poisoned each year but most incidents go unreported because they are attributed to a sorcerer's claims over the victims.

Recently, an Italian outbreak (6), traced to fish caught in

0097-6156/84/0262-0069$06.00/0
© 1984 American Chemical Society

the western Pacific, led to one death and over fifty hospitalizations. Some islands in the western Pacific have been abandoned because of local ciguatera problems. Ciguatera has been a major obstacle in the development of lagoon resources and aquaculture in the southern Pacific. A bottom fishery in Samoa is required to discard all red snappers, as much as 50% of the catch, because of potential toxicity (7). Ciguatera has prevented the development of commercial fisheries in many parts of the Caribbean (8). Some outbreaks affect many individuals. For example, the president of a well known company in Venezuela recently wrote that he and several associates speared 30 snappers in the waters of Isla Los Roques (9). The fish were shared with friends, poisoning over 200 and causing several deaths.

Ciguatera is also an American problem and is grossly underreported. My files contain numerous reports of poisonings which occurred while on vacation or from fish obtained in a hot area. The Center for Disease Control publishes an Annual Survey of Foodborne Disease for the fifty states, Guam, Puerto Rico and the U. S. Virgin Islands. The number of confirmed outbreaks listed for 1976 through 1980 ranged from three to 19 per year. The number of outbreaks listed for 1980 (10), the latest available survey, was at the high end, namely 15 and involved 52 individuals. Please note: fifteen outbreaks and 52 individuals.

My records for 1980 include such reports as nine persons suffered ciguatera poisoning in October from eating a grouper in a Montgomery County, Maryland, restaurant (11), the CDC received numerous inquiries from physicians about ciguatera symptoms on behalf of physicians returning from Puerto Rico and the U. S. Virgin Islands (12), the Washington Post on July 21 described a Florida study (probably Don DeSylva's study) reporting that 1,500 south Florida residents contract ciguatera each year, and Dr. Joe Mac Millan of the College of the Virgin Islands established that during one period 30-40 persons per week were receiving emergency room treatment for ciguatera poisoning at just one of the hospitals on St. Thomas (13). The CDC survey for 1980 includes no reports for Maryland, Florida, Puerto Rico, or the U. S. Virgin Islands. The fifteen outbreaks listed each occurred in Hawaii. Similar underreporting has been noted annually.

It has been our observation that most people in high risk areas do not seek help for ciguatera poisoning. Hence, they are not included in any compiled statistics. They know that the gastrointestinal symptoms and headaches may be treated with over-the-counter medications and that little can be done for the neurological and cardiovascular symptoms. The incidence in these areas can be high. Over 80% of the resident adults we surveyed in the U.S. and British Virgin Islands had been poisoned at least once. Victims are not only made ill but often cannot work during the early stages of the recovery period. Ciguatera continues to be a serious health and economic problem.

Physicians outside the "hot areas" seldom recognize the symptoms of ciguatera poisoning. As a consequence inappropriate time, resources, diagnostic tests, and treatments are used often leaving a frustrated, seriously-depressed, sometimes emotionally-disturbed sick patient whose financial resources have been compromised. An article "Physicians: Know Thy Ciguatera Symptoms" published in the November 1982 issue of the American Journal of Public Health (14) is an attempt to educate physicians.

Clinical data, now supported by research data generated by several research groups represented in this Symposium, suggest that ciguatera is caused by several different toxins. We have sometimes noted significant differences between the list of symptoms reported by victims of different outbreaks in the Caribbean, suggesting a different set of toxins. We have also recorded symptoms of victims in the Cayman and Virgin Islands which, to the best of our knowledge, have not been noted in victims elsewhere in the world. For example, many of these victims have reported blood in the urine, difficulty in urinating, and pain on urination along with other symptoms commonly associated with ciguatera. Dr. van der Sar of the Netherlands Antilles, who has treated many ciguatera victims, disclosed that cardiovascular irregularities are additional symptoms often observed in St. Martin but never in Aruba, Bonaire or Caracao (15). The isolation and characterization of more than one ciguatera toxin, coupled with the progress being made in understanding the physiologic properties of the toxin in such laboratories as those at Southern Illinois University at Carbondale, makes me optimistic about the future.

The chemical studies of ciguatera have been particularly frustrating--shortages of research materials, purification difficulties, instability of some isolated toxin fractions, extremely small quantities present in ciguatoxic fish, lack of a reliable highly-sensitive assay to follow the isolation and purification of toxins, the complex mixtures present in toxic fractions and the complex structures of the toxins. I admire the accomplishments Professor Scheuer and others have made in spite of these obstacles.

The discoveries that polycyclic ethers may be responsible for ciguatera as well as some of the other known seafood toxins is an exciting development. Polycyclic ethers, both natural and synthetic, have become "cutting edge" tools in the investigation of certain living membrane phenomena. The fascinating physiologic effects of polyclyic ethers selected for study has been attributed in part to their ability to complex one or more of the physiologically important cations and to affect cation transport through membrances.

On the side, this causes me to wonder if the diterpenoid toxins of Rhododendrons, known as grayanotoxins, which mimic the physiologic effects of ciguatoxins, also chelate metal cations. These diterpenoids, although not cyclic polyethers, do contain several oxygen functional groups and may assume conformations which could favor the formation of such complexes.

Polycyclic ethers are biosynthesized by a variety of microorganisms. Several were clinical antibiotic candidates in the 1960's and 1970's. The discovery that such substances may have a wide distritution in the marine environment raises intriguing questions. Is the genetic capability of directing the biosynthesis of polycyclic ethers widely distributed? What are the functions of these substances?

In the mid-1970's we isolated cultures of Gram negative marine bacilli from four ciguatoxic surgeon fish speared in the waters of Salt Island in the B.V.I. The cultures appeared identical and elaborated, in submerged laboratory culture, a toxin resembling ciguatoxin. These observations caused our speculation that this bacillus might be the source of ciguatoxin and later to speculate that it may be the source of toxin observed in Gambriadiscus toxicus when attempts to produce the toxin in cultures of this dinoflagellate failed.

Failure to produce ciguatoxins in culture has been attributed to a missing factor. The missing factor may be an appropriate stress or a chemical substance produced by another microorganism. There are numerous examples in both the plant and animal kingdoms of the biosynthesis of substances stimulated by environmental stresses or the presence of another living organism.

Continued research is needed to characterize and understand these toxins, their distribution in the marine environment, their biosynthesis and their function, how to detect them and to minimize their impact on the seafood industry.

Ciguatoxins appear to be stored for long periods in fish and humans. This is a most unusual phenomenon. Animal species and especially man have well developed mechanisms for reducing the toxicity of foreign chemicals and rapidly eliminating them from the body. Thus, it is surprising that fish appear to store ciguatoxins for a lifetime and that humans may also store these toxins for a long time. The slow recovery of victims, which often takes months to years, is evidence of long term storage. Of course, this could be due to other factors such as the slow repair or replacement of biological structures damaged by the toxin(s). Other evidence for long term storage in humans includes the numerous observations of increased sensitivity to ciguatera, lasting several months or longer, among ciguatera victims.

Specific antidotes to ciguatera poisoning might be discovered among chemical derivatives of the toxins or simpler synthetic polycylic ethers. I am suggesting that some of these substances might function as competitive antagonists displacing toxins from critical binding sites and allowing the body to dispose of them.

Where do we go from here? Increased support needs to be provided for seafood toxin research. The progress thus far made in this most challenging research field has been nothing short of phenomenal. This has been accomplished with a very modest investment of research dollars. I compliment the leadership F.D.A. is providing in funding and coordinating research in seafood toxins. This level of support needs to be greatly expanded and jointly funded by F.D.A., National Marine Fisheries, and the N.I.H. Every opportunity to collaborate and communicate with researchers in other countries should continue to be encouraged.

Among the important challenges hanging out there are:

1) To develop highly sensitive, rapid, inexpensive tests to support research activities and researchers for the Seafood industry,

2) To characterize the various seafood toxins, identify factors contributing to their entry into food products, and to develop procedures for minimizing the consumption of toxic seafood,

3) To discover mechanisms of toxicity and to develop effective safe antidotes for the treatment of victims of seafood poisoning, and

4) To develop and utilize an improved information base on all aspects of seafood toxin problems for the seafood industry, health personnel, the public and researchers.

It may not be possible to eliminate ciguatera, PCP and other seafood toxin problems but the kind of research in progress should make it possible to manage these problems, to effectively treat victims, and to minimize economic and health impacts.

Literature Cited

1. Halstead, B. "Poisonous and Venomous Marine Animals of the World"; U.S. Government Printing Office: Washington, D.C., 1967, Vol. II, pp. 63-604.
2. Mansour, N.A., personal communication.
3. St. Louis Post Dispatch, January 22, 1978.
4. Sutherland, S., personal communication.
5. Lili, P., personal communication to Christopher Kohler, Southern Illinois University-Carbondale.
6. Teoboli, A., personal communication.
7. Fields, R., personal communication.
8. Brady, R. W. Proc. 24th Ann. Gulf and Caribbean Fisheries Institute, 1972, p. 100.
9. Comenges, F., personal communication to J. Randall, Bishop Museum.
10. "Foodborne Disease Annual Summary 1980," Center for Disease Control, 1983.
11. Ragelis, E., personal communication.
12. Glaso, R., personal communication.
13. McMillan, J., personal communication.
14. Dembert, M. L.; Pearn, J. H., Am. J. Pub. Health 1982, 72, 1298.
15. van der Sar, A., personal communication.

RECEIVED March 2, 1984

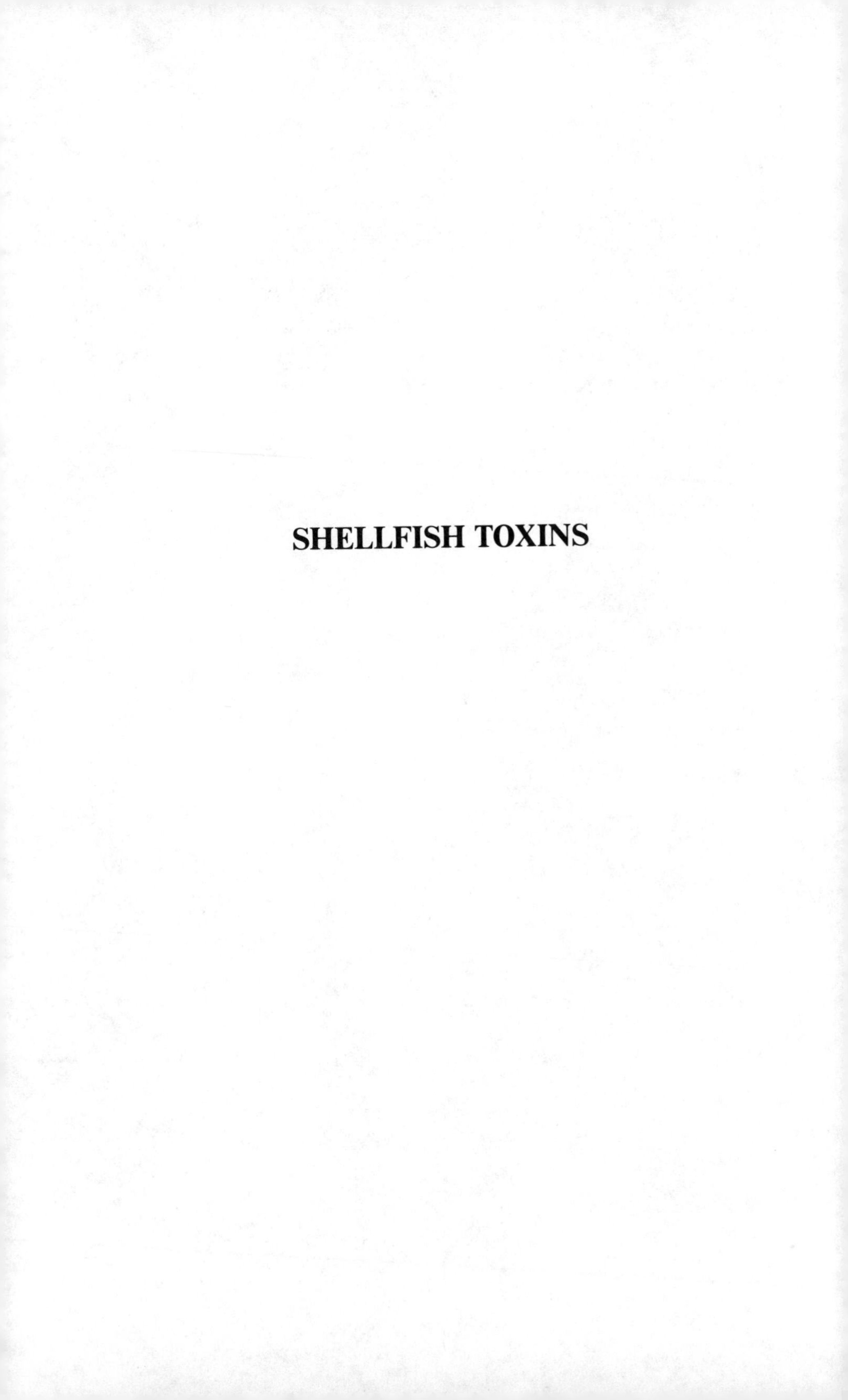

SHELLFISH TOXINS

8

Toxic Dinoflagellates: Taxonomic and Biogeographic Aspects with Emphasis on *Protogonyaulax*

F. J. R. TAYLOR

Departments of Oceanography and Botany, University of British Columbia, Vancouver, B.C., V6T 1W5 Canada

This paper provides a brief summary of taxonomic developments since 1979 and also some features of the global distribution of toxic species. The continued recognition of *Protogonyaulax* as distinct from *Gonyaulax* , with some species referred to *Gessnerium*, seems most reasonable at present. The growing difficulty of distinguishing between *P. catenella* from P. *tamarensis* in areas where they co-exist is discussed, including electrophoretic data, principally from British Columbia isolates. In B.C. they are usually separated distributionally, but intermediate forms occur in intermediate localities. In Japan they may be temporally separated. The allocation of *Gymnodinium breve* to *Ptychodiscus* depends on the presence of a pellicle. Although not seen with TEM it can be seen with light microscopy. Geographic distribution is closely linked to taxonomy for, although some toxin producers appear to be endemic in a restricted sense, closely similar forms occur elsewhere (e.g. *P. brevis*) or the same species may be known by different names in different regions (*Gyrodinium aureolum* ?). *P. tamarensis* appears to be very widely distributed, with both Arctic and tropical forms which must differ significantly in physiology. Some tropical benthic species which produce toxins also occur in temperate regions where they should be tested for toxicity, e.g. *Prorocentrum lima*. Finally, it is noted that many of the toxin producers (all of which are photosynthetic) are close relatives and suggestions are made for the testing of other related taxa.

Several papers dealt with toxic dinoflagellate taxonomy at the Second International Conference on Toxic Dinoflagellate Blooms (1) and this paper summarises developments since then. A paper by Steidinger (2), which appeared after the ACS symposium from which the present papers arise, also discusses this topic and this paper has been revised to reduce duplication. Her paper should be consulted for greater illus-

0097-6156/84/0262-0077$06.25/0
© 1984 American Chemical Society

tration of the species in question and further historical review. In addition to taxonomy the present contribution presents some biogeographical data, since this aspect has been neglected and is intimately linked to questions of species identity, gene pools, and biochemical and physiological variability within similar morphotypes.

Before dealing with the taxa in question, some general points may assist the non-specialist in following recent developments, for dinoflagellate taxonomy is currently in a state of flux which must be annoying and seem capricious to those who simply require a stable name with which to relate their biochemical data. Firstly, much of this has been necessitated by the critical re-examination of species which, because of their association with red tides or toxin production, were often described inadequately or improperly by non-taxonomists. Secondly, because of their importance these species are being subject to much closer scrutiny than others and, as usual, the more that is known, the less simple the picture becomes (discovery of intermediates or biochemical variability within similar morphotypes, for example). Because dinoflagellates have been named by both botanists and zoologists, each of whom follows a slightly different set of nomenclatural rules (International Codes), complications arise which can be resolved eventually. Although sexuality is known in dinoflagellates, it is cryptic, infrequent, unknown in many, and reproductive isolation under natural conditions is very difficult to assess at present. This means that classical species concepts are difficult, if not impossible to apply to such organisms. Genera are subjective constructs, following objective guidelines, which assist in species identifications and the recognition of species groups. The degree of difference necessary for the recognition of genera is usually set by precedent within each group. For example, the presence or absence of chloroplasts is a generic criterion in euglenoids, but has not been used as such in dinoflagellates (except in *Pheopolykrikos*, rejected by most recent taxonomists). It is evident that those working with fossil dinoflagellate cysts accept much smaller differences in generic distinctions than those studying living forms and a current aim is to bring them closer together. Finally, natural populations are mixtures of genotypes and multiple isolates from the same population at the same time should not be expected to be similar in all respects. Differing environmental conditions, both short and long term, will favour particular genotypes within the populations in accordance with basic principles of population genetics.

Protogonyaulax

Three species of this genus have been linked to natural paralytic shellfish poisoning (PSP) events: *P. tamarensis* (which includes *excavata*, attempts to distinguish them by morphological or other means having been abandoned by former proponents; 3), *P. acatenella* and *P. catenella*. In addition there are five other species (see 2, 4, plus the form described as *Gonyaulax kutnerae* Balech - see below) which have not been shown to produce toxins so far. All these were formerly attributed to the genus *Gonyaulax*, but they differ in so many respects (epithecal plate pattern, hypothecal pattern, apical pore, degree of girdle displacement, cyst type), that their generic distinction cannot be disputed. Both Taylor (4) and Loeblich & Loeblich (5) independantly came to this conclusion at the same time but, unfortunately, offered different solutions. The latter authors enlarged the definition of the genus *Gessnerium* (see below) to include all the "*tamarensis* group, minimising variations in the first apical plate. Taylor created a genus especially for them, omitting those species in which the first apical plate homologue (designated 1s in the Taylor homology system, or 1u in Evitt's modification of it)does not usually make contact with the apical pore complex (APC). A typographical error in the generic description of *Protogonyaulax* (it should read " V/VI contacting Z") was corrected later (6). At that time,the cyst of the type of *Gessnerium* was unknown. It was subsequently found to be similar to that of *Protogonyaulax* (7), as is that of *Pyrophacus horologium* (8). Consequently, the distinction between *Protogonyaulax* and *Gessnerium* (not as emended by Loeblich & Loeblich) rests solely on the contact, or lack of it, of the first apical plate. Species referable to *Gessnerium* using this criterion, are indicated in the next section. In *P. fratercula* the contact area may be small because the ventral part of the APC may be very narrow. Balech (9) has argued that this criterion is not reliable for generic distinction because in his newly described form *Gonyaulax kutnerae* the contact was variable. However, the specimen he illustrates, in which the first apical is displaced from the APC, is clearly megacytic (a form in which the cell is greatly enlarged, usually associated with sexuality or immediately prior to division) and the growth of intercalary bands may cause a temporary separation. If such secondary effects, known only for one or two species, are taken into account, the criterion is still usable and useful in maintaining nomenclatural stability (the genus *Protogonyaulax* is now in wide use). In a culture of *P. tamarensis* ,Taylor observed cells with strikingly aberrant tabulations, involving all the primary plate series (cingulars and sulcals were not examined but are assumed to be more conservative), but the majority of cells were of the normal type and this is the condition on which the diagnosis must be based (10).

In this connection it should be noted that an insistence that there should be no overlap whatever, would result in the "sinking" of the genus *Triadinium* (= *Goniodoma*, *Heteraulacus*) as well, because it is basically similar to *Gessnerium* (see Figure 1) and differs only in the plates which are included in the sulcus, a feature which is not always clearly evident due to intergradation. Similarly, the athecate genera *Amphidinium*, *Gyrodinium* and *Katodinium* would all have to be sunk into *Gymnodinium* because of species which exhibit borderline features (see *Gyrodinium aureolum* below).

The morphological distinction between species of *Protogonyaulax* is also becoming increasingly difficult. All share the same basic plate pattern. Their features were summarised and illustrated by Taylor (10). Subsequently,Balech & De Mendiolana (11) described *P. peruviana* (as a *Gonyaulax*), distinguished chiefly by its curving **first apical** plate (which, however, resembles that of *P. phoneus* and some cells of *P. dimorpha*). Later Balech (9) added *Gonyaulax kutnerae* , a large form which he compared to *G. brevisulcatum* Dangeard but which, if contact between the first apical plate and the APC is the usual condition (see above), should be transferred to *Protogonyaulax*, and *Gessnerium* if not.

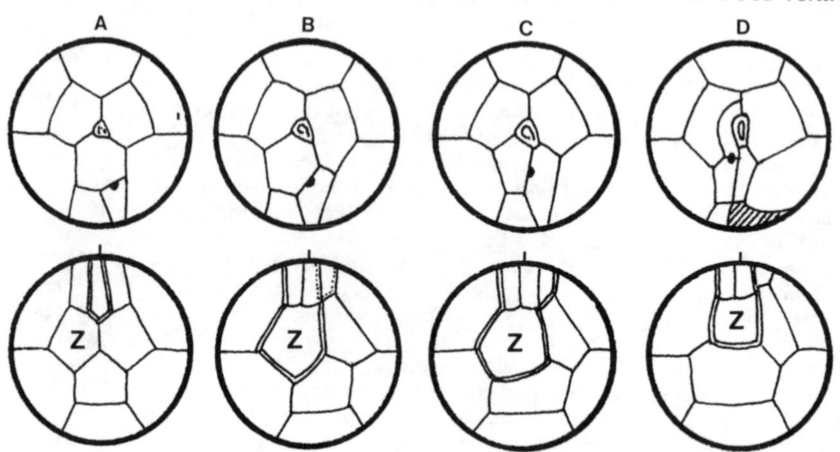

Figure 1. Epi- and hypothecal tabulation patterns. A. Triadinium. B. Gessnerium, Pyrodinium. C. Protogonyaulax. D. Gonyaulax (spinifera).

Each of these species has morphological ideosyncrasies which distinguish it (e.g. the angular anterior sulcal plate, sulcal lists and large size of P. cohorticula; the narrow ventral projection of the APC of P. fratercula; a very narrow sixth precingular in a population from New Zealand which is currently under study), but continued observations are revealing the existence of populations with morphotypes which fall between "species" that were previously readily distinguishable. Unlike the genera discussed above, species are believed to be real biological units which should be discrete morphologically and/or reproductively. Because the Zoological Code does not regulate names below the rank of species or subspecies, zoologists working with dinoflagellates tend to create new species names for forms which differ only minimally, whereas botanists can use varietas and forma for infraspecific variants that differ in small genetic or phenotypic ways respectively. To one using the botanical system it appears that several members of the "Protogonyaulax complex" do not deserve separation at the species level, although further information is required in most instances.

A case in point, which we are investigating, is the distinction of P. catenella from P. tamarensis, the two species which have been distinguished for the longest period. When first described P. catenella appeared to be clearly different because of the plate arrangement on the epitheca, as well as cell shape and chain formation. However, after the realisation that Lebour had erred in the original description of P. tamarensis, drawing the epithecal pattern in optic reversal, and the correction is made (10), the distinction becomes largely a matter of shape. Populations can be found which can be readily attributed to P. catenella (chains of 8 cells or more, cells wider than long with angularly flattened apex and antapex, APC shaped like a lamb chop) or P. tamarensis (cells single or in pairs, longer than wide, more elliptical APC). When brought into culture and grown under identical conditions the distinction often becomes more difficult to make, chain length becoming shorter or no more than pairs in some strains, and cells changing shape. If they simply became indistinguishable we would conclude that they were phenotypic variants resulting from environmental influences belonging to the same morphospecies and this would be a case in which forma might be an appropriate level of distinction. Unfortunately it is not so simple. Some strains maintain a catenella shape even when not in chains. Some make short chains. Some become isodiametric. One of our strains (NEPCC 254 from Nelson Island, British Columbia) was"tamarensoid" when first isolated (12),but now occurs as small, flattened cells (resembling Gaarder's Goniodoma depressum) which could be taken for P. catenella. A ventral pore is absent, which is usually the case with P. catenella, whereas it is of variable occurrence (although constant within a strain) in P. tamarensis. An illustration of the problems involved can be seen in the scanning electron micrographs of Postek & Cox (13), reproduced by Steidinger (2), of a culture sent to them from Washington state as P. catenella, but in which the cells are shown to be longer than wide, with smooth contours (like the variety globosa of P. tamarensis), and possessing a ventral pore. If one were not aware of the potential for change in culture it might be assumed that the culture was misidentified, but this is not clear. The posterior attachment pore on the posterior sulcal plate is found only in the anterior individuals in pairs or chains and appears to be induced by the apical pore of the cell posterior to it. The shape of the posterior sulcal plate and the position of the pore, if present, has been used as an aid in distinguishing these species and P. fratercula (14). However, the shape used for the latter does not agree with the original description by Balech (15) and cells with the posterior sulcal form associated with "catenelloid" cells in Japan has been seen in "tamarensoid" cells from Massachusetts (material of D.M. Anderson).

These problems are seen most clearly in British Columbia and Japan, where both types co-occur (see distributional descriptions below), as well as populations which are difficult to assign because of intermediate features.

In an attempt to resolve this problem Allan D. Cembella, working with the author, has investigated the enzyme compositions of 18 cultured isolates of P. tamarensis and P. catenella, or intermediate morphotypes, using enzyme electrophoresis. The majority of these isolates were from British Columbia or adjacent Washington state waters, but isolates from England, Portugal, the Bay of Fundy (eastern Canada) and New Zealand were available for more remote comparisons. All cultures were grown under identical conditions in enriched natural seawater (NSP-7 medium) and harvested in late exponential phase for application to vertical slab, polyacrylamide gels (PAG). Comparison with starch gel electrophoresis confirmed the superiority of PAG electrophoresis for investigating the biochemical taxonomy of Protogonyaulax, both in terms of number of bands and resolution (Figure 2).

To date we have examined the following enzymes, divided into two functional groups: 1) the pyridine-linked dehydrogenases - alanine dehydrogenase (ADH), glutamate dehydrogenase (GDH), glucose-6-phosphate dehydrogenase (G6PDH), isocitrate dehydrogenase (IDH), malate dehydrogenase (MDH), malic enzyme (ME), and succinate dehydrogenase (SDH); and 2) the non-specific hydrolases - α- and β- acetylesterases (α-,β- EST), propionylesterases (PES), butyrylesterases (BES), and acid phosphatases (AcPH).

An analysis of the banding patterns (16), while providing evid-

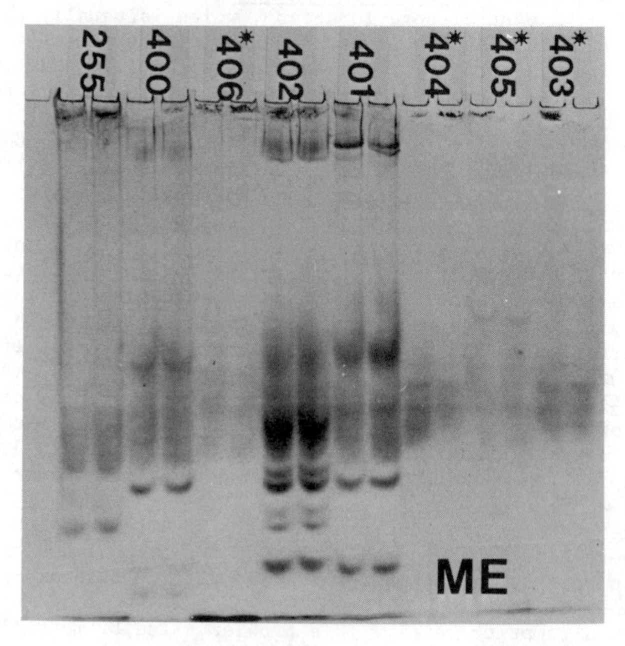

Figure 2. Isozymer banding pattern in PAG of Protogonyaulax tamarensis morphotypes stained for malic enzyme (ME/NADP-dependent malate dehydrogenase).400-406: English Bay isolates (*indicates clonal). 255: Lummi Island (Washington State).

ence for considerable, stable biochemical variability between both clonal and non-clonal isolates, and multiple isolates from the same place and time (English Bay, Vancouver, isolates 400 to 407, see Figure 2), has not supported a simple separation of P. catenella from P. tamarensis, although differences do exist (Figure 3). The presence of such a high degree of electrophoretic polymorphism, even for relatively conservative enzyme systems, such as MDH (NAD - dependent) and ME (NADP - dependent), within a restricted geographical area, is strong circumstantial evidence that Protogonyaulax populations are multiclonal, even when morphologically indistinguishable. In a similar study of five freshwater Peridinium species Hayhome & Pfiester (17) did find a correlation between banding patterns and morphological characters, but the latter (presence or absence of an APC, form of the first apical plate, etc.) were more gross than the distinctions between the tamarensis and catenella morphotypes. Previous attempts to use other biochemical criteria to distinguish Protogonyaulax species, such as luminescence or toxin content. have failed to find consistent correlations with morphotype (3,18).

The distribution of the tamarensis and catenella morphotypes is interesting, both on global (Figure 4) and local (Figure 5) scales P. catenella occurs on the west coast of North America from southern California (La Jolla - F.Haxo and pers.obs.) to south eastern Alaska (19). Its type locality is San Francisco Bay. It is absent south of 32°N, but reappears in southern Chile (20 and pers.obs.). Although it has been assumed to be the source of PSP in the Gulf of Alaska, this has not been confirmed and isolates from cysts collected from several locations in this region were of the tamarensis morphotype (21 and pers. obs.). It is possible that both may occur, but it should be noted that Arctic records so far are restricted to P. tamarensis (see below). P. catenella has not been recorded throughout most of the Atlantic Ocean, but has produced four PSP outbreaks off the west coast of South Africa (22 and pers. obs.). The only other region where P. catenella has been recorded is the east coast of Japan (14, 23). Blooms generally occur when the temperature is close to 20° and the form occurs in both estuarine and open coast localities, including one chain seen in a sample 440km away from the British Columbia coast (pers.obs.).

The P. tamarensis morphotype is found over a much wider area, including both Arctic (24,25) and tropical (26,27) localities, the latter only in the western tropical Atlantic Ocean (Venezuela; northeast of the Amazon mouth, beyond the Guiana Current; and possible Puerto Rico, although some of the latter records appear to be of other related forms). In the North Atlantic it is known from estuarine localities or shallow embayments from Long Island to the Arctic, being particularly common in the Gulf of Maine, Bay of Fundy and Gulf of St.Lawrence in the west, and from Portugal north to Norway and the Arctic near Spitsbergen (25). In the South Atlantic it is known from the central coast of Argentina (28), with an identification attributed to Braarud of material from blooms in Walvis Bay, S.W. Africa (29). In the Pacific Ocean it occurs in British Columbia (10), some coastal locations in the Gulf of Alaska (pers. Obs. and figs. in 21), Unimak Is. in the Aleutians (unpubl. obs., 1965), and bays in north eastern Japan (14,30). A population from the northeast coast of North Island, New Zealand, which appeared after fish and shellfish kills, resembled P. tamarensis, but had an unusually narrow sixth precing-

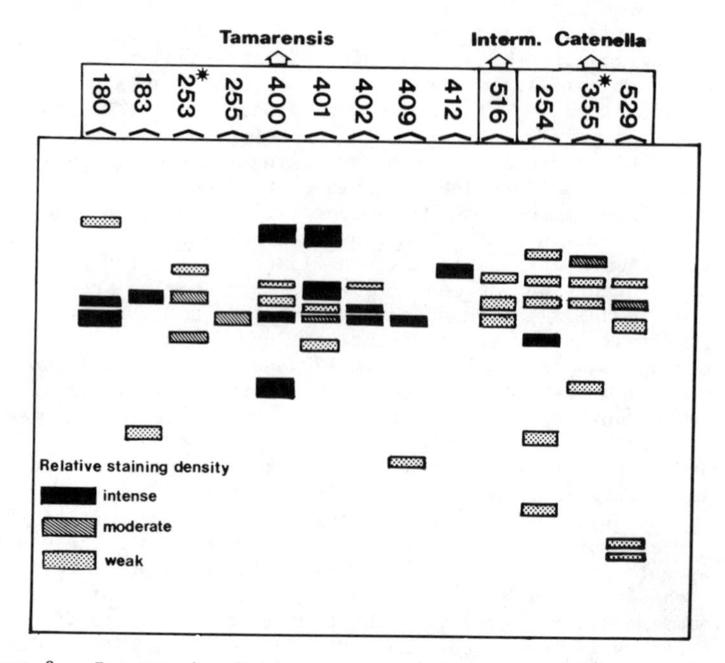

Figure 3. Isozyme banding patterns of Protogonyaulax morphotypes stained for glutamate dehydrogenase (GDH/NAD-dependent). (* indicates clonal).

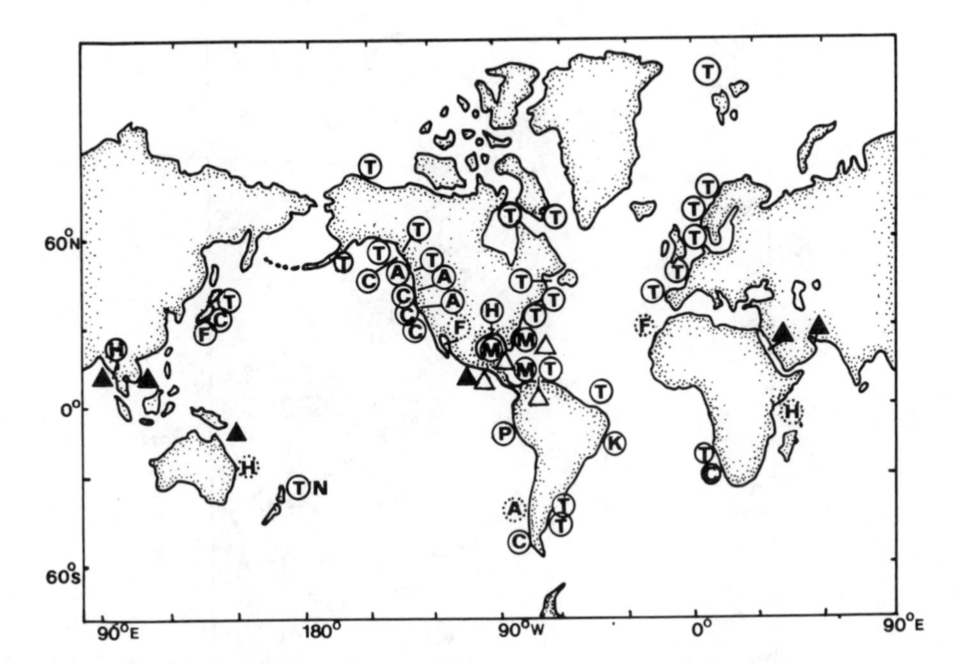

Figure 4. Global distribution of toxic gonyaulacoid species and some close relatives. A.-*Protogonyaulax acatenella*. C.-*P. catenella*. F.-*P. fratercula*. H.-*P. cohorticula*. K.- "*Gonyaulax*" *kutnerae*. M.- *Gessnerium monilatum*. P.-*P. peruviana*. T.-*P. tamarensis* (N- New Zealand isolate). Black triangles- *Pyrodinium bahamense* var. *compressa*. Open triangles- *Pyr. bahamense* var. *bahamense*.

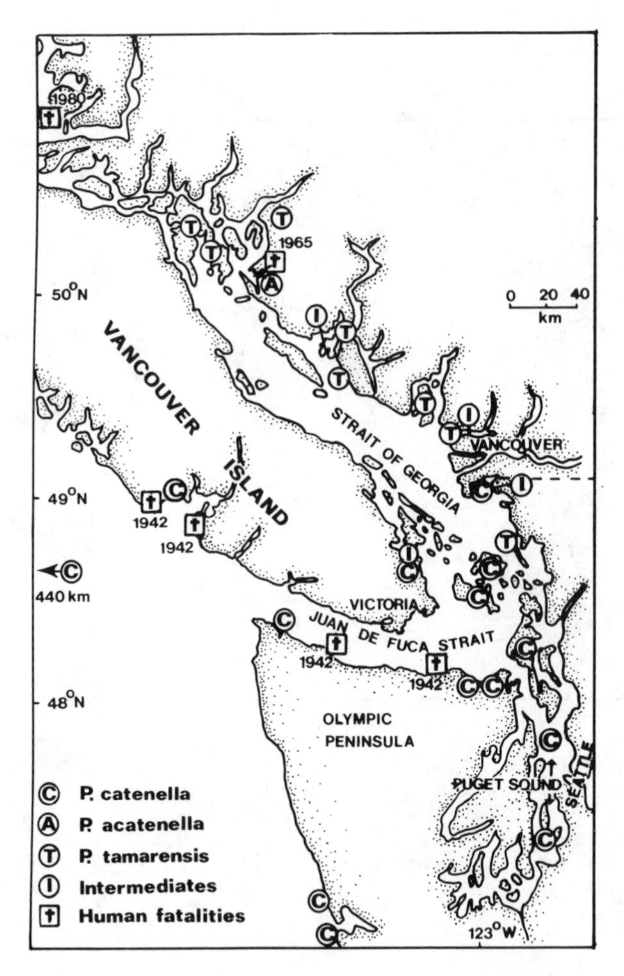

Figure 5. Distribution of protogonyaulax morphotype in southern British Columbia and Washington.

ular which might warrant its separation. It is currently in culture and being studied for toxin production and enzyme spectra. The scattered occurrence of some other closely related taxa are shown in Figure 4.

The distribution of the morphotypes in the coastal waters of British Columbia and adjacent Washington state (Figure 5) is also interesting. Once more, the populations which can be readily attributed to one or the other are spatially separated with little overlap. Records so far indicate that, within this region, the tamarensis morphotype is restricted to bays on the eastern side of the Strait of Georgia (records are not available from the western side), from the northern end to Lummi Island in the south. P. catenella forms occur in Washington state waters from southern Puget Sound (observations by L. Nishitani) to the bays of Juan de Fuca Strait (31, 32) and the open coast. The farthest it has penetrated into the Strait of Georgia is Saanich Inlet on southern Vancouver Island and Boundary Bay, south of Vancouver (33, 34). The Fraser River enters the Strait of Georgia at Vancouver, and much of its flow is deflected northwards. Thus it might be suspected that tamarensis was associated with waters with strong estuarine influence, as it is in many parts of the world (the type locality is the Tamar estuary in southern England). However, there is also appreciable river flow into the Puget Sound region, and yet it does not occur in these areas. Norris & Chew (31) found that some of their catenella isolates grew best within a narrower salinity range (21-30‰) than one from the open coast (20-37‰), but the latter had been maintained in medium of slighly higher salinity than the former.

Within this region both types occur more or less simultaneously, with mid-to late summer maxima. However, in Ofunato Bay, Japan, there is a temporal separation, with P. tamarensis blooming in summer and declining in July, P. catenella blooming from September to November (35).

At a few British Columbia localities intermediates occur (cells isodiametric, single or in pairs, and these are intermediate between the two morphotypes (Figure 5), leading to the enzyme electrophoresis study described above.

In summary, although the occurrence of morphological intermediates, and the decrease in differences in form when cultured under similar conditions may lead one to consider tamarensis and catenella as phenotypic variants (forms) of one species, there are small differences which are maintained in culture. Furthermore, features of their distribution also suggest a genetic basis for some of their differences, such as the absense of the catenella type in the North Atlantic, despite areas of similar water conditions, and the exclusion of tamarensis from the more estuarine parts of Puget Sound. P. tamarensis is usually found blooming in cooler, less saline waters than P. catenella, but the tropical Atlantic form off Brazil and Venezuela (26, 27) is clearly different from the temperate tamarensis physiologically, although it is morphologically similar (slightly smaller). Also, the blooms of P. catenella in the Strait of Magellan region occur in water of 10°C or less and salinities of 25‰ (36).

Other closely similar species of Protogonyaulax, such as the chain-forming P. cohorticula and P. fratercula, have warmer distributions than P. catenella (see Figure 5), as do the chain-formers Gessnerium monilatum and Gymnodinium catenatum, which may also be closely related (4, 5, 10).

P. acatenella is very similar to P. tamarensis (when Lebour's error is corrected) and is distinguished primarily by its higher epitheca than hypotheca. Loeblich & Loeblich (37) added the presence of crests on the hypotheca (additional to the sulcal lists), basing this on their occurrence in a toxic British Columbia bloom (38), although this was not in the original diagnosis and has not been seen in a population since then. In future,it may be necessary to reduce this taxon to infraspecific status as well.

Gessnerium and Alexandrium

As noted above, Loeblich & Loeblich (5) included all those taxa considered to be Protogonyaulax here, in an enlarged definition of Gessnerium , Halim's genus based in error on optically reversed specimens of Gonyaulax monilata (39,40). Arguments have been presented in the previous section for maintaining them as separate, with those in which the first apical homologue does not usually reach the APC, and in which the sulcus involves one of the large, posterior hypothecal plates (Z in Taylor's homology system,4) plus at least two of the plates immediately anterior to it, and with a smooth-walled cyst surrounded by a mucilage layer, remaining in Gessnerium , at least for the present. It has been noted several times (e.g. 2,4)that Alexandrium minutum appears to have fundamentally the same organisation, although its cyst type is unknown (41). As it was published earlier it should have priority. Loeblich & Loeblich (5) asserted that it was "inadequately described" and, lacking a Latin diagnosis, was not available for priority consideration. Although there are problems with Halim's description, this assesment is not accurate. If a species or genus is described in the Animal Kingdom it does not require a Latin diagnosis to be available for use by botanists. The journal in which Alexandrium was published was ecological, and the author does not indicate under which code the genus was published. In other papers he used zoological classification,but he belatedly provided a Latin diagnosis for Gessnerium, apparently to make it acceptable to botanists. His figures of the species are highly detailed relative to descriptions of the time, including details of most of the sulcal plates. However, his terminology was incorrect, referring to the anterior sulcal as a cingular plate and the first postcingular as a sulcal, but these errors are readily corrected. The principal problem is that he appears to have figured the "Z" plate twice, once as a posterior sulcal and once as an antapical, leading to his statement that the genus has two posterior intercalaries. In order to be certain of this,material from the type locality (the harbour of Alexandria) needs to be re-examined. In the meanwhile Gessnerium has the virtue of a type species which can be repeatedly examined. As Steidinger (2) notes, Gessnerium could be officially conserved against Alexandrium, which would obviate further name changes. So far this avenue of action has not been used in dinoflagellate taxonomy and it should be considered in cases like this.

The species which may be referred to Gessnerium in the narrow sense are Gs. monilatum , Gs. balechii and A. minutum if the later genus is conserved. All three have been associated with red water,but only Gs. monilatum has been shown to produce toxins, lethal to fish, annelids, molluscs and crustaceans. The species has only been reported from the western Atlantic, principally from Venezuela (41), through the Gulf of Mexico to the east coast of Florida (40), particularly in coastal lagoons. A closely similar form, not forming long chains, was described by Biecheler (42) from the lagoon at Sète, in southern France, as Goniodoma pseudogoniaulax. The "Z" plate is not apparently in the sulcus, making this referrable to Triadinium, but this arbitrary distinction in such a closely similar form illustrates the point made previously under the Protogonyaulax section: that all these are closely related, making a morphological gradient which is artificially separated into several genera for convenience. Gs. balechii (43) is known only from a Tampa Bay, Florida, bloom which killed fish, apparently due to oxygen depletion.

Pyrodinium

Although this genus has essentially the same tabulation as Gessnerium, it can be satisfactorily separated on the basis of its spiny resting cyst (known to palynologists as Polysphaeridium zoharyi, formerly Hemicystodinium). The thecal ornamentation is also much coarser in this genus and the homologue of the first postcingular plate is on the sulcal side of the left sulcal list.

At present only the var. compressa (44) is known to produce toxins which have produced paralytic shellfish poisoning in the Indo-West Pacific, and also clupeotoxicity in Sabah and Borneo. It has been recorded from various warm, high salinity regions, such as the Red Sea and Persian Gulf (see Figure 4), but has not been observed in the Atlantic Ocean. The var. bahamense, on the other hand, is almost exclusively found in the tropical Atlantic, from the Bahamas to Venezuela, but there is one record of it from the west coast of central America (69). Although the latter is not known to produce toxins there are records of PSP and clupeotoxicity from localities within its range and it is possible that some strains are toxic.

Gonyaulax

Although several species of Gonyaulax (other than those transferred to Protogonyaulax and Gessnerium) have produced red tide blooms, e.g. G. polygramma, G. reticulata, G. spinifera, G. grindleyi (= Protoceratium reticulatum and G. polyedra, only the latter two have been suspected of producing toxins, marine fauna mortalities by the others apparently being due primarily to oxygen depletion. G. grindleyi was associated with PSP and the death of bivalves on several occasions off the west coast of South Africa, but no water soluble toxin could be extracted from it, and the coincident occurrence of P. catenella on some (but not all!) occasions, led to it being discounted as a source of PSP (refs. in 22). The early report of toxicity in a culture of G. polyedra (45) has not been confirmed.

Gambierdiscus, Ostreopsis

The taxonomy of these related forms has recently been thoroughly reviewed (46, see also 2) and so does not require much taxonomic comment here. Fukuyo (46) is correct in asserting that the usual contact of the suture between the fifth and sixth postcingulars (V/VI in Taylor's homology system) is with the Z plate in Gambierdiscus (as in Figure 1A). The present author found only two cells in which it contacted the Y (large antapical) plate in a re-examination of Gambier Island material.

If ciguatera poison is an indication, the distribution of G. toxicus is probably circumtropical, the absence of records from the Indian Ocean being a reflection of the dearth of studies on microbenthos in that region. It was absent in benthic material from Phuket, Thailand, in the Andaman Sea (unpubl. obs.,1973), but this is consistent with its general absense from the close proximity of large landmasses, the east coast of Florida being an exception, presumably due to the closeness of the Gulf Stream to the shore. In the Atlantic the organism has been seen in the Cape Verde Islands, Bermuda, the Florida Keys (refs. in 47,48), and many Caribbean locations, with the most southerly record at Mustique in the Grenadines (49 and Figure 6). Most Pacific records are from French Polynesia (50) and Okinawa (51). An ecological study in the former (50) region confirmed the strong association with macroalgae, but failed to find any clear correlation with nutrients (but it must be noted that, because the water surrounded by the macroalgae constitutes a distinct micro-environment, the samples should have been drawn from within this habitat, rather than the adjacent sea). In a comparative study of 86 localities on 15 Caribbean islands,Taylor & Gustavson (49) found that largest numbers generally occurred in the lee of small, dry islands away from large land masses and were few or absent on the shore of large, volcanic islands with high rainfall. The pattern can be typified by the distribution around southern St. Vincent and the Grenadine islands of Bequia and Mustique (Figure 5). As the distance from the high island of St. Vincent increased the dinoflagellate appeared, first at lee locations, and then protected windward locations. The highest numbers occurred in clumps of the red alga Spyridea, a favoured habitat in the Pacific as well. On Barbados, where sampling was most intense, an apparent decline accompanied the rainy, storm season, with a rise, noticed first at more sheltered locations, afterwards. Maximum numbers were found between 1 and 4m. Contrary to Besada et al.(48), intertidal samples were usually negative. The pattern suggested a strong dislike of land runoff, the organism being limited to the lee of offshore reefs on the windward side of higher rainfall islands, whereas it occurred in sheltered locations all around small islands. Hurricane David caused severe damage to Dominica in late 1979 and two deaths and several illnesses occurred six months afterwards, chiefly from consumption of the liver of Canthidermis maculatus, a predatory triggerfish. A moderate increase of G. toxicus was observed near the island prior to this outbreak.

Both known species of Ostreopsis, O.ovata and O. siamensis, are known to exhibit some toxic activity (51). Like G. toxicus, they are also benthic and common in macroalgae, but they also occur on sand in the vicinity. Although they co-occurred with it periodically, they were often abundant when it was not. (49). They are also known from both the Atlantic and Pacific Oceans and are probably circum-tropical.

Coolia monotis is another related, macro-algal associated species which also spreads into more temperate waters where it is commonly associated with Ulva. It is not very toxic, with only slight hemolytic activity found in a strain from Okinawa (51).

Peridinium

The only toxic freshwater dinoflagellate known is that identified as Peridinium polonicum (52, initially referred to as Glenodinium sp.),

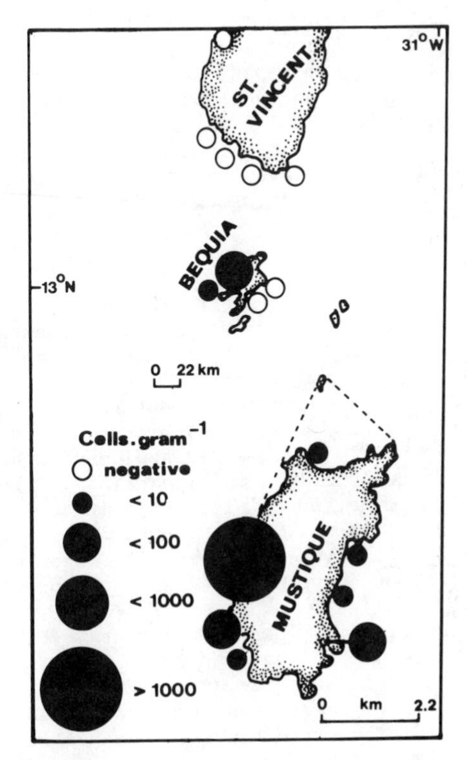

Figure 6. Maximum abundance of Gambierdiscus toxicus found on several species of macroalgae at localities in the northern Grenadine Islands (Mustique is shown to scale and enlarged). Redrawn from Ref. 49.

which produced a fish kill in Sagami Lake near Tokyo in 1962 and 1964. The population described lacked the dorso-ventral flattening noted in the original description from Poland. The species is cosmopolitan in temperate lakes and it is curious that there is only this one report of fish kills and toxin production ("glenodinine",51).

Dinophysis

In addition to the proven production of 35S-methyl okadaic acid (Dinophysistoxin -1) by *D. fortii* (52), with resulting diarrhetic shellfish poisoning (DSP) in Japan, there is strong circumstantial evidence that *D. acuminata* and possibly other *Dinophysis* species have produced DSP in Dutch and Spanish waters (53). These species are very widespread and common in both tropical and temperate waters, but they are not often abundant. They are very difficult to culture and so it is difficult to determine if populations in other areas are toxigenic.

Prorocentrum

One of the worst fatalities on record, the death of 114 people at Lake Hamana (saline) in Japan in 1942, with severe liver injury and hemorrhagic diathesis after the consumption of shellfish, was circumstantially linked with *P. minimum* var.*mariaelebourae* and var.*triangulatum* (which co-occur in blooms), but subsequent testing of these widespread temperate bloomformers has yielded only moderate toxicity (54) and shellfish were only mildly toxic after a bloom which reached concentrations of 1.7 X 10^9 cells.L^{-1} in Oslofjord (55). This species is so common that, if it is the source of the highly toxic "venerupin",toxin must be only in rare strains. Other planktonic *Prorocentrum* species, particularly *P. micans*, have been present as codominants during blooms associated with PSP, but have not been shown to produce acid-extractible toxins.

Two benthic species, *P. lima* and *P. concavum*, which are often found associated with macroalgae (and *G. toxicus*), have been shown to produce toxins, the former okadaic acid and other toxin(s) resembling those associated with ciguatera, and the latter a strong ichthyotoxin (51). *P. lima* is of interest, not simply because it may contribute to the complex spectrum of symptoms associated with ciguatera, but also because it is extremely cosmopolitan, occurring on sand in both tropical and temperate waters. It should be determined if temperate populations also produce the same toxins. Since the fish foodchains differ in colder waters, with less direct grazing on macroalgae, this may be a reason for the absence of ciguatera in temperate regions, even if *P. lima* is toxigenic everywhere, but one might expect its appearance in browsing molluscs. *P. mexicanum* (=*rhathymum*) is mildly hemolytic.

Gymnodinium and Ptychodiscus

The best-known toxin producer attributed to *Gymnodinium* is the fish killer *G. breve* (which also produces toxins which can accumulate in oysters, and has mild hemolytic activity). Recently it was transferred to the genus *Ptychodiscus* by Steidinger (56) because of its general form, with an apical crest (carina) and because the ultrastructure of its peripheral region (amphiesma) differs from the type species of *Gymnodinium*. In *P. brevis* the amphiesma is very compressed, surrounding a highly vesiculated cytoplasmic region. The amphiesmal vesicles appear to be empty, but they are so compressed that it is difficult to be certain. Staining for cellulose in the periphery was negative (57). In a recent review in which she summarised the grounds for the generic transfer Steidinger (2) omitted mention of a feature which is critical if the transfer to Ptychodiscus is to be wholly appropriate. The type, and only other species recognised at present, P. noctiluca, has a prominent, resilient, continuous envelope which appears to be equivalent to the pellicle of other genera, although it has been misleadingly referred to as a theca by some authors (58). The latter term is best limited to the cellulose wall consisting of multiple, closely-fitting plates. There are several genera in which the principal strengthening of the motile stage is a pellicular layer but which differ from Ptychodiscus in other morphological features. Steidinger (pers. comm.) has observed a pellicle-like structure in P. brevis when the cells burst under light microscopic examination. This needs to be documented properly and its precise location and composition determined, if possible. P. noctiluca also requires electron microscopic study. Another species similar to P. brevis has been known as "Gymnodinium Type 65" (and illegitimately named G. nagasaki). It is somewhat toxic to fish and strongly inhibitory to the rotifer Brachionus (59). In ultrastructure it shows resemblances to P. brevis but differs in some morphological features (2). P. brevis appears to be restricted to the Gulf of Mexico although closely similar forms to the "butterfly" form seen in some P. brevis blooms have been seen in northern Spain, Japan, and the eastern Mediterranean (2).

G. catenatum is an interesting species because in general form and chain formation it resembles a Protogonyaulax but it lacks a theca. It has been circumstantially linked with saxitoxin-like PSP from the west coast of Mexico and northwest Spain (62, 63) and is also known to occur in the Gulf of California and the Mar del Plata. In view of the similarity of form and toxin production it has been suggested that it may be a theca-less mutant of Protogonyaulax (64,62,2).

G. flavum has been circumstantially linked to a fish kill off southern California (65). It is very similar in appearance to Gyrodinium aureolum, the different generic attributions being due to their morphology being borderline between Gymnodinium (girdle displacement less than 0.2 the length of the cell) and Gyrodinium (0 2 or more). Technically, G. flavum is limited to the Pacific Ocean and Gyr. aureolum to the north Atlantic, but this author believes that the two may be confused and may even be conspecific. A bloom in Villefranche Bay in the south of France (unpubl. obs.,1972) exhibited features of both.

G. veneficum, isolated from English waters and toxic to fish and mice, has not been linked to any natural toxic events.

Gyrodinium, Cochlodinium

As noted above, Gyrodinium aureolum, has been implicated in the death of fish and benthic invertebrates in the North Sea (principally southern Norway, 61) and the Irish Sea, but only since 1966. It was first described from Massachusetts but has not caused kills in that region.

Cochlodinium resembles photosynthetic Gyrodinium species but has greater girdle torsion (1.5 or more turns around the body). Several species have been circumstantially linked to fish kills, e.g. C. catenatum, C. polykrikoides (= heterolobum) but toxins have not been extracted from the organisms. Severe PSP, with 9 deaths and 159 illnesses, was reported from Venezuela in 1977 (27), associated with an organism that resembled a Cochlodinium sp. in formalin-preserved material. In British Columbia C. citron has been coincident with PSP on several occasions. A culture of it did not yield acid extractable toxins. In Obidos Lagoon, Portugal, C. citron was identified as a codominant with Prorocentrum micans in an outbreak of PSP (66).

Amphidinium

Three species of this athecate genus have been linked to toxin production: the supernatant of cultures at concentrations of 1.5 x 10^6 cells.ml^{-1} of A. klebsii and a species tentatively identified as A. rhychocephalum was toxic to fish (67); and A.carterae, which is known to produce acrylylcholine, choline O-sulphate and an unidentified choline ester (68), as well as a non-choline ichthyotoxin (51). The ichthyotoxin of A. klebsii was present in the diethyl ether fraction, whereas that of A. carterae was butanol-soluble. A toxin-producing culture,sent to the author by John West,was identifiable as A. massartii (42), which is very similar to A. rhychocephalum, but better characterised (they may be conspecific). The Japanese study that found the non-choline ichthyotoxin in A. carterae (51) may have been of another species for the authors pointed to the difference in toxins and a difference in size. None of these species, all of which can be thought of as cosmopolitan sand and macroalgal associants, have been implicated in natural intoxications. However, their occurrence in seaweeds could contribute to the complex spectrum of ciguatera symptoms, although they might also discourage fish from eating the weeds (also possible for Prorocentrum concavum). The ichthyotoxins are not likely to cause fish kills because they appear to be primarily intracellular, being released into the medium on cell lysis (70).

Conclusion

This paper has been limited to the two dozen or so species which have been demonstrated to produce toxins. Not included, for the most part, were red water producing species which are usually harmless, e.g. Gymnodinium sanguineum (= splendens, nelsonii), Noctiluca scintillans(= miliaris), or those causing harm through oxygen depletion associated with bloom decay, respiration demand at low light levels, or bacterial promoting or inhibiting.

It is notable that relatively few genera (10-13) are involved and that several of these appear to be quite closely related. Taylor (4) used a plate homology model to demonstrate that Protogonyaulax, Gessnerium, Pyrodinium, Gambierdiscus, Ostreopsis and possibly Coolia) are gonyaulacoids, more primitive in his view than Gonyaulax. Other close relatives of this "lower gonyaulacoid assemblage" which have not yet been linked to toxicity,but deserve investigation because of their apparent affinities include Triadinium (= Goniodoma, Heteraulacus), Fragilidium (incl. Helgolandinium), Pyrocystis, Pyrophacus and, more remotely, Ceratocorys and Ceratium (64). The gonyaulacoids may be more related to the dinophysoids and prorocentroids than other groups are, with the two latter groups being considered the most primitive dinoflagellates (64, see 71, 72 for alternative views). These affinities have been deduced from morphology, but toxins may also provide clues.

Literature Cited

1. Taylor, D.L.; Seliger, H.H., Eds. "Toxic Dinoflagelate Blooms"; Elsevier North Holland: New York, 1979; p.505.
2. Steidinger, K.A. Progr. Phycol. Res. 1983, 2, 147-88.
3. Schmidt, R.J.; Loeblich, A.R.111. In "Toxic Dinoflagellate Blooms"; Taylor, D.L.; Seliger, H.H. Eds.; Elsevier North Holland: New York, 1979; pp. 83-8.
4. Taylor, F.J.R. In "Toxic Dinoflagellate Blooms"; Taylor, D.L., Seliger, H.H. Eds.; Elsevier North Holland: New York, 1979; pp. 47-56.
5. Loeblich, A.R.111; Loeblich, L.A. In "Toxic Dinoflagellate Blooms"; Taylor, D.L., Seliger, H.H. Eds.; Elsevier North Holland: New York, 1979; pp. 41-6.
6. Taylor, F.J.R. BioSystems 1980, 13,65-108 (correction p. 84).
7. Walker, L.M.; Steidinger, K.A. J. Phycol. 1979, 15, 312-5.
8. Dale, B. Sarsia 1977, 63, 29-34.
9. Balech, E. Bolm. Inst. Oceanogr. S. Paulo 1979, 28(2), 55-64.
10. Taylor, F.J.R. Env. Letters 1975, 9, 103-119.
11. Balech, E.; De Mendiola, B.R.E. Neotropica 1977, 23(69), 49-54.
12. Turpin, D.H.; Dobell, P.E.R.; Taylor, F.J.R. J. Phycol. 1978, 14, 235-8.
13. Postek, M.T.; Cox, E.R. J. Phycol. 1976, 12, 88-93.
14. Fukuyo, Y. Suisan Kenkyu Zosho 1980, 33, 1-8 (In Japanese).
15. Balech, . Bol. Inst. Biol. Mar. (Univ. Nac. Buenos Aires) 1964, 4, 1-49 + 5 pls.
16. Cembella, A.D.; Taylor, F.J.R. Submitted to Biochem. System. Ecol.
17. Hayhome, B.A.; Pfeister, L.A. Amer. J. Bot. 1983, 70, 1165-72.
18. Schmidt, R.J.; Gooch, V.D.; Loeblich, A.R. 111; Hastings, J.W. J. Phycol. 1978, 14, 5-9.
19. Neal, R.A. Ph.D. Thesis, University of Alaska, Fairbanks, 1982.
20. Guzman, L.; Campodonico, I.; Hermosilla, J. Ans. Inst. Patagonia 1975, 6, 173-83.
21. Hall, S. Ph.D. Thesis, University of Alaska, Fairbanks, 1982.
22. Horstman, D.A. Fish. Bull. S.Afr. 1981, 15, 71-88.
23. Hashimoto, Y.; Noguchi, T.; Adachi, R. Bull. Jap. Soc. Fish. 1976, 42, 671-6.
24. Bursa, A. J. Fish. Res. Bd. Canada 1961, 18, 51-83, 563-615.
25. Heimdal, B.R. J. Plankt. Res. 1983, 5, 901-18.
26. Balech, E. Publ. Serv. Hidrogr. Naval Argentina 1971,H. 654, 1-103 + 12 pls.
27. Reyes-Vasquez, G.; Ferraz-Reyes, E.; Vasquez, E. In "Toxic Dinoflagellate Blooms"; Taylor, D.L., Seliger, H.H. Eds.; Elsevier North Holland: New York, 1979; pp. 191-194.
28. Carreto, J.I.; Lasta, L.M.; Negri, R.; Benavides, H. Rep. INIDEP Argentina, 1981, 399, 1-83.
29. Pieterse, F.; Van der Post, D.C. Mar. Res. Lab. S.W. Africa, Invest. Rep. 1967, 14, 1-125.
30. Fukuyu, Y. In "Toxic Dinoflagellate Blooms"; Taylor, D.L.; Seliger, H.H. Eds.; Elsevier North Holland: New York, 1979; pp. 61-64.
31. Norris, L.; Chew, K.K. Proc. 1st Int. Conf. Toxic Dinofl. Blooms; LoCicero, V.R. Ed. 1975, 143-52.
32. Taylor, F.J.R. J. Fish. Res. Bd. Canada 1968, 25, 2241-5.
33. Harrison, P.J.; Fulton, J.D.; Taylor, F.J.R.; Parsons, T.R. Can. J. Fish. Aquat. Sci. 1983, 40, 1064-94.

34. Taylor, F,J. R.; Cembella, A.D. Can. J. Fish. Aquat. Sci., in prep.
35. Fukuyu, Y. Pers. comm. to A.D. Cembella, 1983.
36. Guzman, L.; Lembeye, G. Ans. Inst. Patagonia 1975, 6, 185-95.
37. Loeblich, L.A.; Loeblich, A.R. 111. Proc. Int. Conf. Toxic Dinofl. Blooms; LoCicero, V.R. Ed., 1975, pp. 207-224.
38. Prakash, A.; Taylor, F.J.R. J. Fish Res Bd. Canada 1966, 23, 1265-70.
39. Halim Y. Int. Rev. Ges. Hydrobiol. 1967, 54, 701-55.
40. Howell, J.F. Trans. Amer. Micr. Soc. 1953, 72, 153-6.
41. Halim, Y. Vie Milieu 1960, 11, 102-5.
42. Biecheler, B. Bull. Biol. France Belgique 1952, Suppl. 36, 1-149.
43. Steidinger, K. A. Phycologia 1971, 10, 183-7.
44. Steidinger, K.A.; Tester, L.S.; Taylor, F.J.R. Phycologia 1980, 19, 329-34.
45. Schradie, J.; Bliss, C.A. Lloydia 1962, 25, 212-21.
46. Fukuyo, K. Bull. Jap. Soc. Sci. Fish. 1981, 47, 967-78.
47. Taylor, F.J.R. In "Toxic Dinoflagellate Blooms"; Taylor, D.L., Seliger, H.H. Eds.; Elsevier North Holland: New York, 1979; pp.71-76.
48. Besada, E.G.; Loeblich, L.A.; Loeblich, A.R. 111. Bull. Mar. Sci. 1982, 723-35.
49. Taylor, F.J.R.; Gustavson, M.S. Proc. 7th Int. Sci. Diving Conf. Padova 1984, in press.
50. Yasumoto, T.; Inoue, A.; Oshi, T.; Fujimoto, K.; Oshima, Y.; Fukuyo, Y.; Adachi, R.; Bagnis, R. Bull. Jap. Soc. Sci. Fish. 1980, 46, 1397-1404.
51. Nakajima, I.; Oshima, Y.; Yasumoto, T. Bull. Jap. Soc. Sci. Fish. 1981, 47, 1029-33.
52. Hashimoto, Y.; Okaichi, T.; Dang, L.D.; Noguchi, T. Bull. Jap. Soc. Sci. Fish. 1982, 549-52.
53. Murata, M.; Shimatani, M.; Sugitani, H.; Oshima, Y.; Yasumoto, T. Bull. Jap. Soc. Sci. Fish. 1982, 48, 549-52.
54. Okaichi, T.; Imatomi, Y. In "Toxic Dinoflagellate Blooms"; Taylor, D.L., Seliger, H.H. Eds.; Elsevier North Holland: New York, 1979; pp. 385-8.
55. Tangen, K. Blyttia 1980, 38, 145-58.
56. Steidinger, K. In "Toxic Dinoflagellate Blooms"; Taylor, D.L., Seliger, H.H. Eds.; Elsevier North Holland: New York, 1979; pp. 435-42.
57. Steidinger, K.; Truby, E.W.; Dawes, C.J. J. Phycol. 1978, 14, 72-9.
58. Boalch, G.T. J. Mar. Biol. Ass. U.K. 1969, 49, 781-4.
59. Iizuka, S. In "Toxic Dinoflagellate Blooms"; Taylor, D.L., Seliger, H.H. Eds.; Elsevier North Holland: New York, 1979; pp. 111-4.
60. Noro, T.; Mizuno, J.; Nozawa, Mem. Fac. Fish., Kagoshima Univ. 1981, 30, 179-89.
61. Tangen, K. Sarsia 1977, 63, 123-33.
62. Morey-Gaines, G. Phycologia 1982, 21, 154-63.
63. Campos, M.J.; Fraga, S.; Marino, J.; Sanchez, F.J. ICES ms.rept. 1982, L27, 1-8.
64. Taylor, F.J.R. BioSystems 1980, 13, 65-108.
65. Lackey, J.B.; Clendenning, K.A. Q.J. Fla. Acad. Sci.1963, 26, 263-8.

66. Pinto, J.S.; Silva, E.S. Notas Estud. Inst. Biol. Marit. 1956, 12, 1-20.
67. McLaughlin, J.J.A.; Provasoli, L. J. Protozool. 1957, 4, S7.
68. Taylor, R.F.; Ikawa, M.; Sasner, J.J. Jr.; Thurberg, F.P.; Andersen, K.K. J. Phycol. 1974, 10, 279-83.
69. Tafall, B.F.O. Ans. E.N. Cienc. Biol. (Mexico) 1942, 2, 435-47 + 2 pls.
70. Thurberg, F.P.; Sasner, J.J. Jr. Chesapeake Sci. 1973, 14, 48-51.
71. Bujak, J.P.; Williams, G.L. Can. J. Bot. 1981, 59, 2077-87.
72. Dodge, J.D. Br. Phycol. J. 1983, 18, 335-56.

RECEIVED June 19, 1984

9

Historical Perspective on Paralytic Shellfish Poison

EDWARD J. SCHANTZ

Department of Food Microbiology and Toxicology, University of Wisconsin, Madison, WI 53706

Shellfish have become an important item in the world's food supply. Rapid air transportation with good refrigeration has widened local seacoast markets of past years to many inland cities and areas throughout the world and commercial growing and harvesting of shellfish for this market has become a growing industry. Shellfish such as mussels and clams that feed on microscopic organisms concentrate and retain bacteria, viruses, toxins and poisons associated with the organisms they feed on. One important hazard associated with shellfish consumption is paralytic shellfish poisoning (PSP), caused by a highly lethal neurotoxin which the shellfish retain from feeding on certain poisonous dinoflagellates. Out of about 1200 species of dinoflagellates only a few produce the paralytic poison and occur sporadically along many sea coasts throughout the world. Shortly after eating shellfish that have fed on the poisonous dinoflagellates humans experience a tingling sensation in the lips and finger tips followed by progressive paralysis of the limbs and finally death from respiratory paralysis which may occur within 2 to 24 hours depending upon the dose. If the dose is small enough that one survives 24 hours, the prognosis is good. No antidote is known, but artificial respiration administered soon after symptoms appear is the best treatment and no doubt has saved lives, at least in borderline cases of poisoning.

The purpose of this report is to review and present the important historical developments leading to the present understanding of the chemical nature of the poison including its chemical structure and properties, the mechanism causing death and the present means of control to make shellfish safer for human consumption.

Historical Background and Cause of PSP

As long as medical records have been kept, poisoning of humans from eating clams and mussels has been recorded. The sporadic and unexplainable occurrence of poisonous shellfish was indeed a very puzzling phenomenon. When mussels became poisonous they usually remained so for a few weeks and for no apparent reason soon became

0097–6156/84/0262–0099$06.00/0
© 1984 American Chemical Society

perfectly safe to eat again. Although many theories were proposed for the occurrence of poisonous shellfish throughout the eighteenth and nineteenth centuries, particularly in European countries around the North Sea, the actual cause was not known until about 1927 when outbreaks of PSP occurred along the coast of California. At that time Sommer, Whedon, Meyer and their colleagues (1,2) at the University of California observed a particular dinoflagellate blooming around the California sea mussel beds at the time many people became sick and died after eating the mussels. These investigators suspected that the dinoflagellate upon which the mussels were feeding may be poisonous and found that acidic water extracts of the dinoflagellates and the mussels killed mice in the same manner. The dinoflagellate was identified as Gonyaulax catenella. To test this observation Dr. Sommer and his assistant Lucile Foster placed nonpoisonous mussels in laboratory jars containing seawater cultures of G. catenella and found that the mussels soon became poisonous. When these mussels were placed in jars containing nonpoisonous organisms, the mussels excreted or destroyed the poison within a week or two, thus establishing conclusively the relationship of G. catenella to the poisonous mussels. This organism, like most dinoflagellates, blooms over a period of 2 to 3 weeks and gradually dies off as other organisms bloom. In the natural state mussels become poisonous soon after a poisonous dinoflagellate blooms and they excrete or destroy the poison within 2 to 3 weeks after the bloom has disappeared. About 95% of the poison in a mussel is contained in the dark gland or hepatopancreas and is quite stable to ordinary steaming or cooking which, of course, makes the food poisoning problem more acute. Poisonous shellfish neither look nor taste different from normal ones. Although heavy blooms of dinoflagellates cause a discoloration of the water called a red tide, the average person collecting shellfish for human consumption cannot rely on a change in the appearance of the water. Concentrations as low as 400 or 500 cells per ml are not visible to the naked eye but are sufficient to make the mussels too poisonous for human consumption. Cell counts of around 20,000 or more per ml are necessary to produce a red tide. It would be advisable not to eat shellfish if a red tide bloom is apparent. The organism may or may not be poisonous.

The early work of Sommer and associates established a good mouse assay for the poison in shellfish products which made quantitative work with the poison practical. A mouse unit (MU) was defined as the minimum amount of poison that would kill a 20-gram white mouse in 15 minutes when one ml of an extract of shellfish was injected intraperitoneally. Higher amounts than the minimum kill in shorter time, i.e., death times of 3, 4, 6, and 8 minutes are equivalent to 3.7, 2.5, 1.6, and 1.3 MU, respectively, as illustrated in Figure 1.

The discovery of the relationship of G. catenella to poisonous shellfish by Sommer, Whedon and associates led Canadian investigators Needler (3) and Prakash (4) to the discovery that Gonyaulax tamarensis (var. excavata) caused clams and scallops along the northeast coast of North America and the northeast coast of England to become poisonous.

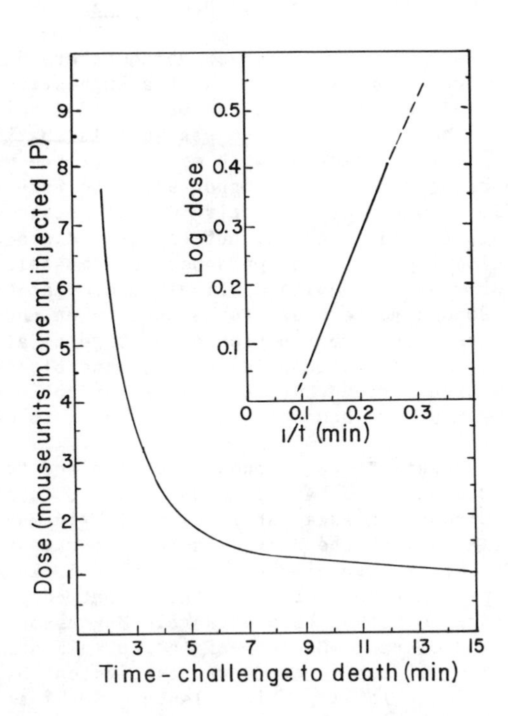

Figure 1. Dose and time to death relationships for saxitoxins. A plot of the log of the dose against the reciprocal of the time results in a straight line relationship. Reprinted with permission from Ref. 37. Copyright 1973, Committee on Food Protection NAS-NRC.

In the early years there was much interest in the amount of poison necessary to cause sickness and death of humans. From accidental cases of poisoning along the coast of California where G. catenella commonly blooms, Sommer and colleagues estimated that death of humans resulted from the consumption of about 20,000 MU of poison. However, along the St. Lawrence estuary and Bay of Fundy the Canadians observed death of persons with a consumption as low as 5000 MU. In these areas shellfish become poisonous by feeding on G. tamarensis.

Purification - Chemical and Physical Properties

Dr. Byron Riegel and Warren Stanger of Northwestern University and Dr. Hermann Sommer undertook in 1944 at the suggestion of the National Academy of Sciences the first organized studies to isolate and characterize the poison from the sea mussels (Mytilus californianis) and in connection with my work at the U.S. Army Biological Laboratories I had the opportunity to take part in the work at this time. Our first objective was to obtain sufficient poisonous mussels for our studies, but the sporadic occurrence of the poisonous dinoflagellate and poisonous sea mussels along the California coast made this quite difficult and meant that organized surveys for poisonous mussels had to be carried on throughout the summer months, the most likely period of occurrence along the California coast. If a poisonous dinoflagellate bloom did occur at the time of low tide period so that above water collections of mussels could be made, we had only about one week to make collections.

Significant amounts of the poison were obtained from the sea mussels, but additional and larger amounts were obtained by the collection of poisonous Alaska butter clams (Saxidomas giganteus). Through arrangements with the United States Department of the Interior and the Alaska Experimental Commission in 1948 I made collections of poisonous butter clams throughout certain areas of southeastern Alaska with the help of Harris Magnusson, John Dassow, J. S. Chambers and Clarence Carlson of the Ketchikan Laboratory. Much of the digging of butter clams was carried out by local Indians who worked for local packing plants. Small amounts also were obtained by culturing the poisonous dinoflagellate in the laboratory at Hooper Foundation, University of California and at the Biological Laboratories, Fort Detrick, Maryland. The culturing at the Biological Laboratories was carried out with axenic cultures of G. catenella and G. tamarensis with help from Luigi Provasoli, Haskins Laboratories, who provided the axenic cultures of these organisms. The original culture of G. catenella was collected in 1944 from a bloom in Monterey Bay, California by Lucile Foster at the Hooper Foundation, University of California and G. tamarensis by Provasoli from the Bay of Fundy. The poisons from the three sources appeared the same, but were kept separate throughout the work. They all produced the same signs of poisoning in mice and all had similar molecular weights of 300 to 400 as measured in crude extracts by the diffusion rate of the poison in a Northrop diffusion cell. Their action on cation exchange resins was also similar. The source from the Alaska butter clam siphons proved to be the best supply because, unlike the mussels, 60 to 70% of the poison in the whole body of the clam was contained in the siphon and remained there for periods of a year or more which made collection possible at low tide any time the weather permitted. Another source of the poison was made available to me shortly after World War II through the courtesy of the Canadian Defense Research Board in 1947. This source was the entrails of commercial scallops collected in the Bay of Fundy. The entrails were from deep sea scallops, which are usually poisonous at all times. The poison in the scallops however had different properties and did not function well on cation exchange resins. As a result I stored them in acid solution about pH 2 for later studies (5).

Riegel and Sommer's early work in 1944 established that the poison from California mussels was a basic substance easily extracted with acidic water pH 2 and some purification was accomplished on Decalso and activated carbon (6). These substances were not very efficient for purification. However the new Amberlite cation exchange resins (Amerlite XE-64) developed by Rohm and Haas Company at about this time proved to be ideal for the quantitative removal of the poison from crude acid extracts of poisonous shellfish containing as little as 1 MU per mg of solids. Extracts of the hepatopancreas of poisonous mussels usually contained 2 to 8 MU per mg of solids whereas extracts of the siphons of poisonous butter clams varied considerably but usually contained around 2 MU per mg of solids. Research on purification methods at the Biological Laboratories and with Riegel's group at Northwestern University resulted in partial purification of the poison on the resin to about 3000 MU per mg of solids. Further chromatography on acid-washed and activated alumina brought the specific toxicity to 5500 MU per mg. Repeated chromatography in various ways did not increase the toxicity and the poison was considered in a highly purified state in 1954, ten years after the work began. Subsequently over 20 grams of highly purified saxitoxin were prepared at the Biological Laboratories mainly from butter clam siphons.

The poison was found to be a white hygroscopic solid very soluble in water, partly soluble in methanol and ethanol, but insoluble in most nonpolar solvents such as ethyl and petroleum ethers. It showed no absorption in the ultraviolet and titration showed two pK_a values at 8.2 and 11.5. The optical rotation was about 130. The molecular formula was found to be $C_{10}H_{17}N_7O_4$ as the free base with a molecular weight of 299 and as the dihydrochloride salt 372. The poison can be reduced with hydrogen at room temperature and pressure in the presence of platinum black to produce a nontoxic derivative. One mole of hydrogen was taken up per mole of poison in this reaction. It also reacted with certain aromatic nitro compounds to form a colored complex in much the same way as creatinine reacts with dinitrophenol in the Jaffe test and with dinitrobenzoic acid in the Benedict-Behre test. The reduced poison (nontoxic) did not react with these reagents. The good correlation of these tests with the poison suggested to us their possible use as a quantitative chemical test to replace the mouse test where mice are hard to get and keep for assay purposes. However, certain impurities contained in clam and mussel extracts interfered sufficiently with the test for its practical use as an assay, but it was valuable for our research work with the purified poison. The specific toxicity determined in several Food and Drug Administration laboratories of the United States and in cooperation with Canadian Public Health Services confirmed the specific toxicity to be 5500 ± 500 MU per mg and one mouse unit equivalent to 0.18 microgram. Our work on the purification and characterization of the poison was published in the Journal of the American Chemical Society in 1957 (7,8) and in the Canadian Journal of Chemistry in 1961 (9). Much interest was shown in the purified poison and small amounts were donated to physiologists, pharmacologists and chemists throughout the world for various studies. As a result in 1964 Dr. M. H. Evans (10) of the Sherrington School of Physiology, London and Dr. C. Y. Kao (11) of the State University of New York discovered that the poison caused death by blocking specifically the inward sodium ion current in nerve and muscle cell membranes, an action similar to that of tetrodotoxin from the puffer fish.

Studies with the Public Health Service and Food and Drug Administration

Through a cooperative arrangement between the Department of Defense and the Public Health Service in 1957 I carried out studies with E. T. Jensen, K. H. Lewis, J. E. Campbell and E. F. Mc Farren on the use of the purified poison for the establishment of a standardized mouse assay procedure for the paralytic poison in shellfish products (12). Many of these studies were carried out cooperatively with Canadian Public Health Agencies involved with PSP. To establish a more nearly accurate and uniform mouse assay between laboratories, we prepared a solution containing 100 µg of the highly purified poison (saxitoxin) in 20% ethanol-water solution at pH 3 with HCl as a reference standard. This solution was dispensed in 5 ml ampoules and furnished free of charge by the Food and Drug Administration to laboratories engaged in the quantitative assay of shellfish products. The use of the reference standard increased the confidence and use of the mouse assay and helped protect against poisonous shellfish getting on the commercial markets. Also, the reference standard made possible the reporting of the poison content of shellfish as µg per 100 grams of meats and established a basis for an official AOAC mouse assay for the poison (12,13). The Food and Drug Administration set a top limit of 80 micrograms of poison per 100 grams of shellfish meats as marketable for human consumption. This limit is well below the 1 to 4 mg deemed necessary in foods to cause sickness and death of humans.

Chemical Structure of the Paralytic Poisons

Our early studies on the chemical structure of saxitoxin were hampered by the fact that the highly polar nature of the molecule prevented good conditions for crystallization. Although we had crystallized saxitoxin as the helianthate in 1957 (7), the molecule was too large to be usable for structure determination by crystallographers in X-ray diffraction studies. In July 1972 I moved to the University of Wisconsin, Department of Food Microbiology and Toxicology and continued the research on saxitoxin with Professors

Frank Strong and Heinrich Schnoes in the Department of Biochemistry. In 1971 Wong et al. (14), working with Rapoport on the structure by degradation of the molecule at the University of California, reported in the Journal of the American Chemical Society a proposed structure for saxitoxin. Although later shown to be incorrect, it did establish that saxitoxin essentially is a tetrahydropurine derivative. During the years that followed questions were raised by some investigators regarding the properties of the reported structure that did not seem in accord with the properties of saxitoxin. As a result we felt that more effort should be placed on obtaining a good crystalline derivative so that crystallographic studies could be used for structure determination. At the laboratory at the University of Wisconsin with Strong, Schnoes and Vartan Ghazarossian we produced a crystalline derivative of saxitoxin with p-bromobenzene sulfonic acid which was very suitable for crystallographic studies. Jon Clardy and colleagues, then at Iowa State University, carried out the crystallographic studies and determined the structure as presented in Figure 2A. This structure for saxitoxin (15) was presented at the first Conference on Poisonous Dinoflagellates at Boston, Mass. in November 1974 and published in the Journal of the American Chemical Society, March, 1975 (16). On the basis of this structure, Tanino et al. (17) with Kishi at Harvard University synthesized d,l-saxitoxin and it was found active in blocking sodium channels in the same manner as natural saxitoxin. Later Bordner et al. (18) working with Rapoport confirmed our structure for saxitoxin with an X-ray analysis of saxitoxin ethyl hemiketal.

Although these studies solved the structure for saxitoxin, the neutral or weakly basic poison from scallops, and that produced in cultures of *G. tamarensis*, was still a mystery. Its poisonous action appeared identical to saxitoxin (19), but because of its weakly basic action on cation exchange resins (Amberlite CG-50) and electrophoresis it was assumed to be of different structure. Approximately 85% of the poison from east coast scallops and *G. tamarensis* cultures was eluted from the cation exchange resin above pH 4 and the remainder of 15% was removed below pH 3. On the other hand 90 to 95% of the poison from the west coast mussels, butter clams and axenic *G. catenella* culture was eluted only below pH 3. The extracts of poisonous scallop viscera I obtained from the Canadians back in 1947 were stored in acid solution and moved to the University of Wisconsin in 1972 along with extracts of cultured *G. tamarensis* cells. After standing in acid solution for over 20 years we found that the poison now chromatographed on the cation exchange resins similar to saxitoxin (20). However the major poison in extracts of freshly collected scallop entrails obtained from the Canadian Department of Human Services in 1978 was eluted from the cation exchange resins above pH 4 in line with the original observations. It was therefore assumed that some conversion took place in the molecule that led to a more basic structure.

Shortly after we were successful in obtaining the structure of saxitoxin, Shimizu and colleagues at the University of Rhode Island purified the poison from east coast scallops and reported that the structure of the major poison from scallops and *G. tamarensis*, based on NMR studies, was 11-hydroxy saxitoxin which he called gonyautoxin II and its epimer gonyautoxin III (21). This work established the similarity of G. tamarensis poison to saxitoxin but in our minds it was questionable if the proposed structure would be weakly basic in character and in line with the properties of G. tamarensis poison. Gregory Boyer in our laboratory at the University of Wisconsin had also purified the poison from freshly collected scallops on Bio-gel P-2 and acidic Al_2O_3 columns and upon elemental analyses found, in addition to the elements in saxitoxin, one mole of sulfur per mole of poison in the form of sulfate. NMR studies showed this group in the 11 position making the correct structure for the major poison from scallops, and also for the poison from G. tamarensis, as the sulfonic acid ester of 11-hydroxy saxitoxin as shown in Figure 2B. This structure was in line with the weakly basic properties of G. tamarensis poison and was published by Boyer et al. in 1978 in the Journal of the Chemical Society (London) Chemical Communications (22).

Shimizu in 1978 published the structure of an interesting saxitoxin derivative, 1-N-hydroxysaxitoxin, which he isolated from scallops and called neosaxitoxin (23) as shown in Figure 2C. Subsequently Boyer in 1980 published the structure of neo-11-hydroxysaxitoxin sulfate and its epimer (24).

Some very interesting derivatives of saxitoxin recently have been isolated by Hall and colleagues of the University of Alaska from dinoflagellates of the genus Protogonyaulax found along the coast of Alaska (25). The spectroscopic studies performed in collaboration with H. K. Schnoes and his associates at the University of Wisconsin suggested that two of these were carbamoyl-N-sulfo-11α-hydroxysaxitoxin sulfate and its 11ß epimer, as shown in Figure 2D. X-ray analysis by S. D. Darling at the University of Akron confirmed the latter structure (26). These compounds were neutral in character. Its properties will be described in more detail by other speakers at this symposium.

The report of the N-sulfocarbamoyl group is the first reported in a natural product and the attenuation of toxicity associated with sulfonation of the carbamoyl group introduces a new aspect of structure-activity relationships in this group of neurotoxins. Saxitoxin has the highest specific toxicity of any of these poisons isolated thus far. Ghazarossian (27) and Ghazarossian et al. (28) had found that removal of the carbamoyl group by treatment of saxitoxin with 3 N HCl at 100°C reduced the toxicity to about 60%. As stated previously in our earlier work (8,27) and that of Koehn et al. (29), reduction of the hydrated ketone at position 12 to a monohydroxy group reduces the toxicity to around 1% of the original.

As mentioned above we obtained most of the poison for our chemical and biological work from the Alaska butter clam. However our search for a dinoflagellate or any other poisonous organism in the water as a source of the poison in the clams was never definitely accomplished. In 1948, and in some years following, along with a survey for toxic butter clams in Southeastern Alaska by the staff at the Fishery Products Research Laboratory at Ketchikan, I collected plankton in areas where the clams were found to be toxic. Mouse assays of the plankton showed no toxicity. Microscopic examination of the water in these areas showed heavy growth of microscopic plankton but little or no evidence of

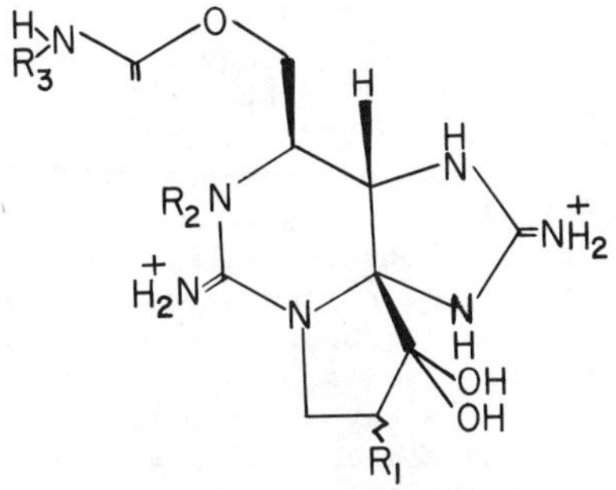

Poison	R_1	R_2	R_3	References
A. Saxitoxin	H	H	H	Schantz et al <u>16</u>
B. 11-hydroxysaxitoxin sulfate	OSO_3^-	H	H	Boyer et al <u>22</u>
C. Neosaxitoxin	H	OH	H	Shimizu et al <u>23</u>
D. Carbamoyl-N-sulfo-11-hydroxysaxitoxin sulfate	OSO_3^-	H	SO_3^-	Wichmann et al <u>26</u>

Including the α and β epimers of these poisons at position 11 there are a total of 12 paralytic shellfish poisons.

Figure 2. Structure of paralytic shellfish poisons

dinoflagellates. Also mussels (Mytilus edulus) showed little or no toxicity. The plankton were collected in No. 25 silk nets. The survey was made in a 50-foot retired navy boat called the "Researcher" assigned to the Ketchikan Laboratory and had living space for eight persons. We carried on board laboratory equipment including mice for assays of clams and a microscope for examination of the water from the Ketchikan area to the area around Juneau. During some of our work at the Ketchikan laboratory we transplanted toxic butter clams in areas where the indigenous clams were not toxic and also transplanted nonpoisonous clams in areas where the indigenous clams were toxic. Only after one year was the toxicity of the transplanted toxic clams significantly lowered. The nonpoisonous clams acquired toxicity equivalent to the indigenous clams in about two months. In 1949 Lucile Foster and Hermann Sommer, in cooperation with the Fishery Products Research Laboratory collected plankton in these waters throughout the summer and identified G. catenella but only in exceedingly low concentrations in areas where they did occur. They showed no particular patterns of occurrence that coincided with toxic clam beds. Examination of dinoflagellate blooms that occurred during this period consisted mainly of species of Noctiluca and were nontoxic to mice. From these findings we believed that the butter clams were toxic, in contrast to the low toxicity of the mussels, because of the slow rate at which the clams destroyed or excreted the poison from the siphon whereas the mussels excreted or destroyed the poison from the hepatopancreas at a much faster rate. Although our surveys and those of Lucile Foster in particular showed only a minute presence of G. catenella along with the fact that the properties of the poison from the clam siphons and from the California sea mussels appeared very similar, it seemed logical to postulate that a poisonous dinoflagellate was the source of the poison. A report of this work was published in the Journal of Protozoology in 1964 (30).

It is interesting too, in light of the recent findings of poisonous dinoflagellates by Hall, Reichardt and Neve at the University of Alaska, that butter clam siphons collected in southeastern Alaska from 1948 to 1958 for our original work contained only saxitoxin. Could it be that the butter clams bind only the strongly basic poisons in the siphon and allow the weakly basic and neutral ones to pass through? If there is a reason for such selection, perhaps it is to ward off predators. However, the main function, in my estimation, for the presence of the poison in shellfish is that it has been a factor in warding off human predators.

Work on the chemical structure and properties of the poisons from Alaska butter clam siphons, the hepatopancreas of California mussels, and axenic cultures of G. catenella carried on in the Biological Laboratories and with Rapoport at the University of California definitely established that the sea mussel did not alter the poison obtained from G. catenella. The poison from all three sources had identical structures (31). The situation with the Alaska butter clam may be different however in light of the different saxitoxin derivatives recently found in the dinoflagellates in that region. Perhaps the clam converts the sulfo and sulfate derivatives to the more toxic saxitoxin.

The absorption of the poison by the hepatopancreas of the mussel and the siphon of the clam is an interesting and unexplained phenomenon. Several years ago (unpublished) I placed ground or sliced hepatopancreas of mussels in solutions of saxitoxin and found no significant absorption of the poison and no more than was adsorbed by mouse liver slices. Under normal conditions in the natural state the hepatopancreas of a small mussel weighing a few grams would accumulate up to 10,000 MU or more from the dinoflagellate. It therefore seemed to me that the intact live hepatopancreas involved something other than just a physical adsorption. Price and Lee, however, at Oregon State University have presented evidence that some of the melanin pigments in the siphon of the clam may be involved in the absorption of the poison (32).

In conclusion I must say personally that my 40 years of work on the paralytic shellfish poisons has been a rewarding experience. The opportunity of working with many colleagues throughout the years has led to the accomplishment of the purification of the paralytic poison and the determination of its structure. The purified poison, saxitoxin, has become a valuable tool for nerve transmission studies for physiologists, and pharmacologists in various laboratories throughout the world. Also it has served as a reference standard in the AOAC Official mouse assay used by the Food and Drug Administration and other control agencies in their work on making shellfish safe for human consumption. There are problems ahead that we hope may be solved. One is a simple test procedure for the poison in shellfish that could be carried out rapidly as a field test by control agencies and commercial fishermen to avoid collecting poisonous shellfish. Several laboratories are working on better analytical methods for the poison. Bates and Rapoport (33) at the University of California and Shoptaugh et al. (34) at the University of New Hampshire have developed techniques for the determination of saxitoxin more sensitive and specific than the mouse test but the procedures and equipment required make them impractical for field use. Other methods based on immunological techniques for the development of simpler methods are being worked on by Chu et al. (35) at the University of Wisconsin and Guire et al. (report at this symposium) at the University of Minnesota. The simple mouse test, although not specific for the poisons, remains the method of choice used by the Food and Drug Administration. Another problem is the development of an antidote for the poison to be used in connection with artificial respiration for the treatment of poisoned persons. In this respect the Klamath Indians of Northern California are reported (36) to have used the sap from the sugar pine tree as treatment of sickness from eating mussels at certain times of the year. Investigations on the use of sap from various parts of the sugar pine obtained through the courtesy of the Forestry Department of the University of California Berkeley, did not relieve the signs of shellfish poisoning in mice.

Funds for the support of the work that I was involved in came from The Army Biological Laboratories; the Public Health Service; the Food Research Institute, University of Wisconsin; and grants from the National Institutes of Health.

Literature Cited

1. Sommer, H.; Whedon, W. F.; Kofoid, C. A.; Stohler, R. AMA Arch. Pathol. 1937, 24, 537.
2. Sommer, H.; Meyer, K. F. AMA Arch. Pathol. 1938, 24, 560.
3. Needler, A. B. J. Fish. Res. Board Canada 1949, 7, 490.
4. Prakash, A. J. Fish Res. Board Canada 1963, 20, 983.
5. Schantz, E. J. Ann. N. Y. Acad. Sci. 1960, 90, 843.
6. Sommer, H.; Monnier, R. P.; Riegel, B.; Stanger, D. W.; Mold, J. D.; Wikholm, D. M.; Kiralis, E. S. J. Am. Chem. Soc. 1948, 70, 1015.
7. Schantz, E. J.; Mold, J. D.; Stanger, W. D.; Shavel, J.; Riel, F. J.; Bowden, J. P.; Lynch, J. M.; Wyler, R. S.; Riegel, B.; Sommer, H. J. Am. Chem. Soc. 1957, 79, 5230.
8. Mold, J. D.; Bowden, J. P.; Stanger, D. W.; Maurer, J. E.; Lynch, J. M.; Wyler, R. S.; Schantz, E. J.; Riegel, B. J. Am. Chem. Soc. 1957, 79, 5235.
9. Schantz, E. J.; Mold, J. D.; Howard, W. L.; Bowden, J. P.; Stanger, O. W.; Lynch, J. M.; Wintersteiner, O. P, Dutcher, J. D.; Walters, D. R.; Riegel, B. Can. J. Chem. 1961, 39, 2117.
10. Evans, M. H. Brit. Pharmacol. 1964, 22, 478.
11. Kao, C. Y.; Nishiyama, A. J. Physiol. (London) 1965, 180, 50.
12. Schantz, E. J.; McFarren, E. F.; Schafer, M. L.; Lewis, K. H. J. Assoc. Off. Anal. Chem. 1958, 41, 160.
13. "Manual of Recommended Practice for Sanitary Control of the Shellfish Industry Part 1," U. S. Public Health Service, 1959.
14. Wong, J. L.; Oesterlin, R.; Rapoport, H. J. Am. Chem. Soc. 1971, 93, 7344.
15. Schantz, E. J.; Ghazarossian, V. E.; Schnoes, H. K.; Strong, F. M.; Springer, J. P.; Pezzanite, J. O.; Clardy, J. Proc. First International Conf. Toxic Dinoflagellate Blooms. 1974, p. 267.
16. Schantz, E. J.; Ghazarossian, V. E.; Schnoes, H. K.; Strong, F. M.; Springer, J. P.; Pezzanite, J. O.; Clardy, J. J. Am. Chem. Soc. 1975, 97, 1238.
17. Tanino, H.; Nakata, T.; Kishi, Y. J. Am. Chem. Soc. 1977, 99, 2818.
18. Bordner, J.; Thiessen, W. E.; Bates, H. A.; Rapoport, H. J. Am. Chem. Soc. 1975, 97, 6008.
19. Narahashi, T.; Browdick, M. S.; Schantz, E. J. Environmental Letters 1975, 9, 239.
20. Ghazarossian, V. E.; Schantz, E. J.; Schnoes, H. K.; Strong, F. M. Biochem. Biophys. Res. Commun. 1975, 59, 1219.
21. Shimizu, T.; Buckley, L. J.; Alam, M.; Oshima, Y.; Fallon, W. E.; Kassai, H.; Miura, I.; Gullo, V. P.; Nakanishi, K. J. Am. Chem. Soc. 1976, 98, 5414.
22. Boyer, G. L.; Schantz, E. J.; Schnoes, H. K. J. Chem. Soc. Chem. Commun. 1978, 889.
23. Shimizu, Y.; Hsu, C.; Fallon, W. E.; Oshima, Y.; Miura, I.; Nakanishi, K. J. Am. Chem. Soc. 1978, 100, 6791.
24. Boyer, G. L. Ph.D. Thesis, University of Wisconsin, Madison, 1980.
25. Hall, S.; Reichardt, P. B.; Neve, R. A. Biochem. Biophys. Res. Commun. 1980, 97, 649.

26. Wichmann, C.; Niemczura, W. P.; Schnoes, H. K.; Hall, S.; Reichardt, P. B.; Darling, S. D. J. Am. Chem. Soc. 1981, 103, 6977.
27. Ghazarossian, V. E. Ph.D. Thesis, University of Wisconsin, Madison, 1977.
28. Ghazarossian, V. E.; Schantz, E. J.; Schnoes, H. K.; Strong, F. M. Biochem. Biophys. Res. Commun. 1976, 68, 776.
29. Koehn, F. E.; Ghazarossian, V. E.; Schantz, E. J.; Schnoes, H. K.; Strong, F. M. Bioorg. Chem. 1981, 10, 412.
30. Schantz, E. J.; Magnuson, H. W. J. Protozool. 1965, 11, 239.
31. Schantz, E. J.; Lynch, J. M.; Vayvada, G.; Matsumoto, K.; Rapoport, H. Biochemistry 1966, 5, 1191.
32. Price, R. J.; Lee, J. S. J. Fish. Res. Bd. Can. 1972, 29, 1659.
33. Bates, H. A.; Rapoport, H. J. Agri. Food Chem. 1975, 23, 237.
34. Shoptaugh, N. H.; Ikawa, M.; Foxall, T. L.; Sasner, J. J. "A Fluorometric Technique for the Detection and Determination of Paralytic Shellfish Poisons"; Carmichael, W. W., Ed.; The Water Environment; Plenum: New York, 1981; p. 427.
35. Chu, F. S.; Fan, T. S. L.; Li, S. W. 97th Ann. Meeting Assoc. Off. Anal. Chem. 1983, p. 77, Abst. 194.
36. Thompson, L. "To the American Indian" 1916 Cummins Print Shop, Eurkea, Calif., p. 28.
37. Schantz, E. J. Seafood Toxicants. Toxicants Occurring Naturally in Foods. National Academy of Sciences, Washington, D.C. 1983; p. 424.

RECEIVED April 23, 1984

10

Cryptic Paralytic Shellfish Toxins

SHERWOOD HALL[1] and PAUL B. REICHARDT[2]

[1] Woods Hole Oceanographic Institution, Woods Hole, MA 02543
[2] Department of Chemistry, University of Alaska, Fairbanks, AK 99701

The saxitoxins from dinoflagellates of the genus Protogonyaulax may accumulate in shellfish and, being potent neurotoxins, pose a serious threat to public health. Six of the saxitoxins are sulfamates which have relatively low toxicity, but are easily hydrolyzed to the respective carbamates with corresponding increases in potency. This paper presents some of the chemistry and public health implications of these toxins.

The twelve saxitoxins known (1) from dinoflagellates of the genus Protogonyaulax are shown in Figure 1. They consist of the parent compound, saxitoxin, 1, (STX) and the eleven derivatives formed by the addition of N-1-hydroxyl, 11-hydroxysulfate, and 21-sulfo groups. The structural relationships among these compounds are summarized in Figure 2. The mouse intraperitoneal potencies of compounds 1-8 are shown in Figure 3, from which it can be seen that each of the sulfamate toxins 2, 4, 6, and 8 is far less potent than its corresponding carbamate (1). The same relationship is observed for compounds 9-12 (2). Since the hydrolysis of the 21-sulfo group is facile, converting sulfamate to carbamate, the sulfamate toxins, when present in seafood, constitute a reservior of latent or cryptic toxicity.

Taxonomy

The dinoflagellates generally recognized to be the source of the saxitoxins in temperate waters were originally assigned to the genus Gonyaulax, within which they form a distinct sub-group. These were referred to collectively by Steidinger (3) as the "tamarensis complex", and subsequently separated into a new genus both by Taylor (4), who named the new genus Protogonyaulax, and by Loeblich (5), who named it Gessnerium. While many authors are still using Gonyaulax and the relative merits of the two new names are being debated, it has turned out that the correct name, according to the rules of biological nomenclature, may be Alexandrium (4).

0097-6156/84/0262-0113$06.00/0
© 1984 American Chemical Society

	R1	R2	R3	R4	
1	H	H	H	H	STX
2	H	H	H	SO_3^-	B1
3	H	H	OSO_3^-	H	GTX 2
4	H	H	OSO_3^-	SO_3^-	C1
5	H	OSO_3^-	H	H	GTX 3
6	H	OSO_3^-	H	SO_3^-	C2
7	OH	H	H	H	NEO
8	OH	H	H	SO_3^-	B2
9	OH	H	OSO_3^-	H	GTX 1
10	OH	H	OSO_3^-	SO_3^-	C3
11	OH	OSO_3^-	H	H	GTX 4
12	OH	OSO_3^-	H	SO_3^-	C4

Figure 1. The twelve saxitoxins from Protogonyaulax.

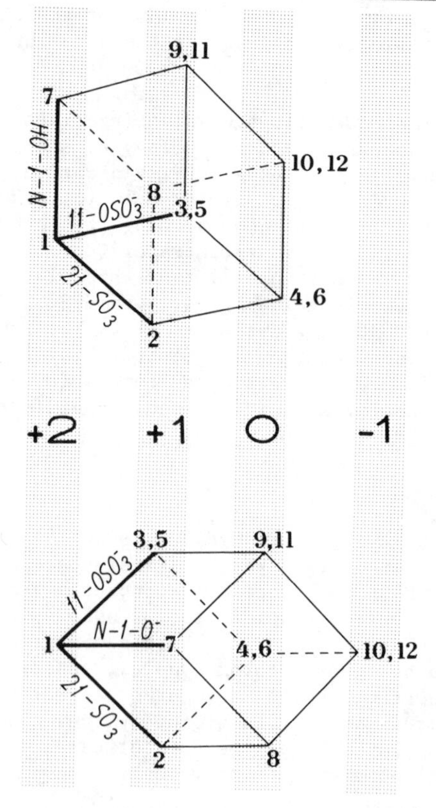

Figure 2. Structural relationships among the saxitoxins and approximate net charge below pH 5 (upper) and in the pH range 7-8 (lower).

Despite the profusion of genus names, the group itself seems relatively well defined. Most of the organisms are similar in morphology, to the extent that it is very difficult to make morphological distinctions among species. Many of the species may, in fact, have been established through historical accident (6). In contrast, the dinoflagellate Pyrodinium bahamense from the tropical Pacific has been shown to contain saxitoxins (7,8) and is clearly distinct from the organisms of the tamarensis group.

The saxitoxins have also been found in the freshwater cyanophyte Aphanizomenon flos aquae (9,10) and, quite recently, in tropical marine coralline algae of the genus Jania (11)

Toxicity

It has been established that saxitoxin, 1, acts through a potent, reversible blockage of sodium conductance in nerve and muscle membranes (12,13). Neurophysiological studies with the various saxitoxin derivatives (14-17) indicate that all act by the same mechanism, simply differing in potency.

The Mouse Assay

The toxicity of extracts and purified toxins has generally been determined using the mouse assay first developed by Sommer (18) and later formalized (19,20), in which 1.0 mL of test solution is injected intraperitoneally into each of several mice and the toxicity of the solution established by correlating the weight of each mouse and the time from challenge to its last gasping breath with a table of dose/death time relationships. The pH of the injected solution must be below 4.5. The solution is diluted as appropriate to give a median death time close to 6 minutes. Used in ths manner, the test does not rely heavily on the shape of the dose/response curve, being more a titration to a six minute death time, and is therefore internally consistent despite variations in sample toxin composition. The values obtained with the standard mouse bioassay are here referred to as mouse intraperitoneal potency (MIP).

Latent Toxicity and Proctor Enhancement

In his meticulous exploration of methods for his thesis research, Proctor (21) noted that the toxicity (MIP) of cell free extracts of Gonyaulax increased following heating at low pH. Hall (1,22) further explored this increase in potency, referring to it as 'Proctor enhancement', and demonstrated that it was primarily due to the hydrolysis of sulfamate to carbamate toxins. It was found that heating at 100°C for 5 minutes in aqueous HCl with a free acid concentration of 0.1 M is sufficient to insure complete hydrolysis and attain maximum MIP.

The Saxitoxins

The x-ray crystal structure (2) of compound 12 (toxin C4),

21-sulfo-N-1-hydroxysaxitoxin-11ß-hydroxysulfate, is shown in Figure 4. This compound bears all three of the functional groups that form the array of Protogonyaulax toxins from the parent saxitoxin. This structure determination unequivocally establishes their position and identity and, since the structural relationships among the toxins have been established by chemical and spectroscopic means (1,2,23,24,25-27), secures the structures of the entire array.

The properties of the structural elements in the array are largely additive. The parent compound, 1, contains two formal guanidinium groups (28) which are strongly basic and deprotonate with pK_as of 8.22 and 11.28 (29). It therefore behaves as a dication when the solution pH is well below 8, binding strongly to carboxylate cation exchange resins. Compounds bearing either 21-sulfo or 11-hydroxysulfate groups (Figure 2) will have a net charge of +1 under these conditions and still bind to carboxylate cation exchange resins, but much less strongly. Compounds bearing both 21-sulfo and 11-hydroxysulfate groups have little or no net charge under these conditions and do not bind to carboxylate resins (1).

Titrations of 7, neosaxitoxin (30), revealed a third dissociable group with a pK_a of 6.75, the other two dissociation constants showing little change. This is clearly due to the presence of the N-1-hydroxyl group, although it is not certain that it is due to dissociation of the proton from it (31). The chromatographic behavior of the compounds bearing this substituent (1) is close to that of those without it when the pH is held well below 6; the apparent net charge relationships are consistent with those in the upper part of Figure 2. As the pH is increased above pH 6, the chromatographic relationships among the compounds shift toward those represented in the lower part of Figure 2, indicating that all six of the N-1-hydroxy compounds undergo a similar deprotonation, although it is likely that substituent effects alter the pK_as slightly.

Compounds bearing the 11-hydroxysulfate substituent epimerize, the conversion from beta to alpha predominating (1,2,23,24,27). This is of particular interest, since the beta epimers are in all cases found to predominate in fresh dinoflagellate extracts (1) suggesting that the nature or circumstances of the toxins within the cell differ significantly from those in vitro.

Compounds bearing the 21-sulfo substituent are easily hydrolyzed to their corresponding carbamates (1,2,22,26,27,32-35). Due to the great increase in potency associated with this conversion and the difficulty of preventing it entirely, it is difficult to accurately determine the potencies of the sulfamates, or mixtures that contain them.

Isocratic chromatography of dinoflagellate extracts on BioGel P2 with 0.1 M acetic acid (22; Figure 5), followed by TLC of the fractions on silica gel using the methods of Buckley (36) permits complete resolution of compounds 1-12. Elution volumes are highly consistent for extracts run under the conditions specified in (22), but differ slightly when mixtures of pure toxins are run. In either case the toxins charged +2, +1, and 0 were found to elute as three well-resolved groups, which inspired the designations A, B and C.

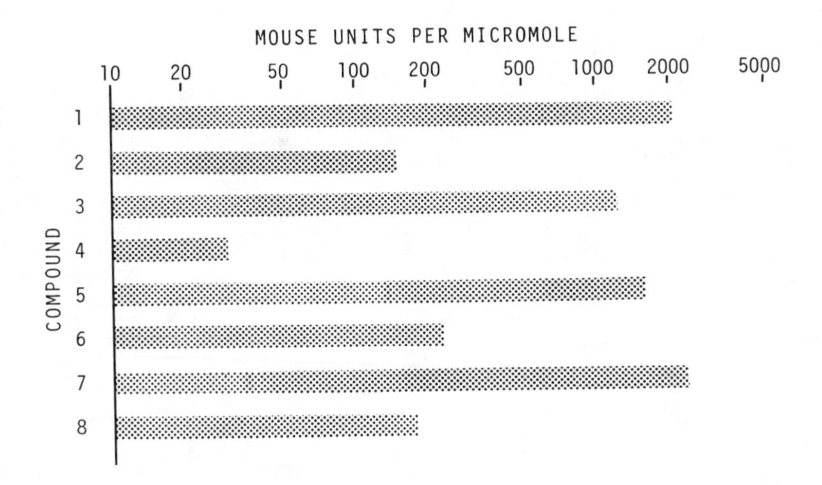

Figure 3. The mouse intraperitoneal potencies of compounds 1-8.

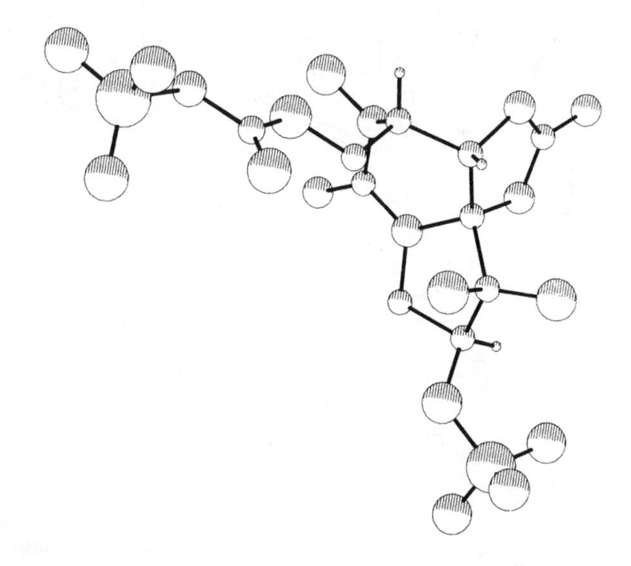

Figure 4. X-ray crystal structure of compound 12 (toxin C4), determined by S. D. Darling at the University of Akron. The calculated positions of hydrogens on carbons 5, 6, and 11 have been shown for clarity.

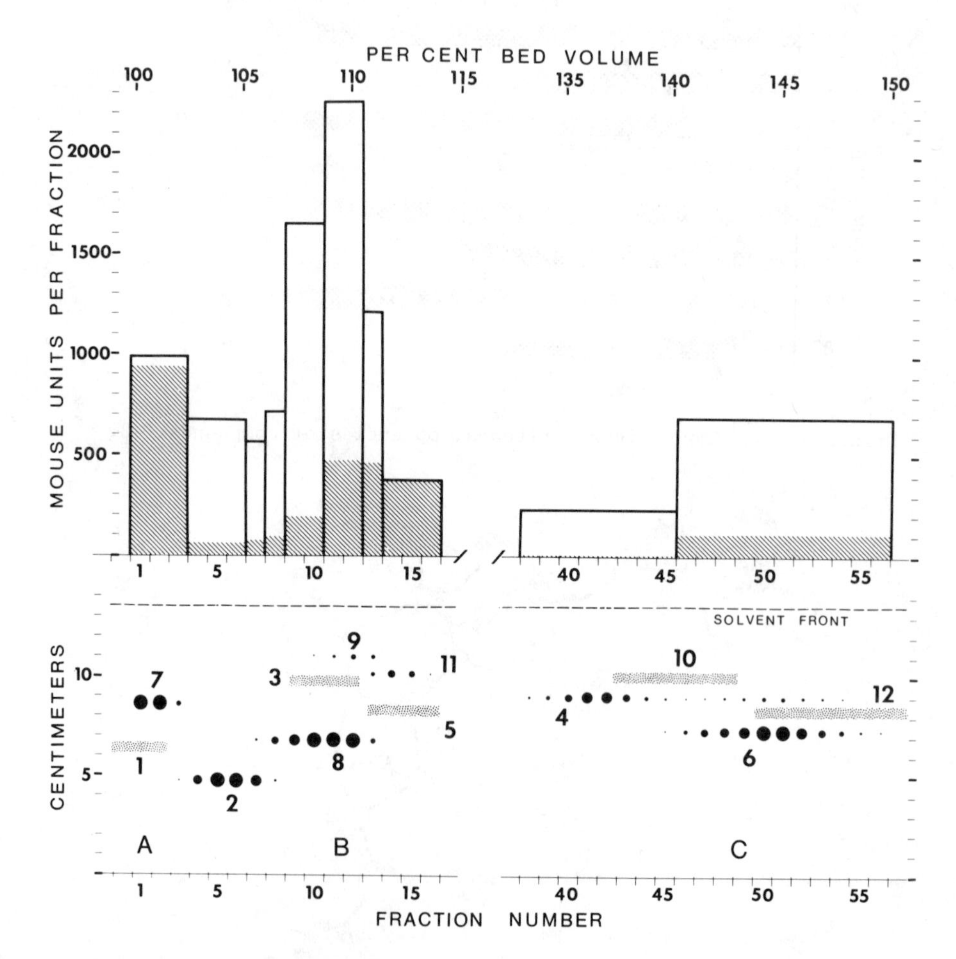

Figure 5. Chromatography of an extract of Protogonyaulax clone PI07, showing the LC/TLC coordinates for the twelve saxitoxins. Conditions are as in (22). Stippled bars in the TLC diagram (lower) show the coordinates of compounds not observed in this run. The toxicity profile (upper), shows the mouse intraperitoneal potency of grouped fractions before (hatched area) and after (outline) hydrolysis of the sulfamates. Adapted from Ref. 22.

In the TLC of the toxins (lower part of Figure 5), R_f values are not presented because they show small variations and are potentially misleading. However, the relative elution sequence has proven to be highly reproducible, such that the two-dimensional composition maps produced by this method can be interpreted with a high level of confidence.

Repeated analyses of the clone for which data are shown in Figure 5 have confirmed that extracts of the organism do not contain compounds 1, 3, or 5. The extracts do contain 10 and 12, but at levels below the TLC detection limits in the run shown. Trace epimerization of toxin C2, probably during application, resulted in the small spots at the R_f of C1 around fraction 50.

The carbamate side chain of saxitoxin can be removed by hydrolysis under vigorous conditions (37,38), yielding decarbamoylsaxitoxin. Work by Sullivan indicates that the same transformation can be brought about by enzymes in certain shellfish (39), implying that a total of six decarbamoyl toxins will be found.

Variations in Toxin Composition

Toxin analyses of clones of Protogonyaulax isolated from several locations along the coast of the Northeast Pacific (1) have shown that, while net toxicity of a clone can vary significantly, toxin composition is a relatively conservative property of a clone and clones from a given region tend to have the same composition. However, composition varies dramatically from one region to another, ranging from that shown in Figure 5, where the sulfamates predominate and 1 is entirely absent, to that where 1 is the main component, and is accompanied only by a trace of 7. The composition of toxins being supplied by dinoflagellates to shellfish will therefore vary substantially with location.

Superimposed upon this spatial patchwork of toxin composition are the changes due to processes in the accumulating organisms, resulting in a compositional diagenesis through time. The actual mechanisms by which the saxitoxins are bound in shellfish are not known. However, the great range in binding properties observed among the saxitoxins both in chromatography (1) and in neurophysiological experiments (15,16) suggests that some of the toxins will be bound more strongly than others, and will thus persist longer in shellfish. It should be recognized that toxins need not be bound to be a hazard to consumers; the toxin content of dinoflagellate cells (1) can be high enough that the undigested cells in a bivalve stomach could make the shellfish unsafe. Toxin composition will therefore start out resembling that of the source organism, possibly with a significant component of ingested cells, and will evolve with the loss of the less strongly bound toxins until only the most strongly bound remain.

It has been observed that while the 11ß-hydroxysulfate epimers predominate in fresh dinoflagellate extracts, the alpha epimers predominate in shellfish and in equilibrated mixtures (1,24). The ratio of alpha to beta hydroxysulfate epimers in shellfish would therefore be expected to increase with time.

In addition to these passive processes, shellfish have been shown to actively modify the saxitoxins. Shimizu has shown (40) that scallops can remove both the N-1-hydroxyl and 11-hydroxysulfate groups from the saxitoxins. Sullivan has shown (39) that enzymes in littleneck clams can remove the sulfamate or carbamate side chain, yielding the decarbamoyl toxins. This activity was not detected in either mussels or butter clams. With both sorts of modification the products are compounds that have higher potency and are likely to be bound in shellfish more strongly.

In summary, the toxin composition of shellfish, while not random, can be expected to vary greatly with location, time, and shellfish species.

Assays and Analyses

Given a food sample which may contain several toxins, methods for determining its toxicity fall into two general classes: assays and analyses. An assay provides a single value, the net response of the assay system to the sample, while an analysis determines the concentration of each toxin.

The value of interest is the human oral potency (HOP): how toxic the food would be if somebody ate it. For an array of toxins that are additive in their effect (not synergistic),

$$HOP_{sample} = \Sigma c_i HOP_i$$

or, the human oral potency of a sample equals the sum of the concentrations of each toxin, each multiplied by that toxin's human oral potency. With analytical methods such as that being developed by Sullivan (41) it should soon be possible to determine the concentration, c_i, of each of the saxitoxins in a sample. From such results the HOP_{sample} could be calculated if the HOP_i of the individual toxins were known. Unfortunately, the latter data are not yet available.

The response of an assay, R, again for toxins with additive effects, is

$$R = \Sigma c_i r_i$$

or, the assay response is equal to the sum of the concentrations, c_i, of each toxin, each multiplied by that toxin's response factor, r_i, in the assay system.

In an ideal assay system,

$$HOP_i / r_i = k$$

or, the ratio of human oral potency to assay response for each toxin is a constant for all of the toxins. If this is the case, then

$$HOP_{sample} = kR = k \Sigma c_i r_i$$

or, the HOP_{sample} can be related to the assay response by a calibration factor which will be independant of variations in sample composition. Since sample composition, the relative proportion of the different toxins, will in general tend to vary, the uniformity of the ratios HOP_i/r_i is an important criterion for an assay. Unfortunately, it is at present a difficult one to apply, again because of the uncertainty in HOP_i for the toxins. The main point is that an assay will be linear with HOP_{sample} only if either the ratios HOP_i/r_i are constant or if the toxin composition of the samples is uniform. The appraisal of results from any assay system must take this potential for nonlinearity into consideration.

Despite the uncertainties in HOP for the toxins, there is reason to suspect that their mouse intraperitoneal potencies (MIP), the r_i for the standard mouse bioassay system, do not bear a uniform relationship to them. Early pharmacological work (42) on the paralytic shellfish toxins was conducted with shellfish extracts. The toxin composition of these extracts is unknown, although their origin and pharmacology make it clear that the active substances were saxitoxins. The extracts were shown to be about 40 times more toxic to mice when administered intraperitoneally than when administered orally. Given the size of this ratio, and the great range of binding behavior found among the saxitoxins, it seems unlikely that they would all have the same ratio of MIP to HOP. Work is in progress to resolve this uncertainty.

Potential Toxicity

The sulfamate saxitoxins have very low potencies relative to their carbamate hydrolysis products. This relationship has been observed in every assay system tried, including the standard mouse bioassay (Figure 3), squid giant axon (17), frog sciatic nerve (16), mammalian brain (16), and single rat sarcolemma sodium channels incorporated into lipid bilayers (15). It seems unlikely that human oral potencies are an exception to this trend.

While the sulfamates themselves probably have low HOP, they can hydrolyze at low rates to the more toxic carbamates under conditions of food storage, preparation, or digestion. From the work of Sullivan (39), it appears that latent sulfamate toxicity can also be potentiated through enzymatic conversion in some shellfish to the corresponding decarbamoyl toxins. The value needed for public health protection is therefore the potential human oral potency, the HOP that product might attain under a worst-case scenario of conversions.

Using analytical methods like those of Sullivan (41), this value could be estimated by using the HOP_i for carbamates in place of those for the corresponding sulfamates when calculating HOP_{sample}.

In an assay that offers acceptable HOP_i/r_i ratios for the carbamates, potential HOP can be estimated by preparing the sample under conditions that insure hydrolysis of the sulfamates. Unfortunately, the conditions specified for sample preparation in the standard mouse bioassay are not sufficiently acidic to insure complete hydrolysis (43). As currently employed in state monitoring laboratories, the mouse assay may substantially underestimate the potential HOP of samples containing the sulfamate toxins.

Acknowledgments

The authors wish to thank S. D. Darling for his tireless efforts in solving the crystal structures and the Institute of Marine Science, University of Alaska, for the facilities that made much of this work possible. High-field NMR data used in structure elucidation were obtained through the generosity of H. K. Schnoes and his associates at the University of Wisconsin. We thank D. M. Anderson for providing facilities used in the latter part of this study, and for comments on the manuscript. Work was supported by the U. S. Food and Drug Administration, the Alaska Sea Grant College Program, cooperatively supported by NOAA Office of Sea Grant, U. S. Department of Commerce, under Grants 04-8-M01-187 and NA81AA-D-00009 and by the University of Alaska with funds appropriated by the State of Alaska, a Graduate Fellowship in Resource Problems from the University of Alaska and a Postdoctoral Fellowship from Woods Hole Oceanographic Institution (to S. H.). This publication is contribution number 5666 from the Woods Hole Oceanographic Institution.

Literature Cited

1. Hall, S. Ph.D. Thesis, University of Alaska, Fairbanks, 1982.
2. Hall, S.; Darling, S. D.; Boyer, G. L.; Reichardt, P. B.; Liu, H.-W. Tetrahedron Lett. 1984, in press.
3. Steidinger, K. A. Phycologia 1971, 10, 183-7.
4. Taylor, F. J. R. In "Toxic Dinoflagellate Blooms"; Taylor, D. L.; Seliger, H. H., Eds.; Developments in Marine Biology No. 1; Elsevier/North Holland: New York, 1979; pp. 47-56.
5. Loeblich, A. R., III; Loeblich, L. A. In "Toxic Dinoflagellate Blooms"; Taylor, D. L.; Seliger, H. H., Eds.; Developments in Marine Biology No. 1; Elsevier/North Holland: New York, 1979; pp. 41-6.
6. Taylor, F. J. R. Environ. Letters 1975, 9, 103-119.
7. MacLean, J. L. Limnol. Oceanogr. 1977, 22, 234-254.
8. Harada, T.; Oshima, Y.; Kamiya, H.; Yasumoto, T. Bull. Japan. Soc. Sci. Fish. 1982, 48, 821-5.
9. Jackim, E.; Gentile, J. Science 1968, 162, 915.
10. Alam, M.; Shimizu, Y.; Ikawa, M.; Sasner, J. J. J. Environ. Sci. Health 1978, A13, 493-9.
11. Kotaki, Y.; Tajiri, M.; Oshima, Y.; Yasumoto, T. Bull. Japan. Soc. Sci. Fish. 1983, 49, 283-6.
12. Evans, M. H. Br. J. Pharmacol. Chemother. 1964, 22, 478-485.
13. Kao, C. Y.; Nishiyama, A. J. Physiol. 1965, 180, 50-66.
14. Kao, C. Y.; Walker, S. E. J. Physiol. 1982, 323, 619-637.
15. Moczydlowski, E.; Miller, C.; Hall, S.; Strichartz, G. Biophys. J. 1984, 45, 286a.
16. Strichartz, G.; Hall, S.; Shimizu, Y. Biophys. J. 1984, 286a.
17. Frace, A. M.; Hall, S.; Brodwick, M. S.; Eaton, D. C. J. Gen. Physiol. 1984, submitted.
18. Sommer, H.; Meyer, K. F. Arch. Pathol. 1937, 24, 560-598.
19. Schantz, E. J.; McFarren, E. F.; Schaeffer, M. L.; Lewis, K. H. J. Assoc. Off. Agric. Chem. 1958, 41, 160-8.

20. Horwitz, W., Ed. "Official Methods of Analysis of the Association of Official Analytical Chemists"; Association of Official Analytical Chemists: Washington, D.C., 1975; pp. 319-321.
21. Proctor, N. H. Ph.D. Thesis, University of California, San Francisco, 1973.
22. Hall, S.; Reichardt, P. B.; Nevé, R. A. Biochem. Biophys. Res. Commun. 1980, 97, 649-653.
23. Boyer, G. L. Ph.D. Thesis, University of Wisconsin, Madison, 1980.
24. Fix Wichmann, C.; Boyer, L. G.; Divan, C. L.; Schantz, E. J.; Schnoes, H. K. Tetrahedron Lett. 1981, 22, 1941-4.
25. Shimizu, Y.; Hsu, C.-P. J. Chem. Soc., Chem. Commun. 1981, 314-5.
26. Koehn, F. E.; Hall, S.; Fix Wichmann, C.; Schnoes, H. K.; Reichardt, P. B. Tetrahedron Lett. 1982, 2247-8.
27. Fix Wichmann, C.; Niemczura, W. P.; Schnoes, H. K.; Hall, S.; Reichardt, P. B.; Darling, S. D. J. Am. Chem. Soc. 1981, 103, 6977-8.
28. Schantz, E. J.; Ghazarossian, V. E.; Schnoes, H. K.; Strong, F. M.; Springer, J. P.; Pezzanite, J. O.; Clardy, J. J. Am. Chem. Soc. 1975, 97, 1238-9.
29. Rogers, R. S.; Rapoport, H. J. Am. Chem. Soc. 1980, 102, 7335-7339.
30. Shimizu, Y.; Hsu, C.-P.; Fallon, W. E.; Oshima, Y.; Miura, I.; Nakanishi, K. J. Am. Chem. Soc. 1978, 100, 6791-3.
31. Hori, A.; Shimizu, Y. J. Chem. Soc., Chem. Commun. 1983, 790-2.
32. Kobayashi, M.; Shimizu, Y. J. Chem. Soc., Chem. Commun. 1981, 827.
33. Onoue, Y.; Noguchi, T.; Maruyama, J.; Hashimoto, K.; Seto, H. J. Agric. Food Chem. 1983, 31, 420-3.
34. Nishio, S.; Noguchi, T.; Onoue, Y.; Maruyama, J.; Hashimoto, K.; Seto, H. Bull. Japan. Soc. Sci. Fish. 1982, 48, 959-965.
35. Harada, T.; Oshima, Y.; Yasumoto, T. Agric. Biol. Chem. 1982, 46, 1861-4.
36. Buckley, L. J.; Ikawa, M.; Sasner, J. J. J. Agric. Food Chem. 1976, 24, 107-111.
37. Ghazarossian, V. E.; Schantz, E. J.; Schnoes, H. K.; Strong, F. M. Biochem. Biophys. Res. Commun. 1976, 68, 776-780.
38. Koehn, F. E.; Ghazarossian, V. E.; Schantz, E. J.; Schnoes, H. K.; Strong, F. M. Bioorg. Chem. 1981, 10, 412-428.
39. Sullivan, J. J.; Iwaoka, W. T.; Liston, J. Biochem. Biophys. Res. Commun. 1983, 114, 465-472.
40. Shimizu, Y.; Yoshioka, M. Science 1981, 212, 547-549.
41. Sullivan, J. J.; Iwaoka, W. T. J. Assoc. Off. Anal. Chem. 1983, 66, 297-303.
42. Müller, H. J. Pharmacol. Exptl. Therap. 1935, 53, 67-89.
43. Hall, S., unpublished data.

RECEIVED June 7, 1984

11

Shellfish Toxicity and Dormant Cysts in Toxic Dinoflagellate Blooms

DONALD M. ANDERSON

Biology Department, Woods Hole Oceanographic Institution, Woods Hole, MA 02543

Several of the dinoflagellates responsible for toxic blooms include a dormant cyst stage in their life cycles. These resistant cells have a variety of potential functions in the overall ecology of the toxic dinoflagellates. As Wall (1) discussed, cysts can theoretically act as "seed" populations to initiate blooms, as a survival mechanism through environmental extremes, as agents for species dispersal, and as means for genetic recombination through sexuality. Two additional functions can be added to this list: cysts can be direct sources of toxicity, and their formation can be a major factor in bloom termination. The purpose of this paper is to examine our present knowledge of the cysts of the toxic *Gonyaulax* species to see whether these hypothetical considerations are valid representations of the actual roles cysts play in toxic dinoflagellate blooms and shellfish toxicity.

Nearly 100 years ago, dinoflagellate resting cysts were first observed in plankton samples, but nearly 50 years passed before their affinity to motile dinoflagellates was demonstrated (2.3). Early researchers investigating paralytic shellfish poisoning episodes (PSP) speculated that a benthic cyst population germinating at specific times could explain the seasonal nature of toxic dinoflagellate blooms (4-7), but it was only in the last few years that true resting cysts were finally identified and described for toxic dinoflagellates (8-10).

It is now generally accepted (10, 30, 39, 40) that the toxic *Gonyaulax* species reproduce asexually through vegetative division (Figure 1), but that this process can switch to sexual reproduction through the formation of gametes which fuse together (Figure 2) to yield a large, swimming zygote (planozygote; Figure 3). This presumably diploid cell swims for up to a week before it is transformed into the thick-walled resting cyst or hypnozygote (Figure 4). Upon germination, the cyst releases one cell (Figure 6) which

0097-6156/84/0262-0125$06.00/0
© 1984 American Chemical Society

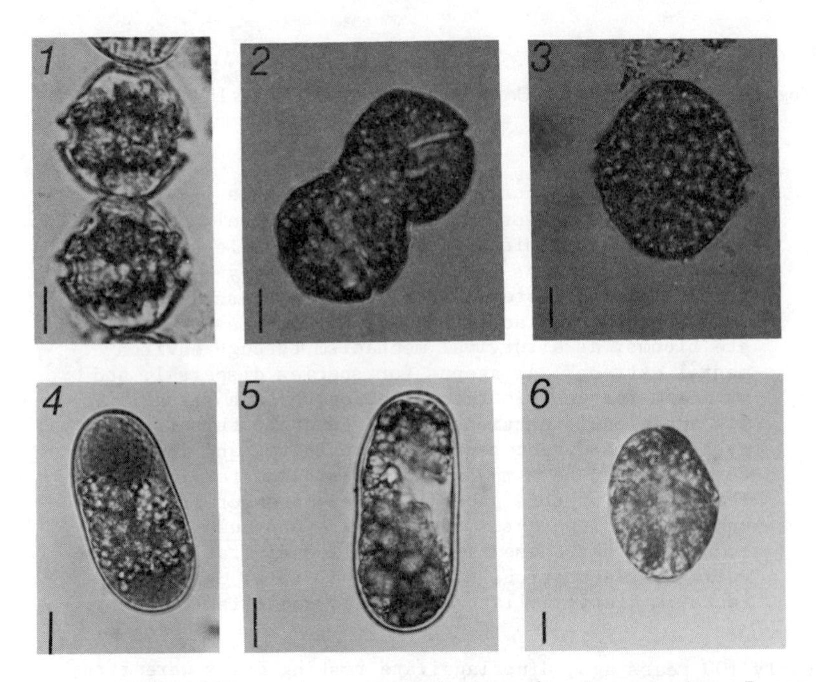

Figures 1-6. Life cycle stages of Gonyaulax tamarensis. Error bars are 10 µm in all cases. Figure 1. Recently divided vegetative cell showing parallel girdle or singulum orientation; Figure 2. Fusing gametes with obliquely-oriented singula; Figure 3. Large, deeply-pigmented planozygote formed from fused gametes; Figure 4. Fully mature resting cyst with starch and lipid accumulations in a central band and pigmented cytoplasm at each pole; Figure 5. Intact, but dead, cyst isolated from fecal pellet of the mussel Mytilus edulis; Figure 6. Recently-germinated cell (planomeiocyte) with faintly-visible pigmented accumulation body lower-left.

divides to yield daughter cells that are haploid and capable of mitotic, asexual division once again.

The discovery of a cyst stage for the toxic Gonyaulax species provided a new set of explanations for various aspects of the toxic bloom phenomenon, all based on the unique characteristics of the resistant, dormant cyst. In an early review by Wall (1), one of the pioneers of studies on this topic, the ecological roles of cysts were presented, but these discussions were largely speculative since so little was known about living cysts at the time. Wall argued that cysts could function: a) as "seed" populations to inoculate overlying waters and initiate blooms; b) as a survival mechanism to permit the species to withstand environmental extremes; c) as agents for population dispersal into new regions; and d) as means for genetic recombination through sexual reproduction. Recent research suggests that cysts are important in two additional ways: e) they represent direct sources of toxicity through their ingestion by shellfish; and f) they can be a major factor in the decline of bloom populations in much the same way that germination is important in bloom initiation.

In the years since these largely hypothetical roles were first proposed, considerable effort has gone into studies of cysts of toxic dinoflagellates. It is the objective of this paper to discuss the importance of cysts in shellfish toxicity episodes with respect to our present state of knowledge. It soon becomes clear from this type of exercise that despite significant progress in certain areas, many of the hypothetical links between cysts and shellfish toxicity remain probable but unverified.

This discussion will focus on two dinoflagellates responsible for PSP (Gonyaulax tamarensis and G. catenella (= Protogonyaulax tamaranesis or excavata (11). The other toxic dinoflagellates known to form cysts (12) are Gonyaulax monilata, Pyrodinium bahamense, and Gonyaulax polyedra (a species that may in fact not be toxic). Studies on the cysts of these latter species are descriptive, however, with no field or laboratory experiments to provide perspective on their ecological importance. Another important toxic species, Ptychodicus brevis (= Gymnodinium breve) has been reported to undergo sexual reproduction, but no resting cyst has yet been observed either in cultures or sediments (13).

Cysts As "Seed" Populations

It seems obvious that cysts would be important in inoculating overlying waters with motile cells through germination. It is, however, essentially impossible to prove that blooms actually originate from cysts since the advection of even a single cell from adjacent waters can theoretically start a bloom. Much of the evidence linking cysts to bloom initiation is thus suggestive but not conclusive, with the most direct evidence coming from shallow estuarine areas where sediments are more easily sampled.

Distributional studies of cysts in sediments provide one indication of inoculum potential. For example, qualitative sediment surveys (i.e., presence vs. absence of cysts) demonstrated a close correlation between sites subject to shellfish toxicity and the presence of G. tamarensis cysts on Cape Cod (10, 14). This association was strengthened by subsequent quantitative cyst surveys documenting large cyst accumulations in sediments of certain key embayments with concentrations falling rapidly to undetectable levels in adjacent areas (15). These distributions suggest "point source" origins for G. tamarensis blooms and supports the contention that cysts provide the initial bloom inoculum. However, the data provide no direct evidence and are only applicable to shallow estuaries at the southern limit of the G. tamarensis geographic distribution in New England, since it is in this region that the blooms and the cyst distributions are patchy and localized. Recent surveys in Maine where cysts are more widespread have not shown a good correlation between cyst distributions and the patterns of PSP outbreaks (16).

Perhaps the most compelling evidence that cysts do function in bloom initiation is based on the observation that newly-germinated G. tamarensis cells are morphologically distinct (Figure 6) from actively-dividing vegetative cells (Anderson and Wall, 1978). In a recent field study, these "planomeiocytes" (17) were observed during the early stages of G. tamarensis blooms, in one instance comprising more than 30% of the initial cell counts (18). This represents the first conclusive evidence that early bloom populations can include recently-germinated cells, but it is also important to recognize that the inocula were relatively small (a few hundred cells L^{-1}). In these instances then, germination of cysts supplied an inoculum, but the magnitude of this input was small relative to the subsequent rapid proliferation of asexually dividing cells.

The link between cysts and bloom initiation becomes more tenuous in regions where the cysts are widespread in estuarine and nearshore sediments (16, 19, 20). For example, given the need for 5-8°C temperatures for germination (18, 21), the fate of cysts deposited in 100 m or deeper coastal water remains an open question since bottom temperatures can be relatively invariant near 4°c (22). Cyst germination from the deep water "seed bed" may thus require resuspension of sediments and advective transport to warmer depths. A further complication is that many cysts are buried by benthic animal activity. Since large members of viable cysts can be found several cm below the sediment surface in deeper coastal waters (Figure 7B), germination (if it occurs) would only follow warming of bottom sediments above 5° C or vertical transport of the cysts to the sediment surface. Unfortunately, the biological and physical processes at the sediment/water interface have not yet been studied in the context dinoflagellate population dynamics.

It is important to recognize that the foregoing discussion is not simply an intellectual exercise attempting to demonstrate the seemingly obvious connection between cysts and bloom initiation. We must recognize that the "easy" answers provided by this dormant life cycle stage do not apply to all situations. In colder temperate waters, it is a reasonable assumption that cyst germination probably introduces the first toxic Gonyaulax cells into the water in some locations, but the subsequent transport of these cells over long distances must be considered a viable "seeding" alternative to the direct input of new cells from underlying sediments.

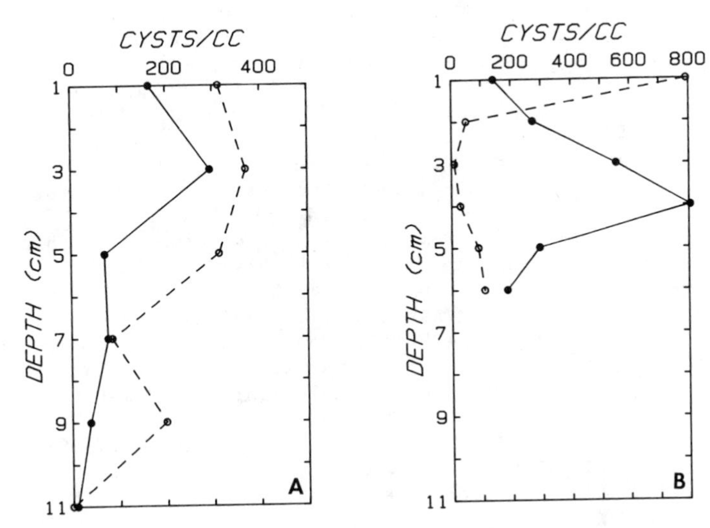

Figure 7. Vertical distribution of Gonyaulax tamarensis cysts in estuarine and nearshore sediments. Each point represents a 1-cm interval above specified depth. A. Two stations in Perch Pond, a shallow Cape Cod salt pond. Note the large number of cysts buried 6-11 cm deep. B. Two coastal stations near Cape Ann, Mass., both approximately 150 m deep. One has a distinct surface maximum in cyst abundance while the other has a peak at 4 cm.

Cysts as Survival Mechanisms

It is well documented that cysts (in general) can be highly resistant to environmental extremes (1, 12, 23). In the context of this discussion, however, the relevant extremes are those to which motile cells would have been exposed had they not encysted. In the dynamic coastal environment, this would include changes in such factors as temperature, salinity, nutrients, and light.

It is important to know the tolerances of motile cells to these parameters in order to evaluate whether encystment truly is a survival mechanism. Unfortunately, the relevant information is incomplete or available only from casual observations. Temperature, for example, is certainly a critical factor for *G. tamarensis* and *G. catenella* given their presence in temperate, waters, yet we do not know how many vegetative cells would survive winter temperatures if encystment had not occured. Most studies have examined temperature effects on growth rate, and these generally indicate that division does not occur below 5°C (5, 24-26). We do know, however, that some motile cells persist at temperatures as low as 0-2°C in laboratory cultures (26, 27), but the duration of this survival has not been determined. Winter temperatures in coastal waters can be highly variable, but they often drop to this same 0-2°C range (5, 28). Summer survival would not be a problem in most areas since *G. tamarensis* growth has been reported at temperatures as high as 24°C (25, 26), a temperature above the maximum for most coastal waters. There is thus the potential for small, motile cell populations to persist in certain temperature regions throughout the year.

Those field studies that include water samples collected during winter months all show undetectable *G. tamarensis* concentrations (5, 18, 21). Given the relatively small volumes of water typically collected and counted, this does not preclude the presence of a few cells (the "hidden flora"), but it does indicate that motile populations are extremely small at best. Furthermore, since the growth rate of *G. tamarensis* is essentially zero at very low temperatures, the appearance of even a few hundred cells L^{-1} in early spring when waters are still very cold suggests that it is excystment and not division of surviving motile cells that initiates the bloom development.

The obvious problem with this discussion is that it is circular - that encystment does occur and thus significantly decreases the size of motile cell populations. Low or non-existent wintertime concentrations of *G. tamarensis* may thus reflect either a low probability of survival or the removal of many motile cells by cyst formation and deposition.

The fate of the toxic *Gonyaulax* species in low-nutrient environments is also poorly understood. On the one hand, *G. tamarensis* cells can persist in older cultures for months without dividing, presumably under impoverished nutrient conditions. On the other hand, encystment has been observed in natural waters at relatively high nutrient concentrations (18). We thus have no firm foundation on which to base the conclusion that toxic *Gonyaulax* species encyst to survive the temperature or nutrient variations of temperate coastal waters. Clearly there are many phytoplankton species that bloom year after year without dormancy - that rely instead on broad environmental tolerances or advection from more suitable waters.

One line of evidence that supports the view that encystment is not primarilly a strategy to survive through short-term stress is that sexuality is not induced in laboratory cultures by a variety of adverse conditions but instead seems to occur under the relatively specific conditions of nutrient limitation (29, 30). For example, decreasing either temperature or light intensity alone has not successfully induced sexuality in *G. tamarensis* cultures (26). In fact, if temperature is dropped from optimal levels to that where growth rate begins to decrease (12°C), cyst formation is actually inhibited under low nutrient conditions that typically induce sexuality.

The only stress that has consistently yielded cysts in *G. tamarensis* cultures to date is nutrient limitation (26). In this context, it is important to recognize that encystment is not a rapid process. The formation and fusion of gametes and the development of a swimming zygote requires up to a week for many dinoflagellates, including *G. tamarensis* (18, 29, 30). It appears that sexual induction is not a response to absolute nutrient starvation but instead occurs in anticipation of impending limitation, perhaps cued by the decreasing size of internal nutrient pools.

The general picture that emerges is that of encystment as a response to a set of conditions that do not represent an immediate threat to survival. Motile cells could persist well beyond the time when cysts form, with a good possibility for additional blooms as conditions improve. There is no doubt that an *indirect* result of encystment is that the species survives in relatively large numbers through a variety of environmental stresses that may severely deplete a motile cell population. Those hostile conditions may, however, occur long after encystment has been completed. The value of encystment as a survival strategy thus rests on the numbers of motile cells that would survive each year if encystment did not occur, relative to the number of cysts that successfully germinate. Unfortunately, the latter number may be relatively small and the former is unknown.

Here again we find ourselves tempted to accept the 'easy answer' that encystment is a necessary mechanism developed by a species to ensure survival through environmental extremes. If it is true that the species would have been capable of survival in many areas without resorting to dormancy, the ecological justification for encystment may then lie in more subtle factors, possibly those associated with genetic recombination during sexuality.

Cysts as Dispersal Agents

There are a variety of ways that toxic *Gonyaulax* cells could be introduced to areas with no previous history of PSP, and most of these involve cysts. The most common claim is that transport of an established motile population by tidal and large-scale circulation patterns permits a species to deposit cysts in new areas as "seeds" for future blooms (1, 9, 10, 31). Within this framework, the cyst is most important in those areas where advection of existing blooms is not a common occurrence. Thus massive PSP outbreaks are often credited with the dispersal of toxic Gonyaulax to new regions.

The most striking example is perhaps the 1972 New England red tide which caused extensive shellfish harvesting closures in Maine, New Hampshire and Massachusetts, the latter two states having no previous history of shellfish toxicity (32). In the more than 10 years since that event, Massachusetts has had PSP-related closures every year (14).

The absence of PSP events in the past does not imply the absence of toxic Gonyaulax cells however, since other factors may be involved including the lack of conditions suitable for dense population growth, low toxicity strains of the causative organism (33), or a shortage of harvestable shellfish resources (14). Nevertheless, there is little doubt that the 1972 bloom introduced G. tamarensis cells into some areas that were previously unaffected in southern New England. The recurrent nature of the toxic episodes and the dangerously high PSP levels measured in popular clamming areas since then both suggest a significant change. The importance of cysts is this spreading event is seen in the subsequent PSP patterns for the southern portion of this region where outbreaks are highly localized and patchy within estuaries, with no offshore populations of G. tamarensis as a source of advected cells (21).

A similar scenario holds for the Seattle area of Washington state. Historically, G. catenella has been a hazard along the open coast of Washington, British Columbia and Alaska, but not within Puget Sound until 1975. A major bloom in 1978 apparently introduced the problem to the southern part of Puget Sound, where it has since persisted year after year (34). One common attribute of both the Seattle and New England PSP problems is that recurrent blooms in previously unaffected areas began in the years immediately following major, catastrophic blooms. There is little doubt then that the persistence of the problem is linked to the deposition of cysts from advected population during the initial bloom events.

The resistant nature of dormant cysts makes them likely candidates for dispersal via dredging operations, shellfish transplants, or boat ballast as well. Each of these is theoretically possible, but there is no direct evidence that demonstrates conclusively that such mechanisms have operated in the past.

Dredging is perhaps the most likely cause for concern, especially in regions where the toxic Gonyaulax cysts are not widespread. In southern New England, for example, the very characteristics of certain estuaries that result in accumulation of cysts (shallow, narrow inlets, reduced tidal flushing, high productivity) are those that create the need for dredging. The advection of sediment resuspended during dredging is a definite concern, as is the distant disposal of larger volumes of sediment at dredge spoil dumping sites. Both of these mechanisms could transport cysts to new environments, but they also serve to provide suitable germination conditions. Evidence is now accumulating that newly-deposited dinoflagellate cysts are buried into the sediments by benthic animal activity (Figure 7), often resulting in significant subsurface maxima (15, 20). Many of these sediments are anoxic just below the surface, and recent experiments indicate that G. tamarensis germination is severely inhibited in the absence of oxygen (although the cysts do remain viable; 35). A dredging operation could thus introduce many cysts to oxygenated surface waters, leading to subsequent blooms in nearby areas. The only data suggesting that this may have happened in the past is associated with the 1972 New England red tide, which began shortly after a major dredging operation in the Merrimac River (36).

The transport of cysts via shellfish transplants or relays is even more difficult to evaluate. Not only is it possible that the sediment on the shells of seed shellfish contains cysts, but ingested cysts may even survive ingestion and germinate following defecation. Many cysts fed to soft-shell clams and mussels are viable following isolation from fecal pellets (35), but experiments have yet to be performed that mimic the conditions associated with prolonged residence in the intestines of shellfish during inter or intra-state transport.

Only one piece of evidence suggests that shellfish seeding may have introduced PSP to an area, and that was in Perch Pond, Falmouth MA which developed PSP in 1976, one year after the salt pond was seeded with quahogs of unknown origin (10). It is also of interest that the Perch Pond strain of G. tamarensis is morphologically distinct from others on Cape Cod (variety tamarensis versus excavata; (10, 37), and thus it is unlikely that advective transport of motile populations introduced the species to that estuary.

In general, it may never be possible to prove that species dispersal is facilitated by dredging, shellfish transplants,or boat traffic. Here again we are faced with mechanisms that are theoretically possible but that may be of minor practical concern relative to the introduction of cysts to new areas through advective transport of established blooms. There is little doubt that this latter mechanism has been, and will continue to be, of major importance to the geographic distribution of the toxic Gonyaulax species.

Cysts and Genetic Recombination

Many studies in recent years have demonstrated a link between sexuality and cyst formation in dinoflagellates (38). It is now evident that both G. tamarensis and G. catenella form cysts following the sexual fusion of gametes (31, 39, 40), but only the latter species has been proven to be heterothallic thus far. The ecological advantage of this process stems from genetic recombination during fusion of self-sterile gametes that should, in theory at least, create heterogeneous cyst and motile cell populations.

It is, of course, difficult to demonstrate that such variability exists and to demonstrate the resulting ecological advantages. Genetic variability has been documented among diatoms and dinoflagellates (41, 42), but no definitive studies have been completed on the toxic Gonyaulax species. The G. tamarensis and G. catenella species assemblages are quite diverse, however, and this provides indirect evidence of recombination. For example, Schmidt and Loeblich (37) found nearly every possible combination of bioluminescence, toxicity, and possession of a ventral pore among nine G. tamarensis isolates, all having identical thecal tabulation. Other workers have demonstrated different toxin content on a per cell basis (33, 43) and different toxin composition between strains of toxic Gonyaulax species (44, 45).

What these differences mean in terms of overall species success is not known, but they imply that there may be other genetic differences that are more difficult to identify that could affect the growth characteristics and environmental tolerances in these species. If this variability were reflected, for example, in the factors affecting germination or motile cell growth rate, we would expect that not all cysts would germinate at the same time under the same conditions or that some motile cells would survive better than others (18). In this way, cyst-forming dinoflagellates might maintain a viable, quiescent seed population in the sediments year after year while optimizing the growth and proliferation of motile cells as well.

These considerations are clearly speculative, but they are suggestive of the tremendous benefits (albeit gradual) that might accrue to sexually-reproducing dinoflagellates. It may be that the ultimate goal of encystment is simply to insure that nuclear fusion occurs. Other benefits of encystment (survival, dispersal, etc.) could be indirect and somewhat coincidental.

Cysts as Direct Sources of Toxicity

Toxicity in deep water scallops throughout the year has been recognized for many years (44, 46, 47). More importantly, the levels of toxicity in the scallop digestive gland were shown to increase dramatically (sometimes by a factor of 2 or 4) during winter months when G. tamarensis motile cell populations were low or undetectable. Bourne (4) postulated that cysts were the toxin source, an opinion also favored by Jamieson and Chandler (47) in a more recent study. It has since been confirmed that G. tamarensis cysts are indeed toxic, and this has led to a proliferation of explanations for toxicity episodes based on cyst ingestion.

The original report of toxicity in cysts claimed that they could be an order of magnitude more toxic than motile cells (48). This contention has since been both confirmed (49) and refuted (20) by studies which compared the toxicities of cysts and motile cells of this species. It is, however, the magnitude and not the existence of toxin in cysts that is in question.

Despite this progress, the evidence linking cysts to shellfish toxicity remains circumstantial and care should be exercised before attributing toxin increases to this mechanism. The major problem is that it has yet to be demonstrated that shellfish can remove toxin from cysts. The feeding studies mentioned earlier (which do not yet include scallops; 35) indicate that many viable G. tamarensis cysts can be isolated from the fecal pellets of mussels and soft-shelled clams fed cyst suspensions. There is certainly some cyst mortality as well (Figure 5), but whether this is also associated with toxin retention by the shellfish has yet to be demonstrated. It is reasonable to expect that the assimilation of toxin from cysts will not be a highly efficient process.

Data are now available that permit a preliminary assessment of the cyst concentrations available for ingestion in deeper waters. White and Lewis (20) mapped the abundance of G. tamarensis cysts in Bay of Fundy sediments in the same regions where toxic scallops have been studied (4, 47). Cyst densities as high as 8000 cc^{-1} were found in the top two cm of sediment, with abundance falling to less than 100 cc^{-1} in rocky, gravelly areas. This latter environment is, however, the one favored by scallops, (47), presumably because the high currents in these areas enhance filter feeding.

If we conservatively assume toxicities of approximately 100 m.u. $cyst^{-1}$ (20) and complete assimilation of toxin (with no depuration), it would require consumption of as many as 100 million cysts to achieve the toxin levels recorded in deep water scallops. In areas of highest cyst abundance, this is equivalent to one scallop ingesting all cysts in the top cm of material over one square meter of sediment. If the lower cyst concentrations reported for the rocky or gravelly bottom are used, the removal must cover an area 100 times larger. (Scallop densities in these regions average 2-3 animals m^{-2}; (50). Arguments by Jamieson and Chandler (47) that areas with high bottom currents would provide a steady supply of cysts advected from high deposition sites nearby may be justified, but this explanation requires that a substantial portion of the cyst population remains resuspended in bottom waters throughout the year. Unfortunately, studies of the dynamics of cyst populations have not reached the stage where this possibility can be evaluated.

Another problem with wintertime toxicity data is that toxin levels are reported per 100 gm of tissue. Studies of the deep sea scallop Placopecten magellanicus indicate that the size of the digestive gland can vary through the year (51). Thus a constant amount of toxin in a gland would look variable when normalized to 100 gm of tissue, with the highest relative toxicity during winter months when the tissues are the smallest. It would appear, however, that this error is small (perhaps 20-30%) relative to the 2 to 4-fold toxin increases typically reported between seasons in scallops (4, 47).

The purpose of the foregoing discussions is to argue for caution in assigning toxic events to cyst ingestion. Once again, this may be an easy explanation that precludes investigations of other mechanisms. It seems highly probable that cysts do contribute to wintertime or deep water shellfish toxicity at some level, but there are other factors that might contribute, including slow toxin depuration or delivery of toxin via fecal pellets, temporary cyst stages, or deep mixing of motile populations.

Cysts as a Factor in Bloom Decline

The prevailing view of the dynamics of many phytoplankton blooms is that they often terminate due to nutrient exhaustion, increased grazing pressure, and/or physical dispersal. Evidence is now accumulating that G. tamarensis blooms have ended when grazing and advection were low and nutrients were above detection limits (18, 21). Instead of persisting without division as nutrients disappear and grazers multiply, asexual reproduction was replaced by sexuality and a substantial portion of the motile population became non-dividing gametes and planozygotes (18). In much the same way that cyst germination may initiate bloom development in many areas, sexuality and cyst formation appeared to dictate the dynamics of bloom decline.

As discussed previously, it is not known how long some portion of the motile populations could remain in the water column if encystment did not occur, but it is evident that it would be longer than is the case with sexuality. The important unknowns in evaluating this process are the magnitude and duration of encystment once initiated. Although up to 35% of all motile cells in the late stages of some G. tamarensis blooms have been planozygotes (the swimmming precursors to cysts), the status of the remainder of the population was not known (18). Other data indicate that those cells that do not encyst can remain in the water for weeks or months after the major surge of encystment (35). This process is undoubtedly mediated by the nutritional environment of the population, with persistance of some fraction of the bloom facilitated by fluctuations in the supply of nutrients.

Nevertheless, the process of encystment dictates when a substantial fraction of a Gonyaulax population leaves the water column, regardless of the ability of motile cells to persist for extended periods. The encystment/excystment cycle can thus define the temporal limits of some blooms. More work is needed to evaluate the importance of this process relative to bloom termination from factors such as grazing or advection that decrease the numbers of motile cells.

Overview

The preceding discussion was intended as an objective evaluation of what has been clearly established (as opposed to what we would like to believe) concerning the relationship between cysts and shellfish toxicity. Clearly a great deal of work is needed if we are to verify the circumstantial evidence that supports most of the attributes ascribed to cysts. The most difficult task is to determine why encystment actually occurs (which is a different question than what initiates it). Certain of the roles of cysts (e.g. survival, seeding) may simply be indirect benefits of a more fundamental drive - that designed to ensure the adaptive benefits accompanying sexuality and genetic recombination for example. Resting cysts are undoubtedly important in many aspects of the PSP phenomenon, but we must not overlook more subtle or complex biological and physical explanations for our observations.

Acknowledgments

This work was supported in part by the Office of Sea Grant in the National Oceanic and Atmospheric Administration through grant NA80AA-D-00077 (R/B-41) to the Woods Hole Oceanographic Institution and in part by the Ocean Assessments Division, Office of Oceanography and Marine Services (NOAA) through grant NA81RAD00012. Contribution number 5541 from the Woods Hole Oceanographic Institution.

Literature Cited

1. Wall, D. Geoscience and Man. 1971, 3, 1-15.
2. Hensen, V. Berichte der Kommission fur Wissenschaftliche Untersuchungen der deutschen meere, Kiel, 1882-1886. 1887.
3. Cleve, P.T. Svenska Vetenskapsakademien, Handlingar. 1900, 32.
4. Bourne, N. J. Fish. Res. Bd. Can. 1965, 22, 1137-1149.
5. Prakash, A. J. Fish. Res. Bd. Can. 1967, 24, 1589-1606.
6. Wall, D. In "Toxic Dinoflagellate Blooms"; LoCicero, V.R. Ed.; Proc. Int'l Conf. (1st). Mass. Sci. and Tech. Found.: Wakefield, 1975; pp. 249-356.
7. Steidinger, K.A. In "Toxic Dinoflagellate Blooms"; LoCicero, V.R. Ed.; Proc. Int'l Conf. (1st). Mass. Sci. and Tech. Found.: Wakefield, 1975; pp. 153-62.
8. Wall, D.; Dale, B. Micropaleontology 1968, 14, 265-304.
9. Dale, B. Sarsia. 1977, 63, 29-34.
10. Anderson, D. M., Wall, D. Journal of Phycology 1978, 14, 224-234.
11. Taylor, F. J. R. In "Toxic Dinoflagellate Blooms"; Taylor, D. L. and H. H. Seliger, Eds.; Proc. Int'l Conf. (2nd). Elsevier/North Holland: New York. 1979; pp. 47-56.
12. Dale, B. In "Toxic Dinoflagellate Blooms"; Taylor, D. L. and Seliger, H. H. Eds.; Proc. Int'l Conf. (2nd). Elsevier/North Holland: New York. 1979; pp. 443-52.
13. Walker, L. M. Trans. Am. Micros. Soc. 1982, 101, 287-93.
14. Anderson, D. M.; Kulis, D. M.; Orphanos, J. A. Estuarine, Coastal, and Shelf Science 1982, 14, 447-58.
15. Anderson, D. M.; Aubrey, D. G.; Tyler, M. A.; Coats, D. W. Limnol. Oceanogr. 1982, 27, 757-65.
16. Thayer, P. E.; Hurst, J. W.; Lewis, C. M.; Selvin, R.; Yentsch, C. M. Can. J. Fish. Aquat. Sci. 1983, 40, 1308-14.
17. von Stosch, H. A. Br. Phycol. J. 1973, 8, 105-34.
18. Anderson, D. M.; Chisolm, S. W.; Watras, C. J. Marine Biology 1983, 76, 179-189.
19. Lewis, C. M.; Yentsch, C. M.; Dale, B. In "Toxic Dinoflagellate Blooms"; Taylor, D. L. and Seliger, H. H. Eds.; Proc. Int'l Conf. (2nd) Elsevier/North Holland: New York. 1979; pp. 47-56.
20. White, A. W.; Lewis, C. M. Can. J. Fish. Aquat. Sci. 1982, 39, 1185-1194.
21. Anderson, D. M.; Morel, F. M. M. Estuarine Coastal Mar. Sci. 1979, 8, 279-93.
22. Spencer, D. W.; Sachs, P. L. Marine Geology 1970, 9, 117-36.
23. Dale, B. In "Survival Strategies of the Algae"; Fryxell, G. A., Ed., Cambridge. 1983. pp. 69-136.
24. Yentsch, C. M.; Cole, E. J.; Salvaggio, M. G. In "Toxic Dinoflagellate Blooms"; LoCicero, V. R. Ed.; Proc. Int'l. Conf. (1st). Mass. Sci. and Tech. Found. Wakefield, 1975; pp. 163-80.
25. Watras, C. J.; Chisholm, S. W.; Anderson, D. M. J. Exp. Mar. Biol. Ecol. 1982, 62, 25-37.
26. Anderson, D. M.; Kulis, D. M.; Binder, B. J. manuscript in prep.

27. Schmitter, R. In "Toxic Dinoflagellate Blooms"; Taylor, D. L. and Seliger, H. H. Eds.; Proc. Int'l Conf. (2nd). Elsevier/North Holland: New York. 1979; pp. 47-56.
28. Yentsch, C. S.; Yentsch, C. M.; Strube, L. R.; Morris, I. Savannah River Project Symposium on Thermal Pollution. Augusta, GA. 1973, 508-17.
29. Pfiester, L. A.; Anderson, D. M. In "The Biology of Dinoflagellates."; Taylor, F. J. R., Ed.; Blackwell Scientific Publications, Ltd. in press.
30. Anderson, D. M.; Lindquist, N. L. Sexuality and cyst formation in the dinoflagellate Gonyaulax tamarensis: II. Time-course measurements of phosphorus limitation. Manuscript under review.
31. Anderson, D. M.; J. Phycol. 1980, 16, 166-172.
32. Bicknell, W. J.; Walsh, D. C. In "Toxic Dinoflagellate Blooms"; LoCicero, V. R. Ed.; Proc. Int'l Conf. (1st). Mass Sci and Tech. Found.: Wakefield, 1975; pp. 447-58.
33. Alam, M. I.; Hsu, C. P.; Shimizu, Y. J. Phycol. 1979, 15, 106-110.
34. Saunders, S.; Sample, T.; Matsuda, R. Paralytic Shellfish Poisoning. METRO (Seattle) Water Qualify Control Dept. 1982.
35. Anderson, D. M., unpublished data
36. Hartwell, A. D. In "Toxic Dinoflagellate Blooms"; LoCicero, V. R. Ed.; Proc. Int'l Conf. (1st). Mass Sci. and Tech. Found.: Wakefield, 1975; pp. 47-68.
37. Schmidt, R. J.; Loeblich, A. R. In "Toxic Dinoflagellate Blooms"; Taylor, D. L. and Seliger, H. H. Eds.; Proc. Int'l Conf. (2nd). Elsevier/North Holland: New York. 1979; pp. 83-8.
38. Beam, C. A.; Himes, M In "Biochemistry and Physiology of Protozoa"; Levandowsky, M. and Hutner, S. H., Eds.; Academic Press. 1980; 2, 171-206.
39. Turpin, D. H.; Dobel, P. E. R.; Taylor, F. J. R. J. Phycol. 1978.
40. Yoshimatsu, S. Bull. Plank. Soc. Jap. 1981, 28, 131-9.
41. Gallagher, J. C. J. Phycol. 1982, 18, 148-62.
42. Brand, L. E.; Murphy, L. S.; Guillard, R. R. L.; Lee, H., Mar. Biol. 1981, 62, 103-10.
43. Maranda, L.; Shimizu, Y.; Anderson, D. M. manuscript in prep.
44. Shimizu, Y. In "Toxic Dinoflagellate Blooms"; Taylor, D. L. and Seliger, H. H. Eds.; Proc. Int'l Conf. (2nd). Elsevier/North Holland: New York. 1979; pp. 321-6.
45. Hall, S. Ph.D. Thesis, Univ. of Alaska, Fairbanks. 1982.
46. Medcof, J. C.; Leim, A. H.; Needler, A. B.; Needler, A. W. H.; Gilbard, J.; Nanbert, J. Bull. Fish. Res. Bd. Can. 1947, 7, 490-504.
47. Jamieson, G. S.; Chandler, R. A. Can. J. Fish. Aquat. Sci. 1983, 40, 313-8.
48. Dale, B.; Yentsch, C. M.; Hurst, J. W. Science 1978, 201, 1223-25.
49. Oshima, Y.; Singh, H. T.; Fukuyo, Y.; Yasumoto, T. Bull. Jap. Soc. Sci. Fish. 1982, 48, 1303-5.
50. Jamieson, G. S., Lundy, M. J. Can. Fish. and Marine. Service Tech. Rep. No. 0915. 1979.
51. Robinson, W. E.; Wehling, W. E.; Morse, M. P.; McLeod, G. C. Fish. Bull. 1981, 79, 449-58.

RECEIVED March 23, 1984

12

Potential Importance of an Endoparasite of *Gonyaulax* in Paralytic Shellfish Poisoning Outbreaks

LOUISA NISHITANI, RALEIGH HOOD, JOHN WAKEMAN, and KENNETH K. CHEW

School of Fisheries, University of Washington, Seattle, WA 98195

Infestation by the dinophycean parasite, *Amoebophrya ceratii*, was a major factor contributing to the decline of two blooms of *Gonyaulax catenella*, the causative organism of many of the outbreaks of paralytic shellfish poisoning in the Pacific Ocean. This finding supports an earlier suggestion that the parasite might be useful as a biological control agent. Another major factor contributing to the decline of the blooms was reduced division rate, perhaps due to low nutrient concentrations. Division rate and nutrient availability may be useful in prediction of the duration of blooms.

Taylor (1) found the endoparasitic dinoflagellate *Amoebophrya ceratii* in phytoplankton samples taken in 1965 from Sequim Bay, Washington (Figure 1). He reported that 30-40% of the *G. catenella* cells in 3 samples taken at intervals over 4 months were infested by this parasite. He suggested that the parasite offered a remote possibility of biological control of *G. catenella*, the causative organism of many of the outbreaks of paralytic shellfish poisoning (PSP) on both sides of the North Pacific (2, 3, 4, 5) and in the Southern Hemisphere (6).

During a 1981 study of a *G. catenella* bloom in Quartermaster Harbor in Puget Sound, Washington (Figure 1), we observed that a rapid decline in the *G. catenella* population was associated with a marked increase in the percentage of cells parasitized by *A. ceratii* (7). This observation of the course of parasitism throughout the development and decline of a *G. catenella* bloom supports Taylor's suggestion that this parasite might be useful as a biological control agent. It also raises the question of whether the percent of natural parasitism could be a useful factor in predicting the onset and duration of PSP outbreaks.

In order to assess its potential value for either purpose, detailed knowledge of the conditions and timing of the various stages

0097-6156/84/0262-0139$06.00/0
© 1984 American Chemical Society

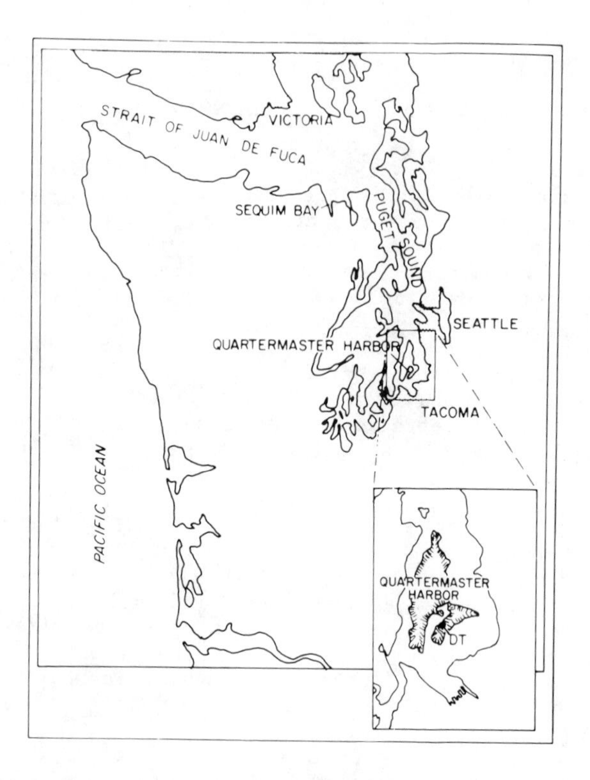

Figure 1. Map of western Washington, showing study area in Quartermaster Harbor.

in the development of *A. ceratii* and its natural role in the phytoplankton will be essential. The developmental stages in the life history of *A. ceratii* have been described in detail by Cachon (8). The parasite, which enters the host cells as a biflagellate dinospore, first attacks the host nucleus (Figure 2), grows through a series of stages, and ultimately consumes the contents of the host cell, leaving only the theca. Eventually the parasite extrudes from the host cell as a long worm-like spiral chain, a vermiform, which divides into hundreds of dinospores (Figure 3).

This paper presents the intial findings of a study of the ecology of this parasite, in particular, its role in controlling *G. catenella* populations, and some general considerations about its potential use as an agent for control of PSP problems.

Methods

Field sampling was done at two stations near the head of Quartermaster Harbor at weekly intervals prior to the initiation of the blooms and much more frequently during the blooms. In 1981 samples for cell counts were taken by Van Dorn bottle at 1, 4, and 9 m depths routinely and at 1 m depth intervals to 9 m during migration studies. Counts were made with an inverted microscope. Samples were taken at 1 and 9 m depths for nutrient analysis by autoanalyzer. Temperature was measured by thermistor probe, and salinity by refractometer. Plankton samples from a bottom-to-surface vertical tow with a 10 µm mesh net were preserved in buffered 2% formalin and stained with acetocarmine for determination of percent of parasitism in *G. catenella* and other species. Samples of bay mussels (*Mytilus edulis*) taken from pilings were analyzed for toxin content by the Washington State Department of Social and Health Services Public Health Laboratory, using the mouse bioassay. In 1983 methods were altered as follows: Water samples for cell counts were taken routinely at 1 m depth intervals to 9 m; samples for nutrient analysis were taken at 1 and 4 m; mussel samples were taken from a cage deployed at a depth of 2 m; and plankton tows were taken with a 20 µm mesh net, and samples were stained by the Feulgen reaction. Plankton samples from 0400 hr were used to derive a division index, which is the percent of non-parasitized cells showing any evidence of division, i.e., nuclei in pre-mitotic configuration, karyokinesis, cytokinesis, or post-cytokinesis. (The 0400 hr was selected on the basis of the finding that the time of maximum division is near dawn.)

Results and Discussion

Role of Parasite in *Gonyaulax* Blooms

1983 Bloom - Gonyaulax Population and PSP in Mussels. A bloom of *G. catenella*, which started in mid-June, peaked in July with PSP levels in bay mussels reaching a maximum toxicity level of 2139 µg toxin/100 g meat (Figure 4). As observed in 1981, the *G. catenella* cells were heavily parasitized by *A. ceratii* during the decline of the bloom.

To evaluate the relative significance of the parasite in the reduction in the *G. catenella* population, it was necessary to compare the changes in abundance of both *G. catenella* and *A. ceratii* with

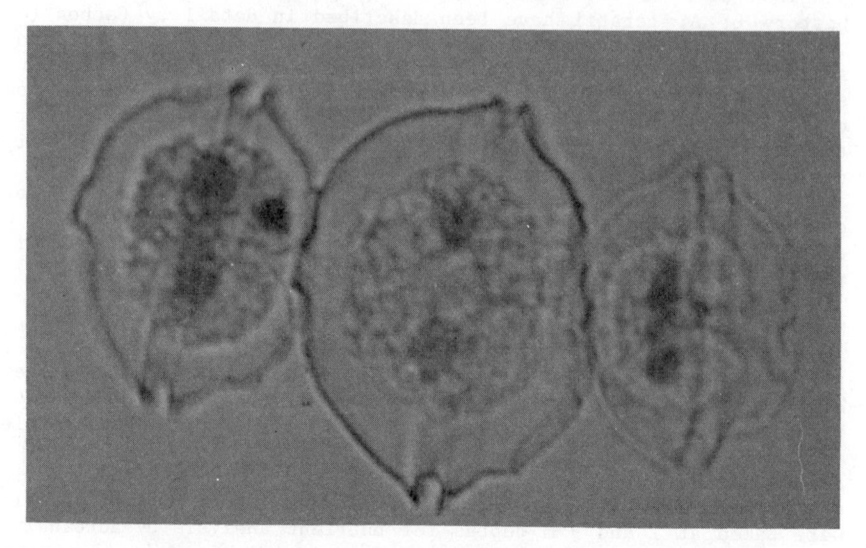

Figure 2. Center of U-shaped nucleus of _G. catenella_ attacked by _A. ceratii_.

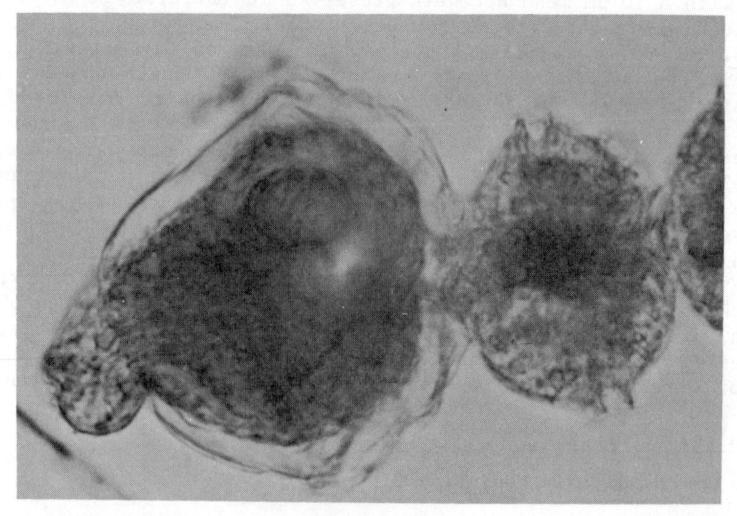

Figure 3. Vermiform of _A. ceratii_ emerging from _G. catenella_ cell.

changes in the division rate of the host species and with the events and trends in the several other environmental factors which could have contributed to the decline of the bloom.

For this evaluation the measurement of the abundance of *G. catenella* in the water column, expressed here in terms of the sum of the cells/l at 1 m intervals to 9 m at one station (DT), can only be used to indicate trends in the population. This summation value was found to be more representative of the population in the 1983 bloom than the density of afternoon samples from 1 m, the value used in 1981, because the cells did not migrate to the top few meters under certain conditions occurring during the 1983 bloom.

The toxin level in bay mussels in cages under floats is often thought to reflect changes in *Gonyaulax* populations more closely than intermittent plankton samples because the more or less continuous sampling of the plankton by the mussels tends to reduce the effect of horizontal patchiness. In this particular bloom, however, the correspondence between the toxicity level in mussels and the abundance of *G. catenella* in the water column was reduced because of two factors: First, on many days the zone of maximum density of *G. catenella* remained below the 2 m level at which the mussel cage was deployed; and, second, it appears that, at the peak of the bloom, cell densities of 3 x 10^5 cells$\cdot l^{-1}$ in the top two meters may have caused the mussels to cease feeding normally, resulting in a drop in toxicity. Other workers have found that the toxin level in mussels increased with increased density of *G. tamarensis* up to 10^5 cells$\cdot l^{-1}$, and at higher densities, pseudofeces were produced, indicating rejection (9).

However, both toxicity levels and the abundance values for *G. catenella* indicate that during the first two weeks of the bloom a sharp increase of *G. catenella* occurred, which was then followed by a rapid decline in the population.

Changes in Division Rate. The division index for 0400 hr samples taken through the bloom was used to indicate whether a reduction in division rate contributed to the decline. (We cannot at present extrapolate from a division index based on samples from one hour to divisions/day.) The index rose during the build-up of the bloom to a value of 22 or more and fell to 10 during a 9-day period in which cell densities peaked and dropped rapidly (Figure 5).

Division indices were compared with data on environmental factors to determine which contributed to the decline in the division rate. Temperatures above the thermocline ranged from 14 to 18°C and salinity from 29 to 31°/oo, both close to the ranges found to be optimal for this species in laboratory experiments (10).

Following the onset of the bloom, concentrations of nitrogen as NO_3, NO_2, and NH_3, declined markedly to values of 1.4 mole$\cdot l^{-1}$, at 4 m depth (Figure 5). (NO_3 values were 0.5 mole$\cdot l^{-1}$ or less for several days following the peak in population densities.) The concentration of PO_4 decreased about 20% to 1.2 mole$\cdot l^{-1}$ at 4 m depth. (Data from the 4 m depth is given because the depth of maximum cell density was 3-5 m during most days and some nights during the bloom and its decline.) The division index fell during the period of low concentrations of nitrogen and orthophosphate and rose again as nutrient concentrations increased.

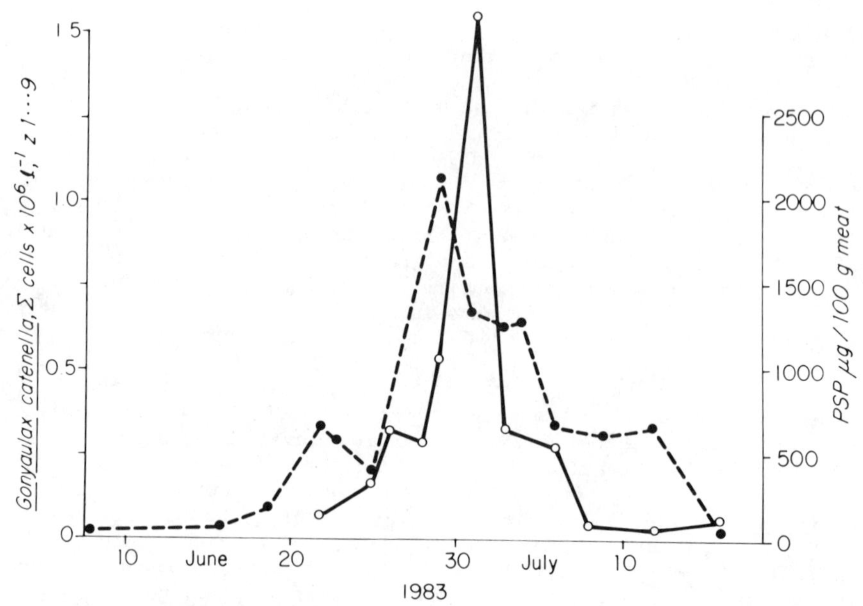

Figure 4. Abundance of G. catenella (0---0) and toxicity in bay mussels (●---●), Station DT, Quartermaster Harbor, Washington, 1983.

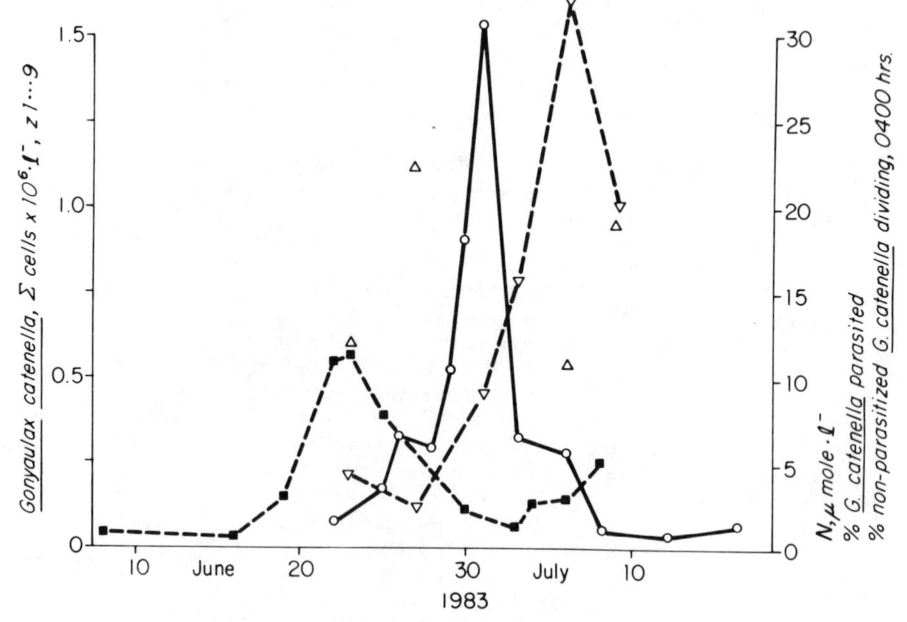

Figure 5. Density of G. catenella (0---0), the percent of G. catenella parasitized by A. ceratii (▽---▽), the percent of non-parasitized G. catenella dividing at 0400 hr (Δ), and concentration of $NO_3 + NO_2 + NH_3$ at z 4 m (■---■), Station DT, 1983.

Physical Dispersion and Biological Removal. The fact that the peak of the bloom and the period of decline occurred between two strong spring tide series suggests that tidal advection of cells was probably not a major factor in the decline of the bloom. Likewise, the wind data available give no indication that there was significant removal of cells in the surface layers by wind-driven advection.

Non-motile stages were extremely rare; hence significant reduction in *G. catenella* densities by settling was unlikely. Although many known and suspected predator species were present during the bloom, no data is available which would indicate the extent to which predation by zooplankton and benthic animals may have reduced the *G. catenella* population.

A potentially significant factor in the decline in abundance of *G. catenella* is the infestation by *A. ceratii* which results in the ultimate removal of the host cells from the water column. The percent of parasitism increased markedly from 2.5% prior to the peak of the bloom to 9.5% at the peak and 32% five days later (Figure 5). During that 9 day period of a 12-fold increase in infestation rate, the *G. catenella* population first increased and then decreased by ~80% in 5 days. During this time the divison rate of *G. catenella* dropped by at least 50%.

1981 Bloom. The June 1981 bloom in Quartermaster Harbor was also characterized by a sharp increase, then a rapid decline in the *G. catenella* population. Cell densities in afternoon samples from the 1 m depth at Station DT rose to 1.7×10^5 cells$\cdot$l^{-1} and toxin content of bay mussels from pilings to 1916 μg toxin/100 g meat (Figure 6).

Some of the conditions associated with the 1981 bloom are very similar to those found in the 1983 bloom. Direct comparison of the factors affecting growth rates is not possible because division indices and 4 m nutrient data are not available for 1981. However, concentrations of N and P at both 1 and 9 m levels decreased markedly as the bloom peaked and began to decline (Figure 7). In the two days following the peak of the bloom, there was a marked drop in temperature throughout the water column, which also may have reduced the division rate.

During the strong spring tide series which coincided with the decline of the bloom, the vertical and horizontal advection indicated by cell counts and hydrographic data probably contributed significantly to the reduction in the population at stations near the head of the bay. A major wind event coupled with the strong tides may have driven a portion of the cells out of the bay.

Significant reduction of the *G. catenella* population by settling was unlikely because non-motile stages were rare. No data on predation effects is available for the 1981 bloom.

The percentage of cells throughout the water column which were infested by *A. ceratii* rose from 2% at the peak of the bloom to 22% five days later during the decline (Figure 7). (In samples from the 1 m depth during and after the decline approximately 50% of the cells were parasitized.)

Comparisons of Patterns in the 1981 and 1983 Blooms. There was a striking parallel between the patterns of the two blooms with regard to populations, nutrients, and parasitism: A sharp decline in

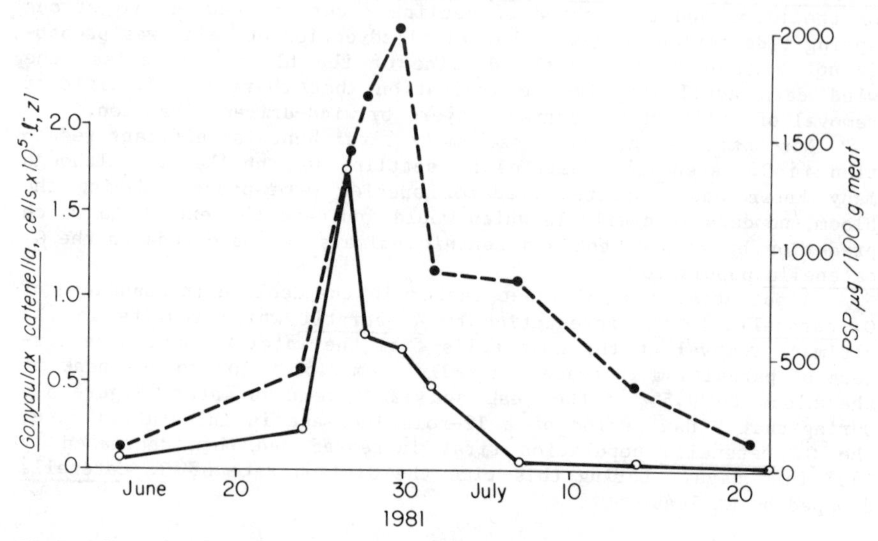

Figure 6. Density of G. catenella (0---0) and toxicity in bay mussels from pilings (●---●), Station DT, 1981.

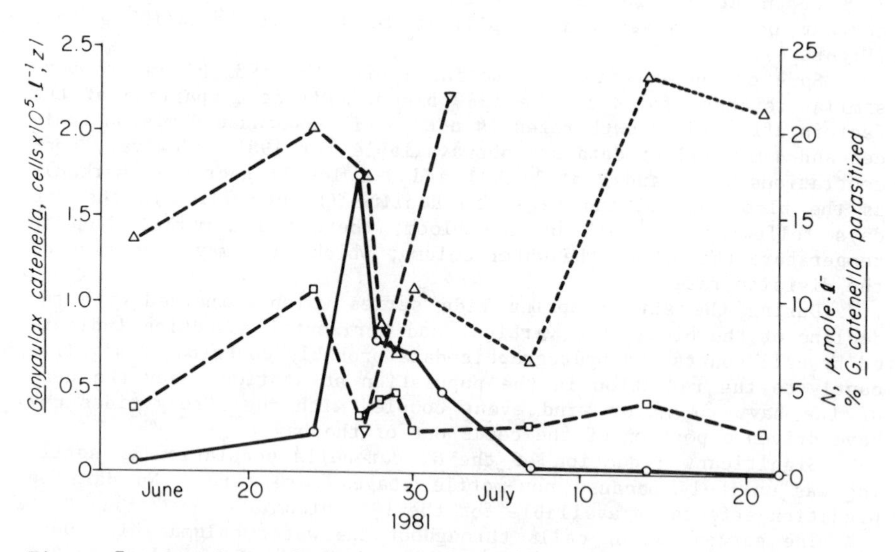

Figure 7. Density of G. catenella (0---0), the percent of G. catenella parasitized by A. ceratii (∇---∇), and concentration of $NO_3 + NO_2 + NH_3$ at z 1 m (□---□) and at z 9 m (Δ---Δ), Station DT, 1981.

nutrients as the G. catenella population increased, followed by both a rapid decline in the population (>95% in <10 days in both blooms), and a marked increase in the percent of cells parasitized by A. ceratii (22 and 32% for 1981 and 1983, respectively). Nitrogen concentrations dropped to levels which may have limited growth, judging from earlier laboratory experiments (10). This combination of events raises questions of whether the increase in parasitism was related to nutrient stress in the host cells and/or to increased density of the host cells. Laboratory experiments may be needed to answer these questions.

In both blooms it is apparent that parasitism was one of several factors contributing to the decline of the population. (In 1983 parasitism alone would not have accounted for the rate of decline of the G. catenella populations even if the division rate had dropped to 0.) Although the relative importance of parasitism will vary with conditions, the magnitude of infestation in these blooms indicates that parasitism must have been a major factor in controlling G. catenella populations.

It should be noted that the possibility exists that the actual destruction of G. catenella cells by A. ceratii may have reached considerably higher percentages than the figures given because relatively minor damage to the host nucleus, undetectable by Feulgen staining, could result in the death of the host cells. If this were the case, the division index values for non-parasitized cells would be correspondingly higher.

Potential Artificial Control of Gonyaulax with Amoebophrya

Present research is aimed toward determining whether biological control of G. catenella by Amoebophrya would be acceptable and practical. In considering the potential use of A. ceratii to control G. catenella populations, two major questions arise: Is it feasible to control microscopic organisms in the vast marine environment? Will the addition of a control agent damage that environment?

It would be highly impractical to attempt to release sufficient parasites to control Gonyaulax in the ocean or in a large-volume portion of an estuary. However, artificial control by means of a parasite might be feasible in certain small bays. (A preliminary estimation suggests that introduction of as little as 5×10^3 l of culture of a non-toxic host parasitized by Amoebophrya into a bay 1 km^2 x 4 m deep might be sufficient to adequately reduce the Gonyaulax density so that harvesting closures would not be required.) Some small bays protected from wind-driven turbulence and strong tidal mixing support repeated major blooms of Gonyaulax (11). Shellfish in such "breeding bays" are often abundant enough to support commercial and extensive recreational harvesting and frequently have high levels of PSP. Artificial control of PSP in such bays might be logistically feasible and could be highly beneficial. Because a portion of the G. catenella population in "breeding bays" can be exported to adjacent channel waters, control by A. ceratii in breeding bays may help to reduce PSP problems in adjacent waters also.

The potential adverse effects of inoculating very large numbers of a parasite, even a naturally occurring one, into a marine habitat to control a pest organism are a major concern. However, other types of control methods, such as application of chemicals or release of predators, are not likely to achieve the same level of target specificity that may be possible with a parasite. Hence, parasite control would appear to be a preferred method of control of *Gonyaulax* populations. Unless it can be shown through extensive tests that *Amoebophrya* is relatively host specific and would damage the food web only minimally, the release of large numbers of this parasite would be unacceptable. Cachon (8) reported *A. ceratii* in 10 genera in Europe. In Sequim Bay, Taylor (1) found it in only 2 cells other than *G. catenella* in plankton which included 22 other species of dinoflagellates. In Quartermaster Harbor samples, *A. ceratii* occurred relatively commonly in *Gymnodinium sanguineum* and very sparsely in 5 other genera. Anderson (12) has recently found *Amoebophrya* in *G. tamarensis* in Massachusetts, a finding which indicates that, if a feasible technique for using it as a control agent is worked out, this method would potentially have application for control of PSP in certain types of environments around the world.

In addition to host specificity, desirable characteristics in a potential biological control agent include a high rate of reproduction, high success rate in seeking out hosts, and ease of culture. It is known that *A. ceratii* can produce hundreds of infestive spores from within one host cell. Information is needed about the effects of environmental conditions and of the density and physiological state of the host population on the rate of success of *A. ceratii* in searching for a host cell, penetrating it, and reproducing within it. The present studies of blooms suggest that in natural circumstances the rate of infestation may be increased when host cells are in a nutrient stressed condition. Additional studies will be needed to determine whether *A. ceratii* can be effective in attacking healthy *G. catenella* cells in pre-bloom abundance. Although Chatton and Biecheler (13), Cachon (8), and the authors have maintained *A. ceratii* for varying lengths of time in the laboratory, techniques for rearing it in large scale cultures have not yet been developed.

In order to assess the potential value of *Amoebophrya* in reducing problems of PSP in shellfish, it will be necessary to determine whether the parasite destroys or retains the toxins as it kills the *Gonyaulax* cells.

Conclusion

The patterns of two blooms of *G. catenella* indicate that parasitism can, at times, play a major role contributing to the decline of blooms, and that the rate of infestation may be dependent on the physiological condition of the host cells. With additional data, percent of parasitism and nutrient concentrations may prove to be useful factors in predicting the duration of PSP outbreaks in some areas.

Amoebophrya ceratii has several characteristics which indicate it may be useful as a biological agent for controlling PSP. However, much work remains to be done to assess the ecological acceptability, biological feasibility, and economic practicality of using *A. ceratii* to control *Gonyaulax* and PSP problems.

Acknowledgments

We gratefully acknowledge the funding of this work by the Washington Sea Grant Program, the National Marine Fisheries Service, and the Pacific Coast Oyster Growers Association. Dr. Edward Haskins offered his kind assistance with the Feulgen staining. We deeply appreciate the dedicated efforts of Jerry Erickson, Janeen Maas, James Nishitani, and Kathleen Turner.

This chapter is Contribution No. 634, Fish. Res. Inst., UW, Seattle, WA 98195.

Literature Cited

1. Taylor, F. J. R. J. Fish. Res. Bd. Canada. 1968, 25, 2241-5.
2. Neal, R. A. Ph.D. Thesis, University of Washington, Seattle, 1967.
3. Hsu, B. C.-C. M.S. Thesis, University of Washington, Seattle, 1967.
4. Sommer, H.; Whedon, W. F.; Kofoid, C.A.; Stohler, R. Arch. Path. 1937, 24, 537-559.
5. Onoue, Y.; Noguchi, T.; Hashimoto, K. Bull. Japan. Soc. Sci. Fish. 1980, 46, 1031-4.
6. Guzman, L.; Campodonico, I.; Hermosilla, J. Ans. Inst. Pat., Punta Arenas (Chile). 1975, VI, 173-183.
7. Nishitani, L.; Wakeman, J. J. Shellfish Res. 1982, 2, 120.
8. Cachon, J. Ann. Sci. Nat., Zool., Paris. 1964, 12 Ser., Tom VI, 1-158.
9. Yentsch, C. M., personal communication.
10. Norris, L.; Chew, K. K. Proc. 1st Intnl. Conf. Toxic Dinoflagellates, Mass. Science Tech. Found., Wakefield, Massachusetts. 1975, p. 143-152.
11. Nishitani, L.; Chew, K. K. Special issue of Aquaculture (in press). Proc. Intnl. Symp. Recent Innovations in Cultivation of Pacific Molluscs, La Jolla, Dec., 1982.
12. Anderson, D., personal communication.
13. Chatton, E.; Biecheler, B. Compt. Rend. 1935, 200, 505-7.

RECEIVED March 2, 1984

13

Biosynthesis of Paralytic Shellfish Toxins

YUZURU SHIMIZU, MASARU KOBAYASHI, ALMOURSI GENENAH, and NAOSHI ICHIHARA

Department of Pharmacognosy and Environmental Health Sciences, College of Pharmacy, University of Rhode Island, Kingston, RI 02881

Biosynthetic routes to saxitoxin and its analogs are discussed, and the current state of progress in the feeding experiment using the dinoflagellates, *Gonyaulax tamarensis*, is summarized. Also described are the biosynthetic and successful ^{15}N-nmr assignment of neosaxitoxin and gonyautoxin-II.

In the past several years we have witnessed tremendous development in the chemistry and biology of paralytic shellfish poisons (PSP). In addition to long-known saxitoxin, more than ten new toxins have been isolated and characterized, and their pharmacological properties have been investigated (1,2,3). New toxin-producing organisms have been discovered, and we now have a better understanding of their life-cycle. Despite all these discoveries, however, we have very little knowledge regarding the biosynthesis of the toxins. Prior to our understanding, there was only one reported experiment on the biosynthesis of PSP, in which ^{14}C-amino acid precursors were fed to the culture of *Gonyaulax catenella*, and the incorporation of some radioactivity was observed in the crude toxin fraction (4).

There are two major facets of the biosynthesis of the toxins: 1) the molecular origin of the tricyclic cyclopentanoperhydropurine skeleton, 2) the origins and biosynthetic sequence of the additional functional groups such as carbamate, O-sulfate, and N-sulfate. The first question is an intriguing one which arouses chemists' imagination. One of the plausible pathways pointed out privately by various researchers is via a normal purine metabolite followed by the Michael type condensation of acrylate or an equivalent moiety (Scheme 1a). Another possible pathway is via a C_7 sugar derivative (Scheme 1b). In fact, some fungal metabolites are known to have amino sugars linked with guanido groups. The third possibility, which was stipulated by our group, is the formation of a nine-membered intermediate by the condensation of an imidazole compound and a C_2 unit (or two C_1 units) followed by prototropic cyclization (Scheme 1c). The last pathway seems to be the most attractive because a sponge metabolite, phakellin, is ostensibly synthesized through an analogous sequence (6).

0097-6156/84/0262-0151$06.00/0
© 1984 American Chemical Society

Scheme 1

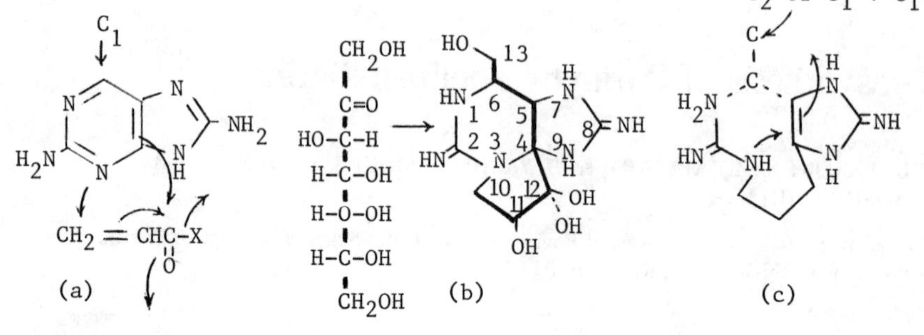

Feeding Experiments using ^{14}C- and ^{13}C-Labeled Precursors (2)

Our first attempt was the feeding of commercially available [guanido-^{14}C]-L-arginine to the cultures of Gonyaulax tamarensis (Ipswich strain). The toxin fraction was isolated and further fractionated to the pure toxins (7). Figure 1 shows an example of the elution pattern of the toxins from a Bio-Rex 70 column. A good correlation between the toxicity and radioactivity was observed. The major toxin, gonyautoxin-III was degraded to locate the radioactivity in the carbamoyl moiety. Thus gonyautoxin-III was converted to saxitoxin by treating with zinc and hydrochloric acid (8), which was then hydrolyzed with 6.7 N HCl to decarbamoylsaxitoxin and carbon dioxide (9). About one-third of the total activity (28%) was found to be associated with the released carbon dioxide and the rest with decarbamoylsaxitoxin (Scheme 2). The result seems to be in

Scheme 2

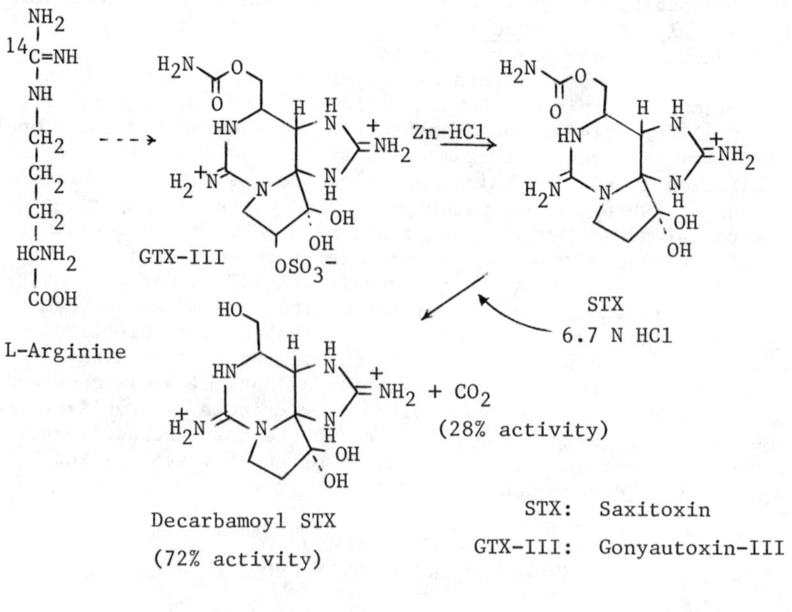

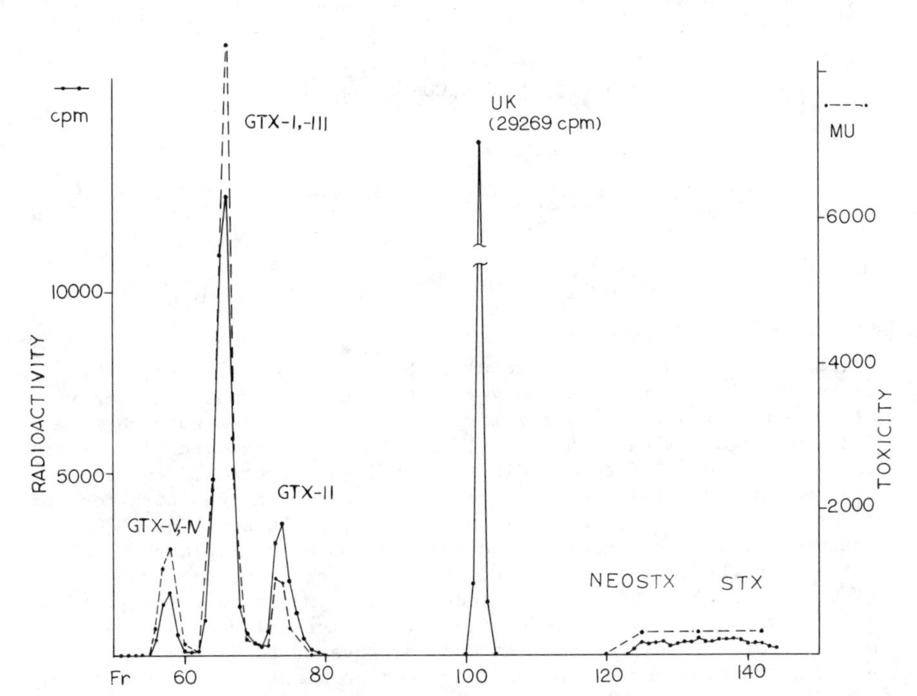

Figure 1. Elution pattern of toxins from Bio-Rex 70 column in ^{14}C-labeled arginine feeding. (Fr.: fraction number; cpm: counts per minute; and MU: mouse units)

agreement with the general understanding that carbamate groups are derived from the ornithine-urea cycle via carbamoyl phosphate.

The next experiment was the feeding of [1-^{13}C]-d,l-arginine, which was synthesized by the following route. Both gonyautoxin-II and neosaxitoxin were isolated from the culture, and their ^{13}C-nmr

Scheme 3

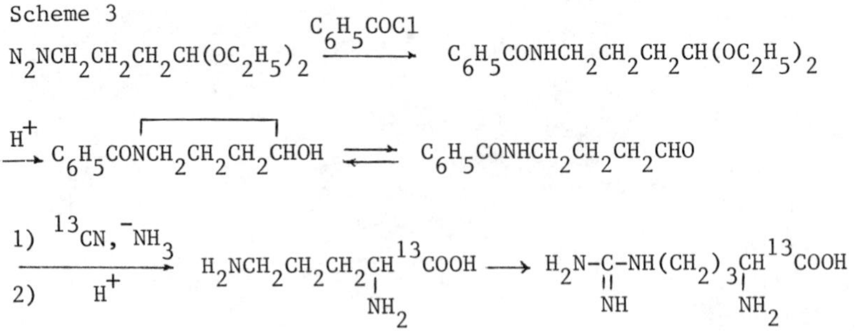

$$N_2NCH_2CH_2CH_2CH(OC_2H_5)_2 \xrightarrow{C_6H_5COCl} C_6H_5CONHCH_2CH_2CH_2CH(OC_2H_5)_2$$

$$\xrightarrow{H^+} C_6H_5\overline{CONCH_2CH_2CH_2CH}OH \rightleftharpoons C_6H_5CONHCH_2CH_2CH_2CHO$$

$$\xrightarrow[2)\ H^+]{1)\ {}^{13}CN,\ {}^-NH_3} H_2NCH_2CH_2CH_2\underset{NH_2}{\underset{|}{CH}}{}^{13}COOH \longrightarrow H_2N\text{-}\underset{NH}{\underset{||}{C}}\text{-}NH(CH_2)_3\underset{NH_2}{\underset{|}{CH}}{}^{13}COOH$$

d,l-ornithine d,l-arginine

spectra were examined. The results, however, indicated that the carbon-13 was incorporated only randomly in both compounds. In view of an apparent difficulty for large organic molecules to penetrate into the organism and reach particular biosynthetic sites in intact forms, we have tried the feeding of small basic metabolic units. The feedings of [1-^{13}C]-acetate resulted in random incorporations of ^{13}C atoms. Similarly, the feeding of doubly labeled [1,2-^{13}C]-acetate afforded the toxin molecules whose C-C spin-spin coupling pattern in the cmr spectra showed no specific groups pattern (Table I and II). Feeding experiments were also tried using [1-^{13}C]- and [2-^{13}C]-glycine, both of which were commercially available. The main purpose of this experiment was to prove or to disprove a possible involvement of the ordinary purine metabolism in the toxin biosynthesis (Scheme 1a). In such a scheme, C-4, C-5, and N-7 of the toxin should be derived from a glycine molecule. Disappointingly the feedings also resulted in general enrichment of all the carbons in both neosaxitoxin and gonyautoxin-II. However, in one feeding experiment with [2-^{13}C]-glycine, we observed extra-enrichment of C-11 and C-12 in gonyautoxin-II, while all the other carbons were also enriched (about 10 times that of the natural abundance) (Scheme 4). This unexpected enrichment of two neighboring carbons from a single-labeled precursor can be explained by assuming that labeled glycine was introduced into TCA cycle via malate. In the cycle, the molecular asymmetricity will be lost at the succinate step, and the labeling will appear on both C-2 and C-3 of succinate. The result seems to support Scheme 1c, since C-4 and C-3 of glutamate correspond to C-11 and C-12 of the toxin molecule in such a scheme.

Despite efforts to pinpoint the molecular origin of PSP, the results have been far from conclusive. The major obstacle is the reluctance of the organism to utilize exogenous organic compounds. Generally phototrophic dinoflagellates such as _Gonyaulax tamarensis_ are known to be very selective in utilization of organic compounds,

Table I. Incorporation ratios of precursors into the toxins and gonyautoxin-III (GTX-III) produced by *Gonyaulax tamarensis*

Feeding Experiments	Precursors (radioactivity in dpm*)	Amounts of toxins isolated in mouse units (mu) and radioactivity: Total Toxins	GTX-III
1 (CO_2 meager)	guanido-^{14}C-L-arginine (2.2×10^7 dpm)	50,000 mu (2.4×10^5 dpm, 1.1%)	21,200 mu (5.3×10^4 dpm, 0.24%)
2 (CO_2 sufficient)	guanido-^{14}C-L-arginine (3.3×10^7 dpm)	21,900 mu (6.1×10^4 dpm, 0.19%)	6,940 mu (9.9×10^3 dpm, 0.09%)
3 (CO_2 meager)	1-^{14}C-glycine (2.2×10^6 dpm)	35,600 mu (1.2×10^5 dpm, 0.56%)	15,500 mu (4.8×10^4 dpm, 0.2%)
4 (CO_2 meager)	2-^{13}C-glycine	71,250 mu (0.5%**)	43,500 mu (2.2 atom excess %**)

*dpm: disintegration per minute

**calculated from the radioactivity of co-fed ^{14}C-glycine.

Table II. Results of feedings of various precursors to _G. tamarensis_ culture

Precursor	Amount of Crude Toxins Obtained	Toxin isolated and amount	Comments
	(in mouse units (mu) and radioactivity (dpm))		
Acetate-1,2-^{13}C	75,000 mu	GTX-II, 12,000 mu	^{13}C-NMR showed no specific incorporation
Ornithine-1-^{13}C	56,000 mu	GTX-II, 11,570 mu	^{13}C-NMR showed no specific incorporation
Arginine-1-^{13}C	35,000 mu	GTX-II, 12,000 mu	0.27% incorporation, no specific incorporation from ^{13}C-NMR
Arginine-1,2,3,4,5-^{14}C (1.45 x 10^7 dpm)	48,000 mu 47,143 dpm	GTX-III, 8,285 dpm	7.9% of RA in CO_2. 92.1% in decarbamoylSTX ratio (1:12.7)
Histidine-C^{14}-(U) (1.1 x 10^7 dpm)	46,000 mu 122,857 dpm	GTX-VIII, 8,357 dpm	1.7% of RA in CO_2. 98.3% in decarbamoylSTX ratio (1:57.8)

Scheme 4

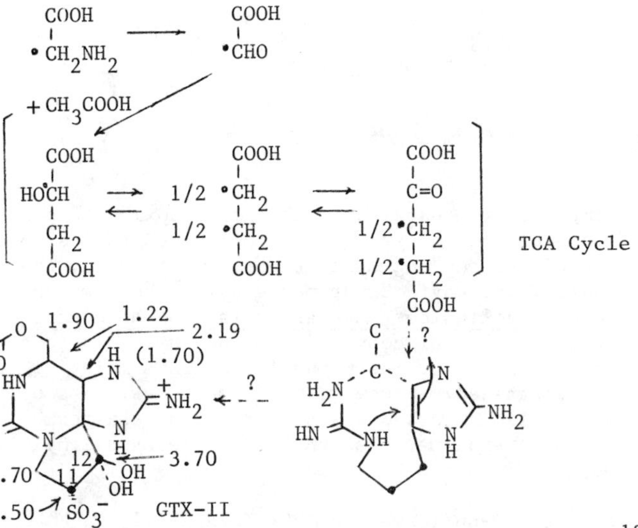

^{13}C-Enrichment pattern of gonyautoxin-II (GTX-II) from the [2-^{13}C]-glycine fed Gonyaulax tamarensis culture, and a possible pathway leading to the enrichment at C-11 and C-12. The numbers depict atom excess % calculated from the relative peak intensities in the spectra of the enriched and unenriched samples based on an average enrichment of 2.2 atom excess %.

if they accept them at all. The second problem may be the compartmentalization of biochemical reactions involved in the formation of the toxins. Under such a circumstance, a precursor, if exogenously fed, reaches the biosynthetic site only after cleaved to basic fragments. The third problem is a technical one. Unlike other microorganisms, the growth of dinoflagellates is very slow, and the maximum attainable population is also low. As a result, the feeding experiment had to be carried out typically in 50 L of culture media to obtain a few milligrams of pure toxins. Although sterile techniques were used throughout the experiment, there was a possibility that small amounts of precursors added to the vast amounts of culture media were quickly metabolized by contaminated or symbiotic bacteria before the dinoflagellate utilized them. Attempts to concentrate the organisms before feeding the precursors or to use the cell homogenate have so far given negative results.

Biochemical Conversion of PSP

When the homogenates of toxic scallops were incubated, other drastic changes in the toxin profile were observed (10); proportionally gonyautoxin-I - IV and neosaxitoxin decreased and saxitoxin increased. In another instance, the analysis of Mytilus exposed to the 1980 red tide at Sonoma County, California, showed the almost exclusive presence of neosaxitoxin in mussels collected just after the red tide and a gradual increase of saxitoxin (Krueger, Meyer and Shimizu, unpublished). These observations suggested the possible

Scheme 5

^{14}C-Labeled neoSTX → ^{14}C-Labeled STX

^{14}C-Labeled GTX-II → ^{14}C-Labeled STX

In Scallop and Clam Homogenates

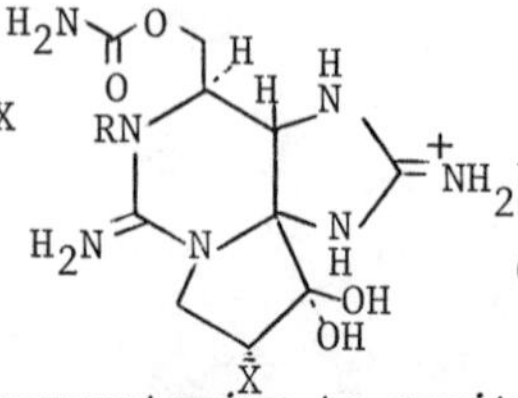

STX: R=H,X=H
neoSTX: R=OH,X=H
GTX-II: R=H,X=OSO_3^-

transformation of neosaxitoxin and gonyautoxins to saxitoxin in the shellfish bodies. The available ^{14}C-labeled toxins obtained by ^{14}C labeled arginine feeding made it possible to study the conversion in a more definitive way. Experiments using the homogenates of Placopecten magellanicus, Mya arenaria, and Mercenaria mercenaria showed the definite bioconversion of neosaxitoxin to saxitoxin. The conversion of gonyautoxins with 11-O-sulfate moieties to saxitoxin was also recognized. However, the same experiments using boiled homogenates also afforded a small amount of saxitoxin, indicating that the conversion may also undergo non-enzymatically or microbially by contaminated bacteria.

Biosynthetic ^{15}N-Enrichment of PSP Toxins (11)

The ^{15}N-enrichment of PSP toxins is important to obtain the ^{15}N-nmr data of the toxins. Such data will be useful in future biosynthetic studies using ^{15}N-labeled precursors. Typically only a few milligrams of individual purified toxins can be obtained out of 50 L of culture medium, and it is simply prohibitive to obtain enough of the toxin for the natural abundance ^{15}N-nmr spectroscopy. The purpose of this experiment was to check the feasibility of biosynthesizing highly enriched toxins by giving this photosynthetic organism an inorganic ^{15}N-precursor.

Gonyaulax tamarensis was grown in enriched sea water. The full-strength Guillard medium (77 L) added with $Na^{15}NO_3$ (99% enrichment, 0.1 g/L). After 33 days, the cells (2 x 10^9) were harvested and processed according to the previously reported method (6). Chromatography of the crude toxic fraction on Bio-Rex 70 afforded gonyautoxin-II (14 mg) and neosaxitoxin (6 mg). For ^{15}N-nmr measurements, the samples were further purified by Bio-Rex 70 and Chelex 100 resin. The spectra were obtained in two modes: proton-decoupled with NOE and proton-coupled with NOE. The observed chemical shifts, coupling constants and parameters used were summarized in Table III. These results are in good agreement with the fully protonated forms of the proposed structures in both compounds excluding alternative structures such as one having a hydroxyl group on N^2. The experiment also opened a way to use ^{15}N-nmr for structure study and biosynthetic work of the toxins even if they are produced only in minute quantities.

Table III. Nitrogen-15 chemical shifts of Neosaxitoxin and Gonyautoxin-II

Position	Neosaxitoxin[a]			Gonyautoxin-II[b]		
	ppm[c]	$^{1}J_{NH}$,Hz	$^{1}J_{NC}$[d]	ppm[c]	$^{1}J_{NH}$,Hz	$^{1}J_{NC}$[d]
N-1	140.1 (s)			80.4 (d)	95.5	$^{1}J_{NC(6)}=7.8$
N^{2}	88.7 (t)[e]	94.2		87.0 (t)[e]	92.8	
N-3	99.0 (s)		$^{1}J_{NC(10)}=6.4$	96.4 (s)		$^{1}J_{NC(10)}=8.4$
N-7	101.0 (d)[f]	90.7		101.0 (d)[f]	99.6	$^{1}J_{NC(5)}=9.4$
N^{8}	70.7 (t)[e]	93.1		70.4 (t)[e]	93.2	
N-9	87.4 (d)[f]	94.7		86.2 (d)[f]	99.1	
N^{14}	76.8 (t)[e]	91.1		76.6 (t)[e]	90.9	

a) Solute concentration 1.8 mM, 5% D_2O/H_2O, pH 4.0

b) Solute concentration 3.3 mM, 5% D_2O/H_2O, pH 4.0

c) (s): singlet; (d): doublet; and (t): triplet

d) Obtained from Carbon-13 NMR Spectra

e) Assignment may be exchangeable

f) Assignment may be exchangeable

Acknowledgments

The work was supported by USPHS grants GM24425 and GM28754.

Literature Cited

1. Taylor, D. L.; Seliger, H. H. "Toxic Dinoflagellate Blooms"; Elsevier/North Holland: New York, 1979.
2. Shimizu, Y. Pure and Appl. Chem. 1982, 54, 1973.
3. Shimizu, Y. "Paralytic Shellfish Poisons"; In "Progress in the Chemistry of Organic Natural Products"; Herz, W.; Griesbach, H.; Kirby, G. W., Eds.; Springer-Verlag: Wien-New York, 1983, in press.
4. Proctor, N. H.; Chan, S. L.; Travor, A. J. Toxicon 1975, 13, 1.
5. Sharma, G.; Magdoff-Fairchild, B. J. Org. Chem. 1977, 42, 4118.
6. Foley, L. H.; Büchi, G. J. Am. Chem. Soc. 1982, 104, 1776.
7. Oshima, Y.; Buckley, L. J.; Alam, M.; Shimizu, Y. Comp. Biochem. Physiol. 1977, 57c, 31.
8. Shimizu, Y.; Hsu, C. P. Chem. Commun. 1981, 314.
9. Ghazarossian, V. E.; Schantz, E. J.; Schnoes, H. K.; Strong, F. M. Biochem. Biophys. Res. Comm. 1976, 68, 776.
10. Shimizu, Y.; Yoshioka, M. Science 1981, 212, 546.
11. Hori, A.; Shimizu, Y. Chem. Commun. 1983, 790-792.

RECEIVED February 6, 1984

Paralytic Shellfish Toxins in Tropical Waters

YASUKATSU OSHIMA[1], YUICHI KOTAKI[2], TAKAKO HARADA[1], and TAKESHI YASUMOTO[1]

[1] Department of Food Chemistry, Faculty of Agriculture, Tohoku University, Tsustsumi-dori, Sendai 980, Japan
[2] Shokei Women's Junior College, Hachiman, Sendai 980, Japan

The dinoflagellate *Pyrodinium bahamense* var. *compressa* and bivalves collected at Palau contained saxitoxin, neosaxitoxin, gonyautoxins V and VI and an unidentified toxin code-named PBT. Chemical structures of gonyautoxins V, VI and PBT were confirmed to be carbamoyl-N-sulfosaxitoxin, carbamoyl-N-sulfoneosaxitoxin and decarbamoylsaxiton, respectively. Occurrence of paralytic shellfish toxins was also evidenced in ten species of crabs belonging to four different families, two turban shells and two top shells collected at Ishigaki Island, Japan. Analyses of the representative species confirmed the presence of saxitoxin, neosaxitoxin, decarbamoylsaxitoxin, gonyautoxins I-III and a new toxin code-named TST. A calcareous red alga *Jania* sp. was proved to produce gonyautoxins I-III and was assigned as the primary source of the toxins in the crabs and gastropods.

Paralytic shellfish toxins in the dinoflagellate *Protogonyaulax* (=*Gonyaulax*) spp. and bivalves of temperate waters have been the subjects of extensive studies. In contrast, information on the occurrence of these toxins in tropical waters has been scarce. Maclean reported the occurrence of poisonings resembling paralytic shellfish poisoning in Papua New Guinea and Borneo and associated the incidence with the concurrent red tide of the dinoflagellate *Pyrodinium bahamense* (1), which was later amended to *Pyrodinium bahamense* var. *compressa* (2). However, no chemical evidence has been presented to prove the occurrence of paralytic shellfish toxins in tropical bivalves until Kamiya and Hashimoto detected saxitoxin (STX) in Palauan bivalves (3). STX was also found in animals of entirely different feeding habit and habitat. Hashimoto and his colleagues found that three xanthid crabs, *Zosimus aeneus*, *Atergatis floridus* and *Platipodia granulosa*, were highly poisonous in certain areas and identified STX in *Z. aeneus* (4-6). STX was also detected in a green turban shell by Yasumoto and Kotaki (7). However, details of toxin composition and the primary source of toxin were left unsolved.

0097-6156/84/0262-0161$06.00/0
© 1984 American Chemical Society

In this paper we summarize our recent findings on paralytic shellfish toxins in tropical waters (8-15) with new assignment of a component previously unreported. It includes the confirmation of paralytic shellfish toxins in the dinoflagellate *Pyrodinium bahamense* var. *compressa* and bivalves exposed to the organism (8), with structural elucidation of three components (9,10). The detailed analyses of the toxin composition of crabs (11-13) and marine snails (13,14) and confirmation of a calcareous red alga *Jania* sp. as the primary source of the toxins (15) are also described.

Materials and Methods

Materials. The dinoflagellate *Pyrodinium bahamense* var. *compressa;* bivalves *Spondylus butleri, Tridacna crocea, Lopha cristagalli, Saxostrea mordax, Modiolus* sp. and *Barbatia* sp.; and a top shell *Tectus* sp. were collected at Palau, Western Calorine Islands, in 1980 and 1981. Eleven species of crab were collected at Kabira reef, Ishigaki Island, Japan, over the period from 1980 to 1982: family Xanthidae *Zosimus aeneus, Atergatis floridus, Platipodia granulosa, Pilmnus vespertilio, Actaea polyacantha, Neoxanthias impressus, Eriphia scabricula, Actaeodes tomentosus;* family Grapsidae *Percmon planissimum;* family Majidae *Schizophrys aspera;* and family Portunidae *Thalamita* sp. Turban shells *Turbo marmorata* and *Turbo argyrostoma* and top shells *Tectus pyramis* and *Tectus nilotica maxima* were collected at Shiraho, Ishigaki Island, in 1980. Eighteen species of alga were collected at Kabira in April and May, 1981: unidentified five species of Cyanophyta; Chlorophyta *Halimeda opuntia* and *Boodles* sp.; Phaeophyta *Sargassum* sp. and *Dictiota* sp.; Rhodophyta *Jania* sp.-1, *Jania* sp.-2, *Gelidiella acerosa, Eucheuma serra, Centrocearas clavulatum, Leveillea jungermunnioides, Hypnea* sp., *Laurencia* sp. and *Ceramium* sp. To examine seasonal variation in toxicity additional samples of *Jania* sp.-1 were collected at the same place in August and December, 1981 and February, 1982. The *Jania* specimens were shaken vigorously in sea water immediately after collection to be freed of sands and other particles. Contaminating algae were then eliminated by forceps to ensure the homogeneity of the specimens.

Bioassay. Toxicity of the materials was measured by the standard mouse bioassay for paralytic shellfish toxins and expressed by mouse unit (MU) as defined by the method (16). For testing the low toxin levels of algal specimens, extracts were treated with a charcoal column prior to injection into mice.

Analysis of toxins. The analytical methods were essentially the same as were used for the toxins of *Protogonyaulax tamarensis* (17). Toxins were extracted with 0.1 N HCl or 75% EtOH acidified to pH 2 and treated with successive columns of charcoal, Bio-Gel P-2 and Bio-Rex 70. Toxins separated by the last column were identified by tlc and electrophoresis. Relative abundance of each toxin was determined by monitoring the eluate from Bio-Rex 70 column by mouse assay. A fluorometric paralytic shellfish toxin analyzer was applied to samples which were too small to be followed by mouse assay. Toxins separated by the ion exchange column (Hitachi gel 3011C) were continuously aromatized by *tert*-butylhydroperoxide and monitored by the fluorometer (18).

NMR spectra. ^{13}C NMR and ^{1}H NMR spectra of toxins were taken with JEOL FX-100 and FX-400 spectrometers in D_2O. Chemical shifts are expressed in ppm downfield from TMS with the use of dioxane (61.4 ppm) and *tert*-butanol (1.23 ppm) as internal standards, respectively, for the ^{13}C and ^{1}H NMR analyses.

Results

Toxicity of the organisms. Dinoflagellate *Pyrodinium bahamense* var. *compressa* was proliferating in Arumizu Bay, Koror Island, 450 cells/ml at the maximum count. *Pyrodinium* comprised more than 97% in the plankton samples collected by net. Toxicity levels of plankton specimens and molluscs in the same area are shown in Table I. Toxin productivity of *Pyrodinium* was comparable to that reported for *Protogonyaulax tamarensis* from Ofunato Bay, Japan (19). Consistent with the distribution of the dinoflagellate, all the shellfishes were highly contaminated by the toxins as shown by an extraordinarily high toxic scores of 1100 MU/g in a specimen of *Spondylus butleri* (8).

Table I. Toxicity of the Dinoflagellate and Molluscs in Palau

Specimens	Toxicities*
Pyrodinium bahamense var. *compressa* **	1.5×10^{-4}
Pyrodinium bahamense var. *compressa*	1.4×10^{-4}
Spondylus butleri	1100
Tridacna crocea	96
Lopha cristagalli	130
Saxostrea mordax	210
Modiolus sp.	160
Barbatia sp.	140
*Septifer bilocularis***	48
Tectus sp.	5.3

* Dinoflagellate: MU/cell, shellfish: MU/g.

** Collected in December, 1980. All other specimens were collected in May, 1981.

Toxic levels of crabs and marine snails are summarized in Table II. Toxins were detected in ten species of crab out of eleven tested. *Z. aeneus*, *A. floridus* and *P. granulosa* were significantly more toxic than others, though individual toxicity varied greatly from 3.6 to 660 MU/g. Considerably high toxicity (80-130 MU/g) was also found in *A. tomentosus*, *E. sebana* and *Thalamita* sp. Toxic scores of other four species were less than 10 MU/g (12). The turban shells and top shells contained toxins in varied degree up to 20 MU/g (14).

Among algal specimens two calcareous algae, *H. opuntia* and *Jania* sp. tentatively coded as type-1, were toxic to mice. However, the

Table II. Toxicity of Crabs and Marine Snails

Specimens	Ratio of toxic specimens*	Toxicities**
Crabs		
Zosimus aeneus	8/8	660
Platipodia granulosa	1/1	110
Atergatis floridus	5/5	490
Neoxanthias impressus	2/3	10
Actaeodes tomentosus	3/5	130
Eriphia scabricula	4/8	180
Pilmnus vespertilio	1/2	6.1
Actaea polyacantha	1/1	ND
Schizophrys aspera	1/1	2.3
Thalamita	4/5	80
Percmon planissimum	2/2	7.4
Gastropods		
Turbo marmorata	2/3	4.2
Turbo argyrostoma	2/4	20
Tectus pyramis	1/1	19
Tectus nilotica maxima	2/2	5.0

* Number of toxic specimens against those tested.
** The maximum toxicity expressed by MU/g. ND refers below 2.0 MU/g.

toxin in *H. opuntia* was judged to differ from paralytic shellfish toxins on the basis of the symptoms in mice and chromatographic properties. On the other hand, toxic components of *Jania* sp.-1 were indistinguishable from reference toxins in both symptomatology and chromatographic behaviors. Increased toxicity of *Jania* specimens after elimination of contaminants indicated that the toxins were genuine products of *Jania* sp.-1. Seasonal observation of toxicity of *Jania* sp.-1 revealed low toxicity through August to December (0.04-0.17 MU/g), moderate in February (0.13-0.69 MU/g) and the highest in April and May (1.3-1.5 MU/g). Presence of this alga was confirmed in the stomach of two crabs *Z. aeneus* and *A. floridus*, and four toxic gastropods (15).

Toxins in the organisms. The compositions and relative abundance of toxins in the organisms tested are shown in Table III. The similar data for cultured *Protogonyaulax tamarensis* are also included in the table as comparison (17). *Pyrodinum* contained STX and neosaxitoxin (neoSTX), gonyautoxins V and VI ($GTX_{5,6}$) and toxin coded PBT, which was later identified as decarbamoylsaxitoxin (dcSTX). Unlike *Protogonyaulax*, the dinoflagellate lacked 11-O-sulfate derivatives such as gonyautoxins I-IV (GTX_{1-4}). In the bivalves, relative ratio of neoSTX, GTX_5 and GTX_6 were lower than in the causative dinoflagellate, indicating bioconversion or preferential accumulation of toxins.

The characteristic feature of the toxin profiles of crabs is the predominance of neoSTX, STX in contrast to the small amount of dcSTX and GTX_{1-3} (11-13). Toxin compositions of gastropods were characterized by the abundance of STX and a new toxin code-named TST (13, 14).

Investigation of toxins in *Jania* sp.-1 by the paralytic shell-

Table III. Composition of Paralytic Shellfish Toxins in Organisms of Tropical Waters

Organisms	Locality	GTX_1	GTX_2	GTX_3	GTX_4	GTX_5	GTX_6	neoSTX	STX	dcSTX	TST
Dinoflagellate											
Pyrodinium bahamense var. *compressa*	Palau	–	–	–	–	++	+	++++	+++	++	–
Protogonyaulax tamarensis	Ofunato	++++	+++	+	++	·	–	++	–	–	–
Pelecypods											
Spondylus butleri	Palau	–	–	–	–	+	–	++	++++	+++	–
Tridacna crocea	Palau	–	–	–	–	+	–	+	++++	++	–
Septifer bilocularis	Palau	–	–	+	–	–	–	+	+++	+	–
Decapods											
Zosimus aeneus	Ishigaki	+	+	–	–	–	–	++++	++	+	–
Atergatis floridus	Ishigaki	–	+	–	–	–	–	+++	+++	+	–
Platipodia granulosa	Ishigaki	–	–	–	–	–	–	–	++++	+	–
Eriphia scabricula	Ishigaki	+	+	+	–	–	–	+++	+++	–	–
Pilumnus vespertilio	Ishigaki	+	+	+	–	–	–	++	++++	–	–
Thalamita sp.	Ishigaki	+	+	+	–	–	–	++	+++	–	–
Gastropods											
Turbo marmorata	Ishigaki	–	+	–	–	–	–	+	+++	+	++++
Turbo argyrostoma	Ishigaki	–	+	+	–	–	–	+	+++	+	+++
Tectus pyramis	Ishigaki	–	+	–	–	–	–	+	+++	+	+++
Rhodophyta											
Jania sp.-1	Ishigaki	++++	+++	+	–	–	–	–	–	–	–

fish toxin analyzer confirmed the presence of GTX_1, GTX_2 and GTX_3 in the ratio 73:31:6 and the absence of STX and neoSTX (15).

Structures of GTX_5, GTX_6 and PBT. The chemical structures of GTX_5 and GTX_6 have not been known despite the frequent occurrence in organisms from cold waters (20). The specific activities of GTX_5 and GTX_6 were 136 and 108 MU/μmol, corresponding to 1/15 of STX and neo-STX, respectively. Upon hydrolysis in mild condition, they were converted to STX and neoSTX by releasing 1 mol of sulfate. As shown in Table IV, NMR spectra of GTX_5 and GTX_6 were essentially the same as those of STX and neoSTX except for the slight difference in chemical shifts for 13-H protons or C-13 and C-14 carbons. This confirms that the conjugation site for the sulfate in both toxins is the carbamoyl nitrogen. Thus GTX_5 and GTX_6 were proved to carbamoyl-N-sulfosaxitoxin and carbamoyl-N-sulfoneosaxitoxin, respectively (Fig. 1)(9). They correspond to the toxins coded B1 and B2 isolated from *Protogonyaulax* by Koehn *et al.* (21).

PBT was proved to be decarbamoylsaxitoxin by the lack of C-14 signal in ^{13}C NMR spectra as well as by the complete agreement of all spectral and chromatographic properties with dcSTX prepared from STX by acid hydrolysis. Production of STX by carbamoylation with chlorosulfonylisocyanate also evidenced the identity between PBT and dcSTX (10).

Discussion

Present work provides a firm evidence that shellfish poisoning involving toxins of STX family occurs in the tropical areas, and *Pyrodinium bahamense* var. *compressa* is the toxin progenitor. Wide distribution of this species in areas such as Papua New Guinea, Brunei, Sabah and Palau and the actual occurrence of poisoning in Fiji (22) and India (23) indicate that the real and potential threat of paralytic shellfish poisoning in tropical waters is as great as in the northern waters.

Multiplicity of the toxin composition was observed in all specimens tested. The predominance of strongly basic toxins over 11-O-sulfated toxins is also a prominent feature common to all tropical specimens, except for *Jania* sp.-1. The wide distribution of dcSTX is also a distinction of toxin composition of tropical specimens.

The present study is the first to demonstrate the production of paralytic shellfish toxins by a macroalga *Jania* sp. and transmission of toxins to crabs and marine snails through food chain. It is debatable whether *Jania* sp. is the sole source of the toxins found in crabs and gastropods. However, the possibility of the presence of a planktonic toxic dinoflagellate such as *Pyrodinium* sp. or *Protogonyaulax* sp. is ruled out because plankton-feeding bivalves collected from the same area were found to be nontoxic (11). Our extensive survey for epibenthic dinoflagellate led to a discovery of a number of toxic species. However none of them produced paralytic shellfish toxins (24,25). Blue-green algae were throughly collected and tested for the toxins because a fresh water species, *Aphanizomenon fros-aquae*, is known to produce paralytic shellfish toxins (26). Yet none of them contained toxins. All these results support our conclusion

Table IV. ^{13}C NMR and ^{1}H NMR Chemical Shifts of STX, GTX_5, neoSTX, GTX_6 and PBT (dcSTX)

No.	STX		GTX_5		neoSTX	GTX_6	PBT (dcSTX)	
C,H	^{13}C NMR	^{1}H NMR	^{13}C NMR	^{1}H NMR	^{1}H NMR	^{1}H NMR	^{13}C NMR	^{1}H NMR
14	159.1 (s)		154.1 (s)				-	
8	158.1 (s)		158.1 (s)				158.0 (s)	
2	156.2 (s)		156.1 (s)				156.1 (s)	
4	82.6 (s)		82.6 (s)				82.5 (s)	
5	57.1 (d)	4.71 (1)	57.2 (d)	4.76 (1)	4.83 (s)	4.83 (s)	56.5 (d)	4.61 (1)
6	53.2 (d)	3.90 (9,5,1)	53.0 (d)	3.88 (9,5,1)	4.13 (6,6)	4.13 (6,6)	55.7 (d)	3.63 (m)
10	42.8 (t)	3.81 (10)	42.9 (t)	3.81 (10)	3.78 (10)	3.78 (10)	42.9 (t)	3.77 (10)
		3.61 (10)		3.61 (10)	3.58 (10)	3.58 (10)		3.54 (10)
11	33.0 (t)	2.41 (m)	*	*	*	*	32.9 (t)	2.39 (m)
12	98.6 (s)		98.6 (s)				98.7 (s)	
13	63.3 (t)	4.07 (11,5)	64.0 (t)	4.12 (12,5)	4.21 (11,6)	4.26 (11,6)	61.4 (t)	3.63 (m)
		4.33 (11,9)		4.42 (12,9)	4.44 (11,6)	4.52 (11,6)		3.63 (m)

* Unobserved due to deuterium replacement at C-11.

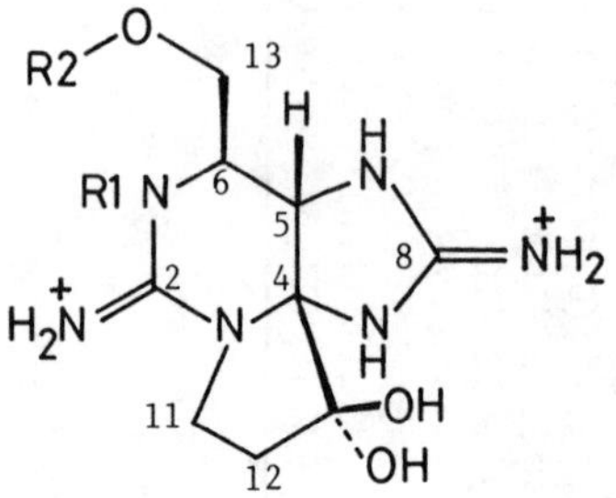

	R1	R2
STX :	-H	$-CONH_2$
neoSTX :	-OH	$-CONH_2$
GTX_5 :	-H	$-CONHSO_3^-$
GTX_6 :	-OH	$-CONHSO_3^-$
dcSTX :	-H	-H

Fig. 1. Structures of toxins in *Pyrodinium bahamense* var. *compressa*.

that on coral reefs, *Jania* sp. is the main source of paralytic shellfish toxins found in crabs and gastropods. Since *Jania* sp.-1 shows spotty distribution, the individual and regional variation observed in the toxicity of animals might be the reflection of the abundance of *Jania* sp. in their habitat. Different preference of crabs and gastropods to the alga might also cause the variation. The discrepancy in the toxin composition between *Jania* sp.-1 and crabs and marine snails suggssts that a reductive cleavage of 11-O-sulfate of GTX_{1-3} takes place in the animals. In preliminary tests, incubation of GTX_{1-4} with the homogenate of the digestive organs of crabs and gastropods indicated the conversion of these toxin to STX. Since the conversion was deterred by the addition of bacteriostatic substances, bacterial role in the conversion of the toxins was suggested. Further investigation is under way to elucidate the mechanism of conversion.

Acknowledgments

Present study was supported by a Grant-in-Aid from the Toyota Foundation.

Literatures Cited

1. Maclean, J. L. In "Toxic Dinoflagellate Blooms", Taylor, D. L.; Seliger, H. H., Eds.; Elsevier North-Holland: New York, 1979; pp. 173-178.
2. Steidinger, K. A.; Tester, L. S.; Taylor, F. J. R. *Phycologia* 1980, 19, 329-37.
3. Kamiya, H.; Hashimoto, Y. *Toxicon* 1978, 16, 303-6.
4. Hashimoto, Y.; Konosu, S.; Yasumoto, T.; Inoue, A.; Noguchi, T. *Toxicon* 1976, 5, 85-90.
5. Konosu, S.; Noguchi, T.; Hashimoto, Y. *Bull. Jpn. Soc. Sci. Fish.* 1970, 36, 715-9.
6. Noguchi, T.; Konosu, S.; Hashimoto, Y. *Toxicon* 1969, 7, 325-6.
7. Yasumoto, T.; Kotaki, Y. *Bull. Jpn. Soc. Sci. Fish.* 1977, 43, 207-11.
8. Harada, T.; Oshima, Y.; Kamiya, H.; Yasumoto, T. *Bull. Jpn. Soc. Sci. Fish.* 1982, 48, 821-5.
9. Harada, T.; Oshima, Y.; Yasumoto, T. *Agric. Biol. Chem.* 1982, 46, 1861-4.
10. Harada, T.; Oshima, Y.; Yasumoto, T. *Agric. Biol. Chem.* 1983, 47, 191-3.
11. Yasumoto, T.; Oshima, Y.; Konta, T. *Bull. Jpn. Soc. Sci. Fish.* 1981, 47, 957-9.
12. Yasumoto, T.; Oshima, Y.; Tajiri, M.; Kotaki, Y. *Bull. Jpn. Soc. Sci. Fish.* 1983, 49, 633-6.
13. Yasumoto, T.; Oshima, Y.; Kotaki, Y. *Toxicon* 1983, Suppl. 3, 513-6.
14. Kotaki, Y.; Oshima, Y.; Yasumoto, T. *Bull. Jpn. Soc. Sci. Fish.* 1981, 47, 957-9.
15. Kotaki, Y.; Tajiri, M.; Oshima, Y.; Yasumoto, T. *Bull. Jpn. Soc. Sci. Fish.* 1983, 49, 283-6.
16. Horwitz, W., Ed., In "Official Methods of Analysis of Association of Official Analytical Chemists" 13th Ed.; A. O. A. C.; Washington, D. C., 1980, 298-9.

17. Oshima, Y.; Hayakawa, T.; Hashimoto, M.; Kotaki, Y.; Yasumoto, T. *Bull. Jpn. Soc. Sci. Fish.* 1982, 48, 851-4.
18. Oshima, Y.; Machida, M.; Sasaki, K.; Tamaoki, Y.; Yasumoto, T. Submitted to *Agric. Biol. Chem.*
19. Kodama, M.; Fukuyo, Y.; Ogata, T.; Igarashi, T.; Kamiya, H.; Matsuura, F. *Bull. Jpn. Soc. Sci. Fish.* 1982, 48, 657-71.
20. Shimizu, Y. In "Toxic Dinoflagellate Blooms"; Taylor, D. L.; Seliger, H. H., Eds.; Elsevier North-Holland; New York, 1979; pp. 321-326.
21. Koehn, F. E.; Hall, S.; Wichmann, C. F.; Schnoes, H. K.; Reichardt, P. B. *Tetrahedron Lett.* 1982, 23, 2247-8.
22. Raj, U. Presentation at the Symposium on Seafood Toxins in Tropical Regions, held at Kagoshima, in September 1981.
23. Bhat, R. V. "Report of Food and Drug Toxicology Research Centre" Hyderabad, India, September 1981.
24. Yasumoto, T.; Oshima, Y.; Nakajima, I.; Bagnis, R.; Fukuyo, Y. *Bull. Jpn. Soc. Sci. Fish.* 1980, 46, 327-31.
25. Nakajima, I.; Oshima, Y.; Yasumoto, T. *Bull. Jpn. Soc. Sci. Fish.* 1981, 47, 1029-33.
26. Alam, M.; Shimizu, Y.; Ikawa, M.; Sasner, J. J. Jr. *J. Environ. Sci. Health* 1978, 13, 439-9.

RECEIVED February 6, 1984

15

Paralytic Shellfish Toxins and Finfish

ALAN W. WHITE

Fisheries and Environmental Sciences, Department of Fisheries and Oceans, Biological Station, St. Andrews, New Brunswick, E0G 2X0 Canada

Paralytic shellfish toxins from dinoflagellates of the genus Gonyaulax have caused recurrent herring kills in eastern Canada and have been implicated in sand lance and menhaden kills in England and the northeastern United States. The kills resulted from toxin transfer through the plankton community. Experiments show that adult and larval stages of various marine fish are sensitive to the toxins. During Gonyaulax blooms zooplankton, molluscan shellfish, and perhaps other components of the food web can become sufficiently toxic to present a danger to fish at several trophic levels. The impact of paralytic shellfish toxins on finfish populations is difficult to assess, but should be considered, especially in view of the trend of intensification and spreading of toxic Gonyaulax and Pyrodinium blooms in many areas of the world. Since finfish, unlike shellfish, are unable to accumulate the toxins in their flesh, there would seem to be little problem in terms of the suitability of fish for human consumption, except possibly in instances where whole fish are consumed without processing.

The impact of paralytic shellfish toxins on the utilization of shellfish resources is widespread and well recognized. The picture is developing that paralytic shellfish toxins also affect finfish resources. Stated simply, paralytic shellfish toxins cause fish kills.

Since 1976, when a herring kill was caused by paralytic shellfish toxins in the Bay of Fundy, our laboratory has been investigating the effects of the toxins on fish and the food web routes through which the toxins reach fish.

Fish Kills Caused by Paralytic Shellfish Toxins

Reports implicating paralytic shellfish toxins in fish kills are summarized in Table I. The first instance in which it was suspected that these toxins were the cause of a fish kill occurred in May 1968

0097-6156/84/0262-0171$06.00/0
© 1984 American Chemical Society

Table I. Fish Kills Associated with Paralytic Shellfish Toxins

Year	Location and Reference	Fish	Toxin Content (μg/100 g) Gut	Toxin Content (μg/100 g) Muscle	Estimated Toxin Dose (μg/kg)
1968	England[1,2]	Sand lance	-	-	-
1976	Canada[3]	Herring	1414	<30	197
1978	U.S.A.[4]	Sand lance	97 (whole fish)	-	970
1979	U.S.A.[5]	Menhaden	185	-	80
1979	Canada[7]	Herring	66-218	<30(33,59)	23-95

off the northeast coast of England (1,2). Sand lance, *Ammodytes* sp., were killed during a toxic bloom of *Gonyaulax tamarensis*. Because sand lance feed exclusively on zooplankton, the suggestion was made that the kill was caused by *G. tamarensis* toxins which had passed through the herbivorous zooplankton community, dominated at the time by the cladoceran *Evadne nordmanni*. Sea bird mortalities were also reported.

In July 1976, an extensive kill of adult Atlantic herring (*Clupea harengus harengus*), involving hundreds of tons of fish, occurred in the Bay of Fundy during the peak of the annual *G. excavata* bloom (3). Paralytic shellfish toxins were detected in the stomach contents of dead and dying fish at the level of 1414 μg per 100 g of material. The only identifiable items in the stomach contents were pteropods in various stages of digestion and degraded algal material which had been ingested by the pteropods. Experiments following the kill confirmed that the amount of toxins measured per fish (21 μg) was sufficient to kill herring rapidly upon oral administration with symptoms similar to those described by fishermen during the kill. The circumstantial evidence left little doubt that the kill was caused by paralytic shellfish toxins and that the toxins had been transmitted to the herring primarily through pteropods.

Another sand lance kill was recorded in June 1978 off Wellfleet (Cape Cod), Massachusetts during a rather unnoticed bloom of *G. excavata* (4). An assay of the fish for paralytic shellfish toxins was positive, 97 μg of toxins per 100 g of whole fish (partly digested). A kill of various sea birds resulted from the ingestion of the contaminated sand lance (4). Further, about 60 large bluefish (*Pomatomus saltatrix*) and a few spiny dogfish (*Squalus acanthias*), skates (*Raja* sp.), and monkfish (*Lophius americanus*) were also killed. Bluefish and dogfish feed on sand lance commonly, presenting the likelihood that the toxins passed from *G. excavata* to zooplankton to sand lance and then to the other fishes.

In July 1979, mortalities of menhaden (*Brevoortia tyrannus*) occurred along the southern coast of Maine during a *G. excavata* bloom (5). Paralytic shellfish toxins were measured in the guts of these fish, up to 185 μg per 100 g of viscera. Since phytoplankton constitute a major part of the diet of menhaden (6), the toxins were probably acquired directly from ingestion of the toxic dinoflagellates.

Also in July 1979, there was a recurrence of herring mortalities in the Bay of Fundy during the annual, summer *G. excavata* bloom (7). Again, paralytic shellfish toxins were detected in gut samples, up to 245 μg per 100 g of material. Stomachs of dead herring contained the cladoceran *Evadne nordmanni* nearly exclusively, which was by far the dominant member of the zooplankton community at the time. The toxins were also detected in zooplankton samples containing *E. nordmanni*, 1800 μg per 100 g wet weight of zooplankton. Symptoms of dying fish were observed during one of the kill episodes and were the same as observed during the 1976 kill.

Fish kills have been reported for the Indo-Pacific region in association with blooms of another dinoflagellate, *Pyrodinium bahamense* (8), which has since been confirmed to contain paralytic shellfish toxins (9). Whether the toxins were involved in the kills apparently remains unknown.

Symptoms of PSP in Fish

During the 1976 and 1979 herring kills in the Bay of Fundy, fishermen reported that fish had lost their equilibrium, swimming either on their sides or upside down, and were gasping at the surface (3,7). Using paralytic shellfish toxins extracted from mass cultures of Bay of Fundy *G. excavata*, we have found the symptoms of poisoning and the time course of their appearance to be similar in Atlantic herring, American pollock, winter flounder, Atlantic salmon, and cod (10). The timing of the appearance of symptoms is the same in these fish regardless of whether the toxins are administered orally or intraperitoneally (Table II). Symptoms become apparent within 5 to 15 minutes of toxin administration and include swimming in an irregular, jerky manner, followed by a loss of equilibrium and swimming on the side or upside down. Generally, fish are unable to swim after 10 to 20 minutes and fall to the tank bottom, breathing in a slow, arrhythmic fashion, with occasional short bursts of wild swimming activity. Death ensues usually within 20 to 60 minutes of toxin administration. In cases in which fish survive the poisoning (some do even after displaying acute symptoms), recovery is complete within several hours with no apparent aftereffects.

Table II. Symptoms of PSP in Fish

Symptoms	Time after Oral or I.P. Administration of Toxins
Irregular, jerky swimming	5 min
Loss of equilibrium	10-15 min
Immobilization, shallow breathing, with occasional hyperactivity	15-30 min
Death	20-60 min

Sensitivity of Fish to Paralytic Shellfish Toxins

Adults. It was previously suspected that fish, like other cold-blooded vertebrates, were relatively insensitive to paralytic shellfish toxins (11). On the contrary, however, we have shown that fish are as sensitive to the toxins as warm-blooded animals (10). The dose responses of herring, pollock, flounder, and salmon to paralytic shellfish toxins extracted from Bay of Fundy G. excavata cultures were determined. The LD50 values are presented in Table III. All of these fishes are sensitive to the toxins and exhibit similar dose responses. Oral LD50 values range from 400 to 755 µg per kg body weight, while intraperitoneal LD50 values range from 4.2 to 12.0 µg per kg body weight. The values for warm-blooded animals are about in these same ranges (12). The results suggest that marine fish in general are sensitive to paralytic shellfish toxins. Japanese puffers of the genus Fugu may, however, be exceptional in this regard because they are able to store saxitoxin (as well as tetrodotoxin) in their livers and gonads (13, 14).

Table III. LD50 Values and 95% Confidence Limits for Paralytic Shellfish Toxins on Marine Fish

Fish	LD50 (µg/kg)	
	Oral	I.P.
Atlantic herring	650 (584, 722)	4.2 (3.4, 5.2)
American pollock	626 (532, 748)	12.0 (8.4, 17.0)
Winter flounder	755 (665, 856)	6.4 (4.9, 8.3)
Atlantic salmon	400 (309, 517)	6.5 (5.4, 7.8)

Reproduced with permission from Ref. 10. Copyright 1981, 'Marine Biology'.

The doses of paralytic shellfish toxins estimated for the fish kills referred to (Table I) are fairly consistent with the oral dose responses determined in the laboratory (10). That most of the estimates of the doses received in the kill episodes are considerably lower than the determined range of oral LD50 values may be accounted for by the fact that the fish assayed from the kills had been subject to decay. The actual, effective doses in these events may well have been higher than measured. Concerning the sand lance kill in 1978 off Cape Cod, Massachusetts, the estimated dose of toxins received by the sand lance (970 µg/kg, 4 and Table I) is somewhat higher than the range of oral LD50 values determined experimentally. High toxin levels in the sand lance may account for the mortalities of other, carnivorous fishes which also occurred in this instance (4). (This particular estimate for sand lance is tenfold greater than reported earlier (10). The present value is correct; the earlier is my error.)

The intraperitoneal LD50 values for fish, as for poikilotherms, are extremely low. This means that fish (with the exception perhaps of Fugu) are unable to tolerate even small amounts of the toxins in their bodies, and hence are unable to accumulate them. This important point will be discussed again below in connection with the effect of the toxins on the use of fish as food.

Larvae. In terms of finfish resources, the most significant aspect of the effects of paralytic shellfish toxins on finfish may well be their effects on larval and juvenile stages. It is an accepted tenet among fisheries scientists that events which occur during the early larval stages are the major determinants of year-class strength. Accordingly, paralytic shellfish toxins may influence the stock size of finfish whose larvae overlap toxic *Gonyaulax* blooms in time and space.

Marine fish larvae are obligate planktivores and thus are liable to acquire paralytic shellfish toxins from the plankton community during *Gonyaulax* blooms, either directly from the dinoflagellates themselves or secondarily from zooplankton. For the most part, the details of the feeding capabilities and preferences of marine fish larvae are unknown. Herring larvae, which have large mouths compared with other larvae, may be exclusively zooplanktivorous. However, other larvae may take *Gonyaulax* (and other phytoplankton) directly, especially during first-feeding stages, as well as feeding on microzooplankton. Regardless of the specific food items selected, any fish larva in the midst of a *Gonyaulax* bloom would likely become exposed to the toxins through feeding.

Recent experiments indicate that marine fish larvae are indeed sensitive to paralytic shellfish toxins (15). First-feeding larvae of Atlantic herring, red sea bream (*Chrysophrys major*), and Japanese anchovy (*Engraulis japonica*) were allowed to feed on cultured Bay of Fundy *G. excavata* cells. Older larvae were fed wild zooplankton which had been fed overnight on *G. excavata* cells and which had accumulated paralytic shellfish toxins. For first-feeding red sea bream larvae feeding on *Gonyaulax*, within just 4 days there was a threefold increase in mortality compared with starved controls. Herring and Japanese anchovy larvae did not feed well on *Gonyaulax* and no effects were seen. For older larvae, within several hours of feeding on zooplankton containing paralytic shellfish toxins there was a 17, 20, and 33% increase in mortality over larvae feeding on non-contaminated zooplankton for herring, red sea bream, and Japanese anchovy larvae, respectively. An increased mortality of first-feeding winter flounder larvae upon exposure to toxic *Gonyaulax* has been reported by Mills and Klein-MacPhee (16).

It is difficult, if not impossible, to assess what these laboratory results signify in terms of the impact of paralytic shellfish toxins and *Gonyaulax* blooms on natural populations of marine fish larvae. The laboratory results show that at least some fish larvae are sensitive to the toxins. Thus, it is suspected that paralytic shellfish toxins play a role as one of the many factors which control the size of fish populations.

Routes of Exposure of Fish to Paralytic Shellfish Toxins

From observations associated with the fish kill events mentioned, coupled with supporting experimental evidence, it is apparent that herbivorous zooplankton are a major biological vector through which paralytic shellfish toxins reach fish (3,7,17,18). It has been demonstrated that a number of herbivorous zooplankters can accumulate (and to some degree retain) the toxins (17,18) and do so rather consistently in nature (Table IV). Further, it appears that the same

Table IV. Maximum Content of Paralytic Shellfish Toxins in Bay of Fundy Zooplankton During *Gonyaulax* Blooms

Year	Toxin Content (μg/g wet zooplankton)	Dominant Zooplankters
1977	60	Podon, Evadne
1978	58	Acartia
1979	18 (maximum?)	Acartia
1980	57	Podon, Evadne
1981	10	Podon, Acartia
1982	66	Acartia, Centropages

complex of toxins present in the dinoflagellates is responsible for killing fish after passage through the zooplankton community (19).

Other components of the marine food web become contaminated with paralytic shellfish toxins, presenting a possible danger to fish. Filter-feeding molluscan shellfish are notorious for their ability to accumulate and store high levels of the toxins and may thus represent a risk to certain groundfish. Carnivorous molluscs can, in turn, acquire the toxins from their filter-feeding relatives. Little is known about whether other organisms such as carnivorous zooplankton, starfish, jellyfish, marine worms, etc. can accumulate the toxins, although toxins have been measured in the gut contents of squid (5). Of course, the toxic dinoflagellates themselves represent a potential toxin source for those fishes such as menhaden which are largely phytoplanktivorous.

The principal routes through which paralytic shellfish toxins reach fish are shown in Figure 1. How likely is it that fish might acquire lethal doses of the toxins through the different routes? Table V lists the approximate number of dinoflagellates and

Table V. Amounts of Foods Necessary to Supply a Potential Lethal Dose of Paralytic Shellfish Toxins to a 100-g Fish

Toxin Source	Amount to Yield 60 μg Toxins
Gonyaulax	1-2 x 10^6 cells (100-200 mL of red tide)
Zooplankton	1 g
Shellfish	1 g

amounts of zooplankton or shellfish that a hypothetical 100-g fish would have to eat during a typical Bay of Fundy *Gonyaulax* bloom in order to acquire the oral LD50 (taken as 600 μg/kg, see Table III). A phytoplanktivorous fish would have to eat about one or two million *G. excavata* cells, which means clearing only about 100 mL or so of red water which can contain many millions of *Gonyaulax* cells per liter. A zooplanktivorous fish would only have to eat 1 g or so wet weight of zooplankton during a toxic *Gonyaulax* bloom (see Table IV). Filter-feeding shellfish commonly acquire 6000 μg toxins per 100 g of

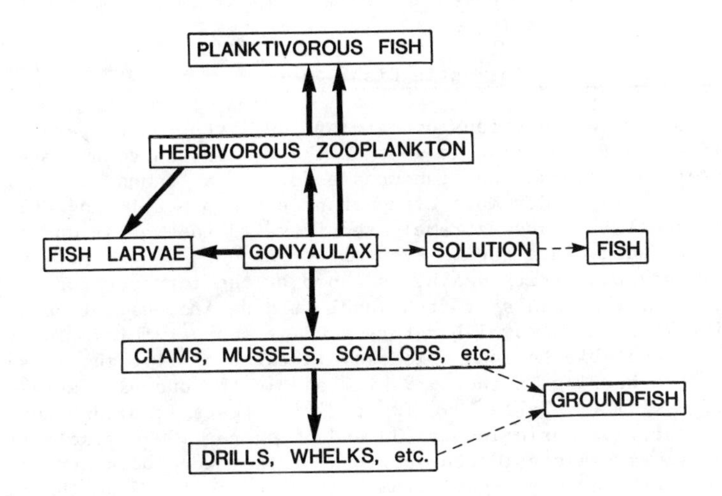

Figure 1. Some routes through which paralytic shellfish toxins reach fish. Solid arrows show known routes. Broken arrows show possible routes. Reproduced with permisssion from Ref. 10. Copyright, 1981, Springer-Verlag.

tissue, or greater, so that it would only take 1 g or so to supply a potential lethal dose to a 100-g groundfish. From these estimates it becomes apparent that fish may be exposed to dangerous amounts of the toxins through several routes. It is unknown if paralytic shellfish toxins exert any effect on fish when dissolved in sea water during dense blooms and red tides of the dinoflagellates, although the lethality in solution of concentrated extracts of G. catenella and Pyrodinium bahamense to juvenile fish has been reported (8, 20).

The toxin content of Bay of Fundy zooplankton has reached high levels in most years since 1977 (Table IV). Yet, curiously, herring kills have not been observed in years other than 1976 and 1979. The occurrence of kills may depend upon the temporal and spatial overlapping of dense Gonyaulax blooms, highly toxic zooplankton, and actively feeding herring.

Paralytic Shellfish Toxins in Fish

One of the major questions concerning the ingestion of paralytic shellfish toxins by fish is whether the toxins can accumulate in fish to levels which make the fish unsafe for use as human food. The answer to this, under most circumstances, is probably no. Finfish, unlike shellfish, are extremely sensitive to the toxins and thus are unable to accumulate them. Even minute amounts of the poisons in a fish's body causes its death, as shown by the intraperitoneal dose studies. Toxin assays of fish muscle tissue from Bay of Fundy herring kills and from laboratory studies support this. The toxins were undetectable by mouse bioassay in three of five muscle samples taken from herring in the 1979 kill and in the one sample taken from the 1976 kill (7, Table I). Two positive tests from 1979 herring showed very low toxin levels (33 and 55 μg per 100 g muscle tissue). These values are considered spurious as they were obtained from decayed fish and may result from leakage of toxins from the viscera. In laboratory studies in which herring, pollock, flounder, and salmon were injected with very high oral doses of toxins (2000 μg/kg), the toxins were undetectable in muscle samples during and after their death (7, Table VI). It seems that fish flesh remains toxin-free, or very nearly so; thus it is highly unlikely that human poisoning would result from its consumption.

Table VI. Paralytic Shellfish Toxins in Fish Tissues Following Oral Dose of Toxins at 2000 μg/kg

Fish	Toxin Content (μg/100 g)	
	Gut	Muscle
Herring	416	<30
Pollock	343	<30
Flounder	837	<30
Salmon	408	<30

Reproduced with permission from Ref. 10. Copyright 1981, 'Marine Biology'.

Paralytic shellfish toxins have, however, been measured in the viscera of dead and dying fish from natural kills (Table I). Although the amounts of toxins measured in fish viscera are relatively low compared with shellfish, it is through this route that sea birds (1,2,4) and possibly carnivorous fish (4) have received lethal doses. Fish viscera was probably the source of the toxins measured in squid (see above). Human poisoning from ingestion either of whole, small fish or of particular internal organs would seem possible, but rather unlikely unless copious amounts of toxic fish viscera were consumed. There are anecdotal reports of non-fatal cases of PSP in humans attributed to ingestion of entire, planktivorous fishes in northern Borneo in 1976 (8) and in the Philippines in 1983 (21). Marine mammals that take large quantities of fish may suffer from PSP through this route, resulting perhaps in kills and strandings. Wyatt (22) provides an interesting account of a massive seal kill in South Africa in the early nineteenth century which may have been caused by red tide toxins in fish.

Impact of Paralytic Shellfish Toxins on Fish and Fish Products

Fish. The major impact of toxic *Gonyaulax* blooms and paralytic shellfish toxins on finfish resources is seen to be the direct effects on fish populations (23). The toxins have been implicated in fish kills in England and eastern North America. It is reasonable to assume that there are similar repercussions on fish of the nearly worldwide toxic blooms of other *Gonyaulax* and *Pyrodinium* species which also contain paralytic shellfish toxins. The occasional losses of fish from such kills would not be expected ordinarily to have a substantial effect on stock size. However, the continual impingement of the toxins on the survival of fish larvae may conceivably have a lasting effect, in terms of year-class strengths, on local stocks. Further, there may be an indirect effect of the toxins on the catchability of fish following a kill. For example, after the extensive herring kill in 1976 off Grand Manan Island in the Bay of Fundy, very few fish were caught in this otherwise productive area for the remainder of the season. This probably resulted from avoidance of the despoiled area by other herring. Set against the background of documented (24) and apparent intensification and spreading of toxic *Gonyaulax* and *Pyrodinium* blooms throughout the world, the problems associated with paralytic shellfish toxins and finfish loom increasingly important.

Fish Products. As explained earlier, it is unlikely that paralytic shellfish toxins have an impact on the utilization of fish products from the point of view of the suitability of fish as food, except perhaps in cases where whole fish are eaten with little processing. Fish simply are unable to accumulate the toxins in their muscle tissues. But the toxins do appear to have an impact on the marketing of fish products, related to consumer wariness of seafood products in general during red tide and PSP incidents. The media blitz surrounding these incidents often leaves consumers unaware of which particular seafood items to be cautious. Consequently, finfish as well as shellfish products have been avoided during these episodes (25).

Literature Cited

1. Adams, J. A.; Seaton, D. D.; Buchanan, J. B.; Longbottom, M. R. J. mar. biol. Ass. U.K. 1957, 36, 169-89.
2. Coulson, J. C.; Potts, G. R.; Deans, I. R.; Fraser, S. M. Brit. Birds 1968, 61, 381-404.
3. White, A. W. J. Fish. Res. Board Can. 1977, 34, 2421-24.
4. Nisbet, I. C. T. Condor 1983, 85, 338-45.
5. Hurst, J. W., personal communication.
6. Bigelow, H. B.; Schroeder, W. C. "Fishes of the Gulf of Maine"; Fish. Bull. 74; U.S. Fish and Wildlife Service: Washington, D.C., 1953; p. 577.
7. White, A. W. Can. J. Fish. Aquat. Sci. 1980, 37, 2262-5.
8. Maclean, J. L. In "Toxic Dinoflagellate Blooms"; Taylor, D. L.; Seliger, H. H., Eds.; Elsevier/North-Holland: New York, 1979; pp. 173-78.
9. Harada, T.; Oshima, Y.; Kamiya, H.; Yasumoto, T. Bull. Jap. Soc. Sci. Fish. 1982, 48, 821-25.
10. White, A. W. Marine Biology 1981, 65, 255-60.
11. Prakash, A.; Medcof, J. C.; Tennant, A. D. "Paralytic Shellfish Poisoning in Eastern Canada"; Bull. 177; Fisheries Research Board of Canada: Ottawa, Ontario, 1971; p. 87.
12. Evans, M. H. Int. Rev. Neurobiol. 1972, 15, 83-166.
13. Kodama, M.; Noguchi, T.; Maruyama, J.; Ogata, T.; Hashimoto, K. J. Biochem. 1983, 93, 243-47.
14. Nakamura, M.; Oshima, Y.; Yasumoto, T. Toxicon 1984, 22, in press.
15. White, A. W.; Fukuhara, O.; Anraku, M., unpublished data.
16. Mills, L. J.; Klein-MacPhee, G. In "Toxic Dinoflagellate Blooms"; Taylor, D. L.; Seliger, H. H., Eds.; Elsevier/North-Holland: New York, 1979; pp. 389-94.
17. White, A. W. In "Toxic Dinoflagellate Blooms"; Taylor, D. L.; Seliger, H. H., Eds.; Elsevier/North-Holland: New York, 1979; pp. 381-4.
18. White, A. W. Limnol. Oceanogr. 1981, 26, 103-9.
19. Hayashi, T.; Shimizu, Y.; White, A. W. Bull. Jap. Soc. Sci. Fish. 1982, 48, 1673.
20. Bates, H. A.; Kostriken, R.; Rapoport, H. Toxicon 1978, 16, 595-601.
21. Hermes, R., personal communication.
22. Wyatt, T. J. Cons. int. Explor. Mer 1980-81, 39, 1-6.
23. White, A. W. Can. Tech. Rep. Fish. Aquat. Sci. 1982, 1064, 1-5.
24. White, A. W. Can. Tech. Rep. Fish. Aquat. Sci. 1982, 1063, 1-12.
25. Jensen, A. C. In "Proc. 1st Int. Conf. Toxic Dinoflagellate Blooms"; LoCicero, V. R., Ed.; Mass. Sci. Technol. Found.: Wakefield, Mass., 1975, 507-16.

RECEIVED February 6, 1984

16

Development of Immunoassays for Paralytic Shellfish Poisoning

A Radioimmunoassay for Saxitoxin

ROBERT E. CARLSON, MARIE L. LEVER, BRIAN W. LEE, and PATRICK E. GUIRE

Bio-Metric Systems, Inc., Eden Prairie, MN 55344

A simple and fast radioimmunoassay for saxitoxin has been developed using an anti-saxitoxinol antibody. The immune serum required for the immunoassay was raised in rabbits using an immunogen conjugate prepared from saxitoxinol and bovine serum albumin with a phosgene based coupling reagent. The antibody has moderate affinity for saxitoxinol ($K_o \sim 5x10^7$ L/mole). It cross-reacts at >90% with saxitoxin but at <1% with neosaxitoxin. This antibody, when used in an anti-rabbit IgG "second antibody" radioimmunoassay format, can detect pmole quantities of saxitoxin. This assay has been shown to be a simple and efficient method for the analysis of saxitoxin in clam extracts. The lack of antibody cross-reactivity to the neosaxitoxin sub-group of the paralytic shellfish poisons limits the general utility of the assay to neurophysiology studies and to certain clam species which preferentially accumulate saxitoxin. However, the radioimmunoassay serves as a good precursor in the development of an enzyme immunoassay for the paralytic shellfish poisons.

Massive dinoflagellate (Gonyaulax and Gymnodinium spp.) blooms in coastal waters are the cause of the phenomenon known as "red tide" (1). Unfortunately, these organisms produce an extremely potent group of tetrahydropurine based (Figure 1) toxins which disable the neuromuscular system by blocking sodium ion transport (12). This group of compounds is readily accumulated in shellfish to concentrations which can be lethal to humans following ingestion of contaminated shellfish meat (13).

The current test for the shellfish toxins is based on a time-of-death mouse assay (14). This test is laborious to perform and relatively expensive (15). Consequently, a variety of assay

0097-6156/84/0262-0181$06.00/0
© 1984 American Chemical Society

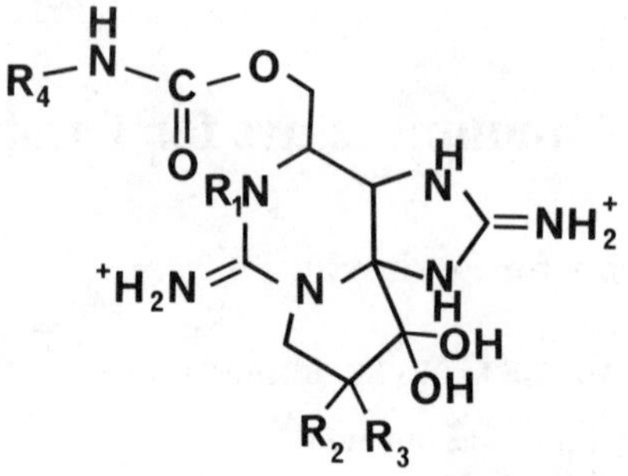

COMPOUND	R1	R2	R3	R4	Ref.
IA - Saxitoxin (STX)	H	H	H	H	2,3
IB - Gonyautoxin2 (GTX2)	H	H	OSO_3^-	H	7
IC - Gonyautoxin3 (GTX3)	H	OSO_3^-	H	H	7
ID - Neosaxitoxin (NEO)	OH	H	H	H	4
IE - Gonyautoxin1 (GTX1)	OH	H	OSO_3^-	H	5,6
IF - Gonyautoxin4 (GTX4)	OH	OSO_3^-	H	H	5,6
IG - Saxitoxinol (STXOL)	H and C12 -CHOH	H	H	H	10

Note: The B and C series of the PSP (R4 - SO_3^-) have not been included in this table because they form their STX, NEO and GTX counterparts (11) on treatment with acid during the preparation of clam extracts. For the structure of tetrodotoxin see (8,9).

Figure 1. The PSP.

methods have been developed as alternatives to the mouse assay. These include; a colorimetric method based on hydrogen peroxide oxidation (16), a fluorimetric method which is also based on hydrogen peroxide oxidation (17,18) and a number of recently introduced high-performance liquid chromatography (HPLC) methods (19,20). Each of these methods has a significant drawback which has prevented it from replacing the mouse assay. The peroxide oxidation methods require strict adherence to a specific protocol (18) and the neosaxitoxin sub-group (Figure 1; Id-f) of the paralytic shellfish poisons (PSP) is not efficiently detected (21). The HPLC methods can require extensive sample preparation. In addition, detection of the PSP, which do not have a significant UV/Vis absorption band, can be a major problem which requires the use of post-column derivatization methods (19,20). However, the HPLC methods do show promise for the development of a laboratory assay for the PSP (22).

An attractive alternative to the above methods is to develop a PSP assay based on the selectivity and high affinity characteristics of PSP binding proteins. Several previous attempts have been made to prepare anti-PSP antibodies (10,23). Koehn, *et al.* were unable to prepare immunogenic PSP-protein conjugates (10) and the antisera prepared by Johnson, *et al.* (23), which was raised in rabbits using a labile PSP-protein conjugate, gave relatively poor anti-PSP antisera (23). The sensitivity of the serological assay which was based on these antisera was marginal (*ca.* 1 nmole/ml minimum detection limit) (24). However, it should be noted that Chu and co-workers (25) have obtained much more impressive results using Johnson's immunogen in combination with an enzyme immunosorbent assay format.

Presumably an assay could be based on the PSP binding proteins from excitable nerve membranes (26). Although the binding characteristics of these proteins would probably be excellent [high K_o (27); good PSP cross-reactivity (based on PSP toxicities, see Figure 1 and references therein)] their isolation in sufficient quantity for application in a routine assay would be expected to be difficult (28). At least one soluble binding protein, which would alleviate the isolation problem, has been identified but the quantity available is extremely limited (29).

This report describes the development of a radioimmunoassay (RIA) for one of the major PSP (saxitoxin, Ia) which is based on our successful production of antibodies to a stable saxitoxin derivative-bovine serum albumin conjugate. The application of the RIA to the analysis of PSP contaminated clam extracts and the problems which must be addressed in the development of a routine immunoassay for the PSP are discussed.

Experimental

Reagents and solvents were of the highest purity commonly available. When possible the PSP derivatives used in this study were checked by NMR. The STX was obtained from the Cincinnati lab of the FDA. The NEO was a generous gift of Dr. S. Hall. The TTX was obtained from Sigma.

Immunogen Preparation. STX was converted to ^3H-STXOL using the method of Koehn, et. al. (10). The STXOL (19 umoles) was dissolved in 250 ul of dry dimethylformamide (Pierce) and an excess of CDI (100 umoles; Pierce) added. After 1 hour, bovine serum albumin (15 mg; Sigma) was added in 10 mM carbonate buffer (10 ml; pH 10.0) and the reaction was stirred overnight. The product was isolated by dialysis against water to remove unreacted STXOL followed by lyophilization of the product. The resulting material was characterized by standard protein, amine and radioactivity analysis procedures and found to contain 8.8 (+/-.5) STXOL/BSA.

Rabbit Immunization. Six rabbits were sub-cutaneously injected with .2 mg of the immunogen in Freund's adjuvent on days 0, 12, 26, 40 and 59 of the immunization schedule (see Table I). Complete adjuvent was used for the first three immunizations followed by incomplete adjuvent for the final two immunizations. The rabbits were bled as needed using ear vein puncture.

Specific Antibody Determination. Serum samples were prepared from each bleed by centrifugation to remove clotted material. 100 ul of the sera was incubated for 30 minutes with sufficient ^3H-STXOL to provide a ca. 20 fold excess of hapten to the anticipated quantity of specific binding sites. The radioactivity of the protein pellet was determined after ammonium sulfate precipitation. After correction for a small amount of non-specific adsorption of label by control sera proteins the mg/ml of specific antibody in the sample was calculated.

Equilibrium Dialysis and Affinity Dilution. Equilibrium dialysis (34) and affinity (35) dilution experiments followed standard procedures.

RIA Procedure. The RIA was developed using an antiserum pool which contained 2.0 mg/ml of anti-STXOL antibody. The protocol was designed for use in a 1.5 ml microcentrifuge tube.
Step 1. Add buffer and test sample to the tube. Total volume 900 ul. The buffer was pH 7.2 phosphate buffered saline. The test samples were STXOL, STX, NEO, TTX or serially diluted clam extract.
Step 2. Add labeled STXOL (32 pmoles, 1.2 eq. to sites of specific antibody used, STXOL specific activity - 138 dpm/pmole).
Step 3. Add control rabbit serum (9 ul).
Step 4. Add anti-STXOL serum (1 ul) and incubate for 30 minutes.
Step 5. Add 60-120 ul of second antibody (anti-rabbit IgG; Sigma Chemical Co.). The exact quantity must be optimized. Incubate for 60 minutes.
Step 6. Centrifuge and determine dpms in a standard aliquot of the supernatant. Plot dpms versus analyte concentration.

Reduction Experiments. NEO (100-1000 pmole) was reacted with an excess (10-1000 equivalents) of sodium borohydride, sodium cyanoborohydride or Zn/acid in buffer at pH 8-10, 3-7 or 2-5 respectively. The borohydride and cyanoborohydride reductions were performed with and without added diglyme for 2-40 hours. The Zn was used with hydrochloric, acetic, phthalic or citric acid with and without the addition of iodide for 2-24 hours. The excess reducing reagent was removed (eg.; using acetone or acidification for borohydride) and the resulting sample evaluated using the RIA.

Results and Discussion

Immunogen preparation. The PSP can be divided into two main subgroups; those compounds which are related to saxitoxin (STX; Figure 1, Ia-c) and those which are related to neosaxitoxin (NEO; Figure 1, Id-f). STX was chosen for these studies predominantly because it is the most readily available of the PSP and because its chemistry has been the most extensively studied of the PSP (for example see 10).

The key step in the preparation of anti-STX antisera was the preparation of a stable STX-protein conjugate for use as an immunogen. The reduction of the ketone (ketone hydrate) on STX gives saxitoxinol (STXOL; Figure 1, Ig (10). This compound, although it is obtained as a mixture of isomers (10,30), is a good choice for the preparation of a hapten-protein conjugate because it introduces a reactive functional group into STX and because the product can be readily radiolabeled by performing the reduction with a tritiated reducing reagent. The radiolabel can be used to facilitate not only the manipulation of STXOL but also the determination of STXOL/protein coupling yield. STXOL has previously been used in an attempt to prepare a hapten-protein conjugate (10). However, even though a hemisuccinate derivative [STXOL~O-C(O)-CH_2-CH_2-CO_2H] could be readily prepared it was not successfully coupled to a carrier protein in significant yield ([STXOL]n-BSA; n<3 (10)).

We chose to take advantage of the fact that phosgene derivatives (eg. carbonyldiimidazole (CDI,31); p-nitrophenylchloroformate (PNPCF,32)) can be reacted with alcohols to give an activated intermediate which will readily couple with nucleophiles (Figure 2). Using this method we have prepared a stable STXOL-BSA conjugate ([STXOL]n-BSA; n=9: presumably coupled through the BSA lysine amino groups) for use as an immunogen. It should be noted, however, that this reaction has proven to be very unreliable.

Using a model system of ^{3}H-STXOL/PNPCF/^{14}C-methylamine we have found that we can reliably obtain a good yield of the STXOL/amine product if STXOL solubility and solvent purity are carefully controlled. However, the same reaction conditions generally fail to produce a useful [STXOL]n-protein (n<2) preparation.

Antibody Preparation and Characterization. Our best [STXOL]n-BSA immunogen (n=9) was used to hyperimmunize a total of 6 rabbits. The formation of anti-STXOL antibodies was monitored by determining the quantity of specific binding protein (presumably IgG) in the serum samples. Table I shows that 5 of the 6 rabbits produced anti-STXOL antibodies within 100 days after initial immunization.

The antisera which were collected at ca. 100 days were first characterized for STXOL affinity (33). Equilibrium dialysis (34) gives a good measure of both antibody-to-hapten affinity (K_o; useful range for RIA development 10^{6-9} L/mole) and antibody

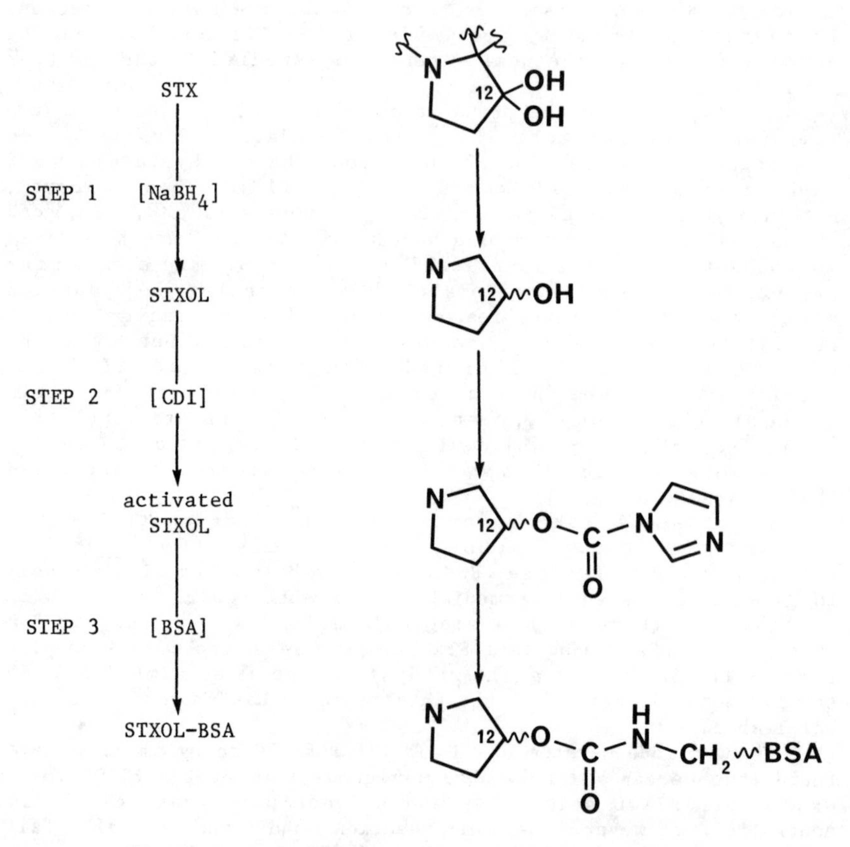

Figure 2. Immunogen preparation chemistry.

TABLE I. DEVELOPMENT OF ANTI-STXOL ANTISERA

Rabbit Bleed (Days post initial immunization)

Rabbit	A(39)	B(49)	C(56)	D(68)	E(75)	F(83)	G(90)	H(104)
#1	.05*	.11	.09	----				
#2	.25	1.54	1.48	1.55	1.74	1.96	1.96	1.80
#3	.20	.69	.67	.82	.85	.79	.74	1.45
#4	.33	.86	.76	1.26	1.38	1.74	1.56	1.84
#5	.14	.49	.30	.42	.61	.90	1.07	1.09
#6	.57	1.41	1.13	1.20	1.50	1.91	2.07	2.65
	Im40**		Im59					

* - mg/ml of specific antibody
**- Immunizations on day 40 and day 59.

homogeneity (relative proportion of antibodies with different hapten binding affinities). Our results (5 rabbits; $K_o = 3\text{-}9 \times 10^7$ L/mole) indicate that the five rabbits have all produced anti-STXOL antibodies with a useful, consistant level of STXOL affinity. In addition, the free versus bound hapten plots of the equilibrium dialysis data (not shown) indicated good antibody homogeneity. This last factor is important because the combination of very high and very low antibody affinities in an antisera can make the development of an RIA with low background interference difficult.

Affinity dilution experiments (Figure 3 (35)) also demonstrated that this antibody has good affinity for STXOL. More importantly, the significant offset of the two curves for STXOL concentrations of 1.5 pmole/ml and 8.7 pmole/ml indicate that an RIA based on this antibody will readily detect and differentiate low pmole/ml concentrations of STXOL.

RIA Development. The essential components of an RIA are (36): 1) competition between a radiolabeled hapten and the analyte for antibody binding sites and 2) separation of the label which is bound to the antibody from label which is free in solution. The assay is based on the expected result that at higher concentrations of competing analyte more label will remain free in solution.

Our RIA was developed to optimize the main criteria of simplicity, sensitivity and reproducibility. The format which was eventually developed was based on the addition of a "second antibody". This procedure separates the bound from the free label by precipitation of the anti-STXOL antibody, which was raised in rabbits, with an antibody produced in goats against rabbit IgG. The source of the labeled hapten was ^{3}H-STXOL.

This assay was developed to be completely performed in a microcentrifuge tube. The basic procedure involves addition of a

TABLE II. CLAM EXTRACT ASSAY

Extract Number	Extract Dilution	pmole Observed	pmole/ ml	ug/100g*	RIA/ Mouse
199	1/10	106	1060	71	.15
	1/100	11	1100		
200	1/10	158	1580	106	.33
	1/100	15	1500		
235	1/10	86	860	57	.21
	1/100	9	900		
0226	1/10	>250			
	1/100	49	4900	330	.25
0302	1/10	>250			
	1/100	~220			
	1/1000	17	17000	1200	.08

* - ug PSP per 100g of clam tissue

serially diluted clam extract sample to a pre-prepared reaction/centrifuge tube which contains antibody and labeled STXOL. After a short incubation the second antibody is added and incubated. Analyte concentration is determined after centrifugation and collection of a standard aliquot of supernatant.

Application of the RIA. The RIA was initially used to evaluate the cross-reaction of the STXOL antisera to STX. The antibody was found to have excellent STX cross-reactivity (93%). Subsequent preparation of a logit/log (37) standard curve for STX (Figure 4) demonstrated that the chosen assay format would give good reproducibility and a desirable order of magnitude linear sensitivity range.

Our first set of clam experiments used control clam and mussel extracts which had been doped with STX and unlabeled STXOL. The results obtained demonstrated that the assay would function properly with STX and STXOL recoveries from a clam matrix solution in the range of 96-103%. Subsequent application of the assay to clam extracts which were known to contain PSP gave the results shown in Table II. In particular, it is significant that the use of an easily managed order of magnitude dilution series gave highly reproducible results except at high PSP concentrations. Unfortunately, a comparison of the results of this assay with the results of the mouse PSP assay (Table II) indicated that the RIA values were only 8-33% of the mouse values.

The reason for these low values becomes apparent when the cross-reaction of our antisera to NEO is considered. We had

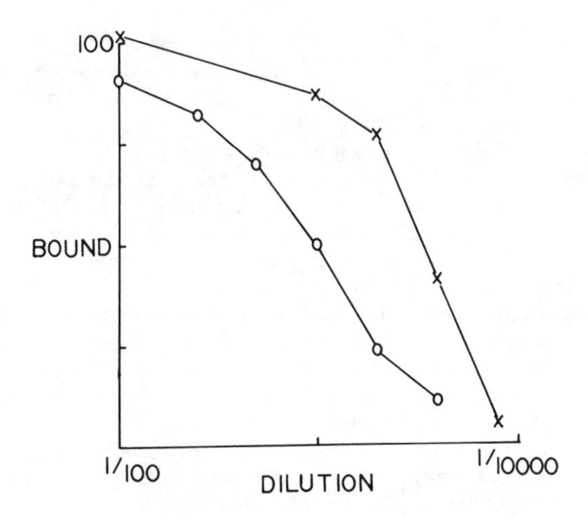

Figure 3. Affinity dilution characterization of an anti-STXOL antisera. STXOL concentration: x-1.5 pmole/ml; o-8.7 pmole/ml.

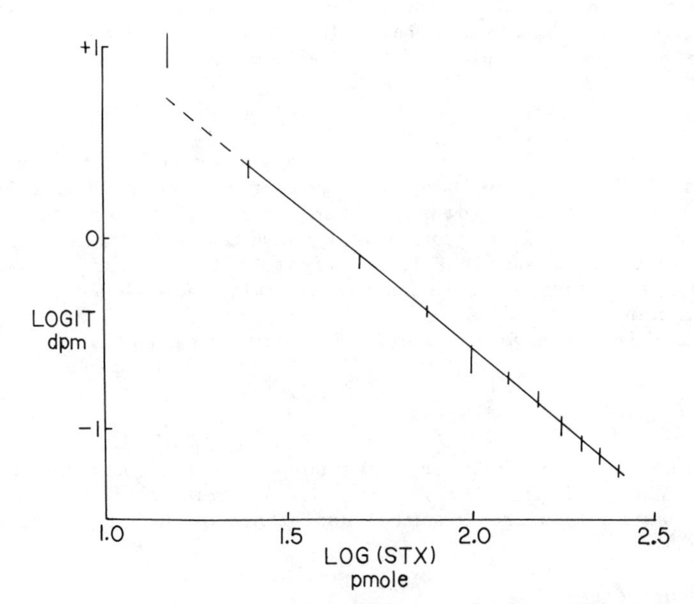

Figure 4. Logit/log RIA calibration curve for STX. Each point is the average of 3–5 determinations.

previously determined that the anti-STXOL antisera had >90% cross-reactivity to STX. However, using the RIA we subsequently determined that NEO has a cross-reactivity of <1%. As a result we would expect that the NEO sub-group of the PSP would be transparent to the RIA.

Continued Development. The RIA will be most useful if it can be used to detect all of the major PSP which are expected to occur in an acidic clam extract (Figure 1).

There are two main approaches to this goal:

The first is to convert all of the PSP into STX (or STXOL) through reductive cleavage of the N1 hydroxyl and C11 sulfate ester groups of the NEO and GTX sub-groups of the PSP (see Figure 1). These reductive processes have been shown to occur both biotically (38,39) and abiotically (40). However, our initial attempts to develop reduction methods using sodium borohydride, sodium cyanoborohydride or zinc/acid as reducing agents have met with limited and variable success. In addition, even after deactivation, these reducing agents and their associated reaction solvents can interfere with the immunoassay.

The second route is to develop an antibody with the required cross-reaction characterisitics. Our antibody binds STX and STXOL but it does not bind to NEO or the related guanidinium neurotoxin tetrodotoxin (Figure 1). These results suggest that the N1,2,3 (as opposed to the N7,8,9) guanidinium ring is a primary recognition site for PSP binding to our anti-STXOL antibody. One possibility for an antigen which would potentially reduce the impact of the N1,2,3 ring on antibody recognition would be the preparation of a conjugate using N1 as the link. This and other potentially useful antigens are being explored.

Conclusion

We have prepared an anti-STXOL antibody and used it to develop an RIA for STX. This antibody does not cross-react with NEO; thus limiting its utility for shellfish assays. However, the RIA can still be of value in neurophysiology studies which use STX (12,27) and in the studies of the PSP content of the Alaska butter clam (*Saxidomus giganteus*) which has previously been shown to contain predominantly STX (41). In addition, the RIA serves as a good precursor in the development of an enzyme immunoassay (42) for the PSP.

Acknowledgments

We would like to acknowledge the support and encouragement of RA Neve′ (State of Alaska Contract No. PS18-0227) and E. Ragelis (USFDA) as well as the technical assistance of RS King (PNPCF model studies).

Literature Cited

1. Schantz, E. J. In "Toxicants Occurring Naturally in Foods"; National Academy of Sciences Printing and Publishing Office: Washington, D.C., 1973; p. 424.

2. Schantz, E. J.; Ghazarossian, V. E.; Schnoes, H. K.; Strong, F. M.; Springer, J. P.; Pezzanite, J. O.; Clardy, J. J. Am. Chem. Soc. 1975, 97, 1238.
3. Bordner, J.; Thiessen, W. E.: Bates, H. A.; Rapoport, H. J. Am. Chem. Soc. 1975, 97, 6008.
4. Shimizu, Y.; Hsu, C.; Fallon, W. E.; Oshima, Y.; Miura, I.; Nakanishi, K. J. Am. Chem. Soc. 1978, 100, 6791.
5. Fix Wichmann, C.; Boyer, G. L.; Divan, C. L.; Schantz, E. J.; Schnoes, H. K. Tetrahedron Lett. 1981, 22, 1941.
6. Shimizu, Y.; Hsu, C. P. J. Chem. Soc. Chem. Commun. 1981, 314.
7. Boyer, G. L.; Schantz, E. J.; Schnoes, H. K. J. Chem. Soc. Chem. Commun. 1978, 889.
8. Woodward, R. B. Pure Appl. Chem. 1964, 9, 49.
9. Tsuda, K.; Ikuma, S.; Kawamura, M.; Tachikawa, R.; Sakai, K.; Tamura, C.; Amakasu, O. Chem. Pharm. Bull. 1964, 12, 1357.
10. Koehn, F. E.; Ghazarossian, V. E.; Schantz, E. J.; Schnoes, H. K.; Strong, F. M. Bioorg. Chem. 1981, 10, 412.
11. For example see Hall, S.; Reichardt, P. B.; Neve', R. A. Biochem. Biophys. Res. Commun. 1980, 97, 649.
12. Evans, M. H. Int. Rev. Neurobiol. 1972, 15, 83.
13. For example see Hudgens, S. Sea Grant Today 1981, 14.
14. For example see Miescier, J. J. Paralytic Shellfish Poison Management Workshop, Calif. Dept. Health Services, Berkeley, CA, 1982, 52.
15. Miescier, J. J., personal communication.
16. Gershey, R. M.; Neve', R. A.; Musgrave, D. L.; Reichardt, P. B. J. Fish. Res. Board Can. 1977, 34, 559.
17. Bates, H. A.; Rapoport, H. J. Agr. Food Chem. 1975, 23, 237.
18. Bates, H. A.; Kostriken, R.; Rapoport, H. J. Agr. Food Chem. 1978, 26, 252.
19. Sullivan, J. J.; Iwaoka, W. T. J. Assoc. Off. Anal. Chem. 1983, 66, 297.
20. Onoue, Y.; Noguchi, T.; Nagashima, Y.; Hashimoto, K.; Kanoh, S.; Mitsuo, I.; Tsukada, K. J. Chromatography 1983, 257, 373.
21. Bose, R. J.; Reid, J. E. In "Toxic Dinoflagellate Blooms"; Taylor, D. L.; Seliger, H. H., Eds.; DEVELOPMENTS IN MARINE BIOLOGY Vol. I, Elsevier/North-Holland: New York, 1979; p. 399.
22. Sullivan, J. J.; Simon, M. G.; Iwaoka, W. T. J. Food Sci. 1983, 48, 1312.
23. Johnson, H. M.; Frey, P. A.; Angelotti, R.; Campbell, J. E.; Lewis, K. H. Proc. Soc. Exp. Biol. Med. 1964, 117, 425.
24. Johnson, H. M.; Mulberry, G. Nature 1966, 211, 747.
25. Chu, F. S., personal communication.
26. Weigele, J. B.; Barchi, R. L. FEBS Letters 1978, 91, 310 and references therein.
27. Narashi, T. Physiol. Rev. 1974, 54, 813.
28. For example see Agnew, W. S.; Levinson, S. R.; Brabson, J. S.; Raftery, M. A. Proc. Natl. Acad. Sci. USA 1978, 75, 2606.
29. Doyle, D. D.; Wong, M.; Tanaka, J.; Barr, L. Science 1982, 215, 1118.
30. Shimizu, Y.; Hsu, C-p.; Genenah, A. J. Am. Chem. Soc. 1981, 103, 605.
31. Bethell, G. S.; Ayers, J. S.; Hancock, W. S.; Hearn, M. T. W. J. Biol. Chem. 1979, 254, 2572.

32. Wilchek, M. Bioc. Int. 1982, 629.
33. Steward, M. W.; Steensgaard, J. "Antibody Affinity: Thermodynamic Aspects and Biological Significance", CRC Press Inc.: Boca Raton, Florida, 1983.
34. Fazekas de St. Groth, S. In "Immunological Methods"; Lefkovits, I.; Pernis, B., Eds.; Academic Press: New York, 1979; p. 1.
35. For example see Wolff, J.; Capraro, H.-G.; Brossi, A.; Cook, G.H. J. Biol. Chem. 1980, 255, 7144.
36. Skelley, D. S.; Brown, L. P.; Besch, P. K. Clin. Chem. 1973, 19, 146.
37. Rodbard, D.; Frazer, G. R. In "Hormone Action (Part B: Peptide Hormones)"; O'Malley, B. W.; Hardman, J. G., Eds.; Academic Press: New York, 1969; p. 3.
38. Shimizu, Y.; Yoshioka, M. Science 1981, 212, 549.
39. Sullivan, J. J.; Iwaoka, W. T.; Liston, J. Bioc. Biophys. Res. Commun. 1983, 114, 465.
40. Shimizu, Y.; Hsu, C. P. J. Chem. Soc. Chem. Commun. 1981, 314.
41. Shimizu, Y. In "Toxic Dinoflagellate Blooms"; Taylor, D. L.; Seliger, H. H., Eds.; DEVELOPMENTS IN MARINE BIOLOGY Vol. I, Elsevier/North-Holland: New York, 1979; p. 321.
42. For example see Oellerich, M. J. Clin. Chem. Biochem. 1980, 18, 197.

RECEIVED February 6, 1984

17

Response of the House Fly to Saxitoxins and Contaminated Shellfish

ALVIN SIGER, BERNARD C. ABBOTT, and MARIA ROSS

Department of Biological Sciences, University of Southern California, Los Angeles, CA 90089

The problem of Paralytic Shellfish Poison as a hazard to human health and life resulting from the consumption of bivalve marine molluscs that have ingested certain toxic dinoflagellates is well known. There is no known antidote and, although the majority of victims recover, there have been many fatalities when toxicity has been high. As a result the health authorities in the several coastal states require monitoring of shellfish and have set criteria for banning the collection of shellfish. The majority of shellfish are harvested by commercial organizations which regularly supply samples for analysis and as a result the number of incidents is kept remarkably low. However there exist long lengths of shoreline where native shellfish populations are harvested recreationally. The control imposed differs between States, varying from regular monitoring with the associated authority to ban collecting in defined local areas, to a State wide quarantine of all beaches for the collection of selected species during certain months of the year. The only legally recognized assay for PSP is that of the Mean Death Time of mice. Not only is this expensive but we believe that it is very innacurate at low levels of toxicity. In addition, there are vast shellfish beds in remote areas of coastline that cannot be opened up for commercial explotation both because of the remoteness from mouse monitoring laboratories, and because of the expense. There is a recognized need for a alternative assay, but in reading the title of the communication a first reaction must be: "why a Bioassay?" and then "why a fly?".

Preference is obviously for a simple chemical assay for PSP. Unfortunately the more specific the chemical test, the narrower is the window of compounds it can assay. The Paralytic Shellfish Poison is not just Saxitoxin (STX) as originally believed, but is a mixture of compounds closely related to STX (1) and the mix varies widely with location and with time (2). It would seem, therefore that a chemical assay should determine at least the ratios of the several compounds, and that the relative toxicity of each of the compounds must be known. An effective assay must evaluate the actual biological toxicity of the shellfish being tested. For the chemical assay this requires the summated toxicity of all the

0097-6156/84/0262-0193$06.00/0
© 1984 American Chemical Society

components and also assumes that there is no synergism. Analysis and identification of the STX family of compounds is essential in the research studies on PSP but we believe that a bioassay giving the summed toxicity of samples will still be needed for some time ahead.

Why not the mouse? There seem to be several cogent objections to the Mouse Mean Death Time bioassay, although the sentiment in support of a method that has been in regular use for a quarter of a century is very strong. Cost is an ever present problem. In order to obtain consistent results the mice used in any one laboratory must be of a defined strain, and of size 19-22 gr. wt (3). Such animals are expensive, and as more intensive monitoring is demanded, costs are rising at a time when budgets are going down. The assay procedure is to inject intrapertoneally 1 ml of the extract under test and observe the time to die. Unfortunately the death throeas are not very pleasant and it has been called the jump test. In an era of society's objection to the imposition of pain to mammals in research and in testing procedures, the government is encouraging the development of alternatives.

But of more concern is what we feel to be an unacceptable error in the monitoring of shellfish that are only weakly toxic. This is due to the "salt effect" observed by Schantz (4) and confirmed by McFarren (5) the following year. It refers to very considerable errors on the low side with weakly toxic shellfish. In calibrating the MTD method known concentrations of pure STX were added to tissue extracts of non-toxic clams, Schantz demonstrated full recovery of the STX added down to the threshold level if the samples were diluted with the low salt extraction buffer. If, on the other hand, the dilution was made with the (high salt) non-toxic clam extract a progressive error was observed as dilution increased, and near the threshold of detection the monitored level of STX was as much as 2-5 times too low. Thus for high levels of PSP the bioassay is quite reliable because the samples are diluted with the buffer, but for extracts that are less toxic and so need no dilution, the assay is made in the presence of full strength clam juice and the error is large. This can be very significant for predictive effects, which require the maximum sensitivity at the lowest levels of toxicity in order to determine if the early onset of a bloom may be detected from the pattern of low level shellfish toxicity.

The common house fly Musca domestica is paralyzed by STX in a way that can be quantally scored (6). This fly is ubiquitous. It can be purchased as pupae for a few cents or can be bred very simply in any laboratory from wild stock for the price of a small volume of hamburger meat. The pupae can be stored for some time if cooled and can be hatched at will within a few days. The adult fly emerges fully grown and will remain healthy for 20-30 days if provided with water, sugar and dried milk.

The method was initially developed as an assay of purity of Leptinotarsin, a protein in the hemolymph of the Colorado Potato Beetle (7) based on an idea of Fraenkel and Hsiao (8). Details of the method are given elsewhere (6). Enough flies for an experiment are chilled and remain immobile in a Petri dish on ice. A volume of the solution being assayed, usually 1.5 µl is injected into the fly with a fine syringe needle. The fly is left for 10 minutes at room temperature and is then quantally scored as Paralyzed or Active. At each dilution of a sample, 10 flies are injected and scored. The probit-log dose response relation is linear for STX with a slope of 4.2 (±0.55) and an ED50 of 360 (+21, -19[S.D.]) picogram (for an injection of 1.5 µl).

This is equivalent to toxin level in a shellfish of 48 µg/mg per 100 gr of shellfish meat. Since, for the toxin and its associated assay protocol there appears to be a unique response it is possible to obtain an estimate of the median ED50 from one set of injections, provided that the dose is not too far from the ED50 value. For STX this represents a practical limit of detection of about 20 µg/100 g of meat.

The fly bioassay is a simple, sensitive and inexpensive method for determining the toxicity of a variety of neurotoxins. It is at least as sensitive as the mouse Median Death Time for PSP and can be used as an alternative with the great advantage that it does not display the salt effect. In fact the response of the fly to toxic shellfish extracts can be accounted for by assuming that there is no "salt effect" as there is in the mouse Median Death Time method, but rather that "non-toxic" shellfish contain a low level (approximately 10 µg/100 gm meat) of STX-equivalent material. This assumption permits a reinterpretation of the response of mice to toxic shellfish and leads to agreement between the fly results and those reported for the mouse on the same shellfish samples. Combinatorial analysis gives the expected error distributions in single dose-response estimations of a sample's toxin content. Chi-square analysis of 126 samples show that the observed error distribution cannot be distinguished from those expected.

Literature Cited

1. Shimuzu, Y. In "Toxin Dinoflagellate Blooms," Taylor, D; Seliger, H., Eds.; Elswert/North Holland; New York, 1979; p.321.
2. Hall, S., Nave, R.A., Reichardt, P.S., and Swisher, J. In "Toxic Dinoflagellate Blooms," Taylor, D; Seliger, H. Eds.; Elswert/North Holland; New York, 1979; p. 345.
3. Official Method of Analysis. Assoc. Offic. Analyst. Chem. 1975, 38, 319.
4. Schantz, E.J., McFarren, E.F., Schafer, M.L., Lewis, K.H. J. Assoc. Off. Agric. Chem. 1958, 41, 160.
5. McFarren, E.J. J. Assoc. Off. Agric. Chem. 1959, 42, 263.
6. Siger, A., Abbott, B.C., Wong, T. Toxicon (in press).
7. McClure, W.O., Abbott, B.C., Baxter, D.E., Hsiao, T.H., Satin, L.S., Siger, A., Yoshino, J.E. Proc. Natl. Acad. Sci. USA 1980, 77, 1219-1223.
8. Hsaio, T.H., Frankel, G. Toxicon 1969, 7, 119-130.

RECEIVED July 20, 1984

18

Determination of Paralytic Shellfish Poisoning Toxins by High Pressure Liquid Chromatography

JOHN J. SULLIVAN and MARLEEN M. WEKELL

Department of Health and Human Services, Public Health Service, Food and Drug Administration, Seafood Products Research Center, Seattle, WA 98174

An improved high pressure liquid chromatographic (HPLC) procedure for the PSP toxins is described. The method involves separation of the toxins on a polystyrene divinylbenzene resin column (Hamilton, PRP-1) in the reversed phase mode using heptane and hexane sulfonic acids as ion-pairing reagents. Detection of the toxins is by fluorescence following post-column alkaline periodate oxidation. The sensitivity of the HPLC method is better than the standard mouse bioassay by at least a factor of four for each of the individual toxins.

Paralytic shellfish poisoning or PSP is a significant toxicological problem affecting shellfisheries in many of the temperate regions of the world. PSP problems arise following the accumulation in bivalve molluscs of a number of dinoflagellate derived neurotoxins (Figure 1). Traditionally, analyses for the presence of the PSP toxins in both shellfish and the causative dinoflagellates have been performed using the mouse bioassay (1). Due to the non-specific nature of the bioassay and a number of factors affecting its precision and accuracy (2), there is a need for alternative assay procedures.

Since the PSP toxins lack native fluorescence, useful UV absorption or adequate volatility, more traditional analytical procedures such as gas chromatography or spectrometry have proven ineffective in assaying for the toxins. A number of chemical assays for the toxins have been developed though with the fluorometric method of Bates and Rapoport (3) proving to be the most useful to date. This assay is based on oxidation of the PSP toxins under alkaline conditions to fluorescent derivatives. The assay is highly sensitive, fairly specific for the PSP toxins and was incorporated into a detection method in the column chromatographic separation of the toxins described by Buckley et al (4).

Following the incorporation of a number of modifications, the fluorometric method was utilized in a post column reaction system (PCRS) with separation of the toxins by high pressure liquid chromatography (HPLC) (5,6). While this system proved quite useful

This chapter not subject to U.S. copyright.
Published 1984, American Chemical Society

for routine toxin assays, a number of problems were evident. Using the procedure as described, it was not possible to achieve complete separation of all the PSP toxins on a single HPLC column. In addition, while the HPLC method was more sensitive than the mouse bioassay for most of the toxins, the detection limits for neosaxitoxin and gonyautoxins I and IV were too high to be useful due to the poor fluorescence of the N-1 hydroxy toxins (see Figure 1). In this paper is described a modified HPLC procedure in which marked improvements in toxin separation and detection can be achieved. The alterations to the previously published method include utilization of an alternate HPLC column and an improved reaction/detection system to achieve adequate sensitivity.

Materials and Methods

Apparatus. The major components in the HPLC system (Figure 2) include a Varian HPLC pump (model 5000, Varian Assoc., Walnut Creek, CA), Altex injection valve (model 210, Beckman Inst., Fullerton, CA), PRP-1 HPLC column (15 cm x 4.1 mm; Hamilton Co., Reno, NV), Kratos post column reaction system (PCRS) (model URS-051, Kratos Analytical Inst., Westwood, NJ), and a Perkin Elmer fluorescence detector (model LS-4, Perkin Elmer Corp., Norwalk, CT). The PCRS was modified slightly to accomplish the two step reaction necessary to achieve oxidation of the toxins and a flow diagram is illustrated in Figure 3. Pump A in the PCRS delivers the oxidant (periodic acid) and base (ammonium hydroxide) with two elevated chromatography columns (60 cm x 1.9 cm) used as reagent reservoirs. Either the oxidant or wash solution (water) is directed to pump A via the six port valve. It is important that the reservoirs and connecting tubing are well matched to assure that both oxidant and base drain at exactly the same rate. The reaction coil (1 ml volume, which is normally mounted internally in the PCRS) was removed from the PCRS and submerged in a thermostatically controlled water bath maintained at 75 C. The HPLC conditions which have proven to be effective in achieving efficient separation and detection of the PSP toxins are illustrated in Table I.

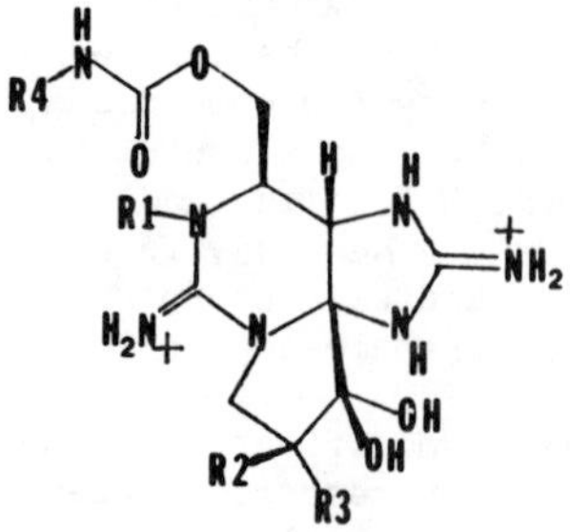

R1	R2	R3	Carbamate Toxins $(R4-H)$	N-Sulfocarbamoyl Toxins $(R4-SO_3^-)$
H	H	H	STX	B1
OH	H	H	NEO	B2
OH	H	OSO_3^-	GTX I	C3
H	H	OSO_3^-	GTX II	C1
H	OSO_3^-	H	GTX III	C2
OH	OSO_3^-	H	GTX IV	C4

Figure 1. PSP neurotoxins: saxitoxin-STX, gonyautoxin-GTX, neosaxitoxin-NEO, and B1-C4 are sulfocarbamoyl toxins.

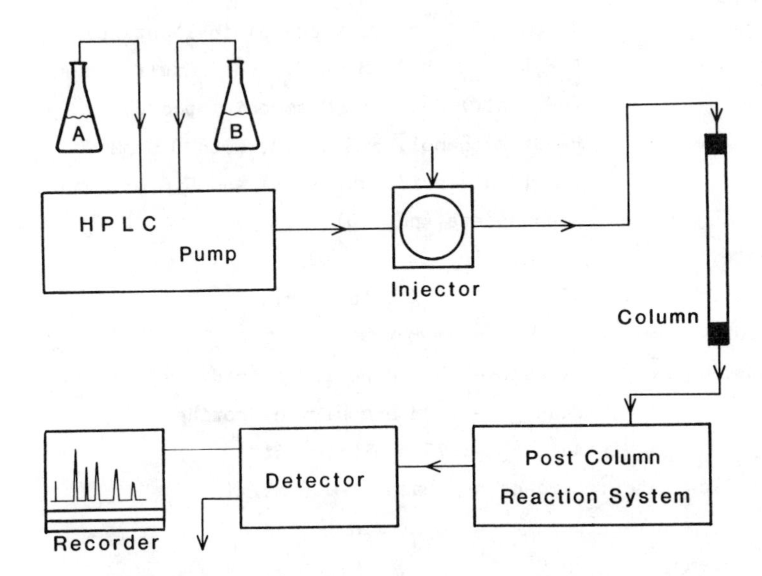

Figure 2. Flow diagram for the HPLC system.

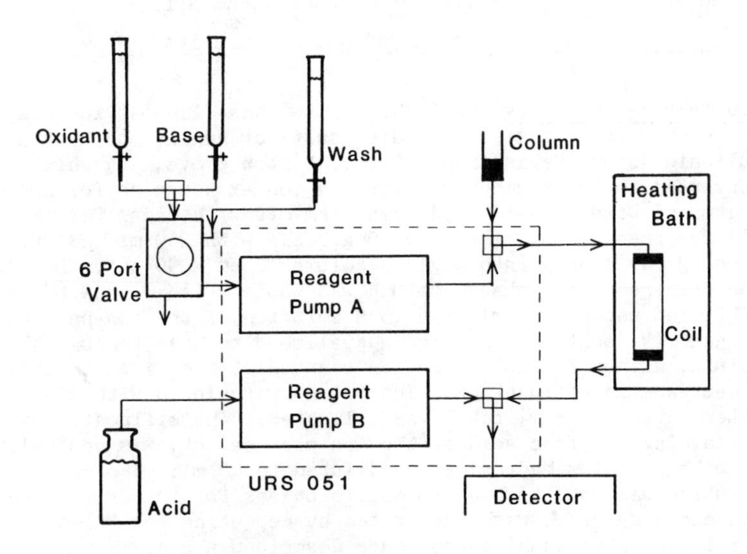

Figure 3. Detail of post column reaction system. Dashed line indicates components enclosed within model URS 051 reaction system.

Table I. HPLC Conditions

Separation	
Column:	15 cm PRP-1 (polystyrene divinylbenzene)
Mobile Phase:	A-Water; with 1 mM ea C_6 and C_7 sulfonic acid (NH_4 salt) and 1.5 mM ammonium phosphate (pH 7.0)
	B-50% Methanol, 50% water; with 0.5 mM C sulfonic acid (NH_4 salt) and 5.0 mM ammonium phosphate (pH 7.0)
Flow Rate:	1.3 ml/min
Gradient Program:	100% A to 100% B in 20 min.
Detection	
Reagents:	Oxidant - .005 M periodic acid
	Base - .5 M ammonium hydroxide
	Acid - .75 M nitric acid
Reagent Flow Rates:	Oxidant and Base - .8 ml/min
	Acid - .4 ml/min
Reaction Temp:	75°C
Reaction Time:	29 sec
Detector:	Excitation 340 nm, 15 nm slits
	Emission 400 nm, 20 nm slits

Mobile Phase and Reagents. The mobile phase ion-pairing reagents are commercially available sodium salts of hexane and heptane sulfonic acids (Regis Chemical Co., Morton Grove, IL) which were converted to their ammonium salts by ion exchange as follows: A column of Dowex 50W-X8 (hydrogen form) (Dow Chemical Co., Midland, MI) was prepared by slurrying 30 g resin with 200 ml .2N HCl and pouring the slurry into a glass column (2 cm x 40 cm). Before use the prepared ion exchange column was washed with 1 L 5% HCl and 1 L distilled water. 20 ml of a .5 M solution of the ion-pairing reagent (Na salt) was applied and allowed to pass through the column, with effluent collection started at the first indication of a decrease in effluent pH. The column was rinsed with distilled water until ca 35 ml total was collected. The effluent, now containing the free acid of the ion-par reagent, was neutralized (pH 7) with ammonium hydroxide and diluted to 50 ml. The resulting .2 M solution was used to prepare mobile phases for the HPLC. The ion-exchange resin was regenerated by repeating the HCl and distilled water washing procedure described above.

Mobile phases were prepared by adding the appropriate quantities of ion-pairing reagent solutions and phosphoric acid to distilled water with subsequent titration to pH 7.0 ($\pm$.05) with ammonium hydroxide. In the case of mobile phase B, pH adjustment was carried out before addition of methanol. The final buffer concentrations (1.5 mM in A and 5.0 mM in B) are expressed as phosphate and all mobile phase solutions were filtered through .45 um filters (type HA, Millipore Corp.) before use.

Reagents to carry out the alkaline oxidation of the toxins were prepared as originally described (5) with the exception that nitric acid was used in place of acetic acid. The reagent concentrations and flow rates are shown in Table I.

Sample and Standards Preparation. Shellfish samples were extracted by the standard bioassay procedure (1). Prior to injection into the HPLC, protein was precipitated with 1.5% TCA at 1°C (60ul) 50% TCA to 2 ml extract). The samples were then centrifuged, an aliquot diluted 1:3 with water and filtered (.45 um).

Standards for use in the HPLC were prepared from individual, purified PSP toxins as previously described (5).

Results and Discussion

The HPLC method as originally described (5,6) lacked adequate sensitivity for several of the PSP toxins and to obtain a complete toxin profile, duplicate injections on two different types of HPLC columns were necessary. To solve these problems, modifications were made that included developing the PSP toxin separations on a different column and the utilization of a more efficient reaction/detection system. Incorporation of these changes enables complete separation of all the toxins with the exception of C1 and C2 which still co-elute (Figure 4). In addition, the improved HPLC

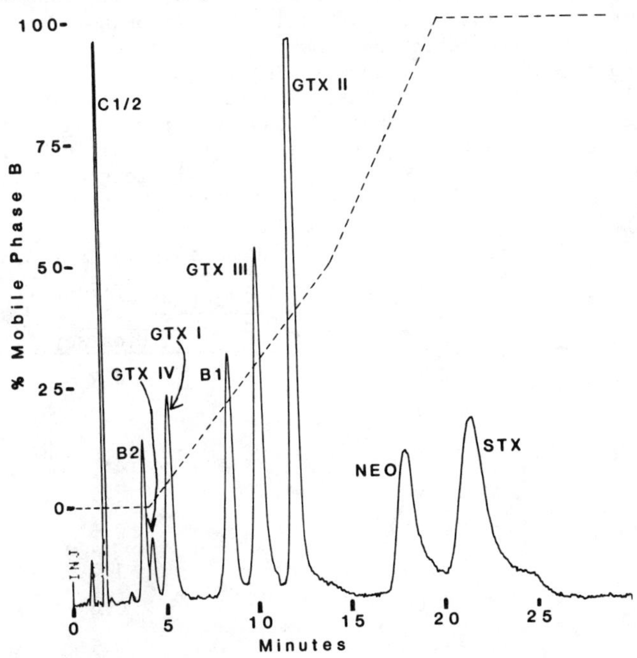

Figure 4. Chromatogram of PSP toxin standard (20 ul). Dashed line is % mobile phase B (see text) and amounts injected (in picomole) are : C½-10; B2-40; GTX IV-8; GTX I-32; B1-20; GTX III-3.4; GTX II-6.6; NEO-40; and STX-10.8.

system exhibits sensitivities that are 10 to 25 times better than the previous method with detection limits that are well below those of the standard mouse bioassay (Table II).

Toxin Separations. A number of columns have been evaluated for suitability using the HPLC method. Of all columns tested to date, the Hamilton PRP-1 column (polystyrene divinylbenzene resin) has proven to be the most useful. Toxin retention on this column is controlled by: 1) methanol concentration, 2) mobile phase ionic strength, 3) chain length of the ion-pair reagent, and 4) mobile phase pH.

Decreased retention on the column is observed for all the toxins by increasing methanol concentration or raising ionic strength of the mobile phase and both of these parameters are utilized to effect elution of the toxins from the column. The effect of altering these two parameters is similar on all the toxins as the relative retention times are not altered significantly. Conversely, altering mobile phase pH or ion-pair reagent chain length has a marked effect on relative retention times and this provides an effective means of controlling toxin separations. As the pH of the mobile phase is decreased, the retention times of the N-1 hydroxy toxins approach those of their non-hydroxylated counterparts (Figure 5). This effect is likely due to deprotonation of the N-1 hydroxy group (pKa ca 6.8) at the higher pH with subsequent neutralization of one charge on the molecule (7). Adjusting the pH of the mobile phase is an effective means of controlling the retention of the N-1 hydroxy toxins and it is apparent that pH values of ±.05 units are important for achieving precision in toxin retention times.

Proper choice of the ion-pairing reagent is also important in achieving adequate separations. The relative retention times of the sulfocarbamoyl toxins, B1 and B2, can be slightly altered by changing the ion-pair reagent chain length. Toxin retention increases for all of the toxins as the chain length is increased with the effect more pronounced for the sulfocarbamoyl toxins.

Table II. Comparison of HPLC and Mouse Bioassay Detection Limits

	Detection Limit (uM)	
Toxin	HPLC	Bioassay
B1	0.055	6.7
B2	0.175	6.7
C1	0.006	59.0
C2	0.006	3.9
GTX I	0.140	0.5
GTX II	0.005	1.0
GTX III	0.005	0.6
GTX IV	0.140	0.5
NEO	0.110	0.4
STX	0.015	0.5

1. Twice baseline noise, 20 ul injection

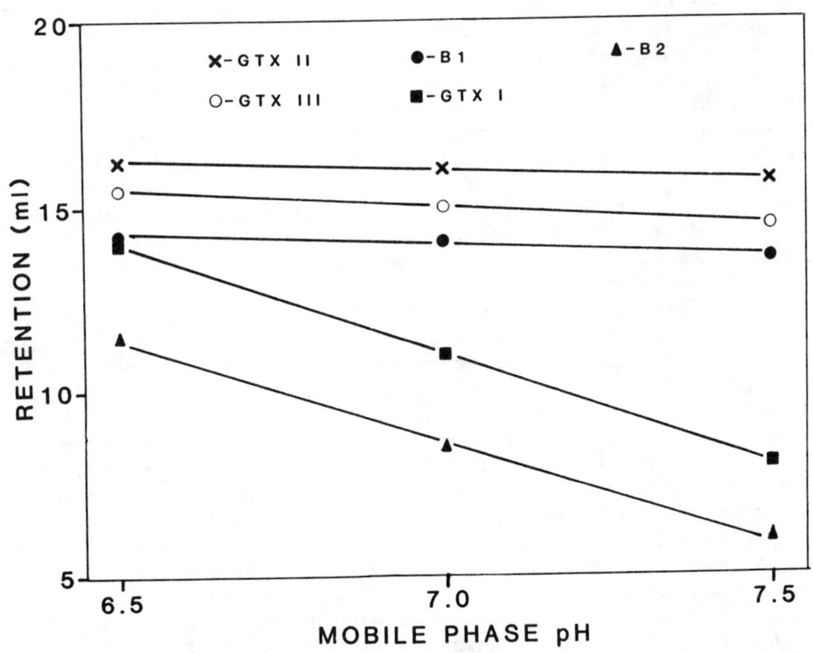

Figure 5. Toxin retention on the PRP-1 column at various mobile phase pH values. 25 min. linear gradient from mobile phase A (2.5 mM ea C6, C7 SO_3; 2 mM amm. phosphate) to mobile phase B (60% MeOH; 5 mM C7 SO_3; 10 mM amm. phosphate).

Combinations of alkyl sulfonic acids from C5 to C8 were tested and the most efficient separations were achieved by running a gradient from 50% C6 plus 50% C7 to 100% C7. Use of the mobile phase compositions illustrated on Table I with the gradient shown on Figure 4 has provided efficient separations for all of the toxins except C1 and C2. Small changes in mobile phase composition and/or gradient profile may be necessary to achieve the desired separations on any particular column or HPLC system.

Toxin Detection. The HPLC method as originally described utilized a PCRS based on Technicon auto-analyzer equipment with fabricated union tees, reaction coil and flow cell. Although this system worked well, it was apparent that improvements could be made by utilizing a system designed specifically for HPLC since band broadening could be greatly reduced. A PCRS that incorporates two pumps such as the one described here is ideally suited for this application. As observed in the original method, periodate concentration, reaction temperature, reaction time and pH all affect fluorescent yield of the toxins. It was found that the optimum conditions of periodate concentration and temperature described in the original method lead to substantial over-oxidation and reduced fluorescence in the improved reaction manifold system. Lower temperature or lower periodate concentrations were necessary using the new manifold and a systematic study was conducted to define the reaction conditions that would lead to optimum fluorescence for all of the toxins. In this study no one set of conditions (periodate concentration, temperature and reagent flow rate) was found to be optimum for all of the toxins. The conditions specified in Table I are therefore a compromise that produces adequate sensitivity for all of the toxins. In general, the low periodate concentration/high temperature reaction specified in Table I increases the response of the N-1 hydroxy toxins relative to their non-hydroxylated counterparts and this is used to favor oxidation of these normally poorly fluorescing toxins.

A further consideration in using this PCRS is precipitate formation in the reaction manifold. Since the flow path volumes are greatly reduced in this system, any precipitation can cause blockage. Precipitation of oxidation reagents will occur in the manifold system if the methanol concentration is too high. Consequently, care must be taken to insure that, if 100% methanol is used to rinse the HPLC column after prolonged use (a procedure that is strongly advised), all methanol must be completely washed out of the system before reagent flow is resumed. In addition, it has been found that under continuous use of the system, precipitates can form in the manifold system and this is manifested by increased back pressure and baseline noise. It is recommended that after 4 to 6 hours continuous use, the system is washed out by opening the pump A purge valve and increasing pump B flow to maximum, forcing nitric acid solution back through the reaction coil in the reverse direction. Precipitation problems are minimized by maintaining a high flow of oxidation reagents relative to mobile phase and the conditions illustrated in Table I are suggested for trouble-free operation of the system.

A major contributing factor to the increased sensitivity of the improved HPLC system over that originally described (5) is the detector. The original method utilized a fluorescence spectrophotometer adapted for HPLC detection by use of a fabricated 40 ul flow cell. The present system utilizes a highly sensitive HPLC fluorescence detector and this contributes greatly to the improved detection limits.

Conclusions

The HPLC system as described here represents a significant improvement over the original system. These improvements have made it possible to obtain a complete toxin profile from a single injection with detection limits well below those of the mouse bioassay (Table II). Analyses of actual shellfish samples (Figure 6) reveals that, as in the original system, no interfering peaks appear in the chromatogram in the region where the majority of the toxins elute, with the exception of extraneous material eluting near the solvent front with retention times similar to C1 and C2. Since the toxicity of these two toxins is extremely low (8), no significant error would occur in total toxicity for the majority of shellfish samples. The HPLC system as described here should prove to be a useful tool for the analysis of PSP toxins in both research and monitoring programs.

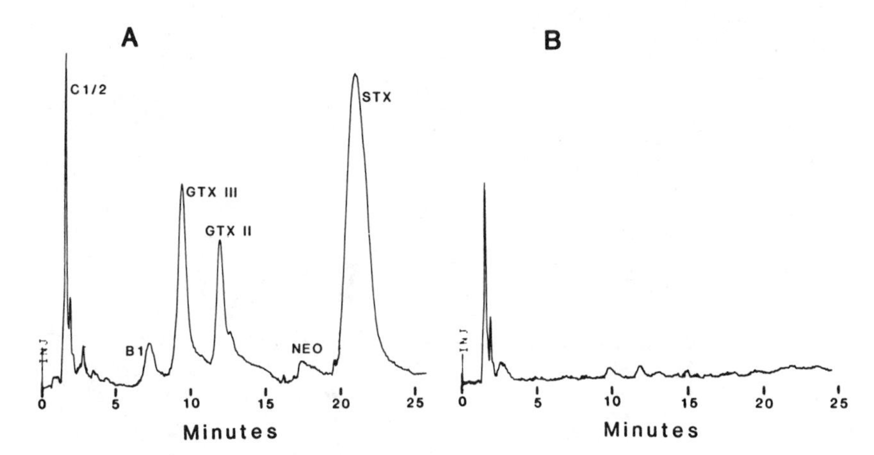

Figure 6. A-Chromatogram of toxic butter clam extract showing the presence of the PSP toxins. B-Chromatogram of extract from non-toxic (bioassay) mussels showing the presence of a trace of GTX II, GTX III, and C½. Conditions as in Table I with gradient shown in Figure 4.

Acknowledgments

The authors thank Sherwood Hall (Woods Hole Oceanographic Inst.) for the supply of purified PSP toxins to standardize the HPLC, and Ms. Linda Vernon for typing of the manuscript.

Literature Cited

1. "Official Methods of Analysis"; Assoc. of Off. Anal. Chem. (13th Ed.), Arlington, VA, 1980.
2. Krogh, P. Nord. Vet.-Med. 1979, 31, 302-08.
3. Bates, H.A.; Rapoport, H. J. Agric. Food Chem. 1975, 23, 237-39.
4. Buckley, L.J.; Oshima, Y.; Shimizu, Y. Anal. Biochem. 1978, 85, 157-64.
5. Sullivan, J.J.; Iwaoka, W.T. J. Assoc. Off. Anal. Chem. 1983, 66, 297-303.
6. Sullivan, J.J. Ph.D. Dissertation, University of Washington, Seattle, WA, 1982.
7. Shimizu, Y.; Hsu, C.; Fallon, W.E.; Oshima, Y.; Miura, I.; Nakinishi, K. J. Am. Chem. Soc. 1978, 100, 6791-93.
8. Fix Wichmann, C.; Niemczura, W.P.; Schnoes, H.K.; Hall, S.; Reichardt, P.B.; Darling, S.D. J. Am. Chem. Soc. 1981, 103, 6977-78.

RECEIVED February 7, 1984

19

Diarrhetic Shellfish Poisoning

TAKESHI YASUMOTO[1], MICHIO MURATA[1,3], YASUKATSU OSHIMA[1], GAYLE K. MATSUMOTO[2], and JON CLARDY[2]

[1]Department of Food Chemistry, Faculty of Agriculture, Tohoku University, Tsutsumidori, Sendai 980, Japan

[2]Department of Chemistry, Baker Laboratory, Cornell University, Ithaca, NY 14853

General aspects of diarrhetic shellfish poisoning including epidemiology, geographical distribution, magnitude of impacts to public health and shellfish industries are described. Succeeding to the previous identification of *Dinophysis fortii* as the origin of shellfish toxins and determination of dinophysistoxin -1 as 35(*S*)-methyl okadaic acid, isolation and structural determination of dinophysistoxin-3 and two novel polyether lactones named pectenotoxin-1 and -2, and identification of *Dinophysis acuminata* as the probable source of okadaic acid are newly reported.

Diarrhetic shellfish poisoning (DSP) is a term proposed by the authors to a shellfish poisoning distinctly different from the paralytic shellfish poisoning (PSP) in both symptomatology and etiology. Unlike PSP, the predominant human symptoms of DSP are gastrointestinal disturbances and no fatal cases have been reported (1). Nevertheless, the high morbidity rate and worldwide distribution of DSP make it a serious threat to both public health and shellfish industries. In a early stage of investigation we established that the origin of the shellfish toxins is the dinoglagellate *Dinophysis fortii* (2). Toxins named dinophysistoxins (DTX_s) were found to be structurally related to okadaic acid, a C_{38} polyether fatty acid first isolated from sponges (3) and then from the dinoflagellate *Prorocentrum lima* (4). A polyether toxin isolated from mussels and coded dinophysistoxin-1 (DTX_1) was identified as 35(*S*)-methyl okadaic acid in the previous study (5). Subsequent study revealed coexistence of numerous toxins with either similar or entirely different skeletons. The purpose of this paper is to present the general aspects of this relatively new type of shellfish poisoning and to report the isolation and structural determination of dinophysistoxin-3, and two novel polyether lactones named pectenotoxin-1 and -2. Detection of okadaic acid and isolation of pectenotoxin-3,-4,

[3]Current address: Suntory Institute For Bioorganic Research, Wakayamadai, Shimamoto-cho, Mishima-gun, Osaka 618, Japan

0097-6156/84/0262-0207$06.00/0
© 1984 American Chemical Society

and -5 as well as identification of *Dinophysis acuminata* as the probable source of okadaic acid are also described.

Materials and Methods

Materials. The following shellfish specimens were collected for toxin analysis from the northeastern part of Honshu, Japan, during the infestation period: the mussel *Mytilus edulis* at Miyagi Prefecture, the scallop *Patinopecten yessoensis* at Aomori Prefecture, the short-necked clam *Tapes japonica* at Fukushima Prefecture, and *Gomphina melanaegis* at Ibaraki Prefecture.

Mouse bioassay. The digestive glands of shellfish were extracted thrice with acetone at room temperature. After removal of acetone by evaporation, the aqueous suspension was extracted thrice with diethyl ether, the combined ether solution was backwashed twice with small portions of water and evaporated. The residue was suspended in 1% Tween 60 solution and serially diluted suspensions were injected intraperitoneally into mice weighing 17-20 g each. The mice were observed for 24 hr and the minimum amount of toxin required to kill a mouse at 24 hr was defined as one mouse unit (1).

Thin layer chromatography. Thin layer chromatography (TLC) was carried out on precoated Silica gel 60 plates (Merck) with the solvent system benzene-acetone-methanol-6N acetic acid (150:80:19:1). Toxins were detected by heating the plates after spraying 50% sulfuric acid.

Instruments. NMR spectra were measured with FX-100, FX-400 (JEOL) and NT-360 (Nicole) instruments, mass spectra with a Hitachi M-80 mass spectrometer, IR spectra with a JASCO A-202 spectrometer, and UV spectra with a Hitachi 124 spectrophotometer.

Gas chromatography. Gas chromatographic analyses were conducted on a Hitachi 163 instrument equipped with hydrogen flame ionization detectors. Okadaic acid and dinophysistoxin-1 were trimethylsilylated with Tri Sil "Z" either intact or after derivatization with diazomethane into methyl esters. A glass column (3 x 800 mm) packed with 2% OV-101 on 60/80 mesh Uniport HP was used for analysis of okadaic acid and dinophysistoxin-1. The column temperature was maintained at 315°C and nitrogen flow rate at 30 ml/min. For fatty acid analyses a glass column (3 x 2000 mm) packed with 10% DEGS on Chromosorb WAW DMCS 60/80 mesh was used. Column temperature was kept at 165°C and the nitrogen flow rate at 30 ml/min. The reference fatty acids were purchased from Wako Pure Chemicals.

Isolation of toxins. The digestive glands of shellfish were extracted with acetone at room temperature. After removal of the acetone by evaporation, the aqueous suspension was extracted with diethyl ether. The ether soluble residue was successively chromatographed twice over silicic acid columns with following solvents: benzene to benzene-methanol (9:1), and diethyl ether to diethyl ether-methanol (1:1). To avoid degradation of dinophysistoxin-3 by contaminant acid, the silicic acid was washed with dilute sodium hydroxide solution and then with water prior to activation at 110°C. Toxic residue obtained in the second eluates was separated into two fractions by passing through Sephadex LH-20 column (28 x 100 mm) using benzene-methanol (1:1) solution. Fractions containing dinophysistoxin-3 (# 24-26, 10 ml each) and those containing other toxins (# 28-31) were combined respectively. Dinophysistoxin-3 was further purified on LiChroprep RP-8 column with methanol-water (20:1) and then on μBondapak C_{18} column with methanol-water (93:7). The toxic fraction containing dinophysistoxin-1 and pectenotoxins was chromatographed first over a Lobar column (LiChroprep RP-8, sizeB) using methanol-water (17:3) as solvent and next on an alumina column (Woelm, basic, activity III) with following solvents: chloroform, chloroform-methanol (9:1), chloroform-methanol (1:1), methanol, and methanol-1% ammonium hydroxide (1:1). Pectenotoxins recovered from the first two eluates were treated on LiChroprep RP-8 column with acetonitril-methanol-water (2:2:3). Final separation of pectenotoxins was achieved on a Develosil column (4 x 250 mm, Nomura Kagaku) using dichloromethan-methanol (98:2) as solvent. Dinophysistoxin-1 was eluted from the alumina column with aqueous methanol containing ammonium hydroxide. Further purification was achieved by repeating chromatography on LiChroprep RP-8 column (5 x 1000 mm) using acetonitrile-methanol-water (3:2:2) as solvent. Separation and purification of toxins were monitored by mouse bioassay, TLC, and a UV-spectromonitor at wave length 220 nm for dinophysistoxin-1 and -3, and at 235 nm for pectenotoxins. Purification of okadaic acid from mussel specimen was carried out in essentially the same manner as employed for purification of dinophysistoxin-1.

Results

General outline of DSP. During the period of 1976-1982, more than 1,300 people were officially diagnosed as DSP cases in Japan. Frequency of signes and symptoms in patients were as follows: diarrhea (92%), nausea (80%), vomiting (79%), abdominal pain (53%), and chill (10%). Incubation period ranged from 30 min to several hr but seldom exceeded 12 hr. Around 70% of the patients developed symptoms within 4 hr. Suffering may last for 3 days, in severe cases, but leaves no after-effect. Causative shellfishes were the mussels *Mytilus edulis* and *M. coruscum*, the scallops *Patinopecten yessoensis* and *Chlamys nipponensis akazara*, and the short-necked clams *Tapes japonica* and *Gomphina melanaegis*. The method of cooking did not alter the toxicity of the causative shellfishes but intoxication could be avoided if the digestive glands were eliminated beforehand. The minimum amount of toxin to induce symptoms in an adult was estimated to be 12 mouse units from the analyses of the left overs of the patient meals. The maximum allowance level of toxins in shellfish meat was set by the government regulation at 5 mouse units/100 g meat. Infestation period ranges from April to September and the highest toxicity of shellfish is overved during May to August, though it may vary locally. Heavy infestation of shellfish normally occurs in northeastern part of Japan. In southwestern areas shellfish toxicity is low and decreases rapidly. The causative organism which transfer toxins to shellfish was identified as *Dinophysis fortii* (2). Mussels and scallops may become toxic beyond the regulation level in the presence of this dinoflagellate at cell density of 200/1, or even lower (2). Gas chromatographic analysis confirmed, for the first time, the presence of okadaic acid in mussels collected in April when *D. fortii* was scarce (3 cells/l) but *Dinophysis acuminata* was relatively abundant (500 cells/l), indicating that *D. acuminata* was also implicated in DSP.

Properties of toxins. Dinophysistoxin-1 was isolated as a white amorphous solid; m.p. 134°C; $[\alpha]_D^{20}$ +28 (c 0.046, chloroform); Rf in TLC 0.42; minimum lethal dose to mouse 160 µg/kg (i.p.). EI mass spectrum gave a dehydrated ion peak at m/z 800, pointing to a composition $C_{45}H_{68}O_{12}$. Overall features of both PMR and CMR spectra of dinophysistoxin-1 closely resembled those of okadaic acid (I) except the presence of an additional methyl in dinophysistoxin-1. Comparison of the spectra of okadaic acid and dinophysistoxin-1 and supplemental spin-spin decoupling measurements enabled us to assign dinophysistoxin-1 to 35(*S*)-methyl okadaic acid (II) (5). A component tentatively named dinophysistoxin-2 was slightly more polar than dinophysistoxin-1 but its characterization was unsuccessful due to the extreme smallness of the sample size.

Pure dinophysistoxin-3 was obtained from the scallop digestive gland as a colorless solid; Rf in TLC 0.57; minimum lethal dose to mouse 500 µg/kg (i.p.); no UV absorption maximum above 220 nm. The PMR spectrum of dinophysistoxin-3 resembled that of dinophysistoxin -1 but contained additional signals assignable to a fatty acid moiety; δ 0.88 (3H, t, terminal methyl), δ 1.25 (*ca.* 25H, br s., methylenes), δ 1.98 (*ca.* 2H, m, methylene neighboring a double bond), δ 2.20 (2H, dd, methylene adjacent to an ester), and δ 5.35 (*ca.* 5H, olefinic protons) ppm. Gas chromatographic analyses of hydrolysis products confirmed the presence of dinophysistoxin-1 and following fatty acids: $C_{14:0}$ (13%), $C_{16:0}$ (29%), $C_{18:3}$ (3%), $C_{18:4\omega3}$ (9%), $C_{20:5\omega3}$ (23%), and $C_{22:6\omega3}$ (23%). Integration of proton signals suggested that dinophysistoxin-3 is a mixture of dinophysistoxin-1 derivatives having one of the above fatty acids in an ester linkage. Another distinction between PMR spectra of dinophysistoxin-1 and -3 was that one oxymethine proton of the former (δ 3.40) was deshielded by 1.36 ppm (δ 4.76) in the latter. This proton was assigned to 7-H because its signal shape and coupling constants (dd, 11.8, 4.2 Hz) was compatible with the axial-axial and axial-equatorial couplings of 7-H to 6-2H. Signals at δ 4.10 (br d, 10 Hz) and δ 4.07 (t, 10 Hz) assignable to 24-H and 27-H respectively (3) remained unchanged. Thus, attachment of a fatty acid at C_{24}-OH or C27-OH was ruled out. Acylation of C2-OH was also denied because the resonance of 39-CH_3 (δ 1.37 ppm) remained unaffected in dinophysistoxin-3. From all these results dinophysistoxin-3 was assigned to 7-O-acyl dinophysistoxin-1 (III). Dinophysistoxin-3 was extremely vulnerable to acid, alkali and exposure to air. Leaving a dried sample, even at low temperature, easily resulted in the loss of toxicity and production of multiple spots on TLC. The presnece of this component in the mussel and two species of the short-necked clams was evidenced by liquid chromatographic analysis of intact toxin and by gas chromatographic analysis of dinophysistoxin-1 and fatty acids in the hydrolyzates.

Pectenotoxins were easily discernible by their strong UV absorption. The presence of 5 components was recognized by TLC and HPLC. Pectenotoxin-1 was isolated as white crystalline solid; m.p. 208-209°

C; $[\alpha]_D^{20}$ +17.1 (c 0.41, methanol); λ_{max} 235 nm (ϵ 12,400, methanol); Rf in TLC 0.43; minimum lethal dose to mouse 250 µg/kg (i.p.). The PMR spectrum exhibited 6 methyls (3 doublets, 3 singlets) and 3 olefinic protons of trans-trans conjugated diene. The CMR spectrum indicated 47 carbons: 1 ketone singlet at δ 213.6; 1 carbonyl singlet at δ 173.3; 4 olefinic carbons at δ 140.5(d), 135.6(d), 130.7(s), and 120.6(d); 3 ketal or hemiketal singlets at δ 108.8, 107.6, and 97.2; 15 oxycarbons (3 singlets, 10 doublets and 2 triplets) between δ 85.5 and 61.3 ppm. A secondary ionization mass spectrum gave a m/z 875 (M^+ +H), indicating a formula $C_{47}H_{70}O_{15}$. The assignment of the structure of pectenotoxin-1 was achieved by single crystal X-ray diffraction techniques. This structure (IV) is completely consistent with the spectral data and an essentially complete assignment of the proton signals is shown in Fig. 1.

Pectenotoxin-2 was obtained as white solid: $[\alpha]_D^{20}$ +16.2(c 0.015, methanol); λ_{max} 235 nm (ϵ 16,000, methanol); Rf in TLC 0.71; minimum lethal dose to mouse 260 µg/kg (i.p.). A secondary ionization mass spectrum gave a m/z 859 (M^+ +H), pointing to a formula $C_{47}H_{70}O_{14}$. Comparison between PMR spectra of pectenotoxin-1 and -2 confirmed that signals due to 43-2H (δ 3.52 and 4.56 ppm) of the former were replaced by a methyl singlet at δ 1.37 ppm in the latter but all other proton signals were in a good agreement. In CMR spectra, the resonances at δ 65.8(t) assignable to C43 of pectenotoxin-1 was transformed to a quartet and appeared at δ 26.0 ppm in the spectrum of pectenotoxin-2. The resonances at δ 83.7(s), 27.0(t), and 31.5 (t) ppm in the spectrum of pectenotoxin-1 were shifted to 80.7(s), 33.8(t), and 36.2(t) ppm in the spectrum of pectenotoxin-2, conforming to the loss of deshielding effect of 43-OH on C18 and the loss of γ-effect of the same function on C17 and C19. Other carbon signals of pectenotoxin-2 agreed well with those of corresponding signals of pectenotoxin-1 within a range of 0.3 ppm. On the basis of these results pectenotoxin-2 was assigned to 43-deoxypectenotoxin-1 (V). Three other components were tentatively code-named pectenotoxin-3, -4, and -5: Rf in TLC 0.49, 0.53, and 0.41, respectively. The secondary ionization mass spectra of pectenotoxin-3 and -4 gave a (M^+ +H) ion at m/z 875, indicating that they share the same formula with pectenotoxin-1. The mass spectrum of pectenotoxin-5 gave a (M^+ +H) ion at m/z 877, suggesting that it is a dihydro derivative of either of pectenotoxin-1,-3, and -4. Further structural elucidation of these components is under way.

Discussion

Aside from 1,300 patients recorded in Japan during 1976-1982, more than 5,000 DSP cases were reported in Spain in a single year of 1981 (6). Sporadic occurrences of DSP is also noted in the Netherlands (7) and Chile (8). In view of the wide geographical distribution of *D. fortii* and its related species, it is likely that DSP occurrence is more frequent and widespread than what is known today. This assumption is deduced from the result of our field survey which indicates that people with mild symptoms do not report to a health office and symptoms are easily mistaken for those caused by bacterial contamination of foods. Apart from the impact to public health, DSP causes a serious economic damage to shellfish industry in Japan due

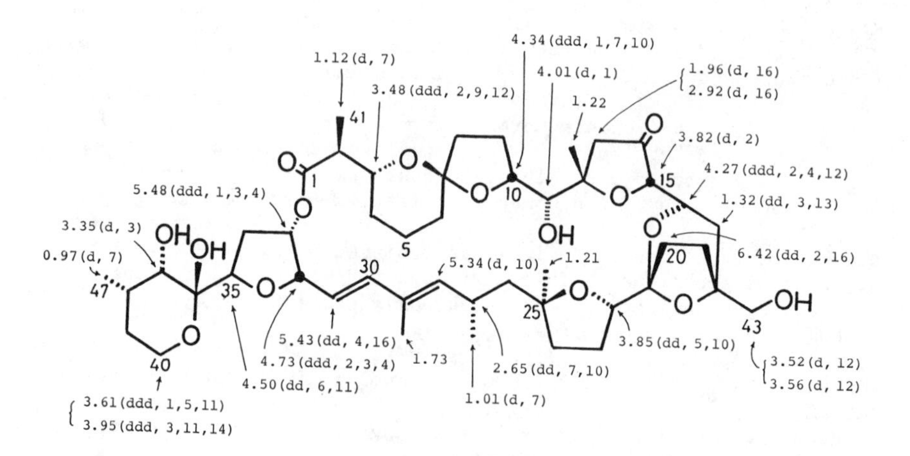

Figure 1. Assignment of proton signals of pectenotoxin-1

I okadaic acid : R_1 = H, R_2 = H

II dinophysistoxin-1 : R_1 = H, R_2 = CH_3

III dinophysistoxin-3 : R_1 = acyl, R_2 = CH_3

IV pectenotoxin-1 : R = OH

V pectenotoxin-2 : R = H

2R,3R,7S,10R,11R,12S,15S,16S,18S,21S,22S,25S,27S,32S,
33S,35R,36R,37R,37S,38S

to its widespread occurrence and long duration of the infestation period. An enormous amount of effort is being paid by both government and private industries to monitor shellfish toxicity. Our work provides the basic information needed to the development of a rapid instrumental assay method.

The present work adds to a current topic of production of bioactive polyether compounds by dinoflagellates. Previous examples are the presence of okadaic acid in *Prorocentrum lima* (4), brevetoxins in *Ptychodiscus brevis* (9-11), ciguatoxin in *Gambierdiscus toxicus* (12,13), and goniodomin in *Gonyodoma* sp. (14). From chemical point of view, it is notable that pectenotoxin-1 and -2 are substantially different from previously described dinoflagellate toxins. Specifically they differ from others in having a longer carbon backbone (C_{40}), a C_{33} lactone ring rather than an open structure, and a novel dioxabicyclo ring. The large oxygen rich internal cavity is grossly similar to the cavities found in the polyether ionophores from terrestrial microorganisms. Coexistence of compounds of such different skeletons raises an interesting question about the biosynthesis of polyether compounds in dinoflagellates. The occurrence of okadaic acid in Japanese mussels exposed to *D. acuminata* coincides with the occurrence of this compound in Dutch mussels collected from an area where the same species of dinoflagellate was abundant (M. Kat and T. Yasumoto, unpublished data). The incidence of gastroenteritis caused by ingestion of mussels in Chile was associated with the occurrence of *Dinophysis acuta*, though details are not available (8). Further study seems necessary to determine toxicity of other species of *Dinophysis*.

Acknowledgment

The authors are grateful to Suntory Institute for Bioorganic Research for undertaking spectral measurements. This investigation was supported by grants from the Ministry of Education, Culture and Science, the Ministry of Agriculture, Forestry and Fisheries, Japan, NIH CA 24487 (JC) and the New York State Sea Grant.

Literature Cited

1. Yasumoto, T; Oshima, Y.; Yamaguchi, M. *Bull. Jpn. Soc. Sci. Fish.* 1978, 44, 1249-55.
2. Yasumoto, T.; Oshima, Y.; Sugawara, W.; Fukuyo, Y.; Oguri, H.; Igarashi, T.; Fujita, N. *Bull. Jpn. Soc. Sci. Fish.* 1980, 46, 1405-11.
3. Tachibana, K.; Scheuer, P.J.; Tsukitani, Y.; Kikuchi, H.; Engen, D.V.; Clardy, J.; Gopichand, Y.; Schmitz, J.J. *J. Am Chem. Soc.* 1981, 103, 2469-71.
4. Murakami, Y.; Oshima, Y.; Yasumoto, T. *Bull. Jpn. Soc. Sci. Fish.* 1981, 47, 1029-33.
5. Murata, M.; Shimatani, M.; Sugitani, H.; Oshima, Y.; Yasumoto, T. *Bull. Jpn. Soc. Sci. Fish.* 1982, 48, 549-52.
6. Campos, M.J.; Fraga, S.; Marino, J.; Sanchez, F.J. *International Council for the Exploration of the Sea Report* 1977-1981, 1982, pp. 1-8.
7. Kat, M. In"Toxic Dinoflagellate Blooms"; Taylor, D.L.; Seliger, H.H., Eds.; Elsevier North-Holland: New York, 1979; pp. 215-20.

8. Guzman, L.; Campodonico, I. *Publ. Inst. Patagonia Ser. Mon.* 1975, No. 9, p. 6.
9. Lin, Y.Y.; Risk, M.; Ray, S.M.; Engen, D.V.; Clardy, J.; Golik, J.; James, J.C.; Nakanishi, K. *J. Am. Chem. Soc.* 1981, **103**, 6773-75.
10. Golik, J.; James, J.C.; Nakanishi, K.; Lin, Y.Y. *Tetrahedron Lett.*, 1982, **23**, 2535-39.
11. Chou, N.H.; Shimizu, Y. *Tetrahedron Lett.*, 1982, **23**, 5521-24.
12. Scheuer, P.J. *Naturwissenschaften*, 1982, **69**, 528-33.
13. Yasumoto, T.; Nakajima, I.; Bagnis, R.; Adachi, R. *Bull. Jpn. Soc. Sci. Fish.*, 1977, **43**, 1021-26.
14. Sharma, G.M.; Michaels, L; Burkholder, P.R. *J. Antibiotics*, 1968, **21**, 659-64.

RECEIVED February 6, 1984

CIGUATERA

20

Mode of Action of Ciguatera Toxins

A. M. LEGRAND and R. BAGNIS

Institut Territorial de Recherches Médicales Louis Malardé, Papeete, Tahiti, French Polynesia

Ciguatera fish poisoning involves one principal toxin, named ciguatoxin (1) and several secondary toxins including maitotoxin (2) and scaritoxin (3), which frequently coexist with ciguatoxin in marine organisms. All these toxins have a strong lethal potency on mice (4-8). Many pharmacological studies concern ciguatoxin, the chief pathogenic compound. Scarcer are the data about scaritoxin, a toxin specific of parrot fishes, while the mode of action of maitotoxin is actually the purpose of a deep research. This review summarizes the present knowledge.

Ciguatoxin

The first report on the pharmacological action of ciguatoxin was by Li (9), he observed in the anaesthetized rat a biphasic cardiovascular response with bradycardia and hypotension followed by tachycardia and hypertension. These biphasic effects resembled those of typical anticholinesterase poisons. In *in vitro* experiments with rabbit intestinal segments, Li observed that ciguatoxin extracts produced a marked inhibition of human and bovine erythrocyte cholinesterases. He concluded that the pharmacological action of ciguatoxin was due to cholinesterase inhibition. The intraperitoneal acute toxicity of several ciguatoxic extracts was found to be closely correlated with their *in vitro* anticholinesterase activity (10). A similar correlation was established between this *in vitro* anticholinesterase activity and pupillary miosis triggered by topical application in rabbits'eyes (11). Yet, later studies showed that the pharmacological action of ciguatoxin cannot result from cholinesterase inhibition. Indeed, Ogura (12) reported a noncompetitive antagonism between neuromuscular action of toxic extracts and acetylcholine. The neuromuscular block induced by ciguatoxin in the rat sciatic nerve-gastrocnemius muscle preparation *in situ* was antagonized by the anticholinesterase physostigmine (13). Rayner *et al* (14) have commented on the differences between the respiratory action of ciguatoxin and typical anticholinesterase effects. Ogura *et al* (15) have not observed with ciguatoxin extracts the typical electroencephalographic activation of anticholinesterase agent while they had obtained the

0097-6156/84/0262-0217$06.00/0
© 1984 American Chemical Society

effect with physostigmine. Further investigation of the *in vitro* anticholinesterase action of ciguatoxin has shown that the effect is non-specific and rather corresponds with a widespread anti-enzymatic activity (16). Red blood cell cholinesterase assays from intravenously intoxicated rats (17) supported the conclusion that ciguatoxin is not an *in vivo* cholinesterase inhibitor (18).

Respiratory and cardiovascular effects of ciguatoxin have been studied. Cheng *et al* (19) reported that respiratory arrest induced by a lethal dose of ciguatoxin is caused by an inhibition of the central respiratory mechanism. Li's results (9) showing biphasic cardiovascular effects, transient hypotension and bradycardia succeeded by hypertension and tachycardia, were confirmed in the rat by Rayner (18). This biphasic response remained unchanged following spinalization, bilateral vagotomy, adrenalectomy and nephrectomy. The hypotension and bradycardia were found to be antagonized by hexamethonium, atropine and hemicholinium, while the hypertension and tachycardia were suppressed by atropine in doses greater than 10 mg/kg, β-blocking agents, prior reserpinization and tetrodotoxin. A similar biphasic response was observed in the rabbit by Laborit *et al* (20). They noted that diethazine (a central anticholinergic agent) hardly affected the responses, but calcium glutamate exerted a protective action. In the pentobarbital or α-chloralose anaesthetized cat, the same complex of effects were observed with slightly increased doses (21). The hypotension and bradycardia were suppressed by hexamethonium and atropine ; the hypertension and tachycardia were partially suppressed by atropine, propranolol and clonidine, a central antihypertensive agent. Moreover, phentolamine was found to be able to antagonize the bradycardia and hypotensive phase as well as the tachycardia and hypertensive response. Further antagonistic action was obtained with prazosin and yohimbine, which are respectively α_1- and α_2-adrenergic blocking agents. Prazosin completly suppressed the action of the toxin while yohimbine only partially antagonized the effects. The results suggest that ciguatoxin has an α-adrenergic mimetic action, specially at α_1-adrenoceptors.

Ciguatoxin effects were investigated also on various isolated organs and tissues. Banner *et al* (22, 23) tested semi-purified extracts of ciguatoxin on isolated toad sciatic nerve-gastrocnemius muscle and guinea-pig phrenic nerve-diaphragm preparations. They concluded that while the action potential of the nerve may be lost from long immersion in a solution of ciguatoxin, the immediate effect was upon the nerve-muscle junction. Similar results were obtained more recently by Miller *et al* (24). Ciguatoxin produced depolarization of the pedal ganglion cells of the sea hare *Aplysia juliana* (25). It increased the permeability of frog skin membrane to sodium ions (26) and depolarized muscle fibers of the frog (27, 28). This depolarization was antagonized by tetrodotoxin and Ca^{2+}-rich medium. Ciguatoxin was also reported to decrease the amplitude of the action potential of the nerve and block the neuromuscular conduction in the sciatic nerve-gastrocnemius preparation of the rat (29). Finally, the effects of ciguatoxin on the resting membrane potential appear to result from replacement of calcium by ciguatoxin at receptor sites which regulate steady-state sodium permeability (28). Consequently, low concentrations of ciguatoxin increase the excitability of excitable membranes while a conduction block could be presumed to result, at higher maintained concentrations, from gradually increasing internal sodium concentration.

In mammalian isolated atria, ciguatoxin induced a biphasic response. Oshika (30) observed in rat and rabbit atria a transitory negative inotropic phase, antagonized by atropine and hemicholinium, followed by a positive inotropic phase antagonized by MJ-1999, guanethidine and pre-treatment with reserpine. In guinea-pig atria, only positive chronotropic and inotropic effects were reported (31). The latter were partially antagonized by propranolol and phentolamine. In the rat myocardium, our studies showed at very low concentrations a positive chronotropic effect in right isolated atria and a negative inotropic effect in paced left isolated atria. Higher concentrations produced a decrease of the spontaneous rate of the right atria while in the left atria a biphasic response was observed, a transitory negative inotropic effect followed by a positive one. Negative inotropic and chronotropic effects were antagonized by atropine. Positive chronotropic response was suppressed by prior reserpinization and propranolol while positive inotropic one was only partially modified by reserpine pretreatment, propranolol and phentolamine (32). These results, as those of Oshika (30) indicate both cholinergic and adrenergic action of ciguatoxin in the isolated rat atria, suggesting a marked release of acetylcholine and a small release of catecholamines at the nerve endings. Yet, other mechanisms in addition to catecholamines release seems to be involved in the positive inotropic response. Preliminary experiments of us showed that the more purified the ciguatoxic extract was, the less important the positive inotropic effect (32). As Rayner and Szekerczes reported an inhibition of the Na^+-K^+ ATP-ase of human erythrocyte ghosts by crude extracts and none by purified extracts (33), thus, the possible presence of a cardiotonic contaminant in partially purified extracts has to be taken in consideration.

Ciguatoxin action was studied on smooth muscle. Miyahara and Shibata (34) investigated the effects of the toxin on the inhibitory mechanism of the guinea-pig *taenia caecum*. They demonstrated that ciguatoxin exerts a prominent calcium-sensitive action on nerve terminals and transmitter release even before affecting the reactivity of the neuro-effector organ. In the guinea-pig *vas deferens*, ciguatoxin caused a potent excitatory effect antagonized by reserpine, guanethidine, phentolamine Ca-free medium and tetrodotoxin (35-37). This action was considered to be due partially to a norepinephrine release and mainly to a supersensitivity of the post-synaptic membrane. Moreover, the authors reported an inhibitory effect on the contractile response to transmural stimulation. They suggest that this effect might result from a strong nerve membrane depolarizing action of ciguatoxin which inhibits the stimulation-induced action potential (35). In guinea-pig *ileum*, ciguatoxin extracts strongly decreased the effect of exogenous acetylcholine and histamine, while in spontaneously active rat *jenunum* preparation they inhibited activity (24). In helical strips of rabbit thoracic aorta, we observed that ciguatoxin slightly potentiated the contraction induced by norepinephrine. On the other hand, a ciguatoxin-induced contraction was obtained. This contraction was completly inhibited by phentolamine and prazosin, partially by yohimbine. The results confirm that ciguatoxin has a marked mimetic activity at the α_1-adrenoceptors.

In conclusion, the complex description of the effects observed on various preparations and their succession following slow dosage increase indicate that ciguatoxin acts at many target sites of slightly differing sensitivity. The reported action of ciguatoxin on the resting membrane potential suggests that the various effects observed may be the result of a direct depolarizing action on excitable membranes.

Scaritoxin

About scaritoxin, the following results were reported. This toxin was found to depress the oxidative metabolic process in the rat brain (20) and to have a depolarizing action on excitable membranes (38). In the guinea-pig atria, scaritoxin caused a marked potentiation of the acetylcholine negative inotropic and chronotropic effects (39). In rat atria, we observed biphasic inotropic and chronotropic effects similar to those of ciguatoxin. Negative inotropic and chronotropic effects were antagonized by atropine. The positive chronotropic response was suppressed by reserpine pretreatment or by propranolol, while the positive inotropic effect was only partially modified by reserpinization, propranolol and phentolamine (32). In the pentobarbital anaesthetized cat, scaritoxin exerted respiratory and cardiovascular effects similar to those of ciguatoxin (21). Although data about scaritoxin are limited, it seems that this toxin has a pharmacological mode of action very close to that of ciguatoxin. It may be speculated that ciguatoxin and scaritoxin are related compounds, scaritoxin resulting from ciguatoxin metabolic transformation in some fish species.

Maitotoxin

Respiratory and cardiovascular effects of maitotoxin have been studied in pentobarbital anaesthetized cats (21). Sublethal doses of maitotoxin induced an important hyperventilation phase, hypertension and a transitory tachycardia followed by slight bradycardia. Higher dosage caused respiratory depression, cardiac arrhythmias and tachycardia leading to cardiac failure. Artificial respiration did not modify the cardiac responses.

Maitotoxin effects were investigated also by *in vitro* experiments. Miyahara *et al* (31) reported a biphasic response in the guinea-pig atria, an initial positive response followed by a progressive decrease of both the rate and the force of contractions. In the rat, we observed a decrease of the spontaneous frequency of right atrial preparations and biphasic inotropic effects in left atrial preparations. Negative effects were not antagonized by atropine. The positive inotropic effect was modified very little by prior reserpinization or prior exposure to propranolol and phentolamine but was sensitive to Mn^{2+} ions. On the other hand, maitotoxin was found to strongly inhibit the Na^{+}-K^{+}ATP-ase from microsomes of cat and human kidneys (40). Takahashi *et al* (41,42) tested maitotoxin in a rat pheochromocytoma cell line, and observed a profound increase in Ca^{2+} influx and a Ca^{2+} dependent release of (3H)-norepinephrine and-dopamine. These effects were not modified by tetrodotoxin or Na-free medium but were inhibited by Mn^{2+}, verapamil, nicardipine and tetracaine, suggesting that maitotoxin activated the voltage-dependent calcium channel. Similar action was observed in the guinea-pig isolated *ileum*, *taenia caeci* and *vas deferens* and in the rabbit thoracic aorta (43-45). The authors reported that maitotoxin had a little effect on the Na^{+}-K^{+}-ATPase from porcine cerebral cortex (45). Recently, we studied maitotoxin effects on the action potential of isolated perfused rat hearts to determine the existence of a direct action on the myocardium. An increase in amplitude and duration of the ventricular action potential plateau was observed at low dosage. Higher doses caused a decrease of the spike amplitude and of the maximum rate of rise of the action potential while the lengthening of the plateau became more marked. These same effects appeared also in hearts from reserpinized animals. They did not develop during perfusion with Mn^{2+}ions, verapamil and low-calcium solution. Addition of Mn^{2+} or verapamil, low-calcium solution or Na-rich medium reversed to a large extent the effects of maitotoxin. These results can be taken as arguments in favor of an action involving calcium movements and/or calcium conductance. Although the mode of action of maitotoxin is not yet completly elucidated, this compound appears to have a specific cellular action. It could become an useful tool for biochemical and pharmacological studies.

The above-reviewed pharmacological results show some slight discrepancies specially with ciguatoxin. These discrepancies may be explained partly by the variability of the sample purity and partly by the presence in the extracts of secondary toxins. The complex description of the effects observed experimentally can explain the polymorphism of the clinical features. These pharmacological data can help the physician to improvise an appropriate treatment in ciguatera fish poisoning.

Acknowledgments

The local data have been obtained with the financial aid of both the French Ministry of Industry and Research and the Territorial Government of French Polynesia.

Literature cited

1. Scheuer, P.J. ; Takahashi, W. ; Tsutsumi, J. ; Yoshida, T. Science (Wash. DC), 1967, 21, 1267-8.
2. Yasumoto, T. ; Bagnis, R. ; Vernoux, J.P. Bull. Jap. Soc. Sci. Fish., 1976, 42, 359-65.
3. Chungue, E. ; Bagnis, R. ; Fusetani, N. ; Hashimoto, Y. Toxicon, 1976, 15, 89-95.
4. Banner, A.H. ; Scheuer, P.J. ; Sasaki, S. ; Helfrich, P. Alender, C.B. Ann. N.Y. Acad. Sci. 1960, 90, 770-87.
5. Yasumoto, T. ; Scheuer, P.J. Toxicon, 1969, 7, 273-6.

6. Chungue, E. Ph.D. Thesis, University of Montpellier, France, 1977.
7. Kimura, L.H. ; Hokama, Y. ; Abad, M.A. ; Oyama, M. ; Miyahara, J.T. Toxicon 1982, 20, 907-12.
8. Hoffman, P.A. ; Grande , H.R. ; Mc Millan, J.P. Toxicon 1983, 21, 363-9.
9. Li, K.M. Science 1965, 147, 1580-1.
10. Banner, A.H. ; In "Animal Toxins" ; Russel, F.E. and Saunders, p.R., Ed. ; Pergamon Press : London, 1967 ; pp.157-65.
11. Kosaki, T. ; Stephens, J. Fedn. Proc. Fedn. Am. Socs Exp. Biol. 1967, 26, 322.
12. Ogura, Y. Report to 2nd Annual Conference on Marine Toxins, University of Hawaii, 1967.
13. Kosaki, T.I. ; Anderson, H.H. Toxicon 1968, 6, 55-8.
14. Rayner, M.D. ; Kosaki, T.I. ; Fellmeth, E.L. Science (Washington. D.C.) 1968, 160, 70-1.
15. Ogura, Y. ; Nara, J. ; Yoshida, T. Toxicon 1968, 6, 131-40.
16. Baslow, M.H. ; Rayner, M.D. Proc. 4th Int. Congr. Pharmac., 1969, p. 180.
17. Rayner, M.D. ; Baslow, M.H. ; Kosaki, T.I. J. Fish. Res. Bd. Can. 1969, 26, 2 208.
18. Rayner, M.D. In "Drugs from the Sea Conference Proceedings", Marine Technology Society : Washington, D.C., 1969, pp.345-50.
19. Cheng, K.K. ; Li, K.M. ; Quintillis, Y.H. J. Path. 1969, 97, 89-92.
20. Laborit, H. ; Baron, C. ; Ferran, C. ; Laborit, G. Agressologie, 1979, 20, 81-96.
21. Legrand, A.M. ; Galonnier, M. ; Bagnis, R. Toxicon, 1982, 20, 311-5.
22. Banner, A.H. ; Helfrich, P. ; Scheuer, P.J. ; Yoshida, T. Proc. 16th Gulf. Caribbean Fish. Inst., 1963, pp. 84-98.
23. Banner, A.H. ; Shaw, S.W. ; Alender, C.B. ; Helfrich, P. South Pacific Commission, 1963, Technic Report 141, 17 p.
24. Miller, D.M. ; Dickey, R.W. ; Tindall, D.R. Fedn. Proc. Fedn. Am. Socs. Exp. Biol. 1982, 41, 1561.
25. Boyarsky, L.L. ; Rayner, M.D. Proc. Soc. Exp. Biol. Med. 1970, 134, 332-5.
26. Setliff, J.A. ; Rayner, M.D. ; Hong, S.K. Toxic. Appl. Pharmac. 1971, 18, 676-84.
27. Rayner, M.D. ; Kosaki, T.I. Fedn. Proc. Fedn. Am. Socs. Exp. Biol., 1970, 29, 548.
28. Rayner, M.D. Fedn. Proc. Fedn. Am. Socs. Exp. Biol. 1972, 31, 1139-1145.
29. Faucomprez, C. ; Ferezou, J.P. ; Bagnis, R. ; Chanfour, B. ; Niaussat, P.M. ; Drouet, J. Bull. Soc. Path. Exot. 1975, 68, 106-15.
30. Ohshika, H. Toxicon 1971, 9, 337-43.
31. Miyahara, J.T. ; Akau, C.K. ; Yasumoto, T. Res. Comm. Chem. Path. Pharmac., 1979, 25, 177-80.
32. Legrand, A.M. ; Bagnis, R. Toxicon 1984, in press.
33. Rayner, M.D. ; Szekerczes, J. Toxic. Appl. Pharmac. 1973, 24, 489-96.
34. Miyahara, J.T. ; Shibata, S. Fedn. Proc. Fedn. Am. Socs. Exp. Biol. 1976, 35, 842.

35. Ohizumi, Y. ; Shibata, S. ; Tachibana, K. J. Pharmac. Exp. Ther. 1981, 217, 475-80.
36. Ohizumi, Y. ; Ishida, Y. ; Shibata, S. Proc. 4th Int. Symp. Vascular Neuroeffector Mechanisms, 1983, pp. 301-4.
37. Ohizumi, Y. ; Ishida, Y. ; Shibata, S. J. Pharmac. Exp. Ther., 1982, 221, 748-52.
38. Rayner, M.D., Personal Communication presented to the South Pacific Commission, 1977.
39. Rentler, J.F. Thesis of Veterinary Surgeon, University of Lyon, France, 1980.
40. Bergmann, J.S. ; Nechay, B.R. Fedn. Proc. Fedn. Am. Socs. Exp. Biol. 1982, 41, 1562.
41. Takahashi, M. ; Ohizumi, Y. ; Yasumoto, T. J. Biol. Chem., 1982, 257, 7287-9.
42. Takahashi, M. ; Tatsumi, M. ; Ohizumi, Y. ; Yasumoto , T. J. Biol. Chem., 1983, 258, 10944-9.
43. Ohizumi, Y. ; Yasumoto, T. Br. J. Pharmac., 1983, 79, 3-5.
44. Ohizumi, Y. ; Kajiwara, A. ; Yasumoto, T. J. Pharmac. Exp. Ther., 1983, 227, 199-204.
45. Ohizumi, Y. ; Yasumoto, T. J. Physiol. (London), 1983, 337, 711-21 .

RECEIVED March 16, 1984

21

Ciguatoxigenic Dinoflagellates from the Caribbean Sea

DONALD R. TINDALL, ROBERT W. DICKEY, ROLLAND D. CARLSON, and GREGORY MOREY-GAINES

Department of Botany, Southern Illinois University, Carbondale, IL 62901

Over 70 separate sites in the British and United States Virgin Islands were surveyed for potential ciguatoxigenic organisms. A culture collection of 65 strains representing 18 species of the most common epiphytic/benthic and planktonic dinoflagellates has been established. To date, nine species have been grown in large-scale culture and screened for toxicity using mice. Five species, namely *Gambierdiscus toxicus*, *Prorocentrum concavum*, *P. mexicanum*, *Gymnodinium sanguineum*, and *Gonyaulax polyedra*, produced one or more toxic fractions which killed mice within 48 hours. Results from mouse bioassays and chromatographic treatments of extracts indicate that *G. toxicus* produces ciguatoxin (and 1 derivative) and maitotoxin; and *P. concavum* produces a scaritoxin-like toxin, a maitotoxin-like toxin, and at least one (up to 2) very potent, unnamed, fast-acting toxins. All of these toxins probably contribute to the ciguatera syndrome in the Caribbean.

Ciguatera is a serious human intoxication that results from eating certain tropical and subtropical fishes associated with coral reefs and adjacent coastal waters. The disease is manifested in humans by a great variety of symptoms (1-4). Those most consistently reported may be generally summarized as follows: (A) moderate to severe gastrointestinal disorders of relatively short duration; (B) moderate to severe neurological disorders that may persist for days, weeks, or months; and/or (C) in some cases, death due to respiratory failure. Ciguatera may be caused by over 400 species of marine fishes, including many that are highly prized for food (1). Banner (5) provided convincing argument supporting the view that ciguatoxin as defined by Scheuer and coworkers (6) was the principal toxin causing ciguatera throughout the world. However, a few authors have suggested that the variety of symptoms displayed and the inconsistent response to certain clinical treatments by patients

0097-6156/84/0262-0225$06.00/0
© 1984 American Chemical Society

suffering from ciguatera result from there being more than one primary toxin causing the disease (3, 7). The occurrence of "secondary" toxins (maitotoxin and/or scaritoxin) in association with ciguatoxin in fishes has been reported (8-14).

Randall (15), following an intensive study of the food habits of ciguatoxic fishes, concluded that all such fishes were tied to the coral reef through the food chain. Furthermore, he speculated that the organism responsible for the production of the toxin was a fine benthic alga or other microorganism. Evidence gathered during the past few years points to certain dinoflagellates that grow in close association with macroalgae and/or other bottom structures as progenitors of ciguatera toxins. The recently discovered species, *Gambierdiscus toxicus* (16) has been reported to produce both ciguatoxin and maitotoxin in the Pacific (17-22). Other species inhabiting the same microhabitats in the Pacific have been found to produce toxins which also may contribute to the ciguatera syndrome. These include *Amphidinium carteri*, *A. klebsii*, *Ostreopsis ovata*, *O. siamensis*, *Prorocentrum concavum* (20) and *P. lima* (20, 23, 24).

Our study of ciguatoxigenic organisms in the Caribbean was initiated in May 1978. The primary objectives of this study were: (A) to conduct an extensive search for potentially toxic organisms in areas of the British and United States Virgin Islands known to support toxic fishes and to select sites as permanent stations for subsequent ecological study; (B) to identify, isolate, and initiate cultures of potentially toxic microorganisms (with emphasis on species of dinoflagellates); (C) to develop large-scale cultures of each species for screening for toxicity; (D) to develop procedures for extraction and purification of toxins from microorganisms; (E) to determine appropriate nerve-muscle preparations for assaying toxins and for determining their sites and modes of action; and (F) to examine distribution, periodicity, growth, and development of toxic organisms in the field and under controlled laboratory conditions.

Preliminary results on toxicity of 3 species of Caribbean dinoflagellates have been reported, namely those on *Gambierdiscus toxicus* (25), *Prorocentrum concavum* (as *P.* cf. *lima*) (26), and *P. mexicanum* (as *P. rhathymum*) (27). A more detailed description of the effects of ether-soluble and water-soluble toxins from the same strain of *G. toxicus* on guinea pig ilea are presented elsewhere in this volume (28, 29). Results on the growth and development of *G. toxicus*, *P. concavum*, and *P. mexicanum* in culture also are presented (30).

The present paper reports on the development of a culture collection of potentially toxic dinoflagellates from the Virgin Islands, large-scale culture and screening for toxicity in nine species, and a more detailed consideration of the toxic properties of *G. toxicus* and *P. concavum*.

Results and Discussion

Study area. Over 70 separate sites in the Virgin Islands were surveyed (Figure 1). These sites represented a wide variety of habitat types, namely live and dead coral reefs, open sand bottoms landward and seaward of reefs, protected coves and bays, areas variously exposed to surf, mangrove stands, and salt ponds.

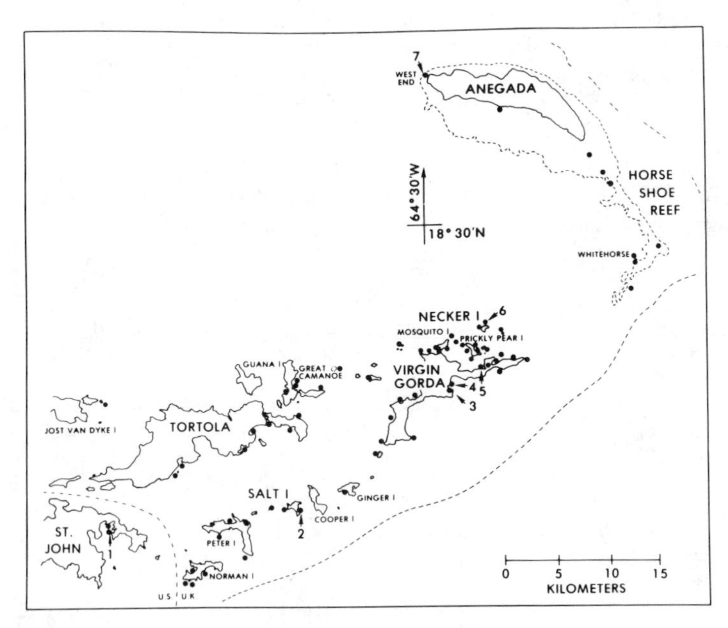

Figure 1. The study area in the British and United States Virgin Islands. Dots indicate sites surveyed prior to establishing permanent collecting stations. Seven permanent stations are indicated by number.

Substrate types characterizing these habitats included surfaces of dead coral, fine and coarse sands, organically rich mud-sand mixtures, wood and steel surfaces on wrecked ships, algal mats, attached macroalgae, and seagrass beds. Water depths at the various sites ranged from 0.2 to 25 meters.

Seven permanent collecting stations were established (Figure 1): (I) Water Creek in Hurricane Hole, St. John; (II) Salt Creek, Salt Island; (III) South Creek in South Sound, Virgin Gorda; (IV) Mattie Point in South Sound, Virgin Gorda; (V) Biras Creek in North Sound, Virgin Gorda; (VI) Northwest Beach, Necker Island; and (VII) West End, Anegada. These stations were selected because of their degree of exposure to open seas, unique bottom topography, history of supporting toxic fishes, and/or abundance of dinoflagellates and other algae.

Collection and culture of dinoflagellates. Large numbers of all conspicuous epiphytic, benthic, and planktonic species of dinoflagellates were collected at each station using plastic bags, syringes, and Van Dorn bottle, respectively. Each collection was transported to the field laboratory in separate polypropylene bottles. At least 30 cells of each species were isolated with micropipetts and inoculated into 10 ml of each of 10 different growth media, namely GPM (31), F (32), ES (33), AG (Carolina Biological Supply Co.), WC (34), ASP and ASP-6 (35), ASP-7 and ASP-12 (36), and NS (31). The resulting cultures were maintained in 30 ml tubes on Plexiglass light tables which were continuously illuminated from below with four 25 watt cool-white fluorescent lamps. Ambient temperature usually was about 29°C. Over 300 of these crude field cultures were successfully transported to Southern Illinois University, Carbondale.

After a 10 to 20 day period of acclimation to laboratory conditions, individual cells were isolated from field cultures supporting best growth. Careful examination of the field cultures revealed that best growth of all species occurred in ES enriched seawater. This medium with a 1.5% soil extract added was selected as our standard medium for stock cultures. Cells were washed several times by serial transfers through drops of sterile medium in Pyrex spot plates. After washing, 30 or more cells of each species were placed in tubes containing 10 ml of fresh, sterile medium. Ten ug/ml of germanium dioxide was added to some cultures to eliminate diatom growth. Once unialgal cultures were obtained, they were transferred to 50 ml volumes of medium in 125 ml flasks. These cultures were used as maintenance stocks and formed the basis of the Southern Illinois University (SIU) culture collection of Caribbean dinoflagellates. To date, we have isolated and cultured 65 strains representing 18 of the most common epiphytic/benthic and planktonic species (Table I). Several cultures of *Gambierdiscus toxicus* were initiated by inoculating 5 ml of medium with single cells from various locations. Cultures obtained in this manner are listed in Table I as clonal cultures. All cultures were grown at 27°C in 3200 lux constant cool-white fluorescent illumination and were transferred weekly, using 10% volume inocula.

Large-scale culture and harvest of dinoflagellates. Large-scale cultures were initiated by first inoculating 2 liters of our

Table I. Southern Illinois University (SIU) Culture Collection of Dinoflagellates From the Virgin Islands. All Stock Cultures are Maintained in ES Enriched Natural Seawater Containing 1.5% Soil Extract. The Letter c Next to Strain Number Denotes Clonal Culture.

Species	SIU Strain Number	Origin
Amphidinium elegans	547	South Sound, Virgin Gorda (VG)
Ceratium furca	682	Hurricane Hole, St. John (SJ)
Cochlodinium polykrikoides	283,489	South Sound, VG
Coolia monotis	263	Hurricane Hole, SJ
C. monotis	390	South Sound, VG
Ensiculifera sp. nov.	415	South Sound, VG
Gambierdiscus toxicus	711, 756c, 763c, 772c	Hurricane Hole, SJ
G. toxicus	350, 453, 467, 566, 570	South Sound, VG
G. toxicus	661, 774c, 775c, 777c	Biras Creek, VG
G. toxicus	509	West End, Anegada
G. toxicus	619, 623, 740c, 741c, 742c, 744c 747c	St. Thomas Lagoon
Gonyaulax diacantha	783	Hurricane Hole, SJ
Gonyaulax grindleyi	403	Hurricane Hole, SJ
G. grindleyi	780, 781	Drake's Channel
Gonyaulax polyedra	278	Hurricane Hole, SJ
Gymnodinium sanguineum	373, 374, 497 499	South Sound, VG
G. sanguineum	437	Biras Creek, VG
G. sanguineum	533	Hurricane Hole, SJ
Gyrodinium fissum	376, 379, 473 474	South Sound, VG
Gyrodinium sp.	665	Biras Creek, VG
Ostreopsis sp.	702, 705	St. Thomas Lagoon
Prorocentrum concavum	364	Salt Island
P. concavum	662	Biras Creek, VG
Prorocentrum lima	700	South Sound, VG
Prorocentrum mexicanum	722	Hurricane Hole, VG
P. mexicanum	228, 262, 273, 276	Salt Island
Scrippsiella subsalsa	86, 404, 724	Hurricane Hole, SJ
S. subsalsa	344	Biras Creek, VG
Scrippsiella trochoidea	556, 557, 558, 582, 587, 685	South Sound, VG

standard medium with cells from dense stock cultures. After approximately 20 days of growth, the 2-liter cultures were each used to inoculate 16 to 18 liters of medium in 20 liter carboys. These cultures were aerated to prevent CO_2 depletion and to provide gentle aggitation. Light and temperature were maintained at 4300 lux and 27°C, respectively. In some cases, 1 or 2 liters of the large cultures were used to inoculate new large-scale cultures. Cultures were harvested by continuous flow centrifugation. The resulting moist cell pellets were removed from the centrifuge chamber and weighed. These products were recorded as yield in grams fresh weight (g f.w.). Nine species have been growth in large-scale culture. Total volume of cultures and yield for each species are included in Table II. Growth rates of selected species in large-scale culture are reported elsewhere in this volume (30).

Table II. Production of Cells and Crude Extracts of Nine Species of Dinoflagellates in Large-Scale Culture. Ages of Cultures Ranged From 20 to 36 Days.

Species (SIU Strain Number)	Culture Volume (l)	Yield (g f.w.)	Crude Extracts (mg)			
			ESAF	ESAP	WSAF	WSAP
Gambierdiscus toxicus (350)	712.4	192.6	1214.3	1129.5	308.9	629.6
Prorocentrum concavum (364)	581.0	191.1	1418.0	481.6	734.7	265.9
Prorocentrum mexicanum (276)	466.5	29.5	210.6	141.7	69.9	171.2
Coolia monotis (263)	32.0	10.3	87.7	81.5	22.7	0.0
Amphidinium elegans (547)	32.0	5.6	66.3	35.6	17.7	1.8
Scrippsiella subsalsa (344)	72.0	46.5	123.0	0.0	33.7	0.0
Gonyaulax grindleyi (403)	54.0	11.7	72.3	4.8	15.8	0.4
Gonyaulax polyedra (278)	16.0	4.7	27.2	1.6	4.6	0.1
Gymnodinium sanguineum (373)	16.0	3.2	33.6	1.2	12.1	0.8

Extraction and partial purification. The cell pellets derived from large-scale cultures were subjected to our standard extraction and partitioning procedure for isolating ciguatera toxins (Figure 2). The final cold acetone treatment of diethyl ether and water soluble fractions usually resulted in four products: (A) ether soluble acetone filtrate (ESAF), (B) ether soluble acetone precipitate (ESAP), (C) water soluble acetone filtrate (WSAF), and (D) water soluble acetone precipitate (WSAP). These acetone soluble and insoluble products were not purified further for this study; however, chromatographic procedures which we have utilized for such purification are presented in this volume (28, 29). The amounts of these fractions obtained from large-scale cultures are shown in Table II.

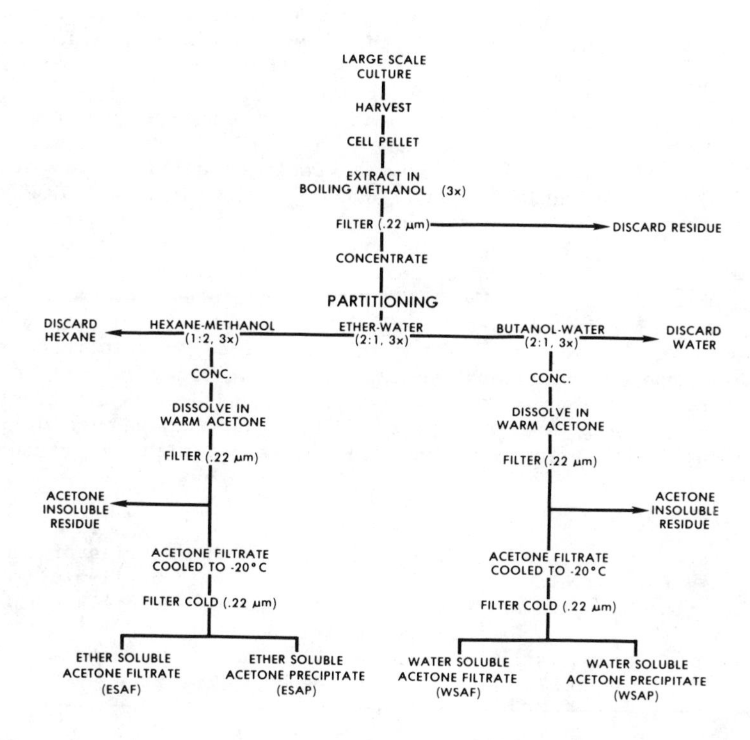

Figure 2. Flow diagram of procedure used for harvesting and extracting dinoflagellates and initial separation of crude toxins.

Screening for toxicity. Toxicity of the crude acetone soluble and insoluble products was determined using the mouse bioassay. Measured quantities of these products were administered via intraperitoneal injection into outbred Harlan Sprague Dawley ICR (BR) mice weighing approximately 20 g. The carrier was 1% Tween 60 in 0.15M sodium chloride and the injection volume was 0.5 ml per mouse. Extracts which caused death of mice within 48 hours were deemed toxic. However, it was apparent that some fractions were toxic even though death was not observed.

Results of our initial screening of the nine species grown in large-scale culture are included in Table III. Five of the nine species screened produced one or more toxic fractions. Two other species produced fractions which induced toxic signs but did not cause death. Of the seven species which produce compounds capable of causing death or inducing toxic signs in mice only G. toxicus and P. concavum have been sufficiently examined to report them as probable significant contributors to the ciguatera syndrome in the Caribbean. However, we believe that the preliminary findings regarding toxicity of the other species examined are important and we plan to conduct more detailed experiments in order to confirm or refute their contribution to the ciguatera syndrome. Furthermore, all species will be examined in order to determine whether or not they produce toxins which may not be detectable with the standard mouse bioassay.

Table III. Results From Initial Screening of Nine Species of Dinoflagellates for Toxicity. Doses of 5 or 10 mg (when available) Were Injected (ip) Into Two or Four 20 g mice. Fractions Were Recorded as Toxic (T) if They Caused Death Within 48 Hours. Fractions Causing Characteristic Symptoms but Not Death Were Recorded as Slightly Toxic (ST). NT Denotes Nontoxic.

Species (SIU Strain Number)	Crude Fractions			
	ESAF	ESAP	WSAF	WSAP
Gambierdiscus toxicus (350)	NT	T	NT	T
Prorocentrum concavum (364)	T	T	T	T
Prorocentrum mexicanum (276)	NT	NT	T	NT
Coolia monotis (263)	NT	ST	NT	--[1]
Amphidinium elegans (547)	NT	ST	NT	ST[2]
Scrippsiella subsalsa (344)	NT	--	NT	--
Gonyaulax grindleyi (403)	NT	NT[2]	--	--
Gonyaulax polyedra (278)	NT	T[2]	--	--
Gymnodinium sanguineum (373)	NT	T[2]	NT	NT[2]
Controls[3]	NT	NT	NT	NT

[1] not available for testing
[2] one mouse tested with amount extracted (see Table II).
[3] carrier only; 2 mice/each fraction

Toxicity of Gambierdiscus toxicus. Our study has utilized only one Caribbean strain (SIU 350) of G. toxicus (Figure 3a). Numerous mouse bioassays of the crude extracts from this strain revealed that the ESAP and WSAP were consistently toxic. Initially, the absence of toxin in the ESAF was perplexing since most previous reports on G. toxicus (Pacific strains) have stated that ether soluble fractions contained ciguatoxin. However, only Bagnis and coworkers (19) subjected ether soluble extracts to cold acetone treatment. The latter authors obtained ESAF which they reported to contain ciguatoxin and ESAP which they consolidated with WSAP and reported the mixture to contain only maitotoxin. Findings reviewed below suggest that the major toxic component of the ESAP is quite different from that of the WSAP in the Caribbean G. toxicus.

The LD_{50}'s for ESAP and WSAP from SIU strain 350 were determined to be 4.95 mg/kg mouse and 1.1 mg/kg mouse, respectively (28, 29). Column chromatography revealed the presence of three toxic components in ESAP (GT-1, GT-2, and GT-3) (28), and one toxic component in the WSAP (GT-4) (29). GT-1 was eluted from silicic acid columns with 9:1 chloroform-methanol; whereas, GT-2 was eluted with 1:1 chloroform-methanol. Due to low yield, GT-1 was not purified further. GT-2 was ultimately eluted from Sephadex G-15 with water. GT-3 and GT-4 were eluted from silicic acid with methanol and from Sephadex G-15 with water. GT-1 and GT-2 have been shown to have precisely the same effects on guinea pig ileum preparations (reversible; competitive inhibition of histamine and actylcholine responses). Both GT-3 and GT-4 were reported to cause non-reversible inhibition of histamine and actylcholine responses (29). These results indicate that GT-3 and GT-4 are identical, the former representing carryover of water soluble toxin in the initial ether fraction. They appear to be the same as maitotoxin reported for G. toxicus from the Pacific. GT-1 is the same as that previously reported as ciguatoxin from G. toxicus. Although differences in solubility of GT-1 and GT-2 were detected, their identical effects on guinea pig ileum preparations suggest that GT-2 is an altered (more polar) form of GT-1, or ciguatoxin.

Toxicity of crude fractions from Prorocentrum concavum. All four crude fractions extracted from P. concavum (SIU strain 364) (Figure 3b) consistently contained potent toxin(s). The results of mouse bioassay of these fractions are included in Tables IV. Corresponding dose response curves and LD_{50}'s are shown in Figure 4.

ESAF plus ESAP from P. concavum (SIU 364) corresponds to fractions from Pacific strains of G. toxicus which were reported to contain ciguatoxin (17, 19, 21, 22), and of P. lima which contained PL toxin II (scaritoxin-like toxin) and okadaic acid (23, 24). Nakajima and coworkers (20) reported an ether soluble extract from P. concavum to be slightly toxic to mice but provided little additional information. Crude fractions containing ciguatoxin obtained from moray eel (Gymnothorax javanicus) and G. toxicus from the Pacific were reported to have LD_{50}'s of 12.2 mg/kg and 28.1 mg/kg, respectively (37). These compare well with the LD_{50} of 8.3 mg/kg for similarly purified ESAF from P. concavum.

Notwithstanding possible differences in amounts and degree of purification of each of the four fractions from the Caribbean strain of P. concavum, the most potent fraction was the WSAF.

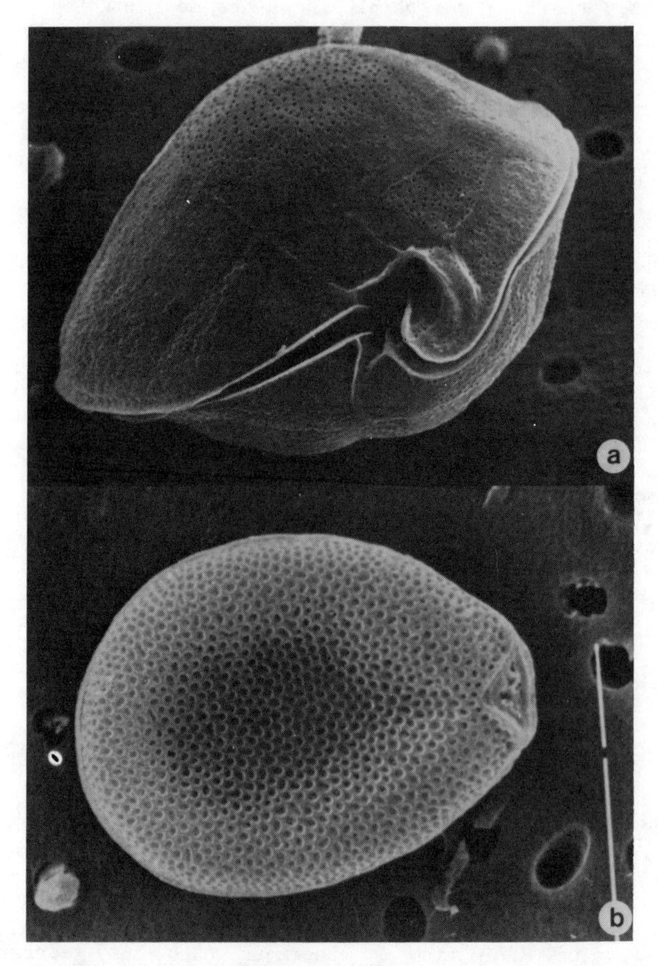

Figure 3. (a) A scanning electron micrograph of Gambierdiscus toxicus (SIU 350) from South Sound, Virgin Gorda (Station 3). (b) A scanning electron micrograph of Prorocentrum concavum (SIU 364) from Salt Island (Station 2).

Table IV. Bioassay of Crude Ether Soluble Acetone Filtrate (ESAF), Ether Soluble Acetone Precipitate (ESAP), Water Soluble Acetone Filtrate, (WSAF), and Water Soluble Acetone Precipitate (WSAP) From *Prorocentrum concavum* Using 10 Female Mice for Each (ip) Injected Dose.

Mice Weight Range (Mean ± S.D.) (g)	Dose mg/Mouse	% Death In 48 Hours	Time To Death Range (Mean ± S.D.)
ESAF			(Hours)
18.0-20.0 (18.9 ± 0.54)	0.100	0	----
19.8-20.9 (20.5 ± 0.32)	0.125	10	30.0
17.8-19.9 (18.7 ± 0.59)	0.150	30	19.7-44.7 (36.4 ± 11.8)
18.0-21.0 (19.6 ± 1.01)	0.200	90	5.3-26.6 (19.1 ± 7.3)
18.4-20.5 (19.7 ± 0.63)	0.250	100	1.1-4.0 (2.0 ± 0.9)
ESAP			
18.9-22.2 (20.8 ± 1.16)	0.200	0	----
19.0-21.7 (20.8 ± 0.73)	0.400	40	4.3-20.0 (12.8 ± 7.3)
19.4-21.9 (20.3 ± 0.78)	0.600	70	1.5-12.0 (4.4 ± 3.4)
19.4-22.0 (21.0 ± 0.86)	0.800	100	1.6-22.5 (6.2 ± 6.2)
WSAF			(Minutes)
17.9-20.9 (20.0 ± 0.79)	0.020	0	----
18.2-19.9 (19.1 ± 0.59)	0.025	30	45-51 (46.3 ± 3.4)
18.4-21.9 (20.5 ± 0.97)	0.030	40	21-45 (32.5 ± 9.0)
18.6-20.8 (20.1 ± 0.65)	0.035	80	23-30 (27.0 ± 3.5)
18.2-20.0 (19.4 ± 0.51)	0.040	70	15-20 (16.9 ± 4.7)
20.0-21.9 (21.3 ± 0.61)	0.045	90	13-19 (15.6 ± 1.9)
20.1-22.3 (21.3 ± 0.77)[1]	0.055	90	9-13 (11.0 ± 1.7)
18.8-21.6 (20.5 ± 0.79)	0.060	100	7-10 (8.2 ± 0.8)
WSAP			
20.7-22.1 (21.3 ± 0.46)	0.025	0	----
18.8-22.1 (20.6 ± 1.01)[1]	0.050	15	180-484 (344 ± 125)
18.3-22.4 (20.1 ± 0.99)	0.075	40	92-1440 (435 ± 580)
19.6-20.9 (20.1 ± 0.35)	0.100	90	43-1260 (258 ± 383)
18.7-20.1 (19.5 ± 0.44)	0.125	100	35-58 (43 ± 7)
18.6-21.0 (20.5 ± 0.67)	0.150	100	26-73 (37 ± 12)

[1] 20 mice assayed

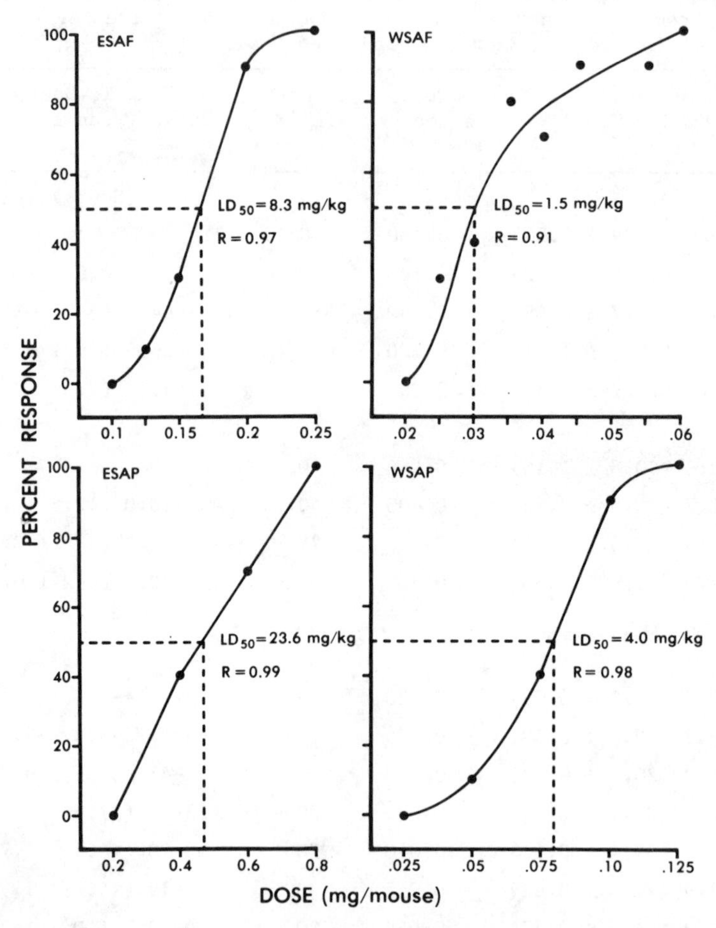

Figure 4. Dose response curves representing per cent mortality (within 48 hours) of approximately 20 g mice (CRE:CD:BR:ICR) injected with specified doses of ether soluble acetone filtrate (ESAF), ether soluble acetone precipitate (ESAP), water soluble acetone filtrate (WSAF), and water soluble acetone precipatate (WSAP) from Prorocentrum concavum (SIU 364). Each dose was administered by ip injection into 10 mice. LD_{50}s were determined using probit analysis. A correlation coefficient (R) is provided for each LD_{50}.

This fraction displayed the greatest killing power; however, the most interesting features of this toxin(s) were its fast-acting and "all-or-none" nature. All mice deaths, regardless of dose, occurred within one hour. In fact, it appeared that there was a critical dose-dependent time to death, after which, surviving mice recovered completely. Although Yasumoto and coworkers (14, 23) also have reported fast-acting water soluble toxins from parrotfish and *P. lima* (PL toxin III), none has been characterized to any extent.

WSAP fractions from Pacific strains of *G. toxicus* usually have been reported to contain maitotoxin (19). As noted above, *P. lima* from the Pacific was reported to have a fast-acting water soluble toxin but not maitotoxin (23). Although the WSAP from the Caribbean *P. concavum* shows some similarities to maitotoxin, we are not prepared to make that designation at this time.

Thin layer chromatography of crude fractions from Prorocentrum concavum. Analytical and preparative thin layer chromatography (TLC) was used to generally characterize and compare the toxic components of the four fractions from *P. concavum*. TLC was performed using Suppelco Redi-Coat silica gel G (0.25 mm) plates and two solvent systems: (A) chloroform-methanol-(6N) ammonium hydroxide (90:9.5:0.5) and (B) chloroform-methanol-water (60:35:8). All plates were developed to 14 cm in sealed glass chambers. Plates were visualized by spraying with 50% aqueous sulfuric acid followed by charring. Retention factors (Rf) were charted and corresponding zones on plates not visualized were scraped and eluted with methanol.

The results of TLC indicated that *P. concavum* produced at least three and possibly as many as five separate toxins (Table V). The three relatively distinct toxic components were designated PC-1, PC-3, and PC-5. Results from TLC did not clearly distinguish PC-2 or PC-4 from PC-5. However, mouse bioassay of these fractions (from preparative TLC) revealed that PC-4 was a fast-acting toxin in contrast to the slow-acting PC-2 and PC-5. Of course, the variation in results of the mouse bioassays could have been a consequence of different amounts and levels of purity of the respective fractions. Regardless, preliminary results from TLC and bioassay suggest that PC-2 may be the same as PC-5, the former being a water soluble carryover in the initial ether fraction.

Solubility and chromatographic behavior of PC-1 were very similar to those reported for SG-1 (scaritoxin) from parrotfish (14), scaritoxin from the turban shell (38), and PL toxin II (scaritoxin-like toxin) from *P. lima* (23). The possibility of PC-1 being a derivative of okadaic acid should not be overlooked.

The fast-acting WSAF from *P. concavum* contained two toxic components (PC-3 and PC-4). Separate bioassay of these fractions showed both to cause fast death and other similar symptoms. At present, we can not speculate on relationships between the two or with fast-acting toxins previously identified from other marine organisms.

Bioassay of both PC-2 and PC-5 revealed similarities to one another and to maitotoxin. However, the Rf's obtained for these fractions (0.45-0.56) were greater than those reported for maitotoxin (0.1-0.3) using a similar TLC system (14).

Table V. A Comparison of The Four Crude Fractions and Their Respective Toxic Components From Prorocentrum concavum Using Analytical and Preparative TLC. Preparative Plates Were Spotted With Amounts of Crude Material Equivalent to the 100% Lethal Doses Determined in LD_{50} Studies. Solvent Systems Used Were: (A) $CHCl_3:CH_3OH:(6N)NH_4OH$ (90:9.5:0.5) and (B) $CHCl_3:CH_3OH:H_2O$ (60:35:8).

Crude Fractions	LD_{50} mg/kg	Solvent Systems	Rf of Toxic Component	Designation and Characterization of Eluted Toxin	
ESAF	8.48	A	0.71-0.86	PC-1	Slow Acting
ESAP	23.38	A	No Migration	PC-2	Slow Acting
ESAP	23.38	B	0.45-0.53	PC-2	Slow Acting
WSAF	1.63	B	0.41-0.48[1]	PC-3	Fast Acting
			0.48-0.56[1]	PC-4	Fast Acting
WSAP	3.72	B	0.48-0.56	PC-5	Slow Acting

[1] two separate but closely associated toxic components; bioassayed separately

Currently, all five toxic components from P. concavum are being subjected to column and high performance liquid chromatography. Products of these treatments will be evaluated and characterized using the mouse bioassay and isolated nerve-muscle preparations.

Summary and Conclusions

One strain of each of nine species of dinoflagellates from the Virgin Islands has been grown in large-scale culture and assayed for toxicity. Five of these, namely Gambierdiscus toxicus, Prorocentrum concavum, P. mexicanum, Gymnodinium sanguineum, and Gonyaulax polyedra produced one or more toxic fractions which killed mice within 48 hours. Two other species, Coolia monotis and Amphidinium elegans produced fractions which caused some signs of toxicity in mice. Recently, a Caribbean strain of Prorocentrum lima was added to the SIU culture collection. Since a Pacific strain of P. lima is known to produce three toxins, the Caribbeana strain probably will prove to be toxic as well. In addition to P. lima, we have 56 strains representing nine species which should be grown in large-scale culture and assayed for toxicity in the near future.

Except for G. sanguineum, the above named species represent cosmopolitan and frequently dominant epiphytic/benthic dinoflagellates in the Virgin Islands. Gymnodinium sanguineum is one of the few common planktonic species in the study area. All of these species are readily accessible to herbivorous fishes throughout the Virgin Island chain.

Our observations strongly suggest that Prorocentrum concavum and Gambierdiscus toxicus are major contributors of toxins to the reef and inshore ecosystems. We believe that these two species produce at least 4 (up to 7) separate toxins which may eventually enter the fish food chain. At present, we can not speculate on the fate of these toxins in the various metabolic systems through which they pass prior to their ingestion by humans. Thus, in addition to our efforts to isolate, and chemically and physiologically characterize toxins extracted from dinoflagellates, we (in conjunction with Dr. C. Kohler, Cooperative Fisheries Research Laboratory, SIU) are examining the same toxins as they pass through primary and secondary consumer fishes. Assuming that dinoflagellates are the progenitors of ciguatera toxins, these "food chain" studies will confirm the actual causal species and will permit us to concentrate our efforts on the study of those toxins which are ultimately transmitted to humans via food fishes.

Finally, it is clearly evident that no one dinoflagellate species should be singled out as the major ciguatera causative organism nor should it be assumed that a single toxin is the major cause of ciguatera in the Caribbean. This is especially important in view of the fact that no species has been shown conclusively to produce ciguatoxin as described from moray eel by Scheuer and coworkers (6) and Tachibana (37).

Acknowledgments

This research was supported in part by the United States Food and Drug Administration (contract #223-79-2287). We wish to thank D. Miller, J. Yopp, and N. Doorenbos for their invaluable assistance in all aspects of this study. Also, we gratefully acknowledge the technical assistance of E. Ragelis, Project Officer, USFDA. Finally, we thank Mr. Myron Hokin, owner of the Bitter End Yacht Club, Virgin Gorda, who graciously provided a field laboratory, living quarters, and boats throughout the course of our field studies. This is but one of many contributions by the Hokin family to scientific studies relevant to the well-being of the people of the Virgin Islands.

Literature Cited

1. Halstead, B. "Poisonous and Venomous Marine Animals of the World"; U. S. Gov. Print. Off.: Washington, D. C., 1967; Vol. 2, 1-1070.
2. Bagnis, R. Rev. Hyg. Med. Soc. 1967, 15, 619-46.
3. Bagnis, R. Hawaii Med. J. 1968, 28, 25-8.
4. Lawrence, D.; Enriquez, M.; Lumish, R.; Maceo, A. J. Amer. Med. Assoc. 1980, 244, 254-8.
5. Banner, A. H. In "Marine Science"; Humm, H.; Lane, C. E., Eds.; Marcel Dekker: New York, 1974; pp. 15-36.
6. Scheuer, P.; Takahashi, W.; Tsutsumi, J.; Yoshida, T. Science 1967, 155, 1267-8.
7. Li, K.; Au, A. Far East Med. J. 1968, 4, 313-5.
8. Li, K. In "Poisonous and Venomous Marine Animals"; U. S. Gov. Print. Off.: Washington, D. C., 1968; Vol. 3, pp.
9. Hashimoto, Y.; Yasumoto, T.; Kamiya, H.; Yoshida, T. Bull. Jap. Soc. Sci. Fish. 1969, 35, 327-32.
10. Yasumoto, T.; Hashimoto, Y.; Bagnis, R.; Randall, J.; Banner, A. Bull. Jap. Soc. Sci. Fish. 1971, 37, 724-34.

11. Bagnis, R.; Loussan, E.; Thevenin, S. Med. Trop. 1974, 34, 523-7.
12. Yasumoto, T.; Bagnis, R.; Vernoux, J. Bull. Jap. Soc. Sci. Fish. 1976, 42, 359-65.
13. Yasumoto, T.; Nakajima, I.; Chungue, E.; Bagnis, R. Bull. Jap. Soc. Sci. Fish. 1977, 43, 69-74.
14. Chungue, E.; Bagnis, R.; Fusetani, N.; Hashimoto, Y. Toxicon 1977, 15, 89-93.
15. Randall, J. J. Bull. Mar. Sci. Gulf Caribb. 1958, 8, 236-67.
16. Adachi, R.; Fukuyo, T. Bull. Jap. Soc. Sci. Fish. 1979, 45, 67-71.
17. Yasumoto, T.; Nakajima, I.; Bagnis, R.; Adachi, R. Bull. Jap. Soc. Sci. Fish. 1977, 43, 1021-6.
18. Bagnis, R.; Chanteau, S.; Yasumoto, T. Rev. Int. Oceanogr. Med. 1977, 45-6, 20-4.
19. Bagnis, R.; Chanteau, S.; Chungue, E.; Hurtel, J.; Yasumoto, T.; Inoue, A. Toxicon 1980, 18, 199-208.
20. Nakajima, I.; Oshima, Y.; Yasumoto, T. Bull. Jap. Soc. Sci. Fish. 1981, 47, 1029-33.
21. Withers, N. Proc. 4th Int. Coral Reef Symp. 1981, 2, 449-51.
22. Shimizu, Y.; Shimizu, H.; Scheuer, P.; Hokama, Y.; Oyama, M.; Miyahara, J. Bull. Jap. Soc. Sci. Fish. 1982, 48, 811-13.
23. Yasumoto, T.; Oshima, Y.; Murakami, Y.; Nakajima, I.; Bagnis, R.; Rukuyo, Y. Bull. Jap. Soc. Sci. Fish. 1980, 46, 327-31.
24. Murakami, Y.; Oshima, Y.; Yasumoto, T. Bull. Jap. Soc. Sci. Fish. 1982, 48, 69-72.
25. Miller, D.; Dickey, R.; Tindall, D. Proc. Fed. Amer. Soc. Exp. Biol. 1982, 41, 1561.
26. Tindall, D.; Dickey, R.; Miller, D. Proc. Fed. Amer. Soc. Exp. Biol. 1982, 41, 1561.
27. Carlson, R.; Tindall, D.; Dickey, R.; Morey-Gaines, G. Proc. Fed. Amer. Soc. Exp. Biol. 1982, 41, 1562.
28. Miller, D.; Dickey, R.; Tindall, D. This volume 1984.
29. Dickey, R.; Miller, D.; Tindall, D. This volume 1984.
30. Carlson, R.; Morey-Gaines, G.; Tindall, D.; Dickey, R. This volume 1984.
31. Loeblich, A. R. III. J. Phycol. 1975, 11, 80-6.
32. Gullard, R. R.; Ryther, J. H. Can. J. Microbiol. 1962, 8, 229-39.
33. Provasoli, L. In "Culture and Collection of Algae"; Watanabe, H.; Hattori, A., Eds. Jap Soc. Plant Physiol.: Hakone, 1968, pp. 63-75.
34. Wilson, W.; Collier, A. Science 1955, 121, 394-5.
35. Provasoli, L.; McLaughlin, J.; Droop, M. Arch. Mikrobiol. 1957, 25, 392-428.
36. Provasoli, L. In "Proceedings of the Fourth International Seaweed Symposium"; DeVirville, D.; Feldman, J., Eds.; Pergamon Press: Oxford, 1963, pp. 9-17.
37. Takhibana, K. Ph.D. Dissertation, University of Hawaii, Honolulu, 1980.
38. Yasumoto, T.; Konno, T. Bull. Jap. Soc. Sci. Fish. 1976, 42, 1399-404.

RECEIVED March 2, 1984

22

Lipid-Extracted Toxins from a Dinoflagellate, *Gambierdiscus toxicus*

DONALD M. MILLER, ROBERT W. DICKEY, and DONALD R. TINDALL

Department of Physiology and Pharmacology, School of Medicine, and Department of Botany, Southern Illinois University, Carbondale, IL 62901

Crude and three diethyl ether extracted, acetone treated, fractions were isolated from large-scale cultures of Gambierdiscus toxicus. Crude extracts at .04 mg/ml inhibited the histamine contraction response in smooth muscle of the guinea pig ileum. Three semi-purified fractions at 5 ng/ml, effectively inhibited the guinea pig ileum preparation. Two of these fractions followed Michaelis-Menten kinetics for a competitive inhibition. The third fraction inhibited in a non-reversible manner. This study has established the presence of three lipid extracted toxins in G. toxicus, outlined a method for their assay in small quantities, and identified at least two of the effects of these toxic extracts in animals.

Ciguatera was the name given to a food poisoning syndrome which affected Spanish people who settled in Cuba during the exploration of the American continents (1-3). The name was derived from the Spanish word "cigua" for a species of turban shell which they thought responsible for the illness. Later this same name was applied to a similar syndrome which occurs in the Pacific. It has come to be defined as a form of food poisoning caused by the ingestion of any one of a number of marine fishes in tropical and subtropical areas. The symptomology in humans has been described (4-8).

Scheuer and coworkers (9) isolated substances from Lutjanus bohar (red snapper), Carcharhinus menisorrah (shark), and Gymnothorax javanicus (moray eel) which they called "ciguatoxin" and implicated in the ciguatera syndrome of the Pacific. The wealth of pharmacological and other data accumulated on the ciguatera syndrome has indicated that more than one toxin is involved (6,10,11). Thus, at present there is no certainty which toxin (e.g., ciguatoxin, maitotoxin, scaritoxin, etc.) plays the major role in the ciguatera syndrome. Indeed, not only may the

0097-6156/84/0262-0241$06.00/0
© 1984 American Chemical Society

causative agent be different in the Caribbean and the Pacific, but some evidence indicates that it may be different from island to island within the same area (12).

A dinoflagellate, Gambierdiscus toxicus has been reported to be the likely progenitor of ciguatoxin in the South Pacific (13-15). Recently, Tindall, (16), Miller, (17) and Carlson, (18) reported on toxic extracts from G. toxicus and two species of Prorocentrum from the Caribbean. The physiological action of toxic extracts from G. toxicus were found to be similar to those induced by toxic extract from Scomberomorus cavalla (Kingfish) from the same region (19).

Attempts to delineate the active components in the genesis of toxic fish and to establish prophylactic measures for ciguatera have been hampered by; the lack of adequate amounts of toxin for testing and purification, the lack of a clear definition of whether the toxic material was a single entity or multiple factors, and the lack of discriminating tests which utilized small amounts of toxin to characterize each toxic component. With these aspects in mind, we have begun a systematic program which includes: (1) the growth, in large-scale culture of eighteen species of "Ciguatera-community" dinoflagellates including G. toxicus, (2) the extraction and purification of large quantities of toxic materials from these large-scale cultures, and (3) the establishment of assay procedures for each toxic component and characterization of each using specific physiological model-systems.

This paper is a report on the lipid extracted toxins from G. toxicus and their effect on isolated guinea pig ileum preparations. This paper also proposes the use of this preparation for quantification of two of the lipid extracted toxins.

Materials and Methods

Culture of Dinoflagellate. The isolate of G. toxicus (SIU 350) used in this study was obtained from the South Sound of Virgin Gorda, British Virgin Islands. The species was brought into unialgal culture and maintained in enriched sea water (20) under continuous light (3200 lux cool white fluorescent) at 27.0° C. The development of large-scale cultures involved transferring cells from stock cultures to a series of two liter Fernbach flasks containing enriched seawater medium. After the early stationary phase of growth had been reached (approximately 15-20 days) each of these cultures were used to innoculate 18 liters of the same medium in 20 liter carboys. Large-scale cultures were grown under continuous light (4300 lux cool white fluorescent) at 27.0° C. Cells were harvested by centrifugation after cultures reached the early stationary phase of growth (30-35 days).

Extraction and Purification of Toxins. Two large-scale cultures were examined: (1) 350F and (2) 350G. Culture F consisted of 53 liters and yielded 13.1 g fresh weight (2.62 g dry mass) of G. toxicus cell pellet. Culture G consisted of 130 liters which yielded 49.5 g fresh weight (9.9 g dry mass). The cells were

extracted by refluxing in boiling aqueous methanol. The methanol extracts were concentrated and subjected to liquid-liquid partitioning followed by cold acetone treatment of the toxic components (Figure 1). Totals of 103.98 and 485.21 mg DM of crude ether soluble acetone precipitates (ESAP) were isolated from 350F and 350G, respectively (sixth and seventh successive large scale cultures). The major portion of the ESAP fraction from 350G was subjected to further purification by sequential treatment on columns of silicic acid, basic aluminum oxide and sephadex G-15. Eluting solvents and their ratios are detailed in the results section.

Thin Layer Chromatography Procedures. Silica gel G (Suppelco, Redi-coat, 5 X 20 cm X 0.25 mm) plates were activated by heating in an oven for 1 hour at 110°C. Toxic products were spotted onto duplicate plates at concentrations corresponding to their previously determined LD_{99} levels. The plates were developed to 14 cm with chloroform, methanol and (6N) ammonium hydroxide (90:9.5:0.5) or chloroform, methanol, and water (60:35:8). Plates were developed and visualized by spraying with 50% aqueous sulfuric acid and charring. Undeveloped plates were scraped, the silica gel fines removed and the extracts concentrated to a residue under a nitrogen gas stream.

Mouse Assays. Outbred female Swiss mice (Harlan Sprague Dawley ICR "BR") weighing 19 to 21 g were used. Doses of toxic extract were suspended in 0.5 ml of 0.1% Tween 60 in 0.15 M NaCl and administered by i.p. injection. Mice were observed for a period of 48 hours.

Guinea Pig Ileum Preparation. Guinea pigs (350-600 gm) were sacrificed by a cervical dislocation. A 2-3 cm segment of ileum was removed and placed in physiological saline solution at 37°C. The saline solution consisted of the following (mM): NaCl, 136.9; KCl, 2.68; $CaCl_2$, 11.84; $MgCl_2$, 1.03; $NaHCO_3$, 11.9; KH_2PO_4, 0.45; glucose, 5.55. If required, the intestinal lumen was evacuated by slowly flushing with saline solution. The dissected ileum segment was attached by a wire hook to a glass rod which extended to the bottom of an organ bath. The other end of the ileum was attached by a silver chain to the lever of the transducer. The apparatus utilized for studying the responses of the ileum to toxin consisted of a: water-jacketed tissue bath, Bionix myograph transducer, Beckmann physiograph, Beckmann temperature bath, and an air pump. Dose response relationships for phasic contractions of ileal segments were determined following the method of van Rossum (21). Time dependent assays were conducted by subjecting the preparation to a single concentration of toxin and then challenging the preparation with the agonist concentration at progressive time periods.

Calcium Substitution Experiments. Saline solutions utilized in this series of experiments contained calcium chloride at 0.5X, 2X,

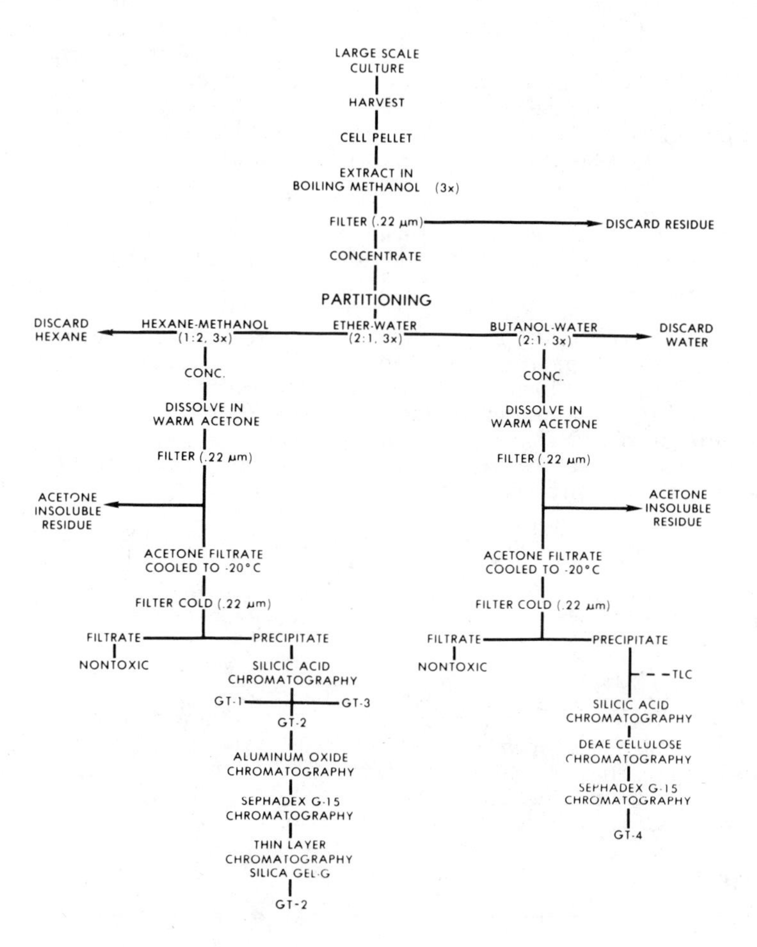

Figure 1. Schematic diagram of separation procedures.

3X and 4X the normal amount. The osmotic concentration was maintained constant by balancing the NaCl concentration.

TTX Addition Experiments. In these experiments the guinea pig ileum was prepared as previously described except that TTX at a concentration of .04 ng/ml was included in the saline solution.

Statistical Treatment of Data. Each dose ratio point was the mean of three determinations. Only points between 20 and 80% on the dose response curves were utilized for the calculation of the dose ratio. After the dose was converted to a log value and the response to percent of the maximum, a best fit line was determined by linear regression.

For the two or more lines involved in a determination, the slopes of the lines were tested for parallelism by computing the "t" value for each line and comparing it with the "t" value determined from the standard tables. If the computed t value exceeded the tabular value, the slopes were considered to differ significantly and the lines were rejected for use. If the computed t value was less than the tabular value, parallel lines were constructed by using weighted slopes to determine a common mean slope. The weighting factors were the reciprocals of the squared standard errors of the slopes.

Results

Separation. The crude ether soluble extract (ESAP) was separated into several fractions by column chromatography and the results are detailed in Table I.

Table 1. Column Chromatographic Separation of 485.21 mg of ESAP from Gambierdiscus toxicus (350G)

DESIGNATION MEDIUM	PARAMETERS	CRUDE	GT-1	GT-2	GT-3
a fraction Silicic Acid	Eluant			*Ch-Me	Me
	Eluant Ratio			(1:1)	100
	Recovered(mg)			220.0	71
b fraction Silicic Acid	Eluant		Ch-Me	Ch-Me	Me
	Eluant Ratio		(9:1)	(1:1)	100
	Recovered(mg)		2.9	128.1	11
c fraction Basic Al-OH	Eluant			Me-Water	
	Eluant Ratio			(1:1)	
	Recovered(mg)			105.9	
d fraction Sephadex	Eluant			Water	
	Recovered(mg)			6.3	

*Ch and Me = Chloroform and Methanol, respectively.

Thin Layer Chromatography. The ESAP from the Caribbean isolate of *G. toxicus* migrated little on TLC using the chloroform, methanol, and ammonium hydroxide system. Prior to treatment with sulfuric acid and charring, five components were visably resolved and all appeared green in color. After charring a total of seven components were visualized, the least polar of which gave an Rf of 0.16. The duplicate plates were divided into four zones, scraped and eluted. These fractions were assayed at 0.4 mg/ml equivalents per mouse. The components falling between RF's 0.0 and 0.1 were found to be toxic.

When the development system was changed to chloroform, methanol and water (60:35:8), separation of the components was enhanced. After charring 16 components were visualized, the duplicate plate was divided into 8 equal zones which were scraped, eluted and tested as above. The components falling between the Rf's 0.33 to 0.47 and 0.47 to 0.61 were found to be toxic. Both of these fractions were observed to co-chromatograph with green pigments.

Crude Extract - Mouse Bioassay. At the level of 5 mg per mouse, the ESAP caused death in all test mice within 60 minutes. Signs of toxicity were noticeable within 15 to 20 minutes following administration, and included in order of occurrence: inactivity and piloerection followed by cyanosis of the tail and feet with concurrent hypothermia, vasodilation in the ears, lacrimation, ptosis of the eyelid on that side of the animal corresponding to the abdominal side of injection, ptosis of the abdomen, asthenia, impairment of hind limb motor ability followed shortly by complete paralysis with the hind limbs extended posteriorly, dyspnea, and near the time of death, clonic convulsions in synchrony with inspiratory gasps. The above signs were observed consistently following administration of ESAP. Several other signs observed, but showing less consistency, included a brief period of hyperactivity shortly after administration, mild diarrhea, and cataracting of one or both eyes coincidental with or just after death.

The LD_{50} for the ESAP was determined by administration of four dosage levels using three mice at each level and was found to be 0.099 mg per 20 gram mouse (4.95 mg/kg mouse, Figure 2). Regression analysis gave a correlation coefficient of 0.95 and the range of activity was 0.2 mg.

Mouse Bioassay of GT-1, GT-2, GT-3. Single doses of 0.1 mg of GT-1, GT-2, and GT-3 respectively, when injected i.p. into mice produced death within 2 to 24 hours.

Guinea Pig Bioassay of GT-1 and GT-2. At less than nanogram concentrations, extracts GT-1 and GT-2 produced an enhancement of histamine stimulation of the ileal preparation. At nanogram concentrations or larger, both caused an inhibition and hence a shift of the dose response curve (Figure 3). Replotting these data for GT-1 and GT-2 into a Michaelis-Menten format (Figure 4) indicates that the action of GT-1 and GT-2 fractions are

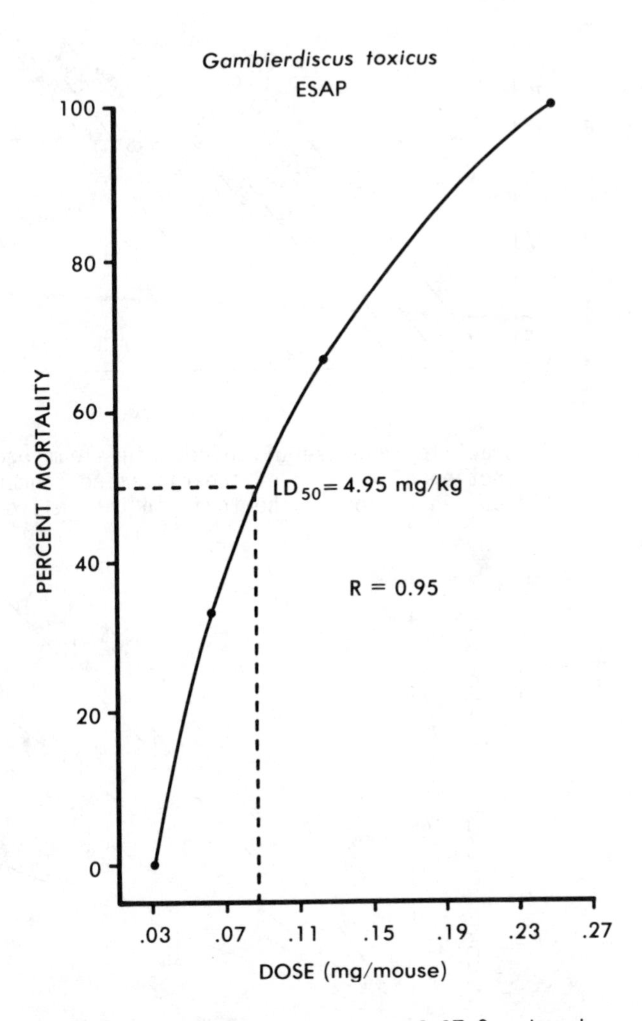

Figure 2. LD-50 determination of GT-2 extract.

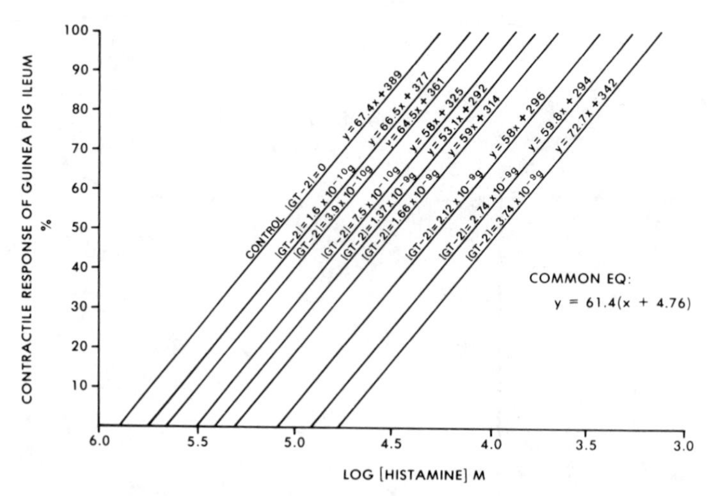

Figure 3. Contractile response of guinea pig ileum preparation as a function of the log of the histamine concentration and the concentration of the toxin in the saline solution.

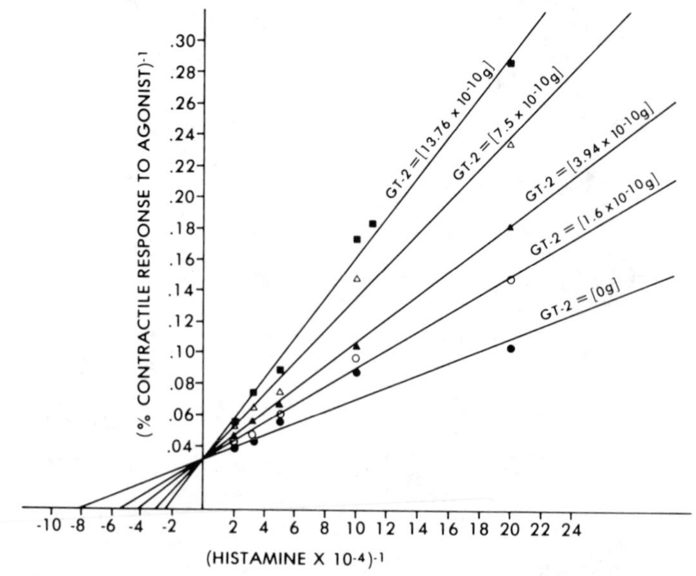

Figure 4. Michaelis-Menten type plot of percent of contractile activity versus the histamine substrate concentration and the toxin concentration in the saline solution.

competitive. Time dependent studies demonstrated that the effects of GT-1 and GT-2 are reversible, after washout, for up to two hours--the time frame for these experiments.

Utilizing the dose-ratio (relative potency) data that we obtained for GT-1 and GT-2, we constructed constrained Schild plots to obtain an estimation of the pA_2 (= -log apparent affinity constant). An estimation of the pA_2 values is indicated by the intercept of a best fit line of slope of -1 (Figure 5).

Guinea Pig Bioassay of GT-3. The application of GT-3 to the guinea pig ileum at 10 ng/ml produced a permanent 50% inhibition in activity. However, this inhibition does not come to a stable level until ninety minutes after the application of the toxin for a fifteen minute interval. We, therefore, expressed the effects of GT-3 by stating both the dose and time to achieve a 50% inhibition from the control as DCI_{50} (Table II).

Table II. Summary of results from mouse and guinea pig ileum assays of ESAP and GT-1, GT-2, and GT-3 from *Gambierdiscus toxicus*

FRACTION (see Table I)	WT (mg)	ACTUAL LD_{50} (mg/mouse)	CALC. MOUSE UNITS X1000	GUINEA PIG ILEUM RESULTS DCI_{50} (ng/min)	or DR=2 (ng)
350F					
ESAP	103.98	.099	1.05	2.8/15	
350G					
ESAP	485.21		3.9	10/10	
GT-1b	2.9				10.6
GT-2b	128.1				8.5
GT-2c	105.9				not done
GT-2d	6.3				5.1
GT-3a	71.5			not done	
GT-3b	11.2			70/15	

Calcium and TTX Experiments. Varying the calcium concentration in the saline of the ileal preparation had no effect on the action of GT-1 and GT-2. GT-3 effects were accentuated at calcium concentrations lower than normal but not appreciably affected at higher than normal calcium concentrations. The inclusion of TTX in the physiological saline at 4 ng/ml did not affect the assay.

Discussion

Effects of Toxic Extracts. We have shown that the crude ESAP of *G. toxicus* had an LD_{50} of 4.96 mg/kg mouse. Recently, Tachibana (22) utilized a dose time relationship to estimate the LD_{50} of crude extract from *G. toxicus* as 28.1 mg/kg. Previous to that

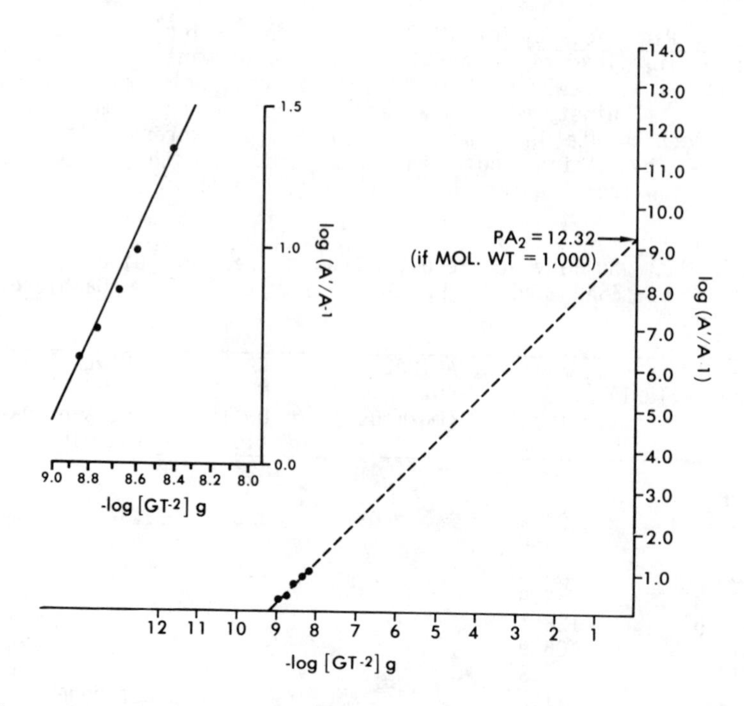

Figure 5. pA_2 determination assuming a molecular weight of 1000.

Scheuer (9) utilized a similar method to obtain an LD_{50} determination of 500 µg/kg.

Reports of effects on mice observed for maitotoxin closely resemble the effects of ciguatoxin, with the exception of diarrhea, hypersalivation, and convulsions before death. The symptoms observed here for ESAP did not include hypersalivation and only occasionally was mild diarrhea observed. The vasodilation in the ears, ptosis of the eyelids and abdomen were observed in this study but not in other accounts. Crude extracts produced irreversible inhibition of the frog nerve muscle preparation, affecting first the synapse, secondly the nerve and lastly the muscle (23). At .04 mg/ml the crude extract completely inhibited the isolated guinea pig ileum preparation.

GT-1, the smallest of the three toxic fractions, was separated on silicic acid with chloroform methanol (9:1). At a concentration of 40 ng/ml it was found to be a reversible inhibitor of the guinea pig ileum. At concentrations less than 8 ng/ml the ileum response to histamine was augmented.

GT-2, separated on silicic acid with chloroform methanol (1:1), constituted the largest fraction. At a concentration of 4 ng/ml, GT-2 was found to be a reversible inhibitor of the guinea pig ileum response to histamine. The separation of all three fractions up to the last step (see d, Table 1) was exactly the same procedure previously utilized by Tachibana (22) Scheuer (2) and others (13, 24). In the last step, when we attempted to utilize a Sephadex column and water for further purification of GT-2, we lost a considerable amount of toxin and also appeared to lose toxicity.

GT-3 was separated from GT-1 and GT-2 on silicic acid using 100% methanol. The separation of this fraction is the same as that used by Tachibana (22) and Yasumoto (13) for maitotoxin. Unlike GT-1 and GT-2, this fraction was found to be an irreversible inhibitor of the guinea pig ileum preparation. The time studies of GT-3 are important in that they characterize GT-3 as a very toxic fraction but one which is very slow in its action.

Table III. Summary of Inhibitory Constant Data

Fraction	CONC (ng)	KI If MW=1000 $x10^{-12}$	pA_2 if MW=1000
GT-1	6.7	5.5	
	4.5	5.49	
(Average)		(5.5)	(11.55)
GT-2	1.6	2.54	
	3.94	3.69	
	7.5	4.06	
	13.7	5.37	
(Average)		(3.91)	(12.2)

Relationship of the Fractions to Ciguatera Syndrome. The Ciguatera syndrome was first described and named in the Caribbean. The isolate of G. toxicus that we have utilized in these studies was isolated from a region of the Caribbean in which Ciguatera is prevalent. Indeed, crude toxic extract from toxic fish from the same area extracted in the same manner gave similar effects on guinea pig ileum as the crude extract in this study (19). Several authors have reported toxic components in the lipid extracts of G. toxicus as being the major contributor to the cause of the ciguatera syndrome (22). We conclude, that GT-1, GT-2 and GT-3 are at least partially responsible for the Ciguatera syndrome in the Caribbean.

Relationship of Fractions to Other Toxins. Our initial extractions of the toxic fractions are the same as that of Tachibana, (22) and others. Thus, the isolation of GT-1a and GT-1b correspond exactly to the initial steps in the isolation of ciguatoxin by Scheuer (9). The isolation of GT-2b and GT-2c and their similar action on the ileum preparation, causes us to conclude that they are either a more polar form of GT-1 or a modification of GT-1 accomplished during the extraction procedures. GT-3, it appears, would most likely correspond to a carry over of a maitotoxin-like fraction from the initial ether-water separation. Without any chemical confirmation, however, these are only tentative identifications. Indeed, toxins extracted from Caribbean isolates of G. toxicus, could be quite different from those extracted from Pacific isolates.

Use of the Ileum Preparations. With respect to the dose response curves and the conclusions that one may try to draw from these, we understand the arguments surrounding the classical receptor theory and that there are limits to both the precision and accuracy that one can expect for the technique. Nevertheless, we infer several things from our data.

Assay of the crude toxin is only competitive in nature within a time frame of fifteen to twenty minutes and after that time the effects are mixed. Therefore, in any determination of toxic activity of crude toxin, the time frame of the assay must be rigidly controlled to obtain consistent results.

Calcium treatment has been recommended for offsetting the effects of "Ciguatera syndrome" in vivo. In our experiments, calcium was ineffective in offsetting the effects of either GT-1, GT-2, or GT-3, on the guinea pig ileum. This would not indicate anything with respect to the action of calcium on the Ciguatera syndrome but it does have implications that calcium is not immediately involved in the actions of GT-1 and GT-2. The fact that it produces an enhancement of the effect of GT-3 at lower than normal concentrations is just the reverse of what we would have expected. Indeed, the guinea pig ileal muscle is known to be quite insensitive to calcium deprivation and also is not affected by TTX. The results with the inclusion of TTX in the medium simply verify that the major action of the toxin in our assay method is directly on the smooth muscles of the ileum.

Significance of the assays. The pA_2 determined from constrained plots of experiments in which a fixed concentration of antagonist is used along with graded doses of agonist upon the same system, is a measure of the affinity of a competitive inhibitor for the implicit receptor site. In general, it is considered to be a truer measurement of the actual affinity when compared to other methods, inasmuch as the experimental design operates upon the null hypothesis. Nevertheless, we recognize that the value which we have determined for the pA-2 of around 11.6 is extremely high. The margin of error in our experimental situation is small, inasmuch as we utilize one microliter of a milligram solution of toxin on one or two liters of physiological saline to perform an assay.

Some reversible systems, *in vivo*, act at very low concentrations. For instance, there is extensive documentation in the literature on insulin to show that it acts at very low serum levels (less than 10-9 M) in a saturable, reversible, dose dependent fashion in recognized target tissues to cause a biological effect. Also, in general, constants derived for competitive inhibitors of anticholinergics and antihistamines tend to be high (25,26). In the same manner, constants measured on calcium channel inhibitors have been exceedingly high (27). As previously mentioned, in any discussion of the effect of an antogonist in a dose response situation there is an implicit protein receptor site. But in fact, there need not be a protein receptor site and as Hollenberg (28) has pointed out, "Most often, the term receptor is used to identify that (probably rate limiting) cellular component which recognized the ligand of interest so as to set in motion the series of events leading to a cellular response." The receptor or accessory receptor sites for the hydrophobic moieties of some competitive antagonists of membrane active agonists might, in fact, be made up of an interface between the hydrophobic area on the receptor protein and the lipid groups in the membrane. The high affinity constants calculated for anticholinergics and antihistaminics are not likely based on interaction on a protein surface. The lipid molecules in a membrane facing the receptor surface are fixed to a large extent in a quasicrystalline form and it perhaps is with these lipid molecules that the antagonist interacts.

In summary, we have isolated three toxins from lipid extracts of *G. toxicus*. We have demonstrated that the guinea pig ileum preparation is an effective assay for these toxins in nanogram quantities. All three fractions effectively inhibited the guinea pig ileum preparation in its response to histamine at nanogram concentrations. The GT-3 toxic fraction is quite different from the other two (GT-1, GT-2) in being very slow acting and nonreversible in nature.

Acknowledgements

This study was supported in part by the United States Food and Drug Administration (contract number 223-79-2287) and the School of Medicine, SIU-C. The animal viscera utilized in these studies were supplied by Dr. William Yau, School of Medicine, SIU-C, Carbondale, Illinois. We also acknowledge the help of Dr. Edward Ragelis, Project Officer (FDA), who supplied the TTX utilized in these studies, discussed the experiments and read the paper. Steven H. Sarfaty and James M. Muenz both assisted in running the ileum preparations.

Literature Cited

1. Hashimoto, Y. In "Marine Toxins and Other Bioactive Marine Metabolites"; Japanese Scientific Societies Press, Tokyo; 1979, pp 91-114.
2. Scheuer, P. J. Acc. Chem. Res. 1977 10,33-9.
3. Baslow, M. H. In "Marine Pharmacology"; 2nd Ed, R. E. Krieger Publishing Co: Baltimore MD; 1977; pp 194-315.
4. Randall, J. E. Bull Marine Sci. 1958, 8, 235.
5. Bagnis, R. Hawaii Med. J. 1968, 28, 25-8.
6. Banner, A. H. In Biology and Geology of Coral Reefs"; Vol III, Academic Press: New York; 1976; p 117.
7. Scheuer, P. J. In "Advances in Food Research"; Chickchester, C. O., Stewart, G. F., Mrak, E. M., Eds; Academic: New York, 1970; Vol XVIII.
8. Lawrence, D. N.; Enriquez, M. B.; Lumish, R. M.; Macer, A. J. Am. med. Ass. 1980; 254-8.
9. Scheuer, P. J.; Takahashi, N.; Tsutsumi, J.; Yoshida, T. Science, 1967, 155, 1267-8.
10. Hessel, D. W; Halstead, B. W.; Peckham, N. H. Add. N.Y. Acad. Sci. 1960, 788-797.
11. Kimura, L. H.; Hokama, Y.; Abad. M. A.; Oyama, M.; Miyakamma, J. T. Toxicon, 1982, 20, 907-12.
12. Sylvester, J. R.; Dammann, A. E.; Dewey, R. A. Mar. Fish. Rev. 1977, 39, 14-6.
13. Yasumoto, T.; Bagnis, R.; Thevenin, S.; Garcon, M.; Bull. Japan. Soc. Scient. Fish. 1977, 43, 1015-9.
14. Yasumoto, T.; Nakajima, I.; Bagnis, R.; Adachi, R. Bull. Japan. Soc. Scient. Fish. 1977, 43, 1021-6.
15. Yasumoto, T.; Nakajima, I.; Chungue, E.; Bagnis, R. Bull. Japan. Soc. Scient. Fish. 1977, 43, 69-74.
16. Tindall, D. R.; Dickey, R. W.; Miller, D. M. Fed. Proc. Fed. Am. Soc. Exp. Biol. 1982, 41, 1561.
17. Miller, D. M.; Dickey, R. W.; Tindall, D. R. Fed. Proc. Fed. Am. Soc. Exp. Biol. 1982, 41, 1561.
18. Carlson, R.; Tindall, D. R.; Dickey, R. W.; Morey-Gaines, G. Fed. Proc. Fed. Am. Soc. Exp. Biol. 1982, 41, 1562.
19. Dickey, R. W.; Miller, D. M.; Tindall, D. R. Fed. Proc. Fed. Am. Soc. Exp. Biol. 1982, 41, 1562.
20. Provasoli, T. In "Culture and Collection of Algae"; Watanabe, H.; Hattori, A., Eds.; Japanese Society of Plant Physiology: Tokyo, 1968; p. 63.
21. Rossum van, J. M. Arch. Int. Pharmacodn. Ther. 1963, 143, 299.
22. Tachibana, K. Ph.D. Thesis, University of Hawaii, Honolulu, Hawaii, 1980.

23. Dickey, R. W.; Miller, D. M.; Tindall, D. R. This volume.
24. Banner, A. H.; Helfrich, P.; Scheuer, P. J.; Yoshida, T. Proc. Gulf Carib. Fish. Inst. 1963, 6.
25. Beld, A. J.; Ariens, E. J. Eur. J. Pharmacol. 1974, 25, 203-9.
26. Ellenbroek, B. W. J.; Nivard, R. J. R.; van Rossum, J. M. Ariens, E. J. J. Pharm. Pharmacol. 1965, 17, 393.
27. Schwartz, A. The Physiologist 1983, 26, 200-5.
28. Hollenberg, M. D.; Cugatrecasas, P. In "The Receptors"; O'Brien, R. D., Ed. Plenum Press: New York, 1980, p. 193-213.

RECEIVED March 16, 1984

23

Extraction of a Water-Soluble Toxin from a Dinoflagellate, *Gambierdiscus toxicus*

ROBERT W. DICKEY, DONALD M. MILLER, and DONALD R. TINDALL

Department of Botany and Department of Physiology and Pharmacology, School of Medicine, Southern Illinois University, Carbondale, IL 62901

A water soluble toxin was extracted from sixth and seventh generation cultures of Gambierdiscus toxicus from the Caribbean. The crude toxin was partially purified using column and thin layer chromatography. Crude and semipurified fractions were tested for toxicity on mice and guinea pig ilea. The LD-50's for crude and semipurified fractions were 1.1 mg/kg mouse and 0.62 μg/kg mouse, respectively. On guinea pig ileum preparations, semiquantitation of this irreversible toxin was accomplished by titrating down to a dosage of toxin which would cause 50% inhibition. The semipurified toxin produced 50% inhibition at 2.8 ng/ml, whereas the crude extract required 5 μg/ml. This water soluble toxin is compared with a similar component found in the lipid extracted fraction.

Ciguatera is a human food poisoning which has for many years endangered public health and hampered maritime economic development throughout subtropical and tropical regions of the world. This form of food poisoning is a serious malady which is occasionally fatal. It is contracted in humans by ingesting any one of numerous species of normally edible inshore and reef associated fishes. The symptoms of intoxication are numerous and diverse, but typically include gastrointestinal and neurological disorders (1-4).

Ciguatoxic fishes are now known to acquire toxicity from at least one, but most probably several species of dinoflagellates which comprise a significant portion of the food base in subtropical and tropical latitudes. Gambierdiscus toxicus was the first dinoflagellate to be linked to the genesis of ciguatera toxins (5-9). However, other investigations conducted in the South Pacific have revealed the presence of several additional toxic dinoflagellate species which may contribute to this complex syndrome (10, 11). Recently, Tindall and co-workers, (12-14)

0097-6156/84/0262-0257$06.00/0
© 1984 American Chemical Society

reported on similar toxins from two species of Prorocentrum and G. toxicus from a ciguatera endemic region of the Caribbean. Several additional species which are prevalent in this Caribbean region have shown similar signs of toxicity, and are currently under investigation (15).

In a study of the effects of lipid extracted toxins (GT-1, GT-2, and GT-3) from G. toxicus on guinea pig ileal smooth muscle preparations, Miller and others (16) concluded that effects of GT-1 and GT-2 could be offset by increasing the agonist concentration used to challenge the preparation after exposure to toxin. In addition, time dependent studies of both GT-1 or GT-2 indicated that their effects were completely reversible after washing within the two hour time frame for the experiments. The results of these experiments were not changed by increasing the calcium concentration in the medium above normal levels.

The most recent reports on the action of the water soluble toxin from G. toxicus concluded that this toxin, referred to as maitotoxin, invokes an unmediated stimulatory (contractile) response on contact with guinea pig ileal smooth muscle preparations. Furthermore, this response was said to be due to an activation of calcium channels in the smooth muscle membranes (17). Previous studies of this toxin from the same source indicated that it produces an inhibitory effect on guinea pig atrial muscle (18, 19).

The present study reports on the effects of the water soluble toxin from G. toxicus from the Caribbean on guinea pig ileal smooth muscle preparations. It also compares the results obtained from the water soluble toxin with those obtained using lipid extraction.

Materials and Methods

Dinoflagellate Culture. The isolate of G. toxicus (SIU-350) used in this study was obtained from the South Sound of Virgin Gorda, British Virgin Islands. The species was brought into unialgal and large-scale culture as previously described (16). Cells were harvested by continuous centrifugation after cultures reached the early stationary phase of growth (30-35 days).

Extraction and Purification of Toxin. Products of two batches of large-scale cultures were extracted: (A) 350F and (B) 350G. Culture F consisted of 53 liters and yielded 13.1 g fresh weight (2.62 g dry mass) of G. toxicus cell pellet. Culture G consisted of 130 liters which yielded 49.5 g fresh weight (9.9 g dry mass). The cells were extracted by refluxing in boiling aqueous methanol (also see 16). The methanol extracts were concentrated and subjected to liquid-liquid partitioning beginning with diethyl ether and water and ending with cold acetone treatment (Figure 1). A total of 414.24 mg dry mass of crude water soluble acetone precipitate (WSAP) was obtained from 350F and 350G, (sixth and seventh successive large scale cultures).

The WSAP extracted from 350G was subjected to further purification by sequential treatment on columns of silicic acid,

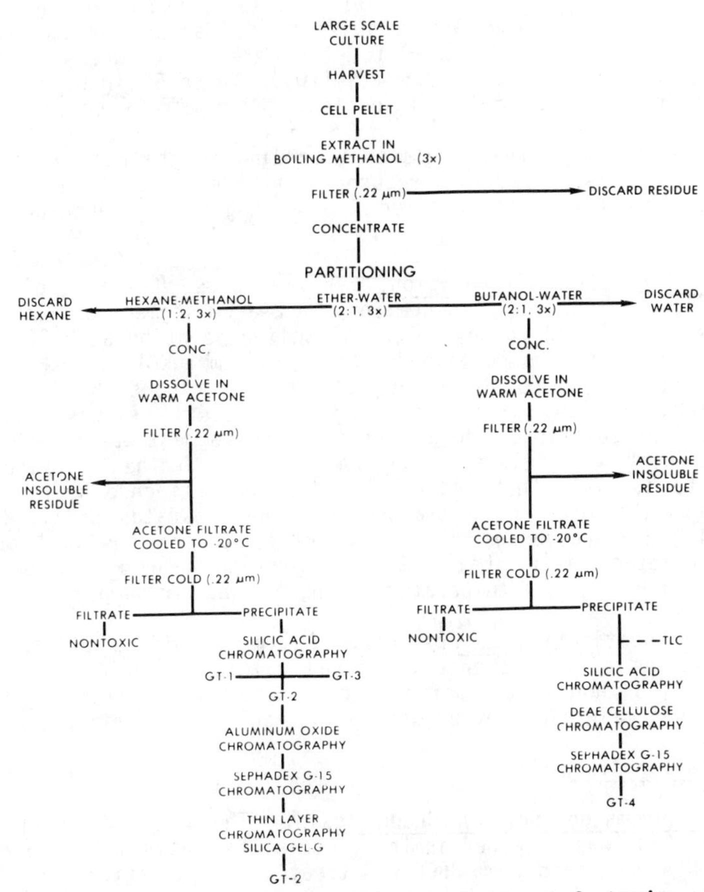

Figure 1. Flow diagram for the isolation of toxins from *Gambierdiscus toxicus*.

DEAE, and Sephadex G-15-120. Eluting solvents in the order of their usage consisted of chloroform-methanol, chloroform-methanol-water mixtures, and distilled water, respectively. The toxin obtained from this chromatographic treatment was designated GT-4.

Mouse Bioassay Procedures. For initial screening and LD_{50} determinations outbred female swiss mice (Harlan Sprague Dawley ICR 'BR') weighing between 19 to 21 g were used. Doses of toxic extract were suspended in 0.5 ml of 0.1% Tween 60 in 0.15 M NaCl and administered by i. p. injection. Mice were observed for a period of 48 hours.

LD_{50} values were calculated according to the method of Weil (20). Four dosage levels were used with three repetitions at each level. The LD_{50} was calculated from moving average interpolation tables.

Guinea Pig Ileum Preparation. Guinea pigs (350-600 g) were sacrificed by cervical dislocation. A 2-3 cm segment of ileum was removed and placed in physiological saline solution at 37°C. The physiological saline contained: NaCl 136.9 mM; KCl 2.68 mM; $CaCl_2$ 11.84 mM; $MgCl_2$ 1.03 mM KH_2PO_4 0.45 mM and glucose 5.55 mM. If required, the intestinal lumen was evacuated by slowly flushing with saline solution. The dissected ileum segment was attached by a wire hook to a glass rod which extended to the bottom of an organ bath. The other end of the ileum was attached by a silver chain to the lever of a transducer. The apparatus utilized for studying the responses of the ileum to toxin consisted of a: waterjacketed tissue bath, Bionix myograph transducer, Beckman physiograph, Beckman temperature bath, and an air pump.

Calcium Substitution Experiments. Saline solutions utilized in this series of experiments contained calcium chloride at 0.25X, 0.5X, 2X, 3X and 4X the normal level. The osmotic balance of the saline was maintained by adjusting the NaCl concentrations.

Results and Discussion

Column Chromatography of Crude Toxin. The WSAP obtained from culture 350F was retained in the crude state for assay. The 266.2 mg of WSAP obtained from 350G was treated on successive columns of silicic acid, DEAE cellulose, and Sephadex G-15 and yielded a single semipurified toxic product (GT-4) of 23.1 mg or 9% of the starting crude extract (Table I).

Thin Layer Chromatography. TLC analysis of the WSAP was accomplished by application to silica gel G plates at a concentration of 0.1 mg and development to 14 cm with chloroform-methanol-water (60:35:8). Prior to sulfuric acid-char treatment, five components were visibly resolved. Following such treatment a total of 10 components were visualized. Duplicate plates were divided into six zones, which were scraped, eluted with methanol and assayed at a concentration of 0.2 mg equivalents

Table I. PURIFICATION OF 266.2 MG OF WSAP FROM *Gambierdiscus toxicus*

SEPARATION MEDIUM	ELUANT	TOXIC ELUATE (mg)*	% RECOVERED
Silicic Acid	Methanol	152.3	57.2%
DEAE Cellulose	Methanol:Water (1:1)	119.5	44.9%
Sephadex G15	Water	23.1	8.7%

*assayed by i. p. injection into mice

per mouse. The component falling within the Rf's 0.0 to 0.2 was found to be toxic.

Further thin layer chromatographic analysis of the toxic fraction indicated a three component system with the major toxic component moving just off of the origin (Rf = 0.05) and two lightly detectable, non-toxic components occurring at Rf's 0.6 and 0.63.

Mouse Bioassay. When administered at 5 mg per mouse in 0.5 ml dose during the initial screening, the WSAP from *G. toxicus* caused death in all test mice within 120 minutes. The toxin had a latency period of approximately 30 minutes after which signs of toxicity were noticeable, and included in order of occurrance: inactivity and piloerection followed by cyanosis of the tail and feet with concurrent hypothermia, vasodilation in the ears ('scarlet ears'), lacrimation, ptosis of the eye lid on the side of injection, ptosis of the abdomen (loss of muscular tone), asthenia, impairment of hind limb motor ability followed shortly by complete paralysis with the hind limbs extended posteriorly (complete prostration), and dyspnea (respiratory distress). Death occurred without convulsions and the eyes became cataracted just prior to or after death.

The LD_{50} of WSAP was found to be 0.022 mg per mouse (1.1 mg/kg) (Figure 2) with a correlation coefficient of 0.97 and a range of activity of 0.036 mg. Interestingly, mice injected with GT-4 tended to die in the second hour after injection; whereas, the lipid extracts, GT-1 and GT-2 caused death within the first hour (16).

Effects of WSAP on Guinea Pig Ileum. In preliminary experiments with a constant infusion of 5 μg/ml of WSAP in the physiological saline bath the following were observed: A) a slow tonic contraction of the smooth muscle segment resulting in a shift of the base line of approximately 30%, B) a latency period of approximately 10 to 15 minutes before there was any significant

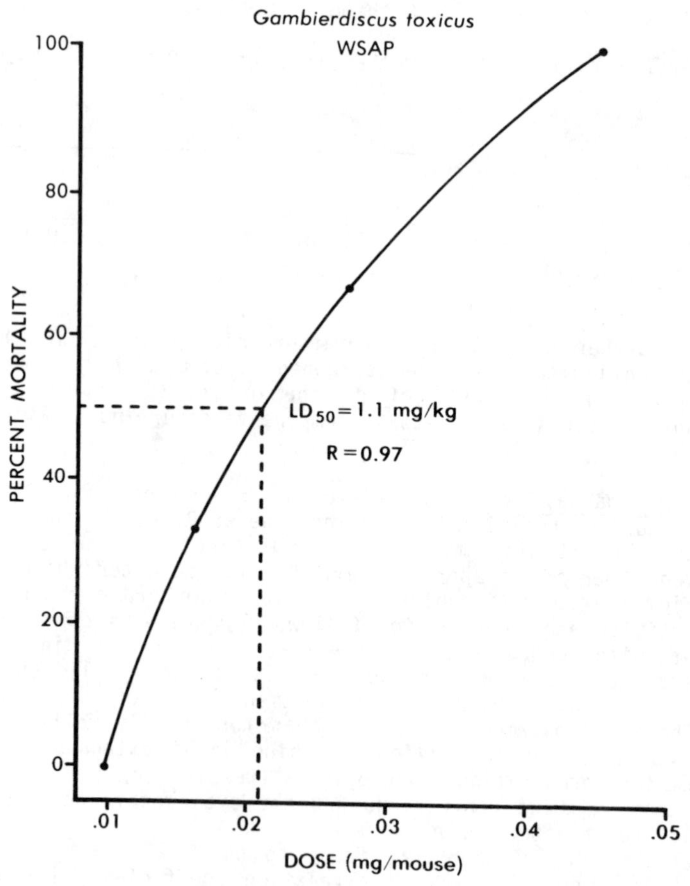

Figure 2. LD_{50} determination of water soluble acetone precipitate.

decline in the maximum response elicited upon challenge, C) a progressive decline of the preparation in response to challenge by agonist, and D) loss of all response to histamine (Figure 3). The death of the preparation consistently followed within 25 to 30 minutes under these conditions.

Effects of GT-4 on Ileum. A constant infusion of GT-4 at nanogram levels produced exactly the same effects as previously described for the WSAP. Parallel assays with GT-4 and WSAP revealed that 2.8 ng/ml of GT-4 caused the same rate of decline of activity as 5.0 μg/ml of WSAP. These parallel assays of WSAP and GT-4 provided an efficient means for the estimation of a purification factor. Purification of the toxin was 1,785 fold. Using this value we projected an estimate for the LD_{50} of GT-4 as 0.62 μg/kg mouse. This value comes quite close to the calculated LD_{50} value reported for maitotoxin (0.17 μg/kg), isolated from *G. toxicus* from the Pacific.

Establishment of Assay Procedures. Further attempts to reduce the dose of GT-4 in the perfusing saline down to a concentration where a linear response of dosage versus percent inhibition could be obtained resulted in the same events described above. The dose-response plots shifted to the right, as experienced with GT-1 and GT-2 (16), but they differed in that the maximum became progressively smaller regardless of the challenging agonist concentration utilized. Eventually the maximum response approached zero, indicating the death of the preparation. The inhibitory effects of GT-4 could not be reversed by washing with clean saline (Figure 3).

These results emphasized the irreversible effects of GT-4 and pointed out the long time frame required for the maximum effect to occur. Accordingly, the experimental design was modified by controlling the time of exposure to the toxin in order to determine its effects after a long time interval. Assays conducted in this manner revealed that the ileal response to the toxin was non-linear for a period of up to 60 minutes following the latency period after exposure to toxin. The results from higher toxin doses deviated less from linearity and stabilized sooner than did lower doses. Further studies revealed that the non-linear responses following the latency periods were a function of the concentration of agonist used to challenge the preparation just prior to and/or after exposure to toxin (Figure 4).

The higher the residual agonist concentration remaining associated with the preparation membranes following the control challenges and/or the higher the challenge levels after exposure to toxin, the closer to linearity the dose response relationship became. Alternately, the lower the residual and/or challenge agonist levels, the more accentuated the departure from linearity; presumably because neither the base line or final effects of the toxin had stabilized. However, when the successive preparations were exposed to the same toxin concentrations, and tested after 60 to 90 minutes the percent inhibition was within 20%.

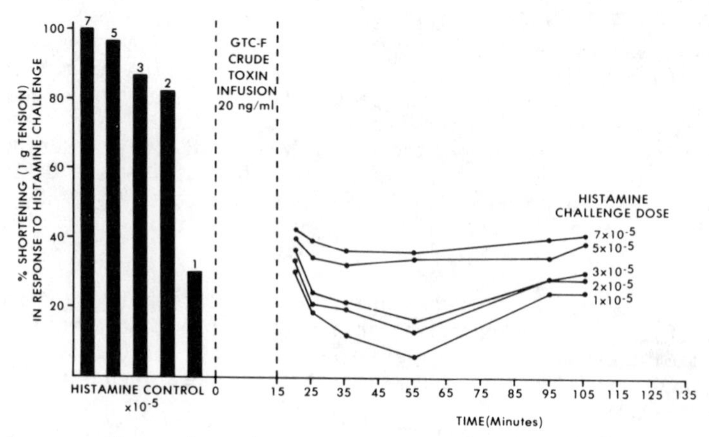

Figure 3. Time dependent response of guinea pig ileum preparation to crude toxin.

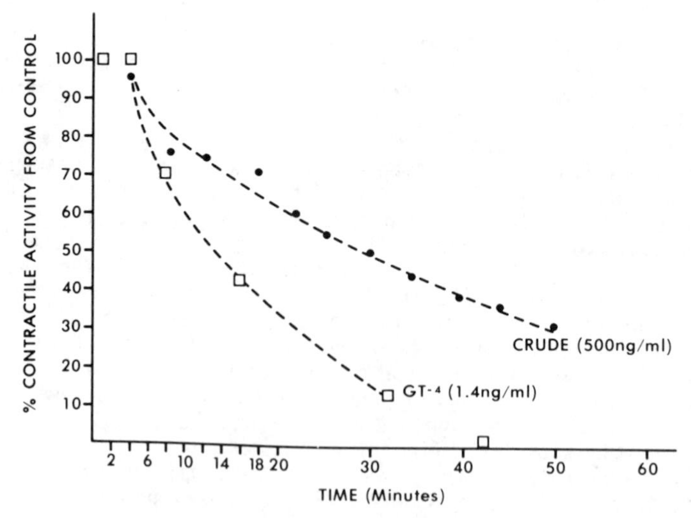

Figure 4. Percent activity remaining as a function of time with toxin continually present.

The results described above indicated that GT-4 caused an accumulative and irreversible antagonism of the histamine response of the guinea pig ileum. In order to more accurately semiquantitate the action of this toxin, a stringent control of experimental variables was imposed. The exposure to toxin could be controlled by either fixing the concentration and varying the duration of exposure or by fixing the duration of exposure and varying the dosage. Following exposure to toxin, challenging the preparation was limited to one agonist concentration within the linear region of the control agonist dose response curve (such curves typically approach 0 and 100% asymptotically and thus may exaggerate levels of inhibition if challenged near these extremes). Finally this challenge dose was given at a time that would be within the stabilized region of the toxin dose response profile. This time interval was set at 90 minutes for this study.

When the duration of exposure to GT-4 was limited to 10 minutes and the percent inhibition for an agonist concentration equivalent to the 50% control response was measured at 90 minutes as a function of toxin concentration, a linear response was obtained (Figure 5). Figure 5 shows that the exposure of the preparation to 2.8 ng/ml, 7.0 ng/ml and 70 ng/ml for 10 minutes resulted in inhibitions of 30, 36, and 66% of the contractile response from control levels, respectively. In addition, exposure of the preparation to 2.8, 7.0 and 70 ng/ml for 15 minutes resulted in 45, 55, and 98% inhibition of the contractile response, respectively. We can express the effects of GT-3 by stating both the dose and time to achieve a 50% inhibition from the control or a DCI_{50} value.

Comparison of GT-3 and GT-4. Time dependent assays of the ether extracted GT-3 reported by Miller *et al.* (16) indicated that it had the same time course of action and irreversible effects as described for GT-4. When the ileal preparation was exposed to 10 ng/ml of GT-3 for 15 minutes a 50% inhibition from control response was observed. Upon contact with the preparation, the toxin caused a slow tonic contraction. Washing the segment with clean saline following the 15 minute exposure period caused a gradual relaxation of the preparation to near the base line levels observed in the control. However, when challenged by agonist at any time following the latency period, and irreversible loss of activity was evident.

Due to the identical behavior of GT-3 (16) and GT-4 on thin layer chromatography in addition to the ileum assays, these toxins are considered to be very closely related, if not identical. GT-3 likely represents a water soluble carry-over in the initial diethyl ether partition. In light of this observation, the effects of the crude ESAP on the guinea pig ileum previously reported (16) are quite understandable. We surmise that the first phase of immediate, but reversible inhibition was ellicitied by GT-1 and GT-2 which have already been shown to be competitive in nature (16), and the second, irreversible phase was caused by GT-3.

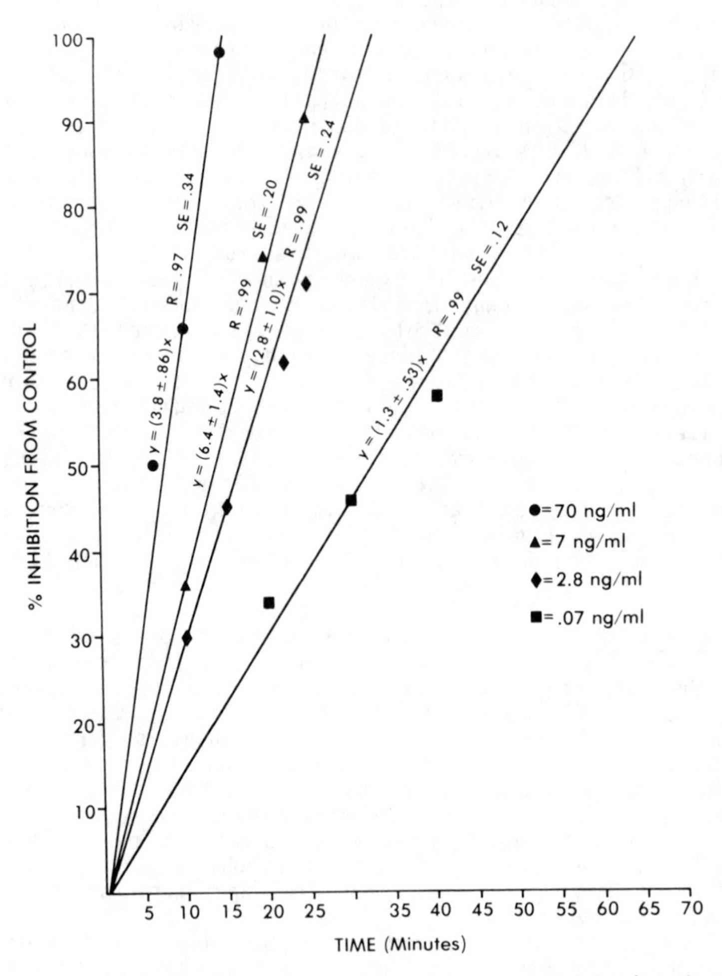

Figure 5. Percent inhibition as a function of time toxin is applied and concentration.

Effects of Calcium. In light of the recent reports that a water soluble toxin (maitotoxin) extracted from a Pacific isolate of *G. toxicus* causes an activation of calcium channels (17, 21), experiments were conducted with altered calcium in the medium. Time dependent studies were conducted utilizing the 10 minute application and 90 minute assay procedures with 2.8 ng/ml of GT-4 added to ileum preparations containing altered calcium concentrations of 0.25, 0.5, 1.0, 2.0 and 4.0 X the normal level (Figure 6). The results indicated a slight decrease in inhibition at higher calcium levels but a drastic increase in inhibition at lower calcium levels.

Possible Modes of Action. The long latent period before the irreversible effects of GT-4 suggests a mode of action other than that of a typical receptor complex. Other possible explanations include: A) some kind of covalent modification of the receptor site is taking place and the presence of an initial time interval during which a nonlinear dose response relationship is observed and whose extent of deviation appears to be a function of agonist concentration; B) the toxin may have to partition or intercalate itself into the membrane before being able to affect the receptor site; C) the toxin may partition intracellularly, somewhat similar to what local anesthetics have to do before being effective, D) the toxin may combine into multiunit complexes to form ionophores similar to other polyethers; and E) indeed, the toxin may even be operating upon an intermediate reaction in a string of reactions which finally lead to the contractile activity.

Acknowledgments

This study was supported in part by the United States Food and Drug Administration (contract number 223-79-2287), the Department of Botany, and the School of Medicine, SIU-C. The animal viscera utilized in these studies were supplied by Dr. William Yau, School of Medicine, SIU-C. We also acknowledge the help of Dr. Edward Ragelis, Project Officer, USFDA, who discussed the experiments and read the paper. Stephen H. Sarfaty and James M. Muenz assisted in conducting the ileum experiments.

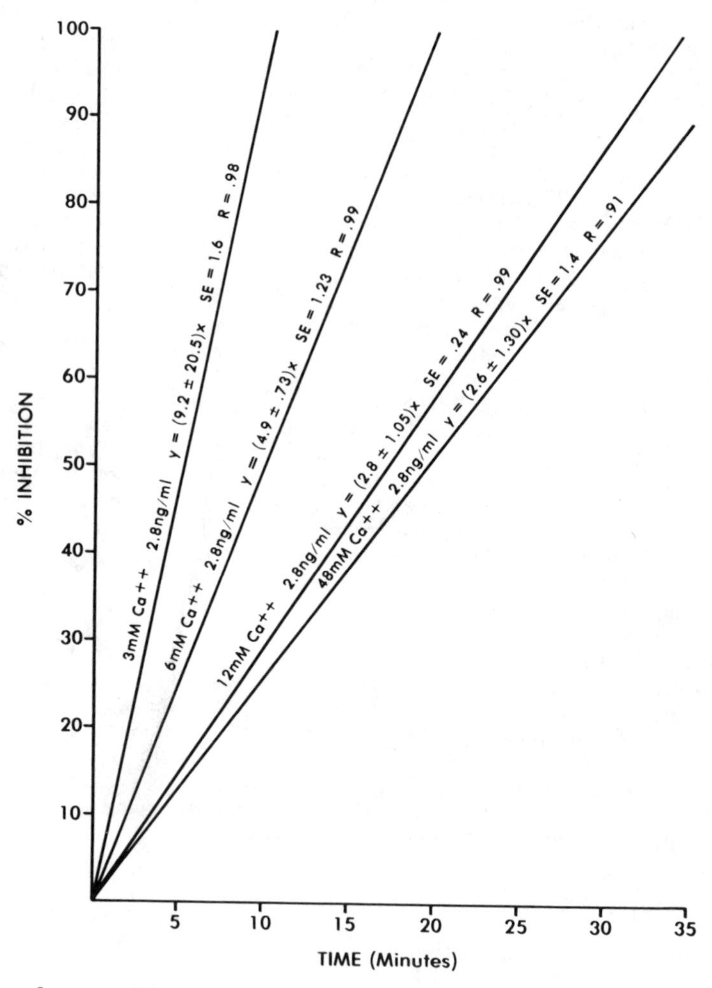

Figure 6. Effect of high and low calcium on the percent inhibition at a constant concentration of toxin.

Literature Cited

1. Randall, J. E. Bull. Marine Sci. 1958, 8, 236-7.
2. Scheuer, P. J. In "Advances in Food Research"; Chickchester, C. O., Stewart, G. F.; Mrak, E. M.; Eds; Academic: New York, 1970; Vol. XVIII, p. 144-61.
3. Banner, A. H. In "Biology and Geology of Coral Reefs"; Academic: New York, 1976, Vol. III, p. 117-213.
4. Lawrence, D. N.; Enriguez, M. B.; Lumish, R. M.; Maceo, A. J. Am. med. Ass. 1980, 244, 254-8.
5. Yasumoto, T.; Bagnis, R.; Thevenin, S.; Garcon, M. Bull. Japan. Soc. Scient. Fish. 1977, 43, 1015-9.
6. Yasumoto, T.; Nakajima, I.; Bagnis, R.; Adachi, R. Bull. Japan. Soc. Scient. Fish. 1977, 43, 1021-6.
7. Yasumoto, T.; Nakajima, K.; Chungue, E.; Bagnis, R. Bull. Japan. Soc. Scient. Fish. 1977, 43, 69-74.
8. Bagnis, R.; Hurtel, J.; Chungue, E.; Inoue, A.; Yasumoto, T. Rev. Int. Oceanogr. Med. 1979, 45, 2.
9. Bagnis, R.; Chanteau, S.; Chungue, E.; Hurtel, J. M.; Yasumoto T.; Inoue, T. Toxicon 1980, 18, 199-208.
10. Yasumoto, T.; Inoue, A.; Ochi, T.; Fujimoto, K.; Oshima, Y.; Fukoya, Y.; Adichi, R.; Bagnis, R. Bull. Japan. Soc. Scient. Fish. 1980, 46, 1397-1404.
11. Nakajima, T.; Oshima, Y.; Fukiyoko, Y.; Yasumoto, T. Bull. Japan. Soc. Scient. Fish. 1981, 47, 1029-33.
12. Tindall, D. R.; Dickey, R. W.; Miller, D. M. Fed. Proc. 1982, 41, 1561.
13. Miller, D. M.; Dickey, R. W.; Tindall, D. R. Fed. Proc. 1982, 41, 1561.
14. Carlson, R.; Tindall, D. R.; Dickey, R. W.; Morey-Gaines, G. Fed. Proc. 1982, 41, 1562.
15. Tindall, D. R.; Dickey, R. W.; Carlson, R. W.; Morey-Gaines, G. This volume 1984.
16. Miller, D. M.; Dickey, R. W.; Tindall, D. R. This volume 1984.
17. Ohizumi, Y.; Yasumoto, T. Br. J. Pharmac. 1983, 79, 3-5.
18. Miyahara, J. T.; Akau, C. K.; Yasumoto, T. Res. Comm. Chem. Path. Pharmac. 1979, 25, 177-80.
19. Shimizu, Y.; Shimizu, H.; Scheuer, P. J.; Hokama, Y.; Oyama, M.; Miyahara, J. T. Bull. Japan. Soc. Scient. Fish. 1982, 48, 811-3.
20. Weil, C. S. Biometrics 1952, 8, 249.
21. Takahashi, M.; Ohizumi, Y.; Yasumoto, T. J. Bio. Chem. 1982, 257, 7287.

RECEIVED March 16, 1984

Ecology of Toxic Dinoflagellates from the Caribbean Sea
Effects of Macroalgal Extracts on Growth in Culture

ROLLAND D. CARLSON, GREGORY MOREY-GAINES, DONALD R. TINDALL, and ROBERT W. DICKEY

Department of Botany, Southern Illinois University, Carbondale, IL 62901

The association of three toxic species of dinoflagellates (Gambierdiscus toxicus, Prorocentrum concavum and P. mexicanum) with selected species of macroalgae (Chaetomorpha linum, Dictyota dichotoma and Turbinaria turbinata) was examined in the Virgin Islands. Maximum concentrations of each of these dinoflagellates were 1.5×10^4, 9.5×10^4 and 1.5×10^6 cells/g f.w. of macroalgae, respectively. Relationships between these dinoflagellates and macroalgae were studied in bacterized and bacteria-free cultures, by enriching culture media with aqueous extracts of macroalgae and soil. In bacterized cultures, Chaetomorpha and soil extracts significantly increased growth rates and cell yields of G. toxicus. All extracts stimulated the growth of P. concavum, but extracts of macroalgae inhibited the growth of P. mexicanum. This indicates that the dinoflagellate-macroalgal association is differentially beneficial to various dinoflagellate species. Under bacteria-free conditions, the growth of P. concavum was significantly different than that observed in the presence of bacteria, indicating that bacteria play an important role in mediating these effects.

Since the discovery that Gambierdiscus toxicus Adachi & Fukuyo is a probable source of ciguatera toxins (1), studies in tropical coral reef areas in the Pacific Ocean have revealed that several species of dinoflagellates regularly occur in close association with benthic macroalgae (2-4). Some of these species produce toxins lethal to mice (2, 4), and thus may also be progenitors of ciguatera toxins. Recently, most of the same species have been found in the Virgin Islands (5), an area well known to produce toxic fish (6, 7). Subsequently, several of these species have proved to be toxic as well (5, 8-10).

0097-6156/84/0262-0271$06.00/0
© 1984 American Chemical Society

Gambierdiscus toxicus has been the focus of several laboratory studies. Optimal temperature, light and salinity requirements have been reported for G. toxicus in culture (11-15). However, the role of macroalgae in the growth of this and other associated toxic epiphytic dinoflagellates has not been examined quantitatively. The present study was undertaken to determine in situ field cell densities and laboratory growth characteristics of three dominant toxic benthic/epiphytic dinoflagellates (Gambierdiscus toxicus, Prorocentrum concavum Fukuyo and P. mexicanum Tafall) from the Virgin Islands. Extracts from macroalgae commonly harboring these dinoflagellates were tested in culture to determine their possible growth regulating properties.

Materials and Methods

Unialgal cultures of G. toxicus, P. concavum and P. mexicanum were established by micropipette isolation of cells from macroalgae collected in the British Virgin Islands. Gambierdiscus toxicus (SIU Strain #350) was isolated from Chaetomorpha linum (Muller) Kutzing near South Sound, Virgin Gorda on January 14, 1980. Prorocentrum concavum (SIU Strain #364) was isolated on July 10, 1980 and P. mexicanum (SIU Strain #273) on January 11, 1980 from Cladophora sp. near Salt Island. In agreement with Steidinger (16), we believe P. mexicanum to be the same species described more recently as Prorocentrum rhathymum Loeblich, Sherley & Schmidt (17). Cells were inoculated into ten different culture media. ES-enriched natural seawater (18) promoted the best growth of all three species, and was selected as the basic medium for the present study.

The macroalgae selected for aqueous extractions were Chaetomorpha linum (Chlorophyta), Dictyota dichotoma (Hudson) Lamouroux (Phaeophyta) and Turbinaria turbinata (Linnaeus) Kuntze (Phaeophyta). Macroalgae were bagged in situ, shaken to remove epiphytes, blotted dry and weighed, freeze-dried in a Curtis Cryo-vac and shipped to the laboratory at SIU in vacuum-evacuated packets. The volume of the epiphytic cell suspension was measured and aliquots were fixed in Lugol's Iodine Fixative for subsequent cell counts. In the laboratory, aqueous extracts of freeze-dried macroalgae were made by adding 25 mls distilled water to 5 g of alga, finely ground in a 60 mesh Wiley Mill. After vigorous stirring for 10 mins, the mixture was centrifuged at 2,500 rpm for 15 mins in a IEC PR-2 centrifuge. The supernatant was filtered aseptically through a 0.22 μm Millipore cellulose acetate filter and stored in the dark at 4 C. Extracts of Turbinaria (TU) and Dictyota (DI) were used at concentrations of 0.1% and 1.0% by volume, whereas Chaetomorpha extracts (CH) were used at 0.01%, 0.1% and 1.0% concentrations.

Soil extract (SE) was prepared by mixing and autoclaving equal weights of packaged soil (Michigan Peat) and distilled water (19). The mixture was then centrifuged and filtered through a succession of glass fiber and cellulose acetate filters to remove particles. The final filter was a 0.45 μm Nuclepore cellulose acetate filter. The extract was then re-autoclaved to prevent bacterial conversion of any of its components. Soil extract was used at a concentration of 1.5% by volume (20).

Each extract was added to two media, both of which were enriched with ES enrichments (18, 21): one made with natural seawater (Florida Gulf Stream Seawater, Carolina Biological Supply Co.) and designated ESNW; the other made with artificial seawater (21, 22), and designated ESAW. To prevent precipitation during sterilization, 0.12 g $NaHCO_3$ and 1.44 mls 1 N HCl were added to each liter of seawater immediately before autoclaving. After autoclaving, when the seawater base was cool, ES enrichments were added aseptically.

Cells were aseptically inoculated into 50 mls of medium in 125 ml Erlenmeyer flasks to an initial concentration of approximately 25 cells/ml. All treatments and controls were represented by triplicate flasks. Cultures were grown at 27°C under continuous 3,200 lux cool-white fluorescent illumination in Percival Model I-35 LL culture chambers. Because cells usually aggregated and adhered to the inside surface of the culture flask, cultures were vigorously aspirated with sterile pipettes before samples were withdrawn for counting, in order to minimize sampling variation. Cell populations were monitored by microscopic counting of two 0.1 ml aliquots of living cells from each flask, using a Palmer-Maloney counting chamber (23). Cell counts were made every 5-7 days for a 28-day experimental period.

Bacteria-free cultures of *P. concavum* were prepared by treatment according to the method of Droop (24) for 9 days with the following mixture of antibiotics: 31.25 µg/ml D-cycloserine, 31.25 µg/ml gentamycin, 137.5 µg/ml ampicillin and 150 µg/ml polymixin B. This mixture proved to be toxic to *G. toxicus* and *P. mexicanum*, and other mixtures were either ineffective or toxic, as well. Experiments with bacterized cultures *P. concavum* were duplicated with bacteria-free cultures of grown in natural seawater medium (ESNW). These cultures were grown under the same conditions as described above but were monitored for 59 days and were regularly tested for bacterial contamination using a sterility test medium (25).

Large-scale (20 liter) cultures of these three dinoflagellates were grown in aerated glass carboys. Each carboy was inoculated with a 2 liter culture grown in a Fernbach flask. ESNW or ESNW with soil extract cultures were grown in a walk-in growth chamber (Sherer-Gillett Model CEL-511-38) at 27°C and under continuous 4,300 lux cool-white illumination, for 25-35 days. Cultures were harvested by continuous-flow centrifugation in a Sharples Model T-1 cream separator, and the fresh weight of the cell pellet was measured.

Results of culture experiments were subjected to analysis of variance using the GLM Procedure of the SAS Statistical Program Package (1979 Edition, SAS Institute Inc., Raleigh, NC). Analyses of final cell densities and culture growth rates were performed using means of the duplicate cell counts of each flask. The probabilities of treatments affecting cell densities or growth rates were evaluated at the $p<0.01$ and $p<0.001$ significance levels, and the particular treatments responsible for these effects were determined at the $p<0.01$ and $p<0.05$ significance levels, using Duncan's Multiple Range Test.

Results

Macroalgae/dinoflagellate associations in the field. Mean cell sizes and maximum cell densities of G. toxicus, P. concavum and P. mexicanum observed in the Virgin Islands are presented in Table I. Both G. toxicus and P. concavum were found in greatest numbers associated with Chaetomorpha linum. Intermediate numbers were found with Dictyota dichotoma. Highest densities of P. mexicanum occurred with Dictyota. Lowest densities of all three dinoflagellate species were found with Turbinaria turbinata. Microscopic examination of the macroalgae showed that G. toxicus and P. concavum were epiphytic; remaining motionless and adhering loosely to the macroalgal surface unless disturbed. Water samples taken near the macroalgae contained few, if any, of these cells. In contrast, P. mexicanum was actively motile and seldom adhered to the macroalgae. Nevertheless, these cells did not appear in water samples, indicating that this species also remains closely associated with macroalgae and the benthos.

Table I. Maximum Cell Concentrations and Cell Sizes of Three Dinoflagellate Species Collected From Three Species of Macroalgae in the Virgin Islands. Cell Concentrations are Expressed as Cells/g Freshweight of Macroalgae

Dinoflagellate Species	Mean Dorsoventral[1] Diameter (μm)	Dinoflagellate Cells/g FW of Chaetomorpha linum	Dictyota dichotoma	Turbinaria turbinata
G. toxicus	77.3 ± 6.6[2] n=176	75,793	45,532	1,617
P. concavum	49.3 ± 6.3 n=199	95,480	16,643	12,505
P. mexicanum	35.7 ± 3.1 n=202	398,030	1,524,215	2,778

[1] Dorsoventral diameter was measured from the sulcus to the opposite (dorsal) side of the cell when viewed from the apex or antapex.
[2] Standard deviation.

Culture of G. toxicus. Table II shows the concentrations of macroalgal extracts used in the present culture studies for all three dinoflagellate species. The macroalgal extract concentrations that supported the highest growth rates and cell densities are presented. Figure 1 shows the effects of different extracts on the growth of G. toxicus. Control or plain ESNW (without added extracts) supported significantly better growth than did control or plain ESAW; however, G. toxicus grew as well in ESAW with soil extract as it did in plain ESNW. The effects of extracts were quite different depending on whether the cells were grown in natural or artificial seawater. Both ESNW with soil extract and ESNW with 0.1% CH supported significantly greater cell yields and growth rates than did plain ESNW (Figures 1 & 2). Growth in extracts of *Dictyota* or *Turbinaria* was no different than growth in plain ESNW. In ESAW, all extracts except CH produced cell densities greater than those in plain ESAW. Growth rates were not so uniformly enhanced: only 1% DI and soil extract promoted significantly increased division rates; however, these increased rates were also greater than those attained in plain ESNW.

Table II. A Summary of Macroalgal Extract Concentrations (Percent by Volume) That Produce the Best Growth of Dinoflagellates in the Culture Experiments Shown in Figures 1-7

	ESNW			ESAW		
Dinoflagellates	CH	DI	TU	CH	DI	TU
G. toxicus	0.1	0.1	0.1	0.1	1.0	1.0
P. mexicanum	0.01	0.1	0.1	1.0	1.0	1.0
P. concavum	0.1	0.1	0.1	0.1	0.1	1.0

Gambierdiscus toxicus cells adhered to the inside surface of the culture flasks and aggregated in mucilaginous strands. The mean dorsoventral diameter of cultured cells was 75 µm. Cultures attained their highest cell densities in 21-28 days (Figure 2). Most of the cells grown in plain ESAW were irregular in shape and were not motile.

Culture of P. mexicanum. Results from similar experiments with *P. mexicanum* are presented in Figure 3. Algal extracts inhibited growth in all cases. Cell densities were increased by soil extract in ESAW, but were unaffected in ESAW. Growth rates in plain ESNW, plain ESAW, ESNW with soil extract, and ESAW with soil extract were not significantly different from one another (Figures 3 & 4).

Prorocentrum mexicanum cells remained motile and regular in shape even in cultures displaying greatly inhibited growth. The mean dorsoventral diameter of cultured cells was 36 µm. Cells did not adhere to the surface of the culture flask and did not produce observable mucilage. Highest cell densities were attained at approximately 14 days of culture age in exponentially growing cultures; cell numbers declined rapidly after 21 days. Although growth rates were low in ESNW with algal extracts, cell division persisted for a longer period; so the highest cell densities were not attained within the 28-day monitoring period.

Culture of P. concavum. Figure 5 summarizes the results of growth experiments using bacterized and bacteria-free (axenic) cultures of *P. concavum*. Bacterized cultures were grown in ESNW and ESAW. However, axenic cultures were grown in ESNW only, and only those extracts which supported the best growth in bacterized cultures were used in the axenic culture experiments.

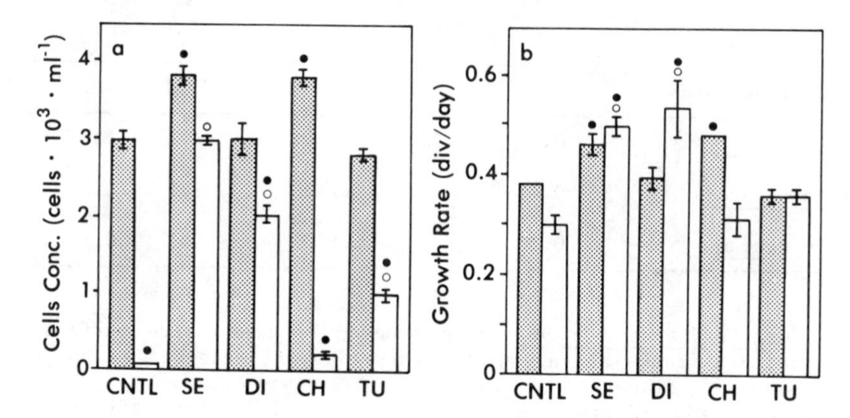

Figure 1. The effects of soil and macroalgal extracts on (a) highest cell concentration and (b) maximum growth rate of *Gambierdiscus toxicus*, grown in natural seawater (ESNW) medium (shaded bars) or in artificial seawater (ESAW) medium (open bars). Vertical brackets indicate standard errors. CNTL=controls (ES medium without added extracts); SE=soil extract; DI=*Dictyota* extract; CH=*Chaetomorpha* extract; TU=*Turbinaria* extract; ●=result significantly ($p < 0.01$) different from ESNW control; o=result significantly ($p < 0.01$) different from ESAW control.

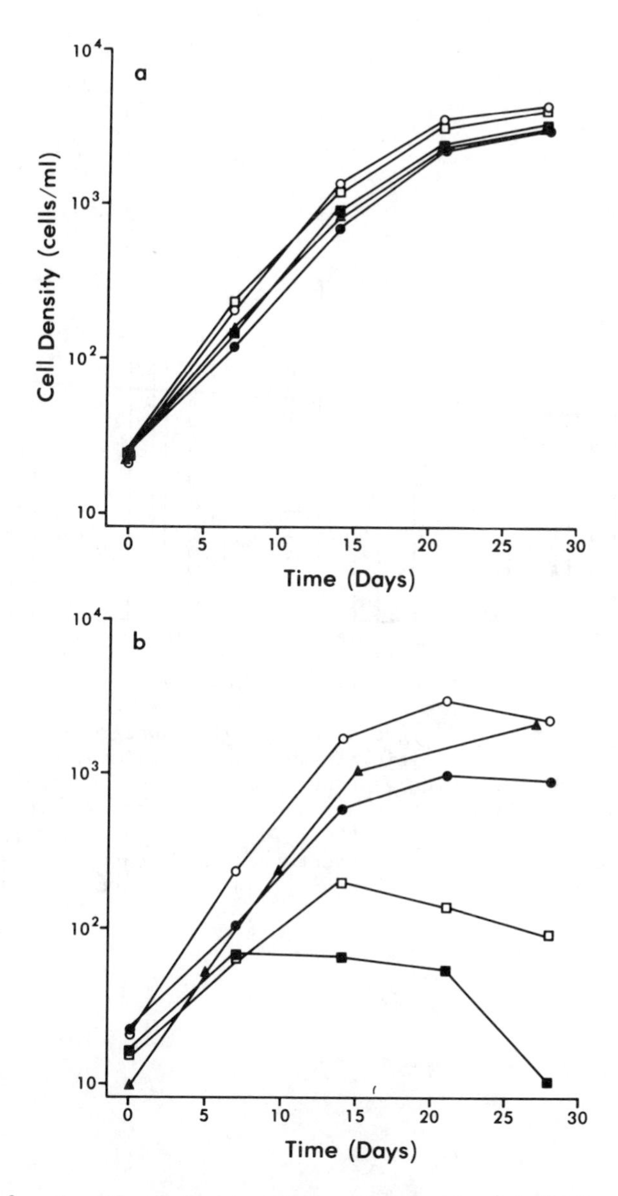

Figure 2. Growth of Gambierdiscus toxicus in (a) ESNW and (b) ESAW; both supplemented with soil and macroalgae extracts. Controls (■), Chaetomorpha extract (□), Dictyota extract (▲), Turbinaria extract (●), and soil extract (○).

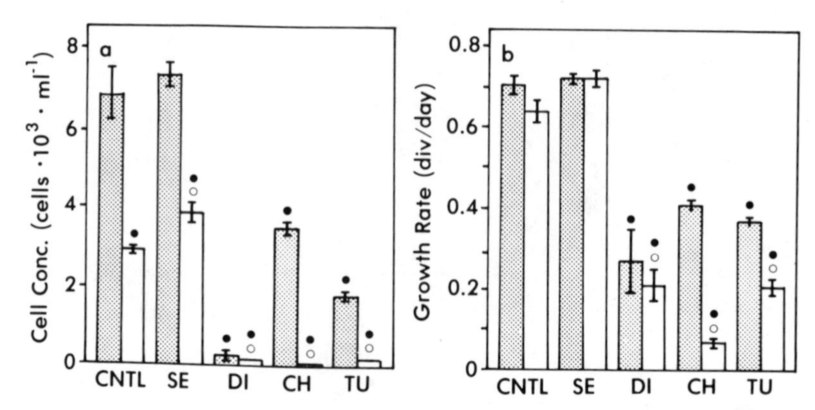

Figure 3. The effects of soil and macroalgal extracts on (a) highest cell concentration and (b) maximum growth rate of *Prorocentrum mexicanum*, grown in natural seawater (ESNW) medium (shaded bars) or in artificial seawater (ESAW) medium (open bars). Vertical brackets indicate standard errors. CNTL=controls (ES medium without added extracts); SE=soil extract; DI=*Dictyota* extract; CH=*Chaetomorpha* extract; TU=*Turbinaria* extract; •=result significantly ($p < 0.01$) different from ESNW control; o=result significantly ($p < 0.01$) different from ESAW control.

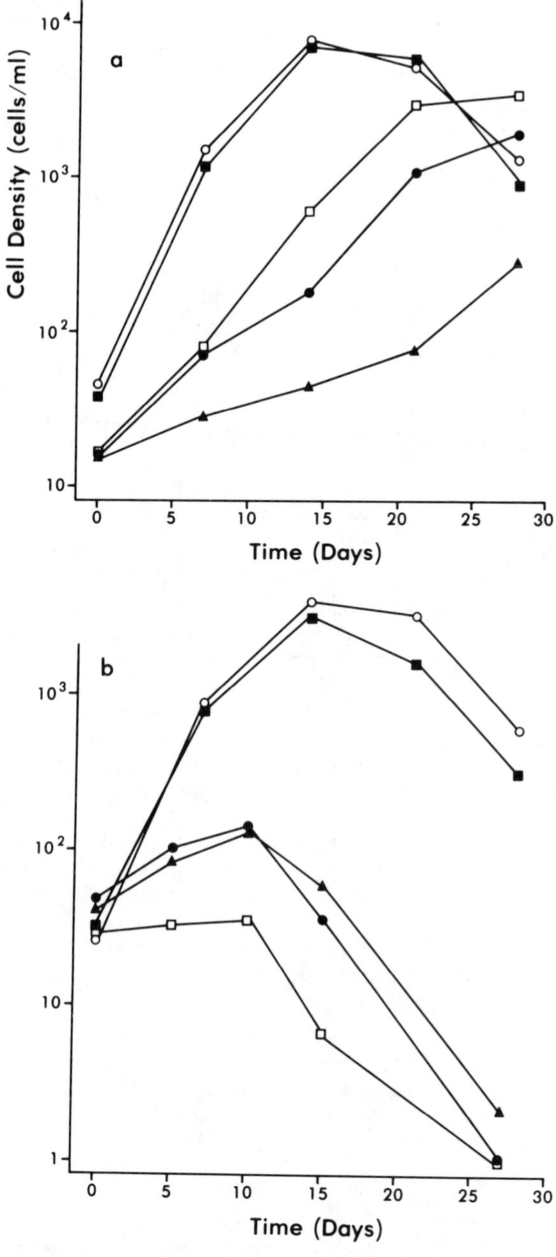

Figure 4. Growth of Prorocentrum mexicanum in (a) ESNW and (b) ESAW; both supplemented with soil and macroalgae extracts. Controls (■), Chaetomropha extract (□), Dictyota extract (▲), Turbinaria extract (●), and soil extract (○).

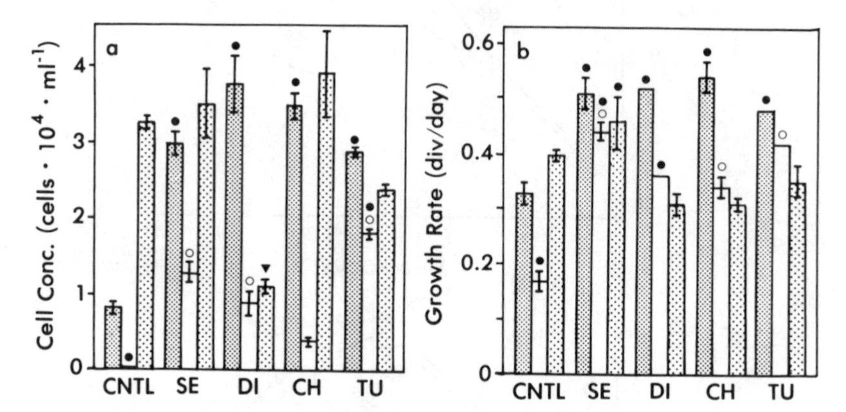

Figure 5. The effects of soil and macroalgal extracts on (a) highest cell concentration and (b) maximum growth rate of *Prorocentrum concavum*, grown in natural seawater (ESNW) medium (shaded bars), in artificial seawater (ESAW) medium (open bars), or in bacteria-free natural seawater (ESNW) medium (lightly shaded bars). Vertical brackets indicate standard errors. CNTL=controls (ES medium without added extracts); SE=soil extract; DI=*Dictyota* extract; CH=*Chaetomorpha* extract; TU=*Turbinaria* extract; ●=result significantly (p 0.01) different from ESNW control; o=result significantly (p 0.01) different from ESAW control; ▼=result significantly ($p < 0.01$) different from bacteria-free ESNW control.

Unialgal cultures of P. concavum had higher cell yields and division rates in ESNW supplemented with any of the four extracts than in plain ESNW (Figures 5 & 6). None of the cultures attained their highest cell densities during the 28-day monitoring period, and the ESNW control (which had a longer generation time than extract-enriched cultures) was still growing logarithmically at the end of this period. Cell densities and growth rates were greater in plain ESNW than in plain ESAW. However, the addition of soil, 1.0% TU or 0.1% DI to ESAW either eliminated or even reversed (with soil extract or 1% TU) the negative effects of plain ESAW. The only artificial seawater treatment that produced growth similar to that in ESNW enriched with soil or macroalgal extracts was ESAW enriched with soil extract.

Since bacterized cultures of P. concavum were not maintained beyond the mid-log phase, axenic cultures were grown for 59 days to allow the culture to reach maximum cell densities (stationary phase). Highest cell densities were obtained in media containing CH extracts, as a result of the extended period of logarithmic growth (Figure 7). The division rate did not decrease. Division rates were greater in soil extract enriched media than in any of the algal extract enriched cultures.

Statistical comparisons of results from bacterized and axenic P. concavum cultures could not be made because growth periods differed. Although the initial cell densities were identical, the axenic cultures required 24 additional days to reach the same cell densities as the bacterized cultures. Only soil extract stimulated division rates above those of bacterized cultures in plain ESNW; these elevated division rates were similar to those attained in extract-enhanced bacterized cultures. Growth in the other extracts was similar to that of bacterized ESNW controls in having longer generation times than bacterized cultures supplemented with extracts.

Cells of P. concavum adhered tenaciously to the surface of the culture flask and were embedded in mucilage. Only the most vigorous aspiration dispersed these aggregations sufficiently to permit reproducible counts. The mean dorsoventral diameter of cultured cells was 45 μm. No morphological differences between axenic and bacterized cells were noted, but irregularly-shaped cells were common in ESAW control cultures.

Large-Scale Cultures. Results of growth in large-scale culture are presented in Table III. Cultures of G. toxicus attained their highest cell densities 21-28 days after inoculation, beginning with an initial cell density of 40-50 cells/ml. Best growth of G. toxicus was achieved in half-filled carboys. Maximum cell densities of P. mexicanum were attained in 19-25 days, from initial cell concentrations of 80-150 cells/ml. Prorocentrum concavum reached maximum cell numbers in 28-35 days in ESNW with soil extract, beginning with cell concentrations of approximately 100 cells/ml. Growth rates of Prorocentrum species were similar to those obtained in 125 ml cultures, but G. toxicus had a shorter period of logarithmic growth in large-scale cultures than in smaller volumes.

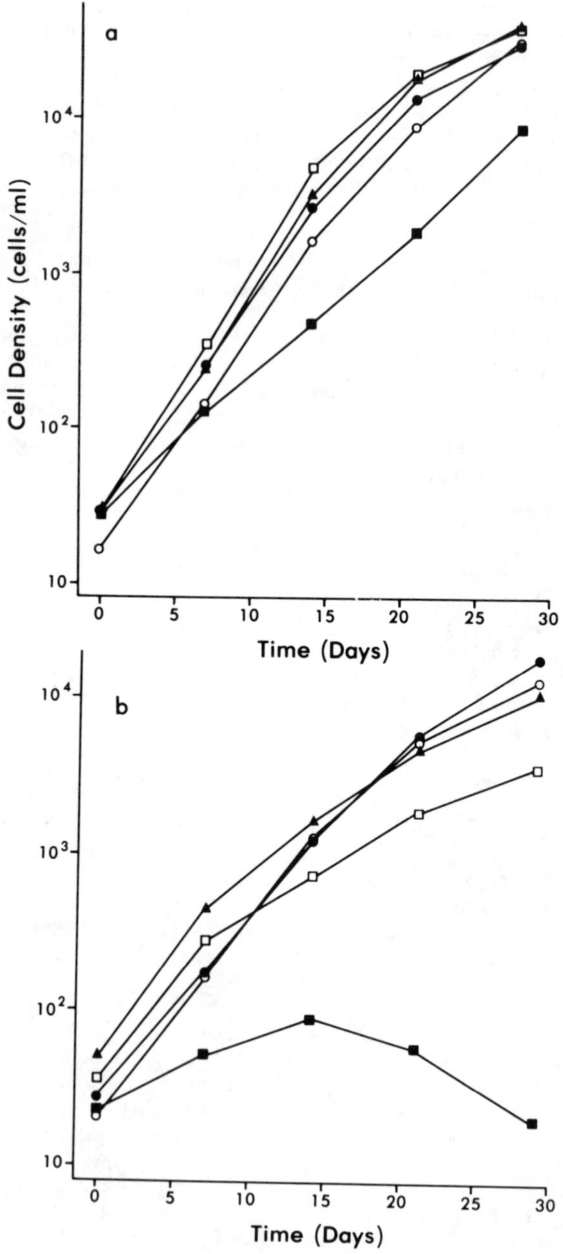

Figure 6. Growth of Prorocentrum concavum in (a) ESNW and (b) ESAW; both supplemented with soil and macroalgae extracts. Controls (■), Chaetomorpha extract (□), Dictyota extract (▲), Turbinaria extract (●), and soil extract (○).

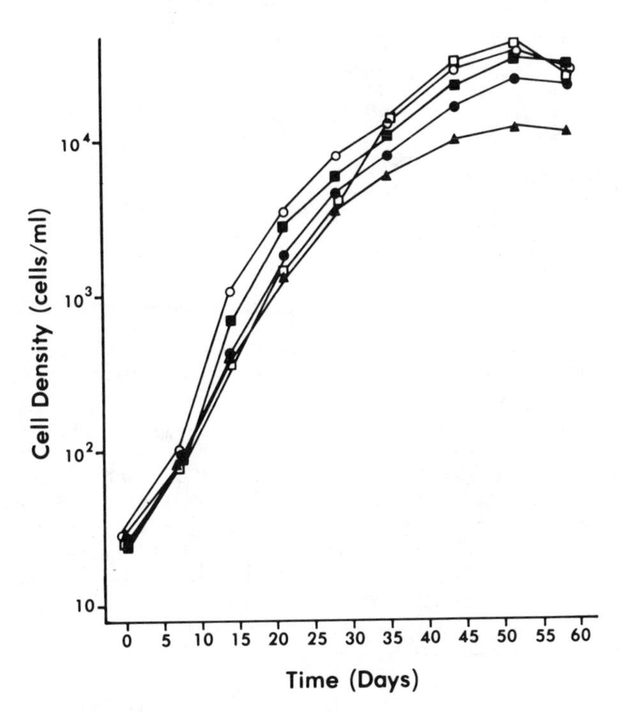

Figure 7. Growth of Prorocentrum concavum in bacteria-free ESNW supplemented with soil and macroalgae extracts. Controls (■), Chaetomorpha extract (□), Dictyota extract (▲), Turbinaria extract (●), and soil extract (○).

Table III. Growth Characteristics of Three Species of Dinoflagellates in Large Scale (20 L Carboy) Batch Cultures

Species	Total Vol. of Cultures (L)	Maximum Cell Conc. (cells/ml)	Culture Age At Harvest (days)	Mean Cell Yield (g FW/L)
G. toxicus	641	2,037	25-35	0.30 ± 0.07[1]
P. mexicanum	304	15,430	24-34	0.12 ± 0.06
P. concavum	355	23,384	25-35	0.49 ± 0.13

[1] Standard deviation.

Discussion

Our field observations demonstrate for the first time that G. toxicus populations in the Caribbean Sea can attain densities as high as those reported from Pacific Ocean areas populated by toxic fish (26, 27). Our field data showing high concentrations of the two Prorocentrum species and G. toxicus indicate a much greater potential for incorporation of toxins into fish feeding upon macroalgal substrates than previously recognized. Prorocentrum mexicanum and P. concavum cell dimensions measured in this study are similar to those reported by others (3, 16, 17). However, Yasumoto (1) and Fukuyo (3) reported finding G. toxicus cells as small as 45 µm in diameter, but we found no specimens with a dorsoventral diameter less than 60 µm. An expanded account of our field studies involving 46 species of dinoflagellates in the Caribbean will be presented elsewhere.

The mean cell yield from harvests of large-scale cultures of G. toxicus was 0.3 g fresh wt/liter of medium. Withers (15) reported a yield of 0.29 g dry wt/liter from her large-scale cultures of G. toxicus. Calculations based on the dimensions of a single cell indicate that cell densities of approximately 10,000 cells/ml would be necessary to obtain a dry weight this large from the culture volumes reported. Other workers have reported maximum cell densities of 4,000 cells/ml, a figure that agrees with our results. On this basis, we speculate that the dry weight yield reported by Withers (15) may be in error.

Growth of the three dinoflagellates in this study was better in ES-enriched natural seawater medium (ESNW) than in ES-enriched artificial seawater medium (ESAW). ESAW supported good growth of P. mexicanum, but G. toxicus and P. concavum grew poorly in it. Unlike Harrison et al. (21), we found ESAW to be a poor medium even for maintenance of G. toxicus. As reported for other species in previous studies (18-20), the addition of soil extract considerably improved the growth of P. concavum and G. toxicus in artificial seawater and also enhanced growth (division rates and cell yields) in natural seawater. The growth-promoting activity of soil extract has been attributed to the chelation of metals (29) and/or the addition of nitrogen, vitamins and other micronutrients (30-32).

In this investigation, soil and macroalgal extracts enhanced the growth of bacterized cultures of G. toxicus and P. concavum. G. toxicus cell yields and growth rates were approximately 25% higher in Chaetomorpha linum extracts than in the controls. These findings are consistent with the highest cell concentrations observed in natural populations in the Virgin Islands. Our results from culture disagree with those reported by Hurtel et al. (13). These authors found that soil extract and other organic substances inhibited the growth of G. toxicus. Furthermore, we did not observe the two cell types differentiated by size (55 μm and 75 μm) and growth characteristics, reported by Bagnis (14) and Hurtel (13). The 79 μm diameter strain of G. toxicus used in the present study showed growth rates and cell yields similar to those of the 55 μm "fast-growing" strain reported by the above authors. The highest division rate (0.55 div/day) reported by Withers (15) is similar to those measured in the present study.

Unlike G. toxicus and P. concavum, bacterized cultures of P. mexicanum were inhibited by algal extracts. Various dinoflagellates species have been reported to respond differently to macroalgal extracts. Growth of Amphidinium carterae Hulburt was inhibited by yellow polyphenol-containing substances liberated from Fucus vesiculosus Linnaeus (33). However, aqueous extracts of common benthic algae from the Norwegian Sea increased the growth rates of several marine dinoflagellate species, although they were not as effective as soil extract (34). The varying responses to macroalgal extracts may account for some of the differences in distribution and numbers of neritic dinoflagellates observed in the natural environment. The epiphytic growth habits of P. concavum and G. toxicus may permit cells to assimilate growth-promoting substances from living or decaying macroalgae. Yasumoto (35) found that G. toxicus was chemotactically attracted to macroalgae, and Withers (15) reported patterns of macroalgal substrate specificity in low level populations of G. toxicus in the Hawaiian Islands. The loose (metaphytic) association between P. mexicanum and macroalgae suggests that macroalgal-derived substances are less important for the growth and metabolism of this species. Our finding that growth in bacterized cultures of G. toxicus and P. concavum was enhanced in the presence of macroalgal extracts, whereas P. mexicanum cultures were inhibited, is consistent with this interpretation.

Loeblich et al. (17) reported that P. rhathymum (=P. mexicanum) was largely non-motile and embedded in copious mucilage in their cultures and in the field. We have not found this either in our cultures or in the field. This may reflect strain variations or differences in culture conditions.

The results of the axenic culture experiments indicate that bacteria may play an important role in the growth of epiphytic dinoflagellates. The lengthened generation times found in bacteria-free cultures of P. concavum are similar to results reported by Hurtel et al. (13) for cultures of G. toxicus. When bacteria were added to axenic cultures of G. toxicus, the growth rates and cell yields increased as compared to those of bacterized cultures (13). Bacteria may serve as intermediaries in the metabolism of macro-algal substances that affect dinoflagellate cell division. Our study of the toxicity of bacteria-free and bacterized cultures of P. concavum suggest that bacteria also may contribute to variations in toxin production (36). However, stimulation of bacterized cultures in the presence of organic substrates may be caused by simple carbon dioxide enrichment from increased bacterial respiration (37, 38). Yentsch (39) found that extracts from Ascophyllum stimulated bacterial respiration. In the present study, such carbon dioxide enrichment cannot account for the increased growth rates in axenic cultures supplemented with soil extract.

Extracts of various macroalgae have been reported to induce cytokinin responses in higher plants (40, 41). Growth enhancement of dinoflagellates in culture may involve similar compounds in algal and soil extracts. Additional studies on the regulation of growth of benthic/epiphytic dinoflagellates in cultures containing extracts from macroalgae and dinoflagellates are in progress.

Acknowledgments

This research was supported in part by the U.S. Food and Drug Administration (Contract #223-79-2287). We thank E. Ragelis, Project Officer, USFDA, for his technical advice and encouragement throughout the course of this study. We also thank L. Hawkins and M. Timmer for their assistance in conducting the culture experiments.

Literature Cited

1. Yasumoto, T.; Nakajima, R., Bagnis, R., Adachi, R. Bull. Jap. Soc. Sci. Fish. 1977, 43, 1021-6.
2. Yasumoto, T.; Oshima, Y.; Murakami, Y.; Nakajima, I.; Bagnis, R.; Fukuyo, Y. Bull. Jap. Soc. Sci. Fish. 1980, 46, 327-31.
3. Fukuyo, Y. Bull. Jap. Soc. Sci. Fish. 1981, 47, 967-78.
4. Nakajima, I.; Oshima, Y., Yasumoto, T. Bull. Jap. Soc. Sci. Fish. 1981, 47, 1029-33.
5. Carlson, R. D.; Tindall, D. R.; Dickey, R. W.; Morey-Gaines, G. Proc. Fed. Amer. Soc. Exp. Biol. 1982, 41, 1562.
6. Halstead, B. W. "Poisonous and Venomous Marine Animals of the World", Darwin: Princeton, 1978.
7. Dickey, R. W.; Miller, D. M.; Tindall, D. R. Proc. Fed. Amer. Soc. Exp. Biol. 1982, 41, 1562.
8. Miller, D. M.; Dickey, R. W.; Tindall, D. R. Proc. Fed. Amer. Soc. Exp. Biol. 1982, 41, 1561.
9. Tindall, D. R.; Miller, D. M.; Dickey, R. W. Proc. Fed. Amer. Soc. Exp. Biol. 1982, 41, 1561.
10. Tindall, D. R.; Dickey, R. W.; Carlson, R. D.; Morey-Gaines, G. This volume 1984.
11. Yasumoto, T. So. Pac. Comm. Rep. 1978, 785, 7-8.
12. Inoue, A. So. Pac. Comm. Rep. 1978, 785, 6-7.
13. Hurtel, J. M.; Chanteau, S.; Drollet, J. H.; Bagnis, R. Rev. Int. Oceanog. Med. 1979, 55, 29-33.
14. Bagnis, R.; Chanteau, S.; Chungue, E.; Hurtel, J. M.; Yasumoto, T.; Inoue, A. Toxicon 1980, 18, 199-208.
15. Withers, N. Proc. 4th Int. Coral Reef Symp. 1981, 2, 449-51.
16. Steidinger, K. In "Progress in Phycological Research, Vol. 2; Round, F. E., Chapman, V. J., Eds.; Elsevier: New York, 1983; pp. 147-88.

17. Loeblich, A. R. III,; Sherley, J. L.; Schmidt, R. J. J. Plankton Res. 1979, 1, 113-20.
18. Provasoli, L. In "Culture and Collection of Algae", Watanabe, H., Hattori, A., Eds., Jap. Soc. of Plant Physiol.: Hakone, 1968, pp. 63-75.
19. Pringsheim, E. G. "Pure Culture of Algae, Their Preparation and Maintenance", Cambridge Univ.: Cambridge, 1946; pp. 1-119.
20. Loeblich, A. R. III. Proc. Biol. Soc. Wash. 1968, 81, 91-6.
21. Harrison, P. J.; Waters, R. E.; Taylor, F. J. R. J. Phycol. 1980, 16, 28-35.
22. Kester, D. R.; Duedall, I. W.; Conners, D. N.; Pytkowicz, R. M. Limnol. Oceanog. 1967, 12, 176-9.
23. Palmer, C. M.; Maloney, T. E. Limnol. Oceanog. (Spec. Publ.) 1954, 21, 1-7.
24. Droop, M. R. Br. Phycol. Bull. 1967, 3, 295-7.
25. Hoshaw, R. W.; Rosowski, J. R. In "Phycological Methods", Stein, J. R., Ed.; Cambridge Univ: Cambridge, 1973, pp. 53-68.
26. Yasumoto, T.; Inoue, A.; Bagnis, R.; Garcon, M. Bull. Jap. Soc. Sci. Fish. 1979, 45, 395-99.
27. Yasumoto, T.; Inoue, A.; Ochi, T.; Fujimoto, K.; Oshima, Y.; Fukuyo, Y.; Adachi, R.; Bagnis, R. Bull. Jap. Soc. Sci. Fish. 1980, 46, 1397-1404.
28. Wilson, W. B.; Collier, A. Science 1955, 121, 394-5.
29. Johnston, R. J. Mar. Biol. Assoc. U. K. 1964, 44, 87-109.
30. Prakash, A.; Rashid, M. A. Limnol. Oceanog. 1968, 13, 598-606.
31. Sweeney, B. M. Amer. J. Bot. 1954, 41, 821-24.
32. Morrill, L. C.; Loeblich, A. R. III. Phycologia 1979, 18, 394-404.
33. McLachlan, J.; Craigie, J. S. Can. J. Bot. 1963, 42, 287-92.
34. Nordli, E. Nytt. Mag. Bot. 1957, 5, 13-16.
35. Yasumoto, T. In "Symposium on Coral Reef Ecosystems", Suva, 1979.
36. Tindall, D. R., unpublished data.
37. Lange, W. Can. J. Microbiol. 1971, 17, 303-14.
38. Richardson, K.; Fogg, G. E. Phycologia 1982, 21, 17-26.
39. Yentsch, C. S.; Reichert, C. A. Botan. Marina 1962, 3, 65-74.
40. Blunden, G. In "Marine Natural Products Chemistry"; Faulkner, D. J.; Fenical, W. H., Eds.; Plenum: New York, 1977; pp. 337-44.
41. Brain, K. R.; Lines, D. S.; Booth, M.; Ansell, G. In "Marine Natural Products Chemistry"; Faulkner, D. J.; Fenical, W. H., Eds.; Plenum: New York, 1977; pp. 345-50.

RECEIVED March 2, 1984

25

Ciguatera in the Pacific: Incidence and Implications for Marine Resource Development

NANCY DAVIS LEWIS

Department of Geography, University of Hawaii, Honolulu, HI 96822

Ciguatera is a public health hazard in the island Pacific. Morbidity caused by this marine biointoxication is apparently on the increase. It also has negative implications for nutrition and marine resource development. A review of the current situation in the Pacific, including Hawaii and Australia, is presented and the implications for small scale commercial fisheries are discussed.

Ciguatera was recorded in the Pacific in the 17th century and it was recognized in the Caribbean at least a century earlier. The pantropical existence of this marine biointoxication and the fact that fish do not seem to be affected by the toxin suggest that it has existed in the marine environment longer than the 500 years documented by the historical record. Island dwellers have had to adapt to the phenomenon and its largely unpredictable distribution in space and time. By and large, they have done so, and where subsistence fishing prevails, they continue to adapt. However, with the increasing importance of small scale fisheries development, marketing to urban consumers, and export, the impact of this long extant fact of Pacific life is exacerbated.

As a geographer I have been concerned with the distribution of ciguatera. In addition I have attempted to assess, within a human ecological framework, the reality of ciguatera as it exists for island dwellers today, its direct effects on health and its more generalized effects on nutrition and marine resource development--all aspects of overall societal health. In this overview I shall present an update on the current situation in the Pacific, including Hawaii and Australia, and then give some examples of the impact that ciguatera has had and may have on marine resource development in the island Pacific. The myriad ways in which islanders have adapted to the ciguatera - beliefs concerning etiology, detection, prophylaxis and cure, and strategies evolved to minimize the effect of ciguatera in subsistence communities, I have dealt with elsewhere (1-2) and are mentioned here only as they relate to ciguatera and marine resource development in the contemporary Pacific.

0097-6156/84/0262-0289$06.00/0
© 1984 American Chemical Society

Incidence in the Pacific

Morbidity statistics for the Pacific Island region, excluding Hawaii and Australia, are recorded by the South Pacific Commission's South Pacific Epidemiological and Health Information Service (SPEHIS) (3). The region is composed of fourteen island states and several territories. Given the wide distribution of the island states, the differences in reporting systems, and the inherent difficulty of obtaining accurate information on ciguatera morbidity under the best of circumstances, the morbidity data that we have can be considered suggestive at best and should not be used for strict epidemiological comparisons. Table I gives the reported annual incidence of ciguatera in the SPEHIS region from 1973-1981. (In calculating the incidence of ciguatera, Papua New Guinea has been excluded. PNG has not reported ciguatera and with over half of the region's population, most of whom are not dependent on fish protein, inclusion would only magnify underreporting. The Trust Territory of the Pacific Islands was used by SPEHIS as the statistical unit for American Micronesia until 1982. The disaggregation of statistics in 1982 highlights important differences in that part of the Pacific.). The last column of the table gives the mean annual incidence for the last three years. Mean annual incidence for the region as a whole during these three years was 98/100,000.

It is important to reiterate that the morbidity figures represent only a small proportion of the total number of individuals actually poisoned each year. There are a host of cultural, economic, and practical factors which influence reporting. In surveys carried out by the author and others in American Samoa and French Polynesia (1,4) there was a large range, from 9%-75%, in the number of individuals who reported to a Western medical facility when experiencing the symptoms of ciguatera. Even if individuals report to Western treatment facilities, they must be diagnosed as having ciguatera and the case must be reported to the proper authority; the path for the transmission of this information from primary health worker to the national register of disease and finally to SPEHIS is for ciguatera, as for some other reportable diseases, fragile and variable. Using the conservative estimate that reporting represents 20% of true incidence, then actual incidence for the area as a whole during this period was 500/100,000 or almost 10,000 individuals poisoned annually. Returning to the last column of Table I, we can see that some countries have annual incidences several times greater than the mean for the region as a whole. French Polynesia, with 633/100,000 is over six times the mean, Tokealu (532/100,000) and Tuvalu (534/100,000) over five times, and Kiribati (315/100,000) over three times (Map 1). The statistics for 1982 are only provisional, but Kiribati (671/100,000), Tokealu (1125/100,000), and Tuvalu (605/100,000), remain high. Higher incidence rates undoubtedly in part reflect better reporting, but the accumulated evidence suggests that ciguatera is a more serious problem in the eastern Pacific, notably French Polynesia, and some of the isolated island arcs of the central Pacific, e.g. Kiribati and Tuvalu. A partial explanation may lie in an inverse relationship with biogeographic diversity (2).

Table I. Ciguatera Morbidity as Reported to South Pacific Commission 1973-1981 (cases/100,000)

Country	1973	1974	1975	1976	1977	1978	1979	1980	1981	79-81 Aver
American Samoa	14	-	-	-	-	-	226	94	95	138
Cook Islands	-	-	-	-	-	-	-	5	11	8
Fiji	1	4	26	5	12	33	21	42	19	27
French Polynesia	482	672	474	489	363	582	470	665	764	633
Guam	-	-	22	17	6	6	9	-	4	7
Kiribati	176	300	314	144	76	68	139	322	485	315
Nauru	-	-	-	-	-	-	14	69	-	42
New Caledonia	-	153	390	481	358	355	137	107	71	105
Niue	159	25	875	103	-	-	-	86	88	87
Solomon Islands	6	4	-	4	3	3	-	2	-	2
Tokelau	-	-	-	500	-	-	875	-	188	532
Tonga	13	66	14	19	48	14	9	7	2	6
TTPI	210	224	171	249	253	223	144	163	112	140
Tuvalu	-	-	-	690	603	960	284	365	948	534
Vanuatu	-	-	37	28	50	52	65	-	30	48
Western Samoa	44	60	10	11	53	117	40	75	82	66

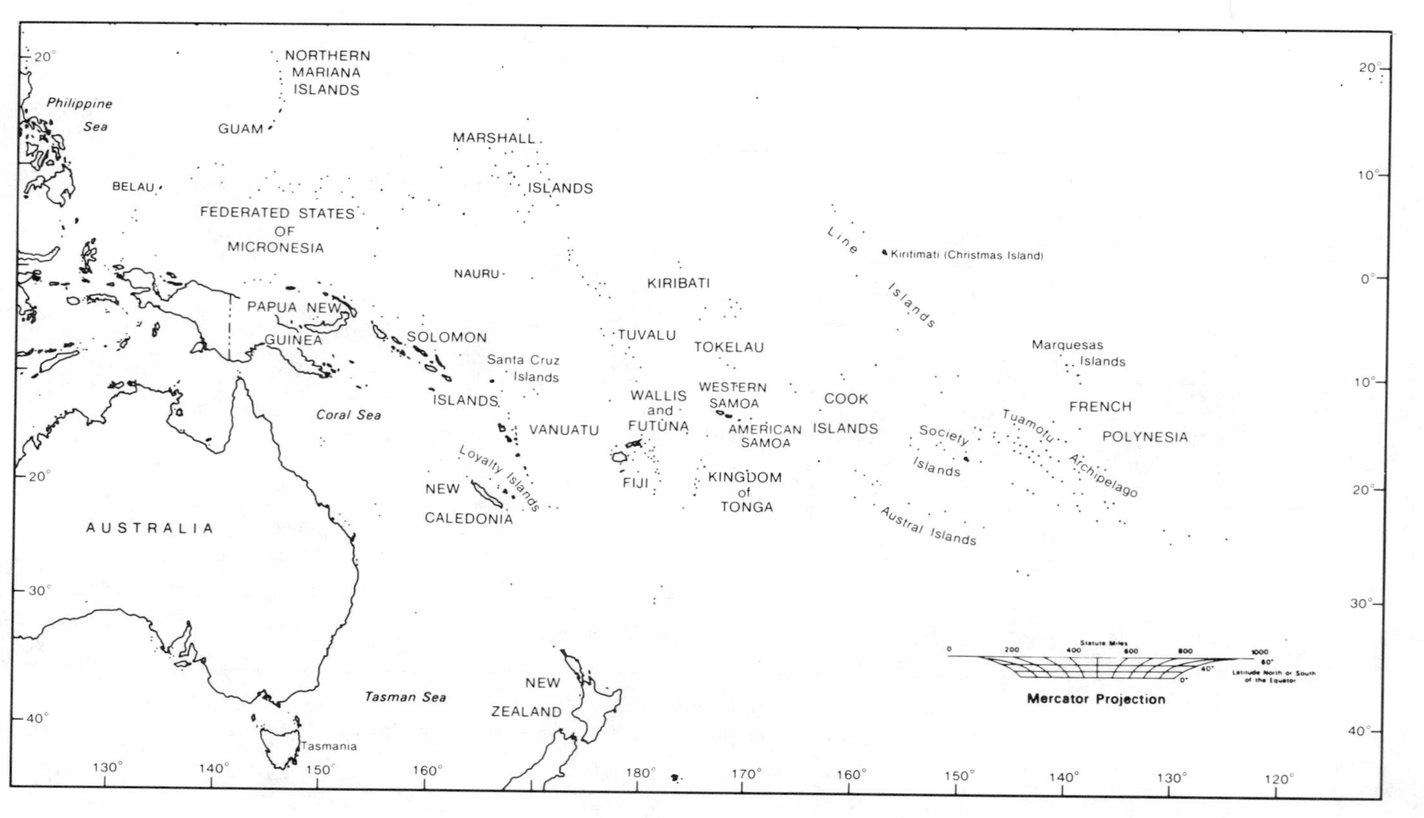

Map 1 The Pacific Islands

In 1982 the incidence for the SPC region as a whole was less than in 1981 - 1604 reported cases during the first ten months of the year, giving an extrapolated annual incidence of 78/100,000. French Polynesia, which has consistently had the highest incidence in the Pacific, did not report any cases for the last four months of 1982, presumably due to the disastrous hurricanes which hit those islands during the 1982-1983 season. Thus, the incidence rate for French Polynesia is artificially low and, consequently, the rate for the region as a whole is depressed. Give the association between ciguatoxicity and disruption of the marine environment, an important area for further ciguatera investigation will be in French Polynesia in the months and years ahead - monitoring for G. toxicus and observing patterns of ciguatera associated morbidity. Fortuitously, French Polynesia is a prime center for ciguatera research. A similar, albeit less extensive natural experiment was provided by hurricane Iwa in Hawaii (November of 1982).

Hawaii

In a recent article on the epidemiology of ciguatera in Hawaii, Anderson et al. (6) noted that between 1975-1981 there were 81 outbreaks involving 203 individuals, giving an average annual rate of 33.8 cases per year or an incidence of 3/100,000 population. The authors estimated that reported cases represented approximately 10% of the actual intoxications. If this is the case, the morbidity rate for the state as a whole may be estimated to be as high as 30/100,000. The species responsible include members of the Carangidae family, papio and ulua which were responsible for 24.7% of the outbreaks, Seriola dumerili, kahala, which was responsible for 21% of the outbreaks, Cheilinus rhodochrous, po'ou, which was responsible for 6.2% of the outbreaks, Mulloidichtys samoensis and Parupeneus multifasciatus, weke, which were responsible for 4.9% of the outbreaks, Parupeneus prophyreus, kumu, which was also responsible for 4.9% of the outbreaks, and other species responsible for 2.5% or less of the outbreaks (one or two outbreaks each). It is worthwhile noting that the species involved include some species not commonly implicated in the ciguatera complex, in areas of low endemicity Mulloidichtys samoensis, Parupeneus multifasciatus, Parupeneus porphyreus, and Cheilinus rhodochrous. In most instances the toxic fish were caught off the leeward coasts of the islands in the main Hawaiian chain. Two outbreaks could be linked to reef or shore modification projects. The first was the construction of the reef runway at Honolulu International airport (1973-1978), where the first toxic fish were caught in April of 1978, four months after the end of construction, the last in November 1980; and the second, the breakwater at Pokai Bay, also on leeward Oahu, which began in 1977 and was completed in 1979. The first case occurred in July, 1978, fifteen months after the beginning of construction. A small bloom of G. toxicus was identified in Pokai Bay in 1978. Initially herbivores were toxic, aholehole (Kuhlia sandvicensis) and palani (Acanthurus dussumieri) and later carnivores, ulua (Carangidae).

Anderson et al. note (6) that Seriola dumerili were responsible for 21% of all outbreaks during the period under study. These are large bottom dwelling jacks, commonly caught in the Northwest

Hawaiian chain. From April 1979 to May of 1981 all kahala passing through the commercial wholesale market were subjected to a radioimmunoassay (RIA) for the detection of ciguatoxin, in a project sponsored by the state of Hawaii, the National Marine Fisheries Service and researchers at the University of Hawaii. Fifteen percent were rejected due to toxicity (40% of those over 18 kg.). The testing program influenced morbidity during that time period--there were no reports of toxic episodes resulting from the consumption of kahala purchased commercially.

In 1982 incidence was low with only eighteen cases reported. Provisional statistics from the State Department of Health for the first eleven months of 1983 indicate an increase in ciguatera poisonings. There have been eighteen outbreaks with fifty-one individuals poisoned. Two of the cases were traced to fish caught off Sand Island, leeward Oahu, and seven people were poisoned in the Spring of 1983 after consuming an ulua, origin unknown, served in a popular fish restaurant. Nine of the outbreaks occurred during the first six months of the year, and eleven from August through November. Seventeen people were poisoned in the November outbreaks. Ten outbreaks occurred on Oahu, seven on Hawaii and one on Kauai. It is tempting to speculate that the increase at the end of the year is related, as expected, to the disruption caused by hurricane Iwa the previous fall. It remains to be seen if ciguatera will continue to increase. Investigators continue to monitor for increases in G. toxicus at sites around the islands and they may be able to establish an association between damage to the marine environment caused by Iwa and increases in dinoflagellate populations.

Following the identification of toxic kahala from the Northwest Hawaiian chain, forty-seven species were tested from waters surrounding both the main Hawaiian islands and the islands and reefs to the northwest. Approximately 16% exhibited some degree of toxicity, the species toxic included Kuhlia sandvicensis, Cheilinus rhodochrous, Myripristis murdjan, M. amaenus, Mugil cephalus, Caranx melampygus, Polydactylus sexfilis, and Caranx ignobilis (7). Researchers from the Hawaii Institute of Marine Biology have begun a long term monitoring project at the construction site of the new Barber's Point deep draft harbor. Transects are run along each side of the dredging site, and counts are taken monthly. G. toxicus has been found, but so far not in epibenthic bloom. Ciguatera is underreported in Hawaii, and therefore, more serious than generally acknowledged.

Australia

For decades toxic fish have been caught along the Great Barrier Reef of northern Queensland, Australia. Toxic species include Lutjanus bohar (red bass), Symphorus nematophorus (chinaman fish) and Lutjanus gibbus (paddle tail) all of which are banned from sale in the state. Since 1976 toxicity has been reported from an area previously considered safe. The main fish incriminated is the narrow banded Spanish mackerel, scomberomorus commersoni, a fish not normally associated with ciguatera in Queensland. (Toxic black kingfish (Rachycentron canadus) and barracuda (Sphyraenidae) have also caused intoxications) (8). The situation is aggravated by the fact that Spanish mackerel is a particularly important sport and commercial species in Australia. The toxic area is Hervey Bay, south of the Great Barrier Reef, the area previously associated with ciguatera in Australia. Ninety percent of all ciguatera intoxications in southern Queensland resulted from the consumption of fish caught in this area. Within Hervey Bay fish caught near Fraser Island are primarily responsible for the intoxications, particularly those from Platypus Bay on the northwestern coast of the island. Interestingly, the most toxic area has minimal coral development. In central Gladstone, north of Hervey Bay, fish appear not to be toxic, yet further north, on the Great Barrier Reef, ciguatera again becomes a problem. A positive association has been noted between lower tidal ranges and ciguatera.

A telephone survey was carried out in southern Queensland, and seven of the three hundred and eighty-three individuals spoken to reported that they had experienced ciguatera recently (9). Telephone surveys are notoriously suspect and "recently" was not defined in the report, but if those individuals has been poisoned within a single year, and the sample was representative of the population at large, then the annual morbidity rate for the region, would be a startling 1828/100,000. Even if incidence is considerably less than that, ciguatera is obviously presenting problems for fishermen and consumers in Queensland. It is common practice for sport anglers to sell their catch to wholesalers, and the fish are then often sold as frozen filets. Buyers in the Maryborough-Rockhampton area are reputed to be wary and the fishing industry throughout Queensland is concerned. Based on observations of the same genus in the Caribbean it has been suggested that *S. commersoni* may be giving up its high seas predatory habit and becoming a reef predatory in part of its range. Researchers in Queensland have constructed a data based on 300 reported cases of ciguatera and are monitoring the distribution of *G. toxicus* along the Queensland coast (8).

South Pacific Commission Region

Fiji. I have presented the morbidity statistics for the South Pacific Commission area but the situation in several of the island nations is worthy of particular attention. In Fiji, ciguatera had not been considered a major public health problem, but as noted by health workers and supported by morbidity statistics, there has been an apparent rise in ciguatoxicity since the mid-1970's. Species confirmed by Raj (10) as being ciguatoxic are listed in Table II. The situation in Fiji is complicated, as it is elsewhere, by the prevalence of other types of marine intoxications, e.g. scrombrotoxicity and those associated with the consumption of the internal organs of bottom species and crustaceans. The ethnic composition of the population in the various island groups also influences the morbidity statistics. Fijian Indians purchase less fish than Melanesians (11). While by and large the islands commercial fishermen are Fijian Indians, the rural/urban distribution of the population, and food preferences suggest that Melanesian Fijians consume more fresh fish than their Indian counterparts. Incidence may be influenced by relative ethnic composition and food preferences in a given area.

Table II. Species Found Toxic in Fiji (10)

Ctenochaetus striatus (balagi)
Caranx sexfasciatus (saqa)
Lethrinus miniatus (dokunivudi)
Lethrinus variegatus (kacika or badamu)
Lutjanus bohar (bati damu)
Macolor niger (regua)
Lutjanus monostigma (kake or damu gumusewa)
Aprion virescens (uto uto)
Gymnothorax sp. (dabea)
Scarus gibbus (ulavi)
Epinephelus fuscoguttatus (dela bulewa)
Cephalopholis argus (donu)
Plectropoma leopardus (donu)
Sphyraena barracuda (oqo)

Kribati. Kiribati, formerly the Gilbert Islands, was known for toxicity in the 1950's and 1960's (12), but toxicity apparently declined until 1974 when these islands experienced a recurdescence. The morbidity rate has been increasing since then, with a minor slump in 1977 and 1978, and a continued increase (13). Species responsible are listed in Table III. Note that herbivores are

Table III. Species Incriminated in Kiribati

Acanthurus gham (te ripa)
Carangoides malabaricus (te mata bareka)
Caranx sp. (te urua)
Lethrinus nebulosus (te morikoi)
Lutjanus argentimaculatus (te ingo)
Cephalopholis argus (te nimanang)
Epinephelus merra (te kuau)
Scarus ghobban (te inai)
Sphyraena barracuda (te baniniua)
Muraenidae (te kairoro)
Pseudobalistes fuscus (te nuonuo)

involved. The situation is apparently most serious in south Tarawa, and within this area, in the village of Naanikai. Population densities in urban Tarawa where 32% of the population lives reach $1137/km^2$. It is one of the most densely populated locations in the Pacific and undernutrition is a recognized problem. As might be expected with high population densities, there has been significant modification of the nearshore marine environment in the urban area. Guthrie (14) reported that in Tarawa there were 2-3 hospitalizations a week during his stay. Marritot and Dalley (13) estimated that the ratio of reported to non-reported cases in the Tarawa area was 4:1; this ratio is probably significantly higher in the outer islands. Ciguatera is a problem in Kiribati - one of the islands nations, which due to its rapidly expanding population and limited resource base, faces the greatest challenges in the 1980's. Ciguatera also threatens the development of the marine resource in the Northern Line Islands, far outliers of Kiribati 1500 km, south of Hawaii. I shall return to the situation there in discussing marine resource development.

Tuvalu and the Tokelaus. Salesa (15) reported that ciguatera was becoming more serious in Tuvalu, particularly on the island of Nukulaelae and to a lesser extent on the main island of Funafuti. Chambers (1980) reported that it was given as a reason for the decline of communal reef fishing on the island of Naunmea to the north. The SEPHIS (605/100,000 in 1982) statistics indicated that it is indeed a health hazard. On Nukulaelae work began on a new seawall in 1976 which was completed in 1980 - this may in part be responsible for the increase in ciguatoxicity. Incidence is also high in Tokealu, but in dealing with statistics from Pacific Island states, it is important to note that due to the small size of many populations variation may be magnified by the small size of the populations. The 875/100,000 incidence rate reported for Tokelau in 1979 (Table I) in actuality represents a total of only fourteen intoxications. The population of Tokelau was 1400. The World Health Organization regional office in Suva requested morbidity information for a technical training meeting held in Suva in January 1983 (17). Data from the Solomons and Vanuatu, island groups where all morbidity reporting is poor, indicated that respectively 3.6% and 13.3%, of the families questioned, had experienced ciguatera within the year.

TTPI. Only in 1982 do we have a disaggregation of statistics for the new states of the former Trust Territory of the Pacific Islands. Various authors (18, 19, 20, 21) reported that a number of islands in American Micronesia had been known to harbor toxic fish: Ponape and Ulithi in the Carolines; Guam, Saipan and Tinian in the Marianas; and Kwajalein, Jaluit and Alinglaplap and most notably Majuro in the Marshalls. Halstead (20) commented that the southern Marshalls were more toxic than the islands to the north. Monotaxis grandoculus was reputedly most toxic. The previously mentioned survey requested by the WHO regional office, revealed that in 1982 in the Marshall Islands, in 56% of the families, at least one member had been poisoned within the last year. While there is reason to question the representativeness of the survey, as is the case in the Queensland telephone survey, it is indicative that individuals are consuming toxic fish in the Marshalls.

Randall's recent survey in the Marshall Islands supports this (22). Surveying the fish populations on Enewetak and Bikini he found that on Enewetak 37% of the fish sampled were toxic to some degree, however only 16.2% were toxic enough to result in human illness; the respective figures for Bikini were 19.7% and 1.4%. Toxic fish included Lycodontis javanicus, Cephalopholis argus, Epinephelus hoedtii, E. microdon, Plectropomus leopardus, Aprion virescens, Lutianus bohar and Lethrinus kallopterus. In 1979 the fisheries officer (23) reported that ciguatoxicity not a serious problem in the Northern Marianas, but more recent reports suggest that there has been an outbreak on the main island of Saipan. SEPHIS data for 1982 give a morbidity rate of 113/100,000 for the Northern Marianas. On Guam, politically separate from the Northern Marianas, ciguatera is reported to be only a sporadic problem. The SEPHIS figures give an average of 7/100,000.

Ciguatera is obviously an exceedingly complex phenomenon. Banner (24) and Withers (25) have reviewed the chemistry, pharmacology and ecology of the causative organism in greater detail. We need to learn not only the requirements of the algal substrate, but more about the toxin-producing organisms and the toxins themselves, including the manner in which they are metabolized by and stored in the fish, as well as the feeding relationships among the myriad tropical species before we will have a grasp of the "life history" of ciguatera and a better understanding of the etiology of the disease. Another research frontier opens up if we are to understand the disease in human populations, one that directs our attention to concepts of health and ill-health, fish as a source of food, risk taking strategies, fishing practices, the utilization of marine resources and a host of other cultural factors. It is to one aspect of this that I now turn.

Marine Resources

The small island states of the Pacific have a restricted resource base and Pacific Island nations are keenly interested in developing their fisheries to feed expanding populations and to contribute to their developing economies. The fisheries of Pacific Island nations are subsistence and small scale commercial endeavors. Subsistence fishing supplies individual families and family units with fish; the small-scale commercial efforts supply urban markets. This latter fishery is developed differentially across the Pacific and is almost without exception underdeveloped. The export of fish caught by local fleets is low. Most fish exported from the South Pacific is caught by foreign fleets operating in the area. In the region as a whole the pelagic catch of Asian - Japanese, Taiwanese, Korean - fleets is many times greater than the catch of the local fishery. With the creation of 200 mile EEZ's, island nations have begun to participate in the pelagic fishery, through licensing agreements and joint venture operations. Further development of this resource, the primary one for many island nations, is central to development plans. Fortuitously, the pelagic fish, most importantly the tuna, are not part of the ciguatera complex. Ciguatera does, however, effect subsistence and small-scale commercial endeavors.

Food Habits

Many related variables effect both nutrition and marine resource development in the Pacific. A number of factors have contributed to a change in traditional dietary patterns, to different agricultural and fishing practices, and to a dependence on imported foodstuffs. Food preferences, particularly for refined carbohydrates, have developed that are not satisfied by local production. Imported Western foods, including canned fish, often have status value and are incorporated in gift exchange. Perhaps nowhere has the colonial administrations' neglect of the development of infrastructure for local production of food had greater impact than on fisheries. Because fresh fish is highly perishable, the lack of refrigeration, ice-making facilities, and regular markets, imposes particularly severe constraints. With the ready availability of canned substitutes, traditional technologies, e.g. the salting and drying of fish, fell rapidly into disuse and were soon lost from the cultural heritage. Furthermore, in large part due to the lack of infrastructure, canned fish is commonly less expensive than fresh. (The price of fresh fish varies greatly by location, season, species, etc., but for example, in Fiji in 1978 a kilogram of fresh reef fish cost approximately $1.40 ($A) whereas a 424 kg. can of mackerel, all of which is consumable, cost $0.43. The canned mackerel was also readily available.)

Urbanization, wage labor, the cost of vessels, fuel and gear, and importantly today near some urban centers, overfishing and reef degradation, are additional factors that influence the development of marine resources and hence the availability of fresh fish or locally caught and processed frozen fish. While it is almost impossible to separate ciguatera from the many confounding variables that affect the consumption of fresh fish except in the most highly toxic locations, it is almost as difficult to measure the impact of ciguatera on the development of marine resources. While also difficult to measure quantitatively, in some cases at least, the impact is more readily ascertainable. It must be reiterated that while the existence of ciguatoxic fish has had a negative impact on the development of small-scale fisheries in the Pacific, it is only one factor of many responsible fore retarded fisheries development. It has played a role however and perhaps most importantly, has been used as an excuse for the lack of development.

Small Scale Commercial Fisheries

The impact of ciguatera on subsistence communities must not be underestimated, and in some very highly toxic locations, e.g. the Gambier Islands, must fresh fish is avoided, however the most wideranging consequence of ciguatera for communities and countries as a whole is its effect on the development of small-scale commercial fisheries and the harvest of the inshore and nearshore bottom resources. While these are resources which are relatively accessible to harvest and within the technological capabilities of island nations, the activities which accompany human habitation near population concentrations, where fisheries development also takes place, may favor the creation of ciguatoxic biotopes. Exacerbating the situation Island dwellers in the Pacific, as elsewhere in the developing world, are moving to population centers. As residence patterns change and subsistence activities, including fishing, are replaced by wage labor, some of the strategies that protect individuals in subsistence situations, as well as general knowledge about fish, fishing, and the marine realm are less widely held and less effective. In urban areas there are specialized fishermen, and individuals may no longer know who is responsible for a particular catch. Nonetheless town dwellers are not necessarily at greater risk of ingesting toxic fish. Some exhibit a wider, less species-specific avoidance, e.g. the avoidance of all "red fish" or "the big fish". This may be an adaptive response to the existence of ciguatera as a health hazard, but not one in terms of overall nutritional status or the development of marine resources.

This generalized avoidance also affects the marketability of new species of fish. Price, or course, is another important component of marketability. The fisheries division in Western Samoa had a catch of red snapper (L. bohar) that did not sell at 20 (A$.24) or 15 or 10 sene but did at 5, and did, in fact, cause intoxications. This incident did nothing to enhance the reputation of "red fish" in Apia, a good example of how one episode can have wide reaching effects.

In the Caribbean a coop marketing scheme reputedly failed because individual fishermen, no longer directly responsible for their catch, began to fish in known ciguatoxic locations because there were plenty of fish and their return for effort greater (26). The temptation to fish toxic areas, or at least take less care in avoiding the capture and sale of toxic fish is one that faces fishermen where fish can not be traced to an individual fisherman or boat. Submitting to this temptation would not be out of character in Polynesia where a fisherman's status may be related to his ability to catch fish and contribute to the system of social exchange. In my research, especially where middlemen or wholesalers were involved, the actual responsibility for safety of an individual fish, devolved to the fish purchaser who was supposed to be wise enough to avoid toxic species. The knowledge of the area from which any individual fish came was of course denied him, and if the fish were fileted, even the species was uncertain, thus important components of the decision-making process were missing.

In urban areas there is also less time for fishing, and often more cash available to purchase canned substitutes. Not uncommonly there is a paucity of fresh locally caught fish in the market place, e.g. Suva, Fiji, and Rarotonga, Cook Islands. These factors of course also have implications for nutrition and economic development and they underline the fact that ciguatera can not be studied in isolation but must be viewed as one factor in island dwellers' complex adaptation to the ever changing reality of the contemporary Pacific.

Pacific Examples

The Tuamotus. The trade in reef fish between the Tuamotus, atolls north of the Society Islands in French Polynesia, and Tahiti, may be the most extensively developed export of reef fish at the local level threatened by cigatoxicity. Over 60% of the entire population of French Polynesia resides on Tahiti and the majority in the Papeete urban zone. Due to overfishing and environmental change associated with population increase, the resources around the island of Tahiti are depleted. In 1976 49% of the fish sold in the Papeete market was from the Tuamotus. Trade links between the two island groups go back to the trade in stone axes in precontact times and continue today in the export of fish from almost a dozen atolls for sale in the Papeete market. In 1976 at least 52% of the intoxications reported in Tahiti were from fish caught in the Tuamotus (20). Fish banned from sale in the Papeete market are listed in Table IV.

Table IV. Fish Prohibited from Sale Due to Toxicity in Papeete Market

Fish
(Sphyraena barracuda) Ono
(Plectropoma leopardus) Tonu
(Epinephelus microdon) Hapuu, from Tahiti
(Lutjanus monostigma) Taivaiva
(Lutjanus rivulatus) Haputu
(Lutjanus bohar) Haamea
(Balastidae) Les oiri
(Ctenochaetus striatus) Maito

From: Service Santé Publique, Papeete, Tahiti, 1978.

In Papeete, as elsewhere in the Pacific, fish banned from sale in the municipal market are commonly sold along the side of the road. While the fish are potentially toxic, the demand for "fresh" reasonably priced fish is high - and it sells.

Inhabitants on Manihi gave toxicity as one reason for the loss of their export trade. While there had been a recognized outbreak in th early 1970's, in this instance, other reasons, perhaps as important include changes in inter-island schooner schedules. On a nearby atoll, where local inhabitants recognize ciguatera to be a problem (as does the author, she was poisoned there), the fish that are exported are caught up to 48 hours before shipping, poorly iced once put aboard ship and sold six days later in Papeete. Spoilage is high and ciguatera is not the only cause of fish-related illness associated with the ingestion of fish caught in the Tuamotus. The export of reef fish also has health implications for the home community. The livers of certain species are considered a delicacy. The fish are gutted and, at least in the case of parrotfish, the livers, which can contain toxin concentrations up to 100 times that of the flesh (21), are collected and served to the family.

Bottom Fishery. The bottomfish resource is one that is receiving considerable attention in the Pacific - the harvest on the outer reef slope at depths of 100-300 meters of snappers (Lutjanidae), groupers (Serranidae), emperors (Lethrinidae), and jacks (Carangidae). A primary reason for developing this heretofore underutilized resource is that it will take the pressure off the reef and lagoon fishery. A secondary reason given in that these fish are not ciguatoxic. The first reason given is a valid and important one, the second, is largely but not always true. In 8 of the 14 areas with SPC bottom fishery pilot projects, the catch has included potentially toxic species. One reason for this is because during the course of the fishing trials the fishermen were also fishing in shallower waters (28) but this would be common practice in any fishery. Western Samoa is making particularly effective use of the bottom resource. Much of the fishing for bottom-dwelling species takes place in the Apolima Channel between the islands of Upolu and Savai'i. One medical practitioner on the island of Savai'i attributed a recent apparent increase in ciguatera morbidity in Western Samoa to this new fishery. The relatively low level of toxicity experienced in Western Samoa (66/100,000) indicates that it is not yet a serious public health hazard. The situation is also complicated by the possible existence of different toxins in the internal organs of some bottom dwelling species.

Species Identification. Bottom species reputation as toxin-free was in part attributable to the fact that they had not previously been important food species. In Hawaii, where the incidence of ciguatera had been on the increase, large bottom-dwelling kahala (Seriola dumerili) from the Northwest Hawaiian Islands were in part responsible. The most common and widely spread of the carangids caught in the SPC projects was Seriola rivoliana, which was initially misidentified as S. dumerili, the toxic Hawaiian species. Correct identification of reef and bottom species is important not only for data collection and management, but also in order to segregate the most potentially toxic species. Pristipomoides microlepis (Opakaopaka), in important food species in Hawaii, showed a relatively high rate of rejection in RIA testing, 15% (7) yet this species has been responsible for only one outbreak in Hawaii (6). Fisheries personnel and public health officials must be cognizant of the risks involved in the bottom fishery, but it must be stressed that the majority of species caught in this new fishery have not been ciguatoxic. With care, ciguatera need not hamper the development of this resource. Species caught during bottom-fishing trials that may occasionally be toxic include Lutjanus bohar, Lethrinus chryostomus, Variola louti, Caranx sexfasciatus, and miscellaneous Serranidae, Lutjanidae, and Acanthuridae (28).

Northern Line Islands. The Northern Line Islands, Christmas, Fanning and Washington, 1500 km. south of the Hawaiian chain play an important role in the development plans of the small nation of Kiribati. Christmas Island is the largest pure coral island in the world, and represents over half (363.63 km^2) of the total land area of Kiribati, yet only 2.2% of the population resides there. Isolated in the mid-Pacific and uninhabited at the time of European discovery, the Line Islands have a long history of use by metropolitan powers. Fanning was the location of the trans-Pacific cable station, Christmas has had various military and transport uses including atmospheric atomic testing. Since the mid-1930's Burns Philp has run copra plantations on Fanning and Washington . Today about 1500 I-Kiribati live on Christmas and work for the government in public works, fisheries, at the agricultural research station, the hotel etc. B.P.'s has just sold their rights to Fanning and Washington. Presumably the 400 plus inhabitants on each of the two atolls are still producing copra, as few alternatives exist. The islands are the projected sites of resettlement schemes from the main chain and perceived to be central to the development of marine resources for the nation as a whole (29). The government has established a small tourist industry on Christmas, catering to sport fishermen and orinthologists - 18 species of seabirds nest on Christmas and its lagoon islets.

In the 1950's and 1960's a well documented pattern of toxicity developed, sequentially, in an almost step like fashion, in the Line Islands. Banner and Helfrich (19) reported that all of the Line Islands (including Kingman reef and Palmyra) with the exception of Washington Island were known for their toxicity. Kingman reef, to the north, first reported intoxications in 1934, Palmyra in 1937, Christmas in 1943, and Fanning in 1945. Toxicity declined in the 1950's and 1960's. Since that time, few ciguatera intoxications have been reported. While there were still few reported intoxications in 1982, individuals were occasionally poisoned, even on Washington. On Christmas, te ingo (*Lutjanus bohar*), te ikanibong (*Lutjanus gibbus*), some large grouper and particularly the eel (Muraenidae) of the outer reef slope were occasionally toxic. Eel is considered a delicacy, special traps are set on the outer reef. The nurse at the hospital in Banana and her family were poisoned by an eel during my short stay during the summer of 1982. On Fanning Island, ciguatera endemicity was also low, but eel caught close to shore, te ingo (*Lutjanus bohar*), te karon (*Chelinus undulatus*), te baweina (*L. monostigmus*), and large ulua, large barracuda, and some large groupers could be toxic. Te ikanibong (*L. gibbus*) used to be toxic but was now believed to be safe to eat. On Washington island toxicity was reputedly very low, occasionally large te ingo (*L. bohar*) and te bukitakeiau (*Variola louti*) caused intoxications. Eel were reputed safe. On all three islands areas considered to harbor toxic fish were near or in the pass and/or in the proximity of the location where a vessel had gone on the reef.

The direction that development of these islands will take is unclear at the present time because of the political situation. Ciguatera is not a severe problem in the Line Islands, but it is one that must be kept in mind as development schemes are implemented, not only because of the increased population dependent on the abundant marine resource, but because of the reef modification that may take place. At the present time, the government is planning to blast a channel through the reef at the new village site on Christmas Island. This could provide yet another valuable natural experiment. *Turbinaria*, one of the macroalgal substrates preferred by *G. toxicus* is abundant at the site.

Kiribati has in fact already begun to export fish from Christmas Island and the fish are flown to Honolulu for sale in retail fish markets. Most of the fish that are exported at the present time are milkfish (*Chanos chanos*) from the interior lagoon, and spiny lobster from the reef flat - neither of which are ciguatoxic; but other reef species make up part of the export. Those responsible for the export of reef fish are careful not to include species that may be toxic, aware of the implications that the sale of a ciguatoxic fish could also affect the newly established sport fishery on the island. In Hawaii alone there are approximately 125,000 recreational fishermen - and attracting this market as well as fishermen from farther away with reasonable air fares and package tours could make an important contribution to the economy of the island. While ciguatera endemicity is low on Christmas Island, monitoring for ciguatera should be included in development plans.

Additional Implications

Ciguatera has other implications for resources and development which I can only touch on: the implications for fisheries development in general, the loss of export markets, the actual cost of the banned fish, the economic cost of the time lost from gainful employment by individuals suffering from ciguatera, costs associated with monitoring for ciguatoxic fish, the implementation of regulations prohibiting the sale of certain species, the cost associated with treatment, and the negative effect on health and future marketability of the inclusion of toxic fish in new food products - weaning foods, fish cakes etc.

Tourism. It also has a potential impact on tourism in general and the hotel and restaurant business. Not long before I arrived in 1978, most guests at the expensive Kia Ora in Rangiroa were poisoned by a meal of ono (*Sphyraena barracuda*). I have mentioned the recent poisoning in an Oahu restaurant. In the United States, the existence of ciguatoxic fish, and the potential sale of toxic fish in markets and restaurants has particular ramifications because of the litigiousness of our population. To date, cases in Hawaii have been settled out of court, but several of the settlements have been substantial. In terms of resource development and use, it is this aspect of the ciguatera problem that may be the most daunting, at least for the U.S. market.

Other Locations and Populations

Institutionalized individuals may be at greater risk of intoxication. In the late 1970's in Vanuatu, inmates of the French jail and hospital and inmates in the jail in Suva were poisoned, both after the consumption of servings of a large fish believed to be barracuda. In Kiribati, 54 high school students were poisoned in April of 1980 after being served portions of barracuda. Certain large fish may be cheap, perhaps because they are suspect, and thus find their way to the institutional kitchen.

Ciguatera as a factor in population movement is of interest to the geographer. It was given as the major reason for population movement from the south coast to the north coast of Hiva Oa in the Marquesas in the 1950's (19), although in the late 1970's population densities were again higher on the southern coast of the island, and it was also used, in this case as an excuse, for the removal of I-Kiribati from drought prone Sydney in the Phoenix Islands (12). Today for islanders living in isolated locations and dependent on the export of fish as their main source of external income, ciguatoxicity and the loss of markets could be major factors in out-migration. The existence of ciguatera also has a potential impact on recovery from natural disaster, although this is probably more important in the long rather than the short term. It may be a factor in population relocation in French Polynesia following the huricanes of 1982-1983.

Conclusion - Ciguatera and General Fisheries Expansion

In the Pacific there are many reasons for encouraging subsistence and small-scale fisheries development - nutritional, cultural, social and economic. Local production in the Pacific is only .1% of global production, and while this is proportionate to the size of the population, it could be much greater given the importance of the resource to the Pacific Islands. While the further development of inshore fisheries in the Pacific may not increase global food resources to any great extent, conservation oriented development can contribute significantly to the health and well being of Pacific peoples and communities. With increased knowledge about the nature of ciguatera and simplier, more widely applicable assay techniques, Pacific Islanders and Pacific nations should be able to continue to adapt to this long extant fact of Pacific life. Ciguatera obviously poses problems for commercial fisheries development but fisheries development can and must proceed. Island governments and planners must incorporate common sense, public health education, and the results of ongoing scientific research which will in turn enable them to monitor the environment, make wise planning decisions and develop artisinal fisheries with a view to protecting consumers from intoxication. They will then be able to fully utilize their local resources to help meet that important but illusive goal of nutritional selfsufficiency. The existence of ciguatera need not, must not, hamper this development. Happily most of the fish in the Pacific are not ciguatoxic.

Literature Cited

1. Lewis, N. Ph.D. Thesis, University of California, Berkeley, 1981.
2. Lewis, N. Under revision for Human Ecology.
3. Annual Reports 1973-1982. South Pacific Commission Epidemiological and Health Information Service, Noumea.
4. Bagnis, R. Bull. WHO 1973, 49:67-73.
5. Dawson, J. Hawaii Med. J. 1977, 36:239-243.
6. Anderson, B.; Sims, J.; Wiebenga, N.; Sugi, M. Hawaii Med. J. 1983, 42, 326-334.
7. Ito, B.; Uchida, R. 1981, National Marine Fisheries Service Pacific Ciguatera Workshop, Honolulu.
8. Gillespie, N. personal communication.
9. Gillespie, N. Aust. Fish., 1980, 39, 28-35.
10. Raj, V. Fiji Times, 1983, Suva.
11. Lindsey, C. 1972, "Fisheries Training in the Region Served by the University of the South Pacific," U.S.P.:Suva.
12. Cooper, M. Pacific Science, 1964, 18, 411-440.
13. Marriott, S.; Dalley, B. 1980, South Pacific Regional Technical Meeting on Fisheries, Noumea.
14. Gutherie, G., personal communication.
15. Salesa, F., personal communication.
16. Chambers, A., personal communication.
17. Olakowski, T., personal communication.
18. Bartsch, A.; McFarren E. Pac. Sci., 1962, 16, 42-56.
19. Banner, A.; Helfrich, P. 1964, Hawaii Marine Lab Tech. Rept. 3.
20. Halstead, B. "Poisonous and Venemous Marine Animals of the

World", 2nd ed., Darwin Press: New Jersey, 1978.
21. Yasumoto, T. 1981, National Marine Fisheries Service Pacific Ciguatera Workshop, Honolulu.
22. Randall, J. Fisheries Bull., 1980, 78, 201-49.
23. Villa-Gomez, J., personal communication.
24. Banner, A. In "Biology and Geology of Coral Reefs"; R. Endean; O. Jones, Eds., Academic Press: New York, pp. 177-213.
25. Withers, N. Ann Rev. Med, 1982, 33, 97-111.
26. Schaefers, E. In "Food and Drugs from the Sea"; Marine Technological Society: Washington, D.C., 1969.
27. Bagnis, R. personal communication.
28. Crossland, J. So. Pac. Fish Newsl., 1980, 21, 26-34.
29. "Five Year Development Plan", Republic of Kiribati, 1978.

RECEIVED March 2, 1984

26

An Enzyme Immunoassay for the Detection of Ciguatoxin

And Competitive Inhibition by Related Natural Polyether Toxins

Y. HOKAMA[1], L. H. KIMURA[1], M. A. ABAD[1], L. YOKOCHI[1], PAUL J. SCHEUER[2], M. NUKINA[2], T. YASUMOTO[3], D. G. BADEN[4], and Y. SHIMIZU[5]

[1]Department of Pathology, University of Hawaii, Manoa, Honolulu, HI 96822
[2]Department of Chemistry, University of Hawaii, Manoa, Honolulu, HI 96822
[3]Faculty of Agriculture, Tohoku University, Sendai, Japan
[4]Department of Biochemistry, School of Medicine, University of Miami, Miami, FL 33101
[5]Department of Pharmacognosy, University of Rhode Island, Kingston, RI 02881

An enzyme immunoassay (EIA) procedure for the detection of ciguatoxin (CTX) and the assessment of structurally related polyether toxins from fish tissue is presented in this study. Ciguatoxin is measured directly from fish tissue with anti-CTX serum previously prepared in a sheep immunized with CTX-human serum albumin conjugate. Fourteen of the 15 fish tissues from clinically documented ciguatera poisoning cases gave EIA ratios of greater than 1.50, defined as positive, while one sample gave a 1.31 ratio, defined as borderline. A total of 76 *Seriola dumerili* consumed by individuals with no incidence of ciguatera poisoning gave a mean and standard deviation ($\bar{X}$ ± S.D.) ratio of 1.00 ± 0.02, defined as negative. Ten other samples of the same fish, also consumed by individuals without poisoning gave a ratio of 1.00 ± 0.22. The results of a preliminary survey of 2 species *Ctenochaetus strigosus* and *Lutjanus kasmira*, demonstrated the presences of CTX and CTX-like toxins in the locations associated with ciguatera poisoning in Hawaii. The percentages of potentially toxic fishes were 14.4% to 15.4% respectively, for the *C. strigosus* and *L. kasmira*. The comparison between liver and flesh tissues of several species of fishes showed significantly ($P < 0.05$ to < 0.01) higher levels of potential toxins in the liver. By competitive EIA using highly purified CTX and polyether containing compounds the structural entity in toxic fish tissue that binds to sheep anti-CTX has been, in part, deduced as polyethers. Inhibition by highly purified CTX was demonstrated with an IC_{50} at 1 to 2 pg amounts. The compound with the least IC_{50} inhibition was monensin (IC_{50} = 70 ng). The EIA as reported has potential for use in a routine laboratory for testing fish samples in ciguatera poisonings, for screening of large number of fishes

0097-6156/84/0262-0307$06.00/0
© 1984 American Chemical Society

and for survey of locations of potential fishing grounds in endemic areas.

Ciguatoxin (CTX) is the major chemical entity associated with ciguatera fish poisoning. CTX, a term given to the low dalton lipid by Scheuer et al. (1) was isolated from livers of Gymnothorax obtained from an endemic area in the Pacific. A recent study by Tachibana (2) in Scheuer's laboratory suggests that CTX consists of numerous cyclic polyether residues similar to okadaic acid (3-5) and monensin (6).

Ciguatera refers to a fish poisoning associated with the ingestion of a variety of fishes of subtropical and tropical regions (7-9). Since not all fishes of the same variety implicated in ciguatera poisoning and even caught in the same area are toxic, the source of this poisoning remained an enigma for a long period. In 1977 Yasumoto et al. (10) demonstrated that the causative factor was a dinoflagellate, thus verifying an earlier suggestion made by Randall (11) on the food chain theory. Adachi et al. (12) identified the dinoflagellate as a new species and called it Gambierdiscus toxicus after the original source of discovery of the organisms, the Gambier Islands in French Polynesia. This same species was found in Hawaii and was characterized by Taylor (13) and Shimizu et al. (14).

Physiological studies show that monensin is a sodium ionophore (15), that induces inotropic effect on guinea pig atria (17). The polyether toxins including ciguatoxin, okadaic acid, and the recently characterized brevetoxin (5, 16) also induce inotropic effect on guinea pig atrial tissue in vitro (17). Additionally, partially purified CTX has been implicated in the depolarization of nerve cells in vitro, which can be reversed by high concentrations of Ca^{++} tetrodotoxin and saxitoxin (18).

A recent area of major concern has been the development of a practical and specific assay for the detection of ciguatoxin directly from fish tissues. Earlier assays, some of which are still used, relied on whole tissues, crude extracts and partially purified ciguatoxin (8, 9, 20). For the mongoose and cat assays, large amounts (10-15% of body weight) of fresh tissues were required, while the mouse assay required concentrated lipid extracts of the fish tissue.

A radioimmunoassay procedure (21) which utilized a purified sheep immunoglobulin isolated from a sheep previously immunized with purified ciguatoxin-human serum albumin conjugate was developed. An extensive two-year evaluation of this procedure has been carried out with Seriola dumerili (kahala, amberjack) in cooperation with the fishing industry in Hawaii (22, 23). This species is commonly implicated in ciguatera poisoning in the Pacific. Though proven to be effective for screening S. dumerili, the radioimmunoassay procedure is essentially economically unfeasible for testing fishes weighing less than 9 kg. Nonetheless, because of its sensitivity and relative specificity, the procedure is useful for surveys of various fishes in undeveloped fishing grounds and for the examination of toxic fishes implicated in ciguatera poisoning (22, 23).

This paper presents recent data obtained from an enzyme-immunoassay similar in sensitivity and specificity to that of the radioimmunoassay. However, compared to the radioimmunoassay, the enzyme-immunoassay is easier to run, economically feasible for screening of all sizes and variety of fishes, and could be used for testing liver in addition to flesh samples. This procedure utilized the sheep-anti-ciguatoxin used for the radioimmunoassay procedure with the antibody coupled with horseradish peroxidase instead of iodine-125. The enzyme-immunoassay procedure and results concerning its sensitivity and specificity are presented. In addition, via competitive inhibition, evidence for the relationship of CTX to other closely related lipoidal polyethers is also presented.

Methods and Materials

Antibody to Ciguatoxin. Sheep anti-ciguatoxin was prepared and purified by DEAE-cellulose chromatography as described in previous reports (21, 22).

Sheep-Anti-Ciguatoxin-Horseradish Peroxidase Conjugate. The purified IgG fraction of sheep-anti-ciguatoxin was coupled to horseradish peroxidase (Type VI, RZ:3.3, Sigma Chemical Co., St. Louis, MO) according to the one-step glutaraldehyde method of Voller et al. (24). The sheep-anti-ciguatoxin-horseradish peroxidase conjugate was divided into aliquots and stored at -20°C until used. Other aliquots were lyophilized and stored at 4°C. Each aliquot was thawed of rehydrated only once and the remaining excess conjugate discarded.

Source of Purified Ciguatoxin and Other Related Purified Polyether Compounds. Partially (LD_{99} = 190 µg/kg mice) and highly (LD_{99} = 0.25 µg/kg mice) purified ciguatoxin obtained from livers of *Gymnothorax javanicus* were prepared by Yasumoto and Nukina, respectively. Brevetoxin (BTX), fractions T34 and T17, was prepared by Baden (17) from *Ptychodiscus brevis* Davies. Maitotoxin was prepared and isolated from *Gambierdiscus toxicus* by Yasumoto. Okadaic acid obtained from *Halichondria okadai* was prepared by Tachibana et al. (3). Monensin was purchased from a commercial source (Calbiochem-Behring Corp., San Diego, CA).

Source of Fish Samples. Samples of *Ctenochaetus strigosus* and *Lutjanus kasmira* were procured by members of the Hawaii State, Aquatic Resources Division. The samples were from an ongoing serial study beginning in September of 1982 to assess the effects of man-made harbor on the CTX levels in nearshore fishes at the entrance of Barbers Point Harbor. A control site, approximately 8 miles northwest of Barbers Point called Ewa was selected for procurement of the same two species. Fish samples were obtained by three methods: (1) spearing; (2) handline; and (3) netting. Eight pieces, four from anterior-dorsal and four from the posterior-ventral sides of each fish were examined. Results of these studies were scored as indicated in the procedures section for the enzyme-immunoassay. Samples of fishes from clinically defined and documented cases of ciguatera poisoning were obtained from the Hawaii State Department of Health (DOH).

Reagents for Enzyme-Immunoassay. Tris buffer contained 0.05 M Tris (hydroxymethyl) aminomethane, pH 7.5 ± 0.05, with 0.1% human serum albumin and 0.01% sodium azide. The H_2O_2-methanol fixative contained 0.3% H_2O_2 in absolute methanol prepared just before use (stable 1 hr. at room temperature). The 4-chloro-1-naphthol substrate was prepared just before use. Twenty-five ml of 0.3% H_2O_2 in Tris buffer without human serum albumin and sodium azide was added to 10 mg of 4-chloro-1-naphthol crystals dissolved in 0.125 ml absolute ethanol. After thorough mixing, the substrate was filtered through Whatman #1 filter paper.

Enzyme-Immunoassay. Fish tissue samples for testing were cut into uniform 3mm thick slices with parallel razor blades mounted on a handle. Four discs were then punched out from each slice with a stainless steel borer, 3-mm in diameter, and each disc was placed in a well of a 96-well polystyrene microtiter plate (Flow Laboratories, Inc., Hamden, CT). Samples were washed once with 0.2 ml Tris buffer. After the wash solution was aspirated, each sample was fixed in 0.2 ml of 0.3% H_2O_2-methanol fixative for 30 min. at room temperature. Samples were then transferred to clean wells and 0.2 ml of a 1:100 dilution of sheep-anti-ciguatoxin-horseradish peroxidase conjugate in Tris buffer was added to each well. The plate was then incubated at room temperature for 1 hr. The sheep-anti-ciguatoxin-horseradish peroxidase was removed by aspiration, and the tissues were immersed for 5 min. in 0.2 ml Tris buffer. Each sample was transferred to clean wells and incubated for 5 min. at room temperature with 0.2 ml of 4-chloro-1-naphthol substrate. The final steps involved removal of the tissue and addition of 0.015 ml of 3 M sodium hydroxide to stop the enzymatic reaction. Absorbance readings at 405 nm of each well were obtained in the Titertek Multiskan (Flow Laboratories, Inc., Hamden, CT).

In each microtiter plate, quadruplicate samples of known toxic (positive) and nontoxic (negative) control fish tissues were tested in parallel with the unknown samples.

Interpretation of Results. The mean absorbance value of the 8 samples from each tissue section of the unknown test fish was divided by the mean absorbance value of 76 negative, nontoxic fish (*S. dumerili*) to give a test: negative control ratio. For evaluation of the samples tested, tentative toxicity ranges of these ratios have been established based on results of clinically documented toxic fishes:

Toxicity Level	Test: Negative Control Ratio
negative	≤ 1.30
borderline	1.30 - 1.49
positive	≥ 1.50

When negative values for each species were available, they were used in the denominator rather than the nontoxic *S. dumerili* value.

Competitive Inhibition in the Enzyme-Immunoassay with Purified Polyether Compounds. Purified ciguatoxin (LD_{99} = 190 µg/kg) at various ng concentrations in Tris buffer was added in 0.1 ml aliquots to H_2O_2-methanol fixed positive and negative fish tissues; then 0.1 ml of a 1:50 dilution of sheep-anti-ciguatoxin-horseradish peroxidase conjugate was added to each well. The samples were then incubated for 1 hr. at 4°C. Controls consisted of positive and negative fish tissues incubated with 0.2 ml of a 1:100 dilution of sheep-anti-ciguatoxin-horseradish peroxidase conjugate. Subsequent steps in the procedure were similar to that of the standard enzyme-immunoassay. Similar experiments using various concentrations of the other polyethers were carried out. These included brevetoxin, maitotoxin, okadaic acid, and monensin.

All statistical analysis were carried out according to the methods of Tallarida and Murray (25).

The percent of inhibition was determined as follows:

$$\frac{\text{Ratio in presence of competitive toxin} - 1.0}{\text{Ratio of toxic tissue in absence of competitive toxin} - 1.0} \times 100 =$$

% activity

(100% - % activity = % inhibition).

Results

Enzyme-Immunoassay Results with Clinically Documented Fishes. The data presented in Figure 1 indicate that the EIA procedure clearly differentiated between clinically documented toxic and non-toxic fishes of several different species: (a) *Parupeneus* sp. (moana), 1; (b) shark, 1; (c) *Sphyraena* sp., 1; (d) *Elagatis bipinnulatus* (rainbow runner), 2; (e) *Caranx* sp. (Jack), 1; (f) *Lutjanus* sp. (red snapper), 2; (g) *Scarus* sp. (parrotfish), 2; (h) *Cephalophalis argus* (grouper), 1; (i) *Cheilinus rhodochrous* (wrasse), 2; (j) *Gymnothorax* sp. (eel), 1; and (k) *Seriola dumerili* (amberjack), 1. The mean and standard deviation ratios for the toxic fishes were 1.83 ± 0.20 for 15 samples. The mean and standard deviation ratios for the non-toxic consumed fishes were 1.00 ± 0.22 for 10 fishes and 1.00 ± 0.02 for 76 *S. dumerili*.

Survey of Ciguatoxin Levels in Two Species by EIA at Two Sites in Hawaii. Results of a preliminary survey of two species of near shore fishes implicated in ciguatera poisoning using the EIA are shown in Table I. The Barbers Point and Ewa sites are both situated on the leeward side of the Island of Oahu. The leeward region has had the highest incidence of ciguatera poisoning in the state of Hawaii. The study at these sites is an ongoing program to assess the effect of dredging and development of a harbor at Barbers Point. As the results show, the frequency of positive and borderline *Lutjanus kasmira* (blueline snapper) is low compared to *Ctenochaetus strigosus* (surgeon fish) from samples at Barbers Point during the sampling periods of September 1982 through March of 1983. This higher frequency of fishes in the positive and borderline categories at the Ewa control site compared to Barbers Point was not anticipated for *L. kasmira*. Similarly, the frequency of positive

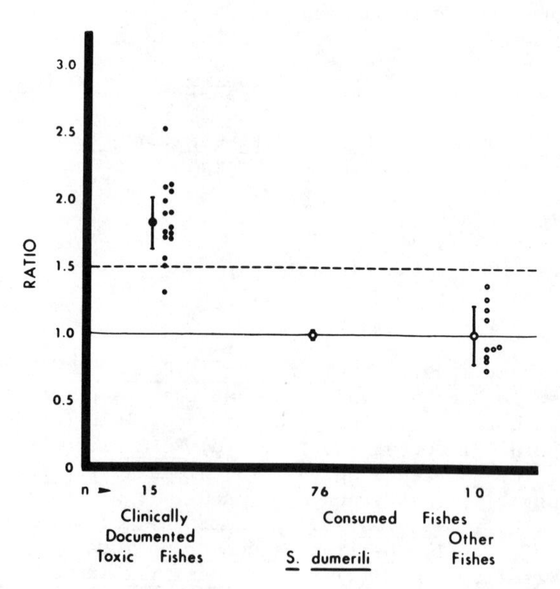

Figure 1. The comparison of the ratios between clinically defined and documented toxic fishes and non-toxic consumed fishes is shown. The large circles are the means and the vertical lines the standard deviation. The smaller circles represent the individual fish samples, except for the 76 S. dumerili.

Table I. Survey of Ciguatoxin Levels in Two Species of Fishes by EIA in Areas Associated with Ciguatera Poisoning in Hawaii

Species	Total Number	Results Number (%)			Location
		Positive	Borderline	Negative	
Lutjanus kasmira (taape)	89	1 (1.1)	2 (2.3)	86 (96.6)	Barbers Point
Lutjanus kasmira (taape)	62	6 (9.7)	11 (17.7)	45 (72.6)	Ewa
Ctenochaetus strigosus (kole)	262	15 (5.7)	18 (6.9)	229 (87.4)	Barbers Point
Ctenochaetus strigosus (kole)	68	6 (8.8)	5 (7.4)	57 (83.8)	Ewa

and borderline C. strigosus was slightly higher at Ewa site than at Barbers Point. The means and standard deviations of O.D. at 405 nm were 0.191 ± 0.038 and 0.197 ± 0.049 for C. strigosus and L. kasmira, respectively.

Competitive Inhibition in the Enzyme-Immunoassay with Purified Polyethers Including Ciguatoxin. Figure 2 represents an inhibition curve in the competitive test between a positive fish tissue and highly purified CTX (LD_{99} = 0.25 μg/kg mouse) in suspension. Approximately 0.002 ng of purified ciguatoxin gave 50% inhibition of the normal binding between sheep anti-CTX and the toxic fish tissue (Lutjanus sp.).

In the non-toxic tissues the ratio (1.11 ± 0.07) in the presence of purified CTX remained essentially the same. This may be attributable to non-specific binding of the immune complexes to the normal fish tissue.

Table II summarizes the concentration in ng/ml of the other toxic polyethers examined which inhibited approximately 50% of the interactions between sheep anti-CTX and toxic fish tissue. The % inhibition was calculated based on the interactions between sheep anti-CTX and toxic tissue ratio as 100%.

Comparison of EIA Results between Flesh and Liver Tissues from the Same Species. Table III summarizes the differences in the ciguatoxin activity between flesh and liver from the same fish sample in species of fishes obtained from the Northwestern Hawaiian Islands. Significant differences in the ratios ($p < 0.05$ to < 0.01) were shown for all four species examined. In each instance the liver showed a higher ratio, especially in the Bodianus and Caranx species. All of these fishes have been implicated in ciguatera poisoning in Hawaii, especially the Caranx species.

Discussion

The analysis of fish tissues for ciguatoxin by a newly developed enzyme-immunoassay procedure (26, 27) has been carried out in this study. Three areas of examinations have been attempted: (1) the examination of clinically defined and documented and non-toxic consumed fish samples; (2) the assessment of freshly caught fishes from the sites in the Leeward part of the island of Oahu where ciguatoxin is found; and (3) competitive inhibition with suspension of purified ciguatoxin and closely related structurally similar polyether toxins.

The comparison of clinically defined and documented toxic with non-toxic consumed fishes by EIA clearly demonstrated a statistically significant difference ($p < 0.001$) between the two populations. This result is similar to that reported for the radioimmunoassay and enzyme-immunoassay procedures (21, 22, 26-28). The assessment of the EIA with each species of fish samples caught in the Hawaiian waters presented positive and borderline frequency values comparable to that reported earlier by Ito and Uchida (28) by the RIA, and more recently by Kimura et al. (27) by the EIA.

An interesting assessment of the EIA is the demonstration that structurally known polyether compounds such as brevetoxin, okadaic

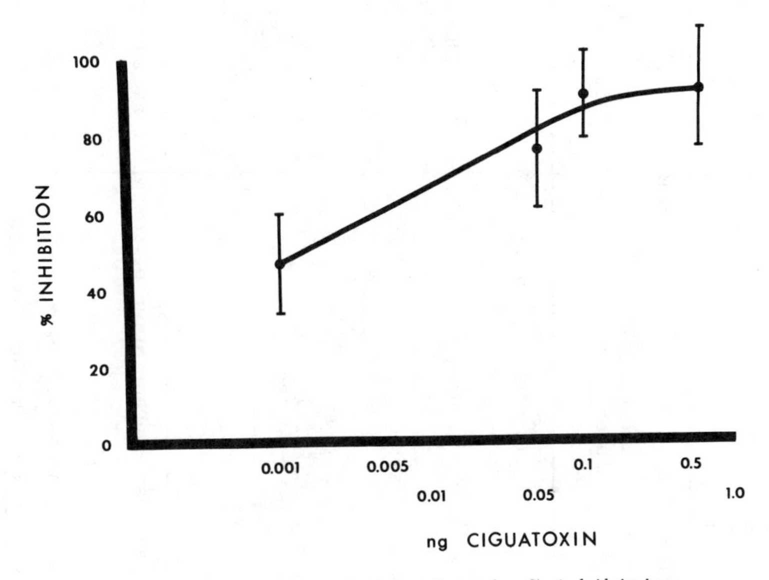

Figure 2. A representative curve for the % inhibition plotted against the concentration of ciguatoxin sample is shown. The vertical lines from each point represent one standard deviation.

Table II. The Concentration of Polyether Compounds Inhibiting 50% of the Reaction between Sheep Anti-CTX and Toxic Fish Tissue: Competitive Inhibition Assay[a]

Compound	Concentration in ng: 50% Inhibition			Toxicity[b] in Mouse IP Administration µg/kg
		Toxic Fishes		
	Lutjanus sp.	*Elagatis* sp.	*Seriola* sp.	
Monensin	165.0	70.0	—	16,800.00, LD_{50}
Brevetoxin				
Fraction 34	7.3	12.0	—	150.00 - 270.00, LD_{50}
Fraction 17	—	11.0	—	140.00 - 210.00, LD_{50}
Okadaic Acid	6.3	3.2	—	> 400.00, LD_{50}
Ciguatoxin[c]				
Preparation A	0.37	0.19	0.85	190.00, LD_{99}
Preparation B	0.001	0.002	—	0.40, LD_{99}
Preparation C	0.002	—	—	0.25, LD_{99}
Maitotoxin[d]	—	—	0.85	?

[a] IC_{50} was obtained by plotting % inhibition against concentration of toxin for each compound examined (Figure 2).

[b] Mouse toxicity for monensin, brevetoxin and okadaic acid was obtained from references (3, 4, 6, 16).

[c] Ciguatoxin preparation A came from Dr. Yasumoto; preparations B and C from Dr. Nukina.

[d] The maitotoxin was isolated from *Gambierdiscus toxicus* by Dr. Yasumoto.

Table III. Comparison between Ratios Obtained in Flesh (A) and Liver (L) Tissues in Four Species of Fish

Species	Number of Samples	Ratio[a] $\bar{x}$ ± S.D.		P Value
	(n)	A	L	
Bodianus bilunulatus	15	1.04 ± 0.29	1.98 ± 0.21	<0.05
Caranx melampygus	13	0.88 ± 0.15	1.47 ± 0.20	<0.01
Caranx ignoblis	82	1.01 ± 0.27	1.39 ± 0.33	<0.01
Polydactylus sexfilis	33	1.08 ± 0.23	1.25 ± 0.22	<0.01

[a]Ratio = $\frac{\text{mean absorbance of each test sample}}{\text{mean absorbance of 0.233 (from the 76 consumed amberjack)}}$

acid, and monensin can inhibit the sheep anti-CTX toxic fish tissue binding (Table II). The IC 50% inhibition concentrations suggest that maitotoxin, okadaic acid and brevetoxin are closely related to ciguatoxin. These data support the contention by Scheuer's group that CTX is a polyether compound (see Tachibana (3)). The inotropic and chronotropic properties associated with a polyether compound (6, 15) are compatible with the properties demonstrated for ciguatoxin by several investigators (17). Thus, the high level of frequency of borderline to positive results in the EIA at present may be attributable to these closely related polyethers produced by dinoflagellates. To increase specificity it is hoped to develop monoclonal antibodies to each of these closely related polyethers. This is presently being carried out in our laboratory.

The critical factors in the enzyme-immunoassay procedure as indicated previously (26, 27) are as follows: (1) the size and shape of the tissue samples should be uniform, and replicate samples from each area should be tested; (2) the fixation step of the fish tissue with methyl or ethyl alcohol containing H_2O_2 is essential (isopropyl, butanol and acetone are not satisfactory); and (3) both fresh, including liver, and cooked fish tissues may be examined by this procedure, provided the tissues are firm and not flaky. Further studies are needed for tissue sampling sites. In this regard, preliminary studies show that liver samples can be tested by the enzyme-immunoassay procedure, unlike the radioimmunoassay method (21). As shown earlier by bioassay procedures, livers tend to show greater toxicity (7, 9).

An earlier attempt by Berger and Berger (29) to develop an enzyme-linked immunosorbent assay (ELISA) procedure with sheep-anti-ciguatoxin was unsuccessful. In part, this may be attributable to the approaches taken and to the insufficient number of samples of documented toxic and non-toxic fishes examined. Except that the putative toxic fish samples came from Tahiti, evidence for their toxicity was not presented in the report (29).

For lack of an acceptable routine procedure for the direct assessment of ciguatoxin in fish tissues, the mouse bioassay with extracted lipids remains the method of choice for most laboratories doing ciguatoxin research. In addition, the radioimmunoassay has proven of value in screening S. dumerili (22, 23), but it is precluded for routine screening because of its cost factor and technical procedure. Nevertheless, it has been of value in the survey and assessment of new fishing grounds (28).

Based on the data presented in this study, it is suggested that the enzyme-immunoassay procedure may be an acceptable choice for the routine direct assessment of ciguatoxin in fish tissues. The procedure is sensitive (can detect approximately 0.01 to 0.005 ng), practical and has specificity for ciguatoxin, and ciguatoxin-like polyethers. The procedure warrants further examination and evaluation. Thus, we are presently using the procedure for routine survey of developing fishing grounds and for analyzing fishes from clinically documented ciguatera outbreaks.

Acknowledgments

Supported in part by the Hawaii State Board of Land and Natural Resources, Division of Aquatic Resources and through the Dingell-Johnson Federal Aid to Fish Restoration Project No. F-17-R, Study No. 111, Job No. 3, Fishery Resource Assessment of the NWHI; and the Hokama - Yagawa Memorial Fund, U.H. Foundation.

The authors acknowledge the excellent technical assisstance of Julie L. R. Y. Hokama, Allan Chun and Lance Shirai. We thank members of the DLNR, Division of Aquatic Resources and the Department of Health for the fish samples.

Literature Cited

1. Scheuer, P. J.; Takahashi, W.; Tsutsumi, J.; Yoshida, T. Science 1967, 155, 1267-8.
2. Tachibana, K. Ph.D. Thesis, University of Hawaii, Hawaii, 1980.
3. Tachibana, K.; Scheuer, P. J.; Tsukitani, Y.; Kikuchi, H.; Van Engen, D.; Clardy, J.; Gopichand, Y.; Schmitz, F. J. J. Am Chem. Soc. 1981, 103, 2469-71.
4. Murakami, Y.; Oshima, Y.; Yasumoto, Y. Bull. Jap. Soc. Scient. Fish 1982, 48, 69-72.
5. Lin, Y. Y.; Risk, M.; Ray, S. M.; Van Engen, D.; Clardy, J.; Golik, J.; James, J.C.: Nakanishi, K. J. Am. Chem. Soc. 1982, 103, 6773-75.
6. Agtarap, A.; Chamberlin, J. W.; Pinkerton, M.; Steinrauf, L. J. Am. Chem. Soc. 1967, 89, 5737-9.
7. Banner, A. H. In "Biology and Geology of Coral Reefs"; Jones, A. O. and Endean, R., Eds.; Academic Press, Inc.: New York, 1976; pp. 177-213.
8. Banner, A. H.; Scheuer, P. J.; Sasaki, S.; Helfrich, P.; Alender, C. B. Ann. N.Y. Acad. Sci. 1960, 90,770-87.
9. Halstead, B. W. "Poisonous and Venomous Marine Animals of the World", Vol. III, U.S. Government Printing Office: Washington, D.C., 1970; pp. 857-84.
10. Yasumoto, T.; Nakajima, I.; Bagnis, R.; Adachi, R. Bull. Jap. Soc. Scient. Fish 1977, 43, 1021-6.
11. Randall, J. E. Bull. Mar. Sci. Gulf Caribb. 1958, 8, 236-67.
12. Adachi, R.; Fukuyo, Y. Bull. Jap. Soc. Scient. Fish 1979, 45, 67-71.
13. Taylor, F. J. R. In "Toxic Dinoflagellate Blooms"; Taylor, D.; Seliger, H. H., Eds.; Elsevier North Holland, Inc.: New York, 1979, pp. 71-6.
14. Shimizu, Y.; Shimizu, H.; Scheuer, P. J.; Hokama, Y.; Oyama, M.; Miyahara, J. T. Bull. Jap. Soc. Scient. Fish 1982, 48, 811-3.
15. Bassett, A. L.; Wiggins, J. R.; Gelband, H.; Pressman, B. C. J. Pharm. Exp. Therap. 1978, 207, 966-75.
16. Baden, D. G.; Mende, T. J. Toxicon 1982, 20, 457-61.
17. Miyahara, J. T.; Akau, C. K.; Yasumoto, T. Res. Comm Chem. Path. Pharm. 1979, 25, 177-80.
18. Rayner, M. D. Fed. Amer. Soc. Exp. Biol. 1972, 31, 1139-45.
19. Bagnis, R. A. Bull. Wld. Hlth. Org. 1973, 49, 67-73.
20. Banner, A. H.; Sasaki, S.; Helfrich, P.; Alender, C. B.; Scheuer, P. J. Nature. London. 1961, 189, 229-30.
21. Hokama, Y.; Banner, A. H.; Boyland, D. B. Toxicon 1977, 15, 317-25.

22. Kimura, L. H.; Abad, M. A.; Hokama, Y. J. Fish Biol. 1982, 21, 671-80.
23. Kimura, L. H.; Hokama, Y.; Abad, M. A.; Oyama, M.; Miyahara, J. T. Toxicon 1982, 20, 907-12.
24. Voller, A.; Bartlett, A.; Bidwell, D. E. J. Clin. Pathol. 1978, 31, 507-520.
25. Tallaria, R. J.; Murray, R. B. "Manual of Pharmacologic Calculations with Computer Programs"; Springer-Verlag: New York, 1981; p. 54.
26. Hokama, Y.; Kimura, L. H.; Abad, M. A. Toxicon 1983, in press.
27. Kimura, L. H.; Hokama, Y.; Abad, M. A. In "Proceedings of the Resource Investigations in the Northwestern Hawaiian Islands"; Grigg, R. W.; Pfund, R. T., Eds.; University of Hawaii Sea Grant College Program: Honolulu, 1983, in press.
28. Ito, B. M.; Uchida, R. N. In "Proceedings Symposium on the Status of Resource Investigations in the Northwestern Hawaiian Islands": Grigg, R. W.; Pfund, R. T., Eds.; University of Hawaii Sea Grant College Program: Honolulu, 1980; pp. 81-9.
29. Berger, J. A.; Berger, L. R. Rev. Int. Oceangr. Med. Tomes 1979, LIII-LIV, 23-32.

RECEIVED February 6, 1984

Effect of Ciguatera-Associated Toxins on Body Temperature in Mice

PHYLLIS R. SAWYER[1], DAVID J. JOLLOW[2], PAUL J. SCHEUER[3], RICHARD YORK[4], JOSEPH P. McMILLAN[5], NANCY W. WITHERS[1], H. HUGH FUDENBERG[1], and THOMAS B. HIGERD[1,6]

[1]Department of Basic and Clinical Immunology and Microbiology, Medical University of South Carolina, Charleston, SC 29425
[2]Department of Pharmacology, Medical University of South Carolina, Charleston, SC 29425
[3]Department of Chemistry, University of Hawaii at Manoa, Honolulu, HI 96822
[4]Hawaii Institute of Marine Biology, University of Hawaii, Kaneohe, HI 96744
[5]Department of Biology and Mathematics, College of the Virgin Islands, St. Thomas, VI 00801
[6]National Marine Fisheries Service, NOAA, Southeast Fisheries Center, Charleston Laboratory, Charleston, SC 29412

Toxic materials extracted from ciguatoxic fish (ciguatoxin) and laboratory cultures of Gambierdiscus toxicus (putative maitotoxin) were chemically distinct yet elicited similar symptomatology in mice. Administration of either toxin intraperitoneally to mice evoked a dramatic and prolonged suppression of body temperature. The extent of the depression was dependent on ambient temperature and could be abrogated by elevation of room temperature. However, the artificial maintenance of normal body temperature by raising the surrounding temperature of mice treated with the dinoflagellate toxin had an adverse effect on the survival rate of the animals.

Ciguatera is a human illness that results from the ingestion of certain tropical and subtropical, coral reef-associated fish (1). The toxin believed responsible for the illness was first isolated from the livers of shark, red snapper and moray eel by Scheuer et al. (2) and named ciguatoxin. Indirect evidence suggests that toxic fish accumulate the toxin over a time period via their diet. While the source of the toxin remained elusive for many years, current evidence suggests that one source of the toxin is the benthic dinoflagellate, Gambierdiscus toxicus, found in association with certain macroalgae of coral reefs (3). Several laboratories have been successful in growing the organism as a unialgal culture under controlled conditions. Extracts of laboratory grown cultures of G. toxicus, however, have not yielded a toxin unequivocally identified as ciguatoxin. Instead, a more polar toxin similar to maitotoxin described in visceral extracts of surgeonfish (4), can be readily isolated from these cultures (5). Maitotoxin's biochemical relationship to ciguatoxin and role in the pathogenesis of ciguatera remain unknown. During preliminary studies, we noted that the gross

0097-6156/84/0262-0321$06.00/0
© 1984 American Chemical Society

symptoms of mice treated with maitotoxin were similar to those of ciguatoxin-treated mice. This manuscript reports our observations on the apparent hypothermia elicited by both toxins and postulates that these toxins possess an interrelationship not previously described.

When mice receive sub-lethal intraperitoneal injections of the fish toxin, they exhibit symptoms of generalized malaise but display no gross distinguishing biological response useful in defining or identifying the responsible toxin. Earlier reports by Doorenbos et al. (6)and Hoffman et al. (7) demonstrated that mice receiving crude extracts of fish toxin manifested depressed body temperature. The initial purpose of this study was to exploit their observation as a useful tool to help distinguish various toxins. Instead, our results showed that the dinoflagellate toxin (putative maitotoxin) also evokes this unusual and dramatic response of decreased body temperature previously thought to be unique to ciguatoxin. In addition, this study evaluates the biological nature of the temperature depression elicited by these two ciguatera-associated toxins.

Methods and Materials

Gambierdiscus toxicus culture. *G. toxicus*, Adachi and Fukuyo (8) was collected from *Dictyota acutiloba* growing at a depth of about 3 meters on the west of Tern Island, Northwest Hawaiian Island Chain. Single dinoflagellate cells isolated by micropipetting were washed exhaustively in sterile seawater. In order to diminish the contamination by diatoms, GeO_2 was added to 80 ppm. The single cell isolate eventually selected for these studies, clone T-39, was cultured in F/2-t medium (9) supplemented with a seaweed extract (an autoclaved, aqueous extract of blended *Acanthophora spicifera* and *Sargassum sp.*; 10). The 10 ml tubes used for the initial cultures were maintained at 25°C to 27°C at a continuous light intensity of 40 u einsteins/m/sec without aeration. The dinoflagellate cells for this study were mass cultured and harvested from 100 liter culture vats by filtration through a 36 u mesh screen.

Extraction of cultured G. toxicus. To prepare the harvested dinoflagellate cells (about 1 x 10^7 cells) for shipment from Hawaii to South Carolina, absolute methanol was added to effect a final concentration of approximately 25% (v/v). Upon receipt of the culture, additional methanol was added to attain a final concentration of 80% (v/v). After extraction at room temperature for 7 days, the suspension was clarified by centrifugation. The supernatant was dried and the resultant solids were weighed and resuspended in absolute methanol. The suspension was filtered and the filtrate designated as the crude dinoflagellate extract used in this study. Subsequent fractionation of this extract by HPLC (Higerd et al., manuscript in preparation) showed that the material responsible for toxicity eluted in a fraction well removed from the fraction that exhibited toxicity when extracts of either Pacific moray eel or Caribbean fish were chromatographed. In all cases, the body temperature depression effect was evident only in those fractions which also exhibited toxicity.

Source of Extracted Fish Toxin. Two different extracts of ciguatoxic fish were used in this study. Partially purified ciguatoxin was derived from the liver and viscera of the Pacific moray eel (*Gymnothorax javanicus*) and was a side fraction of the chromatographically purified ciguatoxin described by Tachibana (11). Crude fish toxin of Caribbean origin was partitioned extract derived primarily from the blackfin snapper (Lutjanus buccanella) as reported previously (7). Each was stored at 4°C in absolute methanol.

Toxicity Tests. Assays for toxicity were conducted on ICR female mice weighing approximately 20 g each. Animals were maintained on Wayne Laboratory Animal diets (Lab-Blox) and water, ad libitum. A known quantity of toxic extract was dried under a stream of dry nitrogen and placed in a vacuum dessicator overnight. The dried extracts were dissolved in phosphate buffered saline containing 5% Tween 80 and administered intraperitoneally. Control animals received an equal volume of the vehicle. Lethality was assessed at 48 h.

Body Temperature Determination. Unless otherwise noted, experiments were carried out at 24°C (room temperature) and no attempt was made to individually cage the treated or control animals. For the elevated temperatures of 28°C, 34°C and 37°C, animals were acclimated to the desired temperature for 24 h. Temperatures of the mice were taken using a YSI temperature probe inserted approximately 20 mm into their rectum. The initial temperatures (recorded as zero time) were taken immediately before administration of the extracts.

Results

A pronounced drop in the body temperature of mice was observed following the injection of a methanol extract of G. toxicus. This extract was used throughout the study and had an estimated LD50 dose of 20 mg/kg for mice (Figure 1).

To quantitate the change in body temperature, eight mice were injected with the dinoflagellate extract (43 mg/g). Control groups of eight mice received the carrier alone and were maintained with the toxin-treated animals at room temperature (Figure 2). The rectal temperatures of the treated and control mice were measured intermittently for 24 h. The average body temperature of the toxin treated mice was significantly lower than the control group at the end of the first hour. In a few mice, body temperatures as low as 24°C were recorded prior to death. In addition, the duration of this apparent hypothermia was striking; a few animals in the population exhibited lowered body temperature for 48 h or longer.

The response of depressed body temperature in mice treated with dinoflagellate extract is dose dependent (Figure 3). When doses as low as 3 mg/kg or 6 mg/kg (15% and 30% of the LD_{50}, respectively) were administered, no drop in body temperature was noted. With injections of extract containing 50% or more of the LD_{50} dose, however, an apparent hypothermia resulted, and the rate at which the temperature dropped was directly dependent on the quantity of extract administered.

To ascertain the biological nature of this response, groups of toxin-treated and control animals were maintained at elevated ambient temperatures (Figure 4). Animals that had received 76% of an LD_{50} dose and were held at room temperature exhibited the

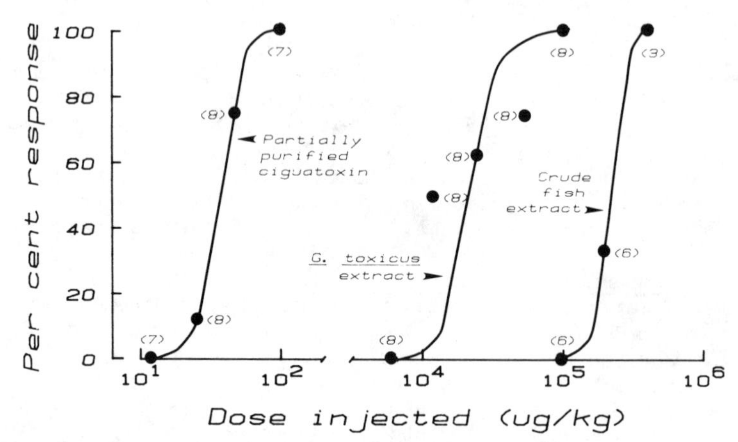

Figure 1. Dose-response curve of the lethality of the three different extracts used in this study. The extracts were administered to mice i.p. in PBS with 5% Tween-80. Lethality was assessed at 48 h. for animals maintained at 24 °C. Extracts were Gambierdiscus toxicus, fish of Caribbean origin, and partially purified moray eel ciguatoxin.

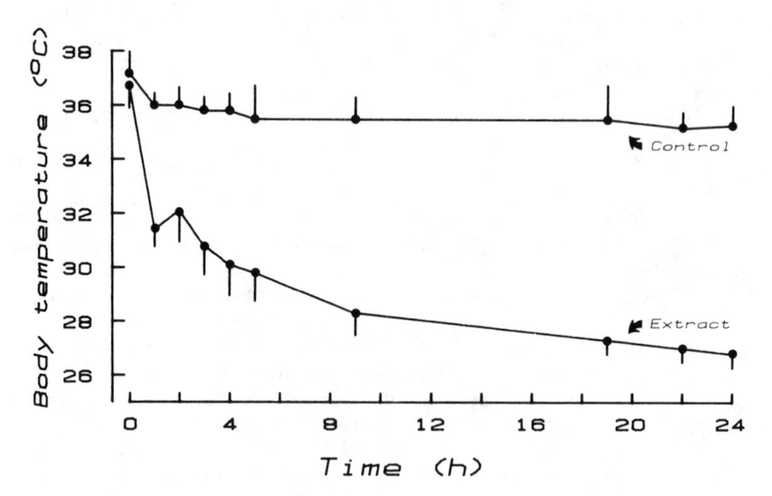

Figure 2. Body temperature response of mice receiving the extract of cultured Gambierdiscus toxicus or the vehicle alone. Each point represents the mean ± S.E. of the rectal temperature measurements on 8 mice.

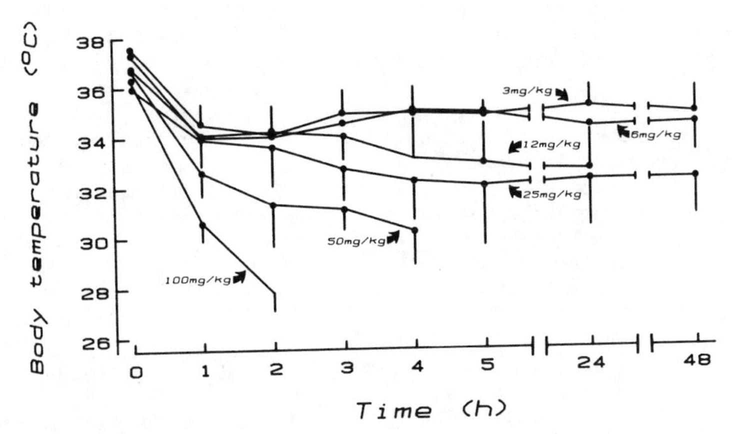

Figure 3. Body temperature response of mice receiving various quantities of cultured Gambierdiscus toxicus extract. Each point represents the mean ± S.E. of the rectal temperature of 8 animals.

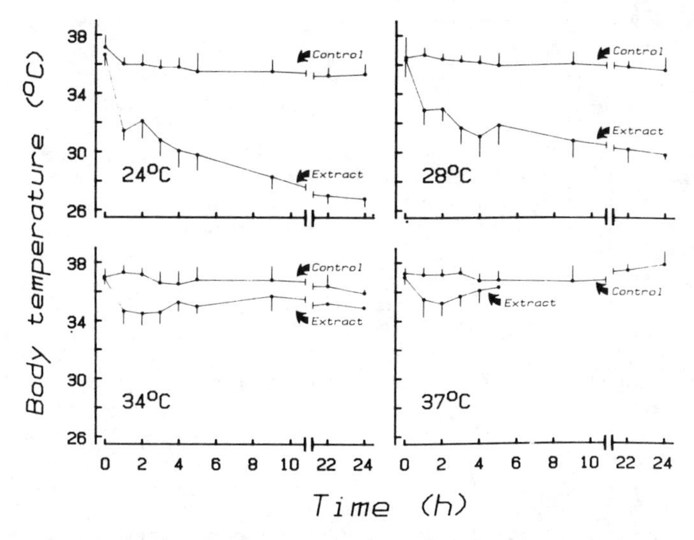

Figure 4. Body temperature of mice administered an extract of Gambierdiscus toxicus or the vehicle alone, and maintained at 24 °C, 28 °C, 34 °C, or 37 °C. Each point represents the mean ± S.E. of the rectal temperature of 8 mice.

expected drop in body temperature. On the other hand, as the ambient temperature was increased to 28°C, 34°C and 37°C, the body temperature difference between the treated and control mice diminished. When treated mice were held at 37°C, no significant alteration of body temperature was observed. Since the extent of the observed hypothermia was dependent on ambient temperature, the resulting temperature response suggested an impairment in the animals' ability to control normal body temperature.

Since crude extracts of ciguatoxic fish have been reported to produce hypothermia in mice (6,7), it was of interest to confirm these earlier reports with well-characterized toxin preparations and to measure the extent and duration of any temperature alteration in light of our results with *G. toxicus* extracts. Crude extracts of ciguatoxic fish from the Caribbean subjected to liquid-liquid partition and to chromatography on silicic acid (7) were used in this study. The LD_{50} of these crude extracts was approximately 250 mg/kg (Figure 1). As was the case with the dinoflagellate extract, the extent and duration of the decrease in body temperature was dependent on dose of administered fish extract (Figure 5). Furthermore, both extent and duration of the observed hypothermia produced by the crude fish toxin were similar to that recorded with the dinoflagellate extract.

Since only a limited amount of fish toxin was available, two temperatures (24°C and 34°C) were used to assess the influence of ambient temperature on the apparent hypothermia induced by crude fish toxin. The hypothermia was dependent on ambient temperature and experimentally demonstrated that mice treated with fish toxin also lost their ability to maintain body temperature (Figure 6).

To determine the effect of lowered body temperature on mortality, groups of toxin-treated mice were maintained at several elevated temperatures and their survival rates monitored. Because of the limited quantity of fish toxin, only the dinoflagellate toxin could be administered to a significant number of animals held at various temperatures. Groups of eight mice treated with one-half of an LD_{50} dose were placed at four different ambient temperatures. The percentage of mice in each group that survived after five hours was 100%, 75%, 62% and 25% when maintained at 24°C, 28°C, 34°C and 37°C, respectively. A similar result was observed when crude fish toxin was administered to a group of three mice. It would appear, therefore, that holding animals at higher temperatures, while ameliorating the drop in body temperature, had an adverse effect on survival.

Discussion

Sufficient quantities of these toxins have not yet been purified to test the hypothesis that both poikilothermia and toxicity are properties of the same molecule. Recently, however, the toxic components in both the dinoflagellate and the Caribbean fish have been partially purified by HPLC on a reversed phase octylsilane column using a methanol-water gradient (Higerd et al., manuscript in preparation). In every instance; the mouse toxicity could not be separated from the hypothermia eliciting property. In addition, a side fraction obtained during the purification of ciguatoxin from moray

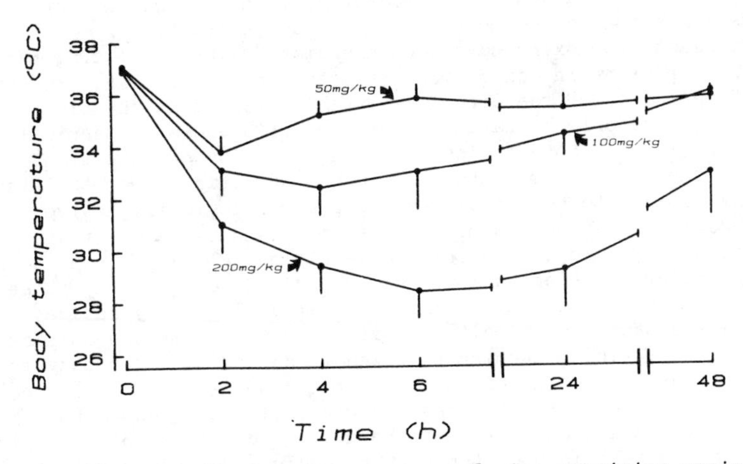

Figure 5. Body temperature responses of mice receiving various amounts of crude Caribbean ciguatoxic fish extract. Each point represents the mean ± S.E. rectal temperature of 6 animals.

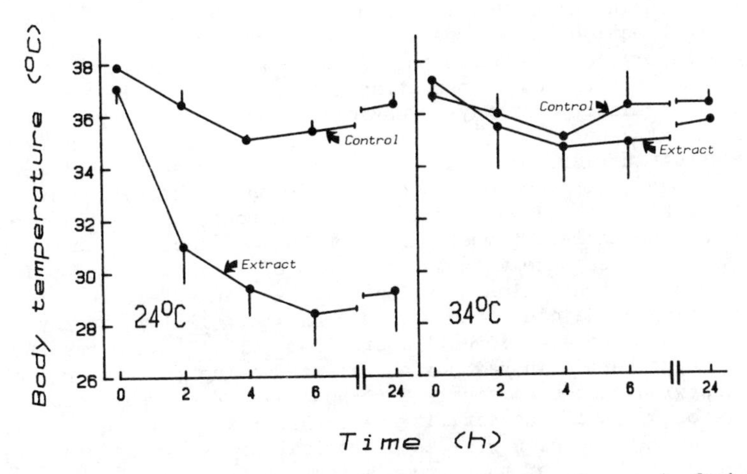

Figure 6. Body temperature of mice receiving the crude fish extract or the vehicle alone. Mice were maintained at either 24 °C or 34 °C. Each point represents the mean ± S.E. of the rectal temperature of 6 animals.

eel liver described by Scheuer (2) was assessed for relative toxicity (Figure 1). When small amounts of this toxin were administered to mice, a marked body temperature depression was observed. It would appear, therefore, that the biological response of apparent hypothermia is a characteristic property of the toxins and not due to some contaminating component of the crude extract.

At least two toxins have been associated with ciguatera, namely ciguatoxin present in toxic fish extracts, and maitotoxin, present in cultured cells of the dinoflagellate implicated as the biological source of ciguatoxin in nature. Despite differences in chemical structure (11,12), both toxins affect the ability of mice to regulate their body temperature in such a dramatic manner that this property appears to be singular among pharmacologically active compounds. The uniqueness of this response and the ease of its measurement may make this quantifiable biological response a useful tool in differentiating various marine toxins, particularly those associated with ciguatera, from other pharmacologically active toxins which otherwise elicit non-discriminatory symptoms in mice.

The underlying mechanism by which these toxins lower body temperature is unclear. Increased peripheral blood flow with augmented convective and radiant heat loss are generally believed to be important factors leading to hypothermia. Additional possibilities mediating the fall in body temperature include decreased metabolic heat production or increased respiratory evaporative heat loss. It may be significant that mice injected with either of these toxins display decreased respiratory rates and no gross evidence of vascular change (including tail vein vasodilation). These toxins could have a direct action on peripheral temperature sensors or on the thermoregulatory foci of the hypothalamus. Additionally, they may exert an indirect effect on neurotransmitters involved in the central thermoregulatory pathways. Obviously, we can only speculate on the mechanism(s) by which these toxins exert this effect. Recent hypotheses in thermoregulation focus on the set-point concept which describes some reference level around which body temperature is regulated. Since the lowered temperature response can be reversed by changing the ambient temperature from low to high, it would appear that the response is not the result of an altered set-point.

Since both ciguatoxin and maitotoxin have an effect on transportation of calcium ions across membranes (12,13), it may be significant that several investigators (14,15,16,17) have reported prolonged hypothermia in various animals following repeated microperfusions with excess calcium. Our knowledge of the physiological mediators of hypothalamus activity is cursory; with further studies, these toxins may become useful as biological probes to help define the biochemical basis of thermoregulation.

It is generally accepted that ciguatoxin, as extracted from toxic fish, is not synthesized *in situ* but is the result of dietary intake. Although *G. toxicus* may be one source of ciguatoxin in the diet, it is difficult to collect large enough numbers of wild cells in order to extract sufficient amounts of toxin for an unequivocal chemical assessment of ciguatoxin content. Alternately, both wild and cultured *G. toxicus* cells produce a toxin that is chemically distinct from ciguatoxin (12). As reported in this study, cultured dinoflagellate toxin, putatively maitotoxin, has biological properties that are indistinguishable from the fish toxin, ciguatoxin. Both toxins exhibit similar dose-response profiles and both toxins elicit the apparent hypothermia to the same extent and for the same duration. This response, whether promoted by the fish toxin or the dinoflagellate toxin, is reversed by increasing ambient temperature. In addition, both toxins evoke equivalent gross symptoms of malaise. These similarities in biological traits suggest that more than a casual relationship exists between ciguatoxin and maitotoxin.

Acknowledgments

We thank Marilyn Orvin for her technical assistance. This report is publication no. 635 from the Department of Basic and Clinical Immunology and Microbiology, Medical University of South Carolina. Research supported in part by Grant Na-80-AA-D-00101 from the National Oceanic and Atmospheric Administration.

Literature Cited

1. Withers, N. W. Ann. Rev. Med. 1982, 33, 97-111.
2. Scheuer, P. J.; Takahashi, W.; Tsutsumi, J.; Yoshida, T. Science 1967, 155, 1267-8.
3. Yasumoto, T.; Nakajima, I.; Bagnis, R.; Adachi, R. Bull. Jpn. Soc. Sci. Fish. 1977, 43, 1021-6.
4. Yasumoto, T.; Bagnis, R.; Vernoux, P. Nippon Suisan Gakkaishi 1976, 42, 359-65.
5. Yasumoto, T.; Nakajima, I.; Oshima, Y.; Bagnis, R. In "Toxic Dinoflagellate Blooms"; Taylor, L.; Seliger, H. H. Ed.; Seliger, North Holland, 1979; pp. 65-70.
6. Doorenbos, N. J.; Granade, H. R.; Chen, P. C.; Morgan, J. M. In "Food-Drugs from the Sea"; Mar. Tech. Soc., 1976; p. 414.
7. Hoffman, P. A.; Granade, H. R.; McMillan, J. P. Toxicon 1983, 21, 363-9.
8. Adachi, R.; Fukuyo, Y. Bull. Jpn. Soc. Sci. Fish. 1979, 45, 67-71.
9. Guillard, R. R. L. In "Culture of Marine Invertebrate Animals"; Smith, W. L. ; Chanley, M. H. Ed.; Plenum Publication Corp., New York, 1975; pp. 29-60.
10. Withers, N. W. In "Symposium on Northwestern Hawaiian Islands"; Sea Grant Program Publication, Honolulu, 1983; in press.
11. Tachibana, K. Ph.D. Dissertation, U. Hawaii, Honolulu, 1980.
12. Takahashi, M.; Ohizumi, Y.; Yasumoto, T. J. Biol. Chem. 1982, 257, 7287-9.
13. Rayner, M. D. Fed. Proc. Am. Soc. Exp. Biol. 1971, 31, 1139-45.
14. Feldberg, W.; Saxena, P.N. J. Physiol. (London) 1970, 211, 245-61.
15. Myers, R. D.; Veale, W.L.; Yaksh, T.L. ibid. 1971, 217, 381-92.
16. Myers, R. D.; Brophy, P. D. Neuropharmacology 1972, 11, 351-61.
17. Myers, R. D.; Buckman, J. E. Am. J. Physiol. 1972; 226, 1313-8.

RECEIVED March 2, 1984

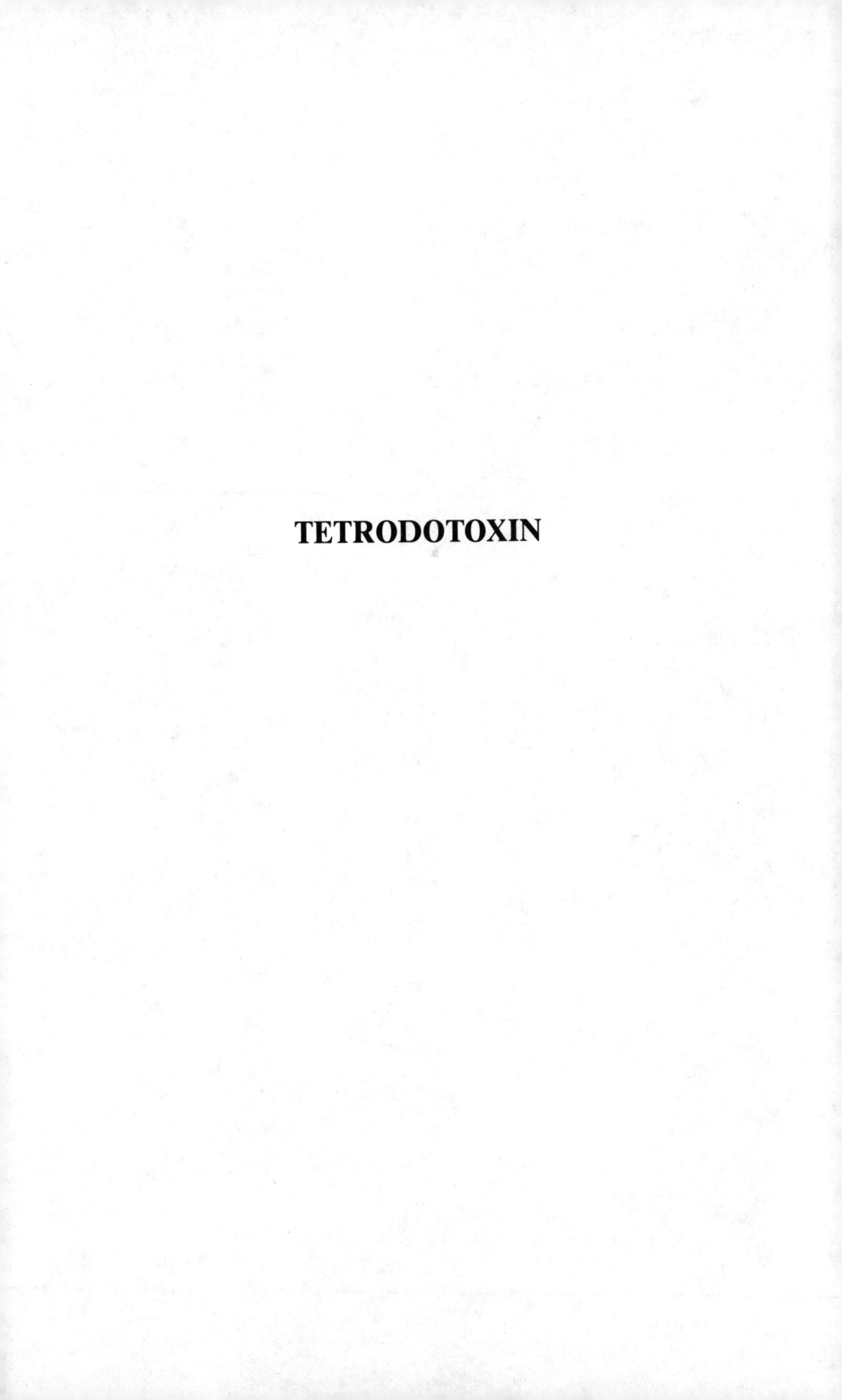

TETRODOTOXIN

28

Occurrence and Origin of Tetrodotoxin

HARRY S. MOSHER and FREDERICK A. FUHRMAN[1]

Departments of Chemistry and Physiology, Stanford University, Stanford, CA 94305

The wide, but species selective, distribution of tetrodotoxin in pufferfish, salamanders, frogs, octopus, and two Japanese shellfish, together with the observation that pufferfish raised in captivity are non-toxic, presents an intriguing problem concerning the origin of tetrodotoxin. Is tetrodotoxin produced by the animal itself; is it part of the food chain; or is it produced by an infectious, symbiotic microorganism? The pertinent observations concerning the occurrence of tetrodotoxin are examined in light of these possibilities. The possibility of infection of selected species by an as yet unknown tetrodotoxin-producing, symbiotic microorganism can be reconciled with most of the observations and should be investigated further.

The occurrence and distribution of tetrodotoxin (TTX) presents us with a perplexing problem that increases in complexity as new observations are made. This puzzle is best appreciated by considering three questions. First, how can an animal such as a pufferfish or a salamander survive with a concentration of tetrodotoxin in its body which is fatally toxic to other animals that do not contain tetrodotoxin? Second, how can we account for the peculiar distribution of tetrodotoxin in various animals? It occurs in such diverse animal species as pufferfish, a goby, _Atelopid_ frogs, _Taricha_ salamanders, an octopus, and has recently been found in two Japanese shellfish. Yet, it is not known to be widely distributed in other species of fish, salamanders, frogs, etc. Third, how can one explain the observation that pufferfish cultured on fish farms in Japan are non-toxic? We propose to examine the currently existing information bearing on these questions and then consider possible explanations.

[1]Current address: P.O. Box 313, Pebble Beach, CA 93953

0097-6156/84/0262-0333$06.00/0
© 1984 American Chemical Society

Occurrence, Isolation and Structure of Tetrodotoxin

The toxicity of pufferfish has been recognized in many early writings (1,2). The history of the pufferfish toxin, tetrodotoxin, has been recounted elsewhere (1-8), but a brief review will serve to introduce the problem we are discussing.

Pufferfish. Toxic species are members of the order Tetrodontiformes (Plectognathi) and include the sharp-nosed puffers (Canthigasteridae), porcupine fish (Diodontidae) as well as the true puffers or fugu (Tetraodontidae). Many human fatalities continue to occur from eating these fish, including an occasional fatality in Florida (*cf.* 1,6,8). The toxin occurs principally in the ovaries, liver and, in some species, the skin.

A crude toxic principle, which we now know to be only about 0.2% pure, was obtained by Yoshizumi Tahara (9,10) in 1909. He named this tetrodotoxin. In 1950 Aikira Yokoo (11) obtained pure crystalline tetrodotoxin. The purification was further developed by Tsuda and coworkers (12) so that the production of the necessary amounts of tetrodotoxin for structural and chemical studies could be achieved. From 1000 kg (1 ton) of pufferfish ovaries they obtained 8-10 g of tetrodotoxin.

The structure shown below was determined and reported upon at the IUPAC Natural Products Symposium in Kyoto on April 13, 1964, by three groups--Tsuda *et al.* (13) (Sankyo Company, and the University of Tokyo), Goto *et al.*, (14) (Nagoya University), and Woodward (15) (Harvard University). This structure is based on extensive degradation studies and a total of five X-ray crystallographic studies on tetrodotoxin derivatives, the final one on tetrodotoxin hydrobromide (16). This structure has many unusual features and is unprecedented among both

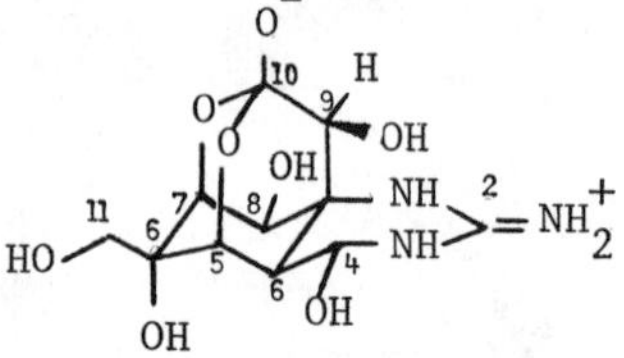

Crystalline Tetrodotoxin

natural and synthetic products (15). The structure of tetrodotoxin is pertinent to the topic of the origin of tetrodotoxin because it has no obvious relationship to known animal or plant natural products; no precursor can be readily identified.

Goby. Tetrodotoxin has been isolated in pure form from a fish of a completely different order, a goby (*Gobius criniger*) from the Ryukyu islands south of the main Japanese islands (17). Three species of California goby contain a toxin that resembles tetrodotoxin pharmacologically, but the toxin has not been identified (18). In gobies the highest concentrations of toxin are in the ovaries and liver.

Salamander. In the late 1930's Professor Twitty at Stanford University discovered that eggs of the California newt or salamander *Taricha torosa* contained a toxic substance which was subsequently named tarichatoxin. In 1963 we isolated from these eggs, collected near Stanford University, a crystalline substance which proved to be indistinguishable from tetrodotoxin (2). The identical tetrodotoxin structure was proposed in 1964 at the same Kyoto meeting where the structure of tetrodotoxin from the pufferfish was announced (2,15). The eggs of another salamander, *T. rivularis*, also yielded crystalline tetrodotoxin, as did another population of *T. torosa*, from the California Sierra foothills, which have a very different type of terrain and flora. Extracts of adult salamanders from ten different species related to *T. torosa*, collected around the northern Pacific perimeter, were injected into mice and produced symptoms resembling those of tetrodotoxin poisoning (19). The clear presumption is that these other closely related salamanders contain tetrodotoxin, but this point has not been rigorously proven. In *Taricha torosa* the highest concentrations of toxin occur in the skin, ovaries, muscle and blood (19). One human fatality has been reported from ingestion of *Taricha granulosa* (20).

Atelopid Frogs. It is now becoming apparent that tetrodotoxin is more widely distributed among animals than at first supposed. A study of toxic frogs which belong to the genus *Atelopus* from Costa Rica, Panama, Columbia, and Ecuador was begun in 1965. These studies ultimately led to the isolation of crystalline tetrodotoxin from the skin of *Atelopus varius* (21) (the harlequin frog of Costa Rica) in 1975 and from the skin and eggs of *A. chiriquiensis* (from Costa Rica) in 1977 (22). We suspect that the other toxic *Atelopus* frogs also contain tetrodotoxin or tetrodotoxin-like compounds, but this was never explicitly proven.

Octopus. The isolation and identification of tetrodotoxin from the Australian blue-ringed octopus *Hapalochlaena maculosa* (23) in 1978 expands the diversity of animals proven to contain tetrodotoxin. In contrast to the pufferfish, salamander and frog, the toxin is found in the posterior salivary venom glands of the octopus and is used for paralyzing prey. Details of the final proof of tetrodotoxin in this octopus are of interest because they illustrate and emphasize the difficulties associated with positively establishing the presence of tetrodotoxin. Following human fatalities attributed to the bite of this octopus (24), a potent toxin called maculotoxin was obtained, but the active component was not identified. Several groups of investigators then sought to establish the nature of the toxic principle (24-29). Maculotoxin was readily distinguished from saxitoxin chromatographically, but there was uncertainty concerning the relation of tetrodotoxin and maculotoxin on both pharmacological and chemical grounds. Croft and Howden (28) in 1972 published results from which they concluded that "maculotoxin, isolated in a chromatographically pure state...was chemically similar but distinct from tetrodotoxin." Jarvis *et al.* (25) in 1975 found the chromatographic and pharmacological properties of tetrodotoxin and maculotoxin to be indistinguishable. In 1977 Savage and Howden showed that the salivary gland extract from *H. maculosa* contained two toxins; one was maculotoxin "having chromatographic properties similar to tetrodotoxin, but with differing infrared spectra, and a second more polar toxin, hapalotoxin." Then in 1978 Sheumack, Howden, Spence, and

Quinn (23) starting with the glands from 250 H. maculosa, following the scheme outlined below, obtained 1.8 mg of crystalline tetrodotoxin. This procedure appears to be straightforward and deceptively

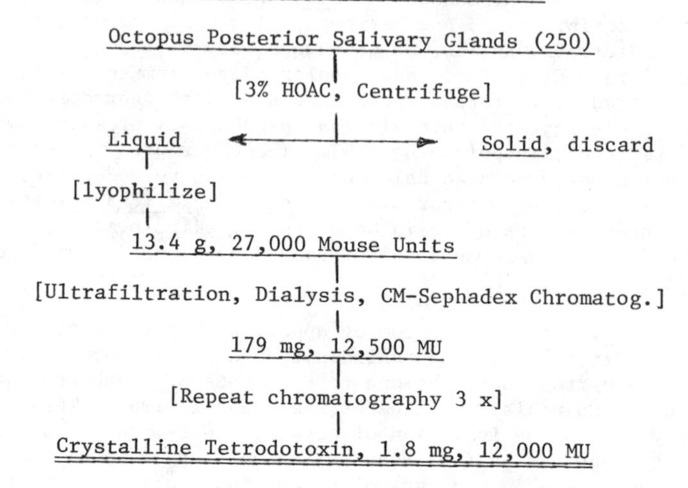

Flow Diagram for Isolation of Tetrodotoxin from the Blue Ring Octopus Hapalochlaena maculosa (23)

easy. However, if so, why did it take knowledgeable and experienced researchers some eight years to establish the identity of maculotoxin and tetrodotoxin? Clearly it is not easy to do, and the presence of impurities, especially closely related toxic impurities, makes the task even more difficult. This work emphasizes the great need for a relatively simple, sensitive test for detecting tetrodotoxin in extracts. Work along these lines has been published (30,31) but will not be reviewed here.

Shellfish. Extended distribution of tetrodotoxin has been uncovered by investigations of recent incidents of fatal and near-fatal shellfish poisonings in Japan. The Japanese ivory shell (Babylonia japonica) is usually non-toxic, but toxic ivory shells appeared in several places in Japan and were identified as a source of tetrodotoxin (32, 33). The toxicity of the ivory shell is complicated by the previous isolation of surugatoxin (34) from the same animal. The structure of surugatoxin, which was proven by X-ray analysis, is totally different from tetrodotoxin. It was found only in the midgut gland and clearly arose from the food chain. In 1981 tetrodotoxin was also isolated from another gastropod, the trumpet shell (Charonia sauliae) (35).

Starfish. Finally, tetrodotoxin has been provisionally identified in the starfish *Astropecten polyacanthus* (36). This starfish was shown to be a food source for the trumpet shell, and the reasonable assumption is that the tetrodotoxin in the digestive gland of the trumpet shell came from this starfish. Although this may be so for the trumpet shell and starfish, Y. Hashimoto (37) concluded that it was very unlikely that tetrodotoxin in the pufferfish was obtained from the environment.

The following outline summarizes the animals in which tetrodotoxin has been proven to occur:

TETRODOTOXIN (TTX) DISTRIBUTION

1) Fish:
   Pufferfish, *Fugu* (Tetraodontidae)
   Goby (*Gobius criniger*)

2) Amphibians:
   Newts, *Taricha* (Salamandridae)
   Frogs, *Atelopus varius*, *A. chiriquiensis* (Atelopidae)

3) Octopus, *Hapalochlaena maculosa* (Octopoda)

4) Shellfish (Gastropoda)
   Ivory Shell, *Babylonia japonica*
   Trumpet Shell, *Charonia sauliae*

5) Starfish (Echinodermata)
   Astropecten polyacanthus

Naturally Occurring Tetrodotoxin Analogs

Currently we know very little about naturally occurring tetrodotoxin analogs or precursors. Chiriquitoxin, isolated along with tetrodotoxin from the Costa Rican frog *Atelopus chiriquiensis* (21,22), is known to possess the same basic skeleton as tetrodotoxin but differs by an uncharacterized substituent at C-6, as shown in the structure below. It also was shown, based on isolation experiments, that

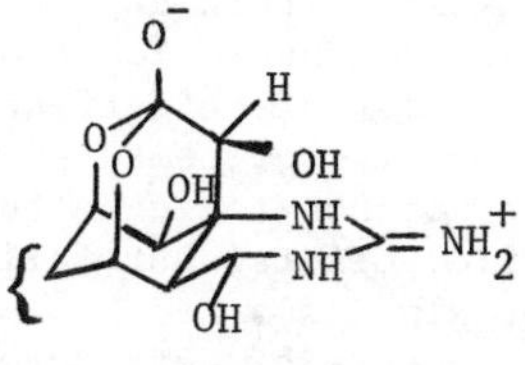

O^- H OH OH NH NH $= NH_2^+$ O O OH

121 mass units

Chiriquitoxin

tetrodotoxin and chiriquitoxin occur in the eggs of *A. chiriquiensis* (22). They were not extracted by water but were extracted with dilute acetic acid. Presumably the toxins were present in a bound form, from which they were readily liberated by dilute acetic acid.

Most recently Kodama et al. (38) have described the separation of a non-toxic high molecular weight fraction from toxic pufferfish livers which, on incubation at 37° for 30 minutes with RNase T_2, liberates a toxin indistinguishable from tetrodotoxin. This points to an inactive form of tetrodotoxin bound in some unknown way as part of RNA present in the toxic liver. These important experiments may contain a clue as to the origin of tetrodotoxin.

Saxitoxin is only one of ten to twelve closely related naturally occurring toxins produced by dinoflagellates of the genus Gonyaulax (39). Could it be that tetrodotoxin is also but one member of a larger group of related toxins?

Possible Origin of Tetrodotoxin

What is the origin of tetrodotoxin in this diverse collection of animals? We must consider four possibilities:

A. Endogenous. Tetrodotoxin is produced by each of the animals in question. It may be an accidental metabolic product; it may have evolved because it serves some survival purpose from an evolutionary standpoint; or it may be a control substance which serves some useful physiological function in these organisms.
B. Exogenous. Tetrodotoxin is not produced by the animal but arises from the food chain in some manner, for instance, similar to the way saxitoxin is concentrated by shellfish that ingest the toxic Gonyaulax dinoflagellates.
C. Symbiotic Microorganism. Tetrodotoxin is the product of some microorganism which lives symbiotically with these particular animals.
D. Multiple Origins. Tetrodotoxin arises from some combination of these three.

Observations to Be Rationalized

The most important observations which must be rationalized in attempting to account for the origin and distribution of tetrodotoxin may be enumerated as follows:

1. The absence of any obvious common factor or relationship among the groups of different animals in which tetrodotoxin has been found, on the one hand, and the limited number of toxic species within these groups, on the other hand.
2. The variability of toxicity within a particular species with respect to individuals, geographic location, and season (1,40).
3. Decrease in toxicity of eggs of pufferfish during development (41) and the lack of toxicity of cultured pufferfish (42,43).
4. Development of toxicity of non-toxic cultured puffers when fed the livers of toxic wild puffers, but failure to develop toxicity when fed pure tetrodotoxin (42).
5. The unique chemical structure of tetrodotoxin (12,16).
6. The ability of tetrodotoxin-containing animals to tolerate a concentration of tetrodotoxin in their bodies that would be fatal to other animals and the resistance of these animals to administration of tetrodotoxin (44-46).
7. The generation of tetrodotoxin from non-toxic, high molecular weight extracts of livers taken from toxic pufferfish by incubation with ribonuclease (RNase T_2), followed by extraction with dilute acetic acid (38).

Compatibility of Theories with Observations

How do these observations measure up to the proposed origins of tetrodotoxin? We will examine each of these separately.

Endogenous Origin. The diversity of animals which have been shown to contain tetrodotoxin and the variability of toxicity between individuals of a given species, as well as variability with respect to location, argue against the self-production of tetrodotoxin by the various animals. If the production of tetrodotoxin is a genetic characteristic of the various species, it is strange and unusual that it should emerge in these diverse groups of animals--fish, salamanders, frogs, octopus, gastropods--but not in closely related species within each group. The lack of toxicity of cultured pufferfish in contrast to wild puffers is also hard to explain on the endogenous origin theory. However, it is possible that there could be some stimulus or nutrient in nature, which is lacking in captivity, that causes the production of tetrodotoxin by the animal and thereby accounts for its variability. Several different diets were tried by Matsui et al. (43), but none resulted in toxicity of cultured pufferfish. The fact that feeding pure tetrodotoxin failed to induce toxicity in non-toxic puffers while feeding livers from toxic puffers did (42), definitely favors an exogenous source for the origin of tetrodotoxin and is a powerful observation against the self-production of tetrodotoxin in these animals.

It is difficult to reconcile the unique chemical structure of tetrodotoxin with that of an animal product. Its structure is not related to that of other animal products by any readily recognized biosynthetic scheme. It is not a terpenoid, not obviously formed from amino acid or carbohydrate units, and apparently not constructed from acetate or propionate units. Nor does it resemble any of the various plant alkaloid patterns. It thus appears to be a very unlikely animal product to result from known biogenetic pathways. In this connection the metabolic incorporation of radioactive precursors using T. torosa and T. granulosa salamanders was studied by Shimizu et al. (47). They observed significant isotopic incorporation into amino acids and steroid metabolites, but they found no such incorporation associated with tetrodotoxin.

The experiments by M. Kodama et al. (38) in which tetrodotoxin was liberated from non-toxic, high molecular weight extracts of toxic pufferfish livers by digestion with RNase T_2 (37° for 30 minutes), followed by extraction with dilute acetic acid, can be accommodated by the endogenous, exogenous, or symbiotic microorganism theories. The idea of a non-toxic RNA precursor to tetrodotoxin is extremely interesting and should be investigated further. As a first step, it should be established that acetic acid alone will not liberate tetrodotoxin.

Exogenous Origin. The theory that tetrodotoxin arises in the environment and is ingested by these animals is favored by the finding that cultured puffers are non-toxic. It is also favored by the feeding experiments of Matsui (42), in which non-toxic puffers became toxic upon ingesting toxic livers of wild puffers. The observation that feeding pure tetrodotoxin to these non-toxic puffers did not induce toxicity is explicable by assuming that tetrodotoxin is not absorbed or is destroyed before absorption, whereas, when present in a bound or protected form in the toxic livers, it is incorporated into the host. Certainly the variable toxicity of puffers with respect to individuals, location, and season is readily accountable by assuming an external environmental source. This is the proven origin for saxitoxin in shellfish and for ciguatoxin in fish. The circumstantial evidence that pieces of tetrodotoxin-containing starfish were found in the midgut gland of the toxic trumpet shell (35) is convincing evidence that the food chain is a viable route in this case even if it is not in other cases.

However, the diverse eating habits and food supplies for such widely different creatures as fish, frogs, salamanders, and octopus make the postulate of an environmental source for tetrodotoxin less than attractive. The case for such a source is further weakened by the total lack of any evidence for a tetrodotoxin-containing dietary component for any of these animals, let alone any common source for the postulated component. The one suggestion in this regard is that the source may be a special non-toxic, tetrodotoxin-associated, high molecular weight substance which Kodama et al. (38) suggest may be an RNA component.

Symbiotic Microorganism Origin

How does the tetrodotoxin-producing microorganism theory measure up to these observations? This theory is easily reconciled with the strange variety of tetrodotoxin-containing animals and with their variable toxicities by assuming a corresponding variability in the extent and intensity of infection. The lack of toxicity of cultured pufferfish is accountable, based on this theory, by assuming a lack of infection. It is the practice in fish hatcheries to treat the fish in the larval or fry stages with antibiotics to reduce loss by ordinary infections. The theorized tetrodotoxin-producing microorganisms may be controlled by such a treatment. The feeding experiments in which non-toxic cultured pufferfish did not become toxic when given pure tetrodotoxin but did become toxic when fed livers of wild, toxic puffers are understandable on the basis that pure tetrodotoxin was not absorbed (or was destroyed in the digestive tract) while the livers of the infected wild puffers introduced the tetrodotoxin-producing microorganism into the otherwise non-infected cultured fish.

The unique chemical structure of tetrodotoxin does not seem to follow any established biogenetic pattern in either plants or animals, whereas very unusual structures are frequently encountered in the products from microorganisms. For instance, saxitoxin has an unique chemical structure (39); it is known to be produced by the plant-like dinoflagellate of the genus Gonyaulax. Thus it is much easier to imagine tetrodotoxin arising biogenetically from some microorganism than directly from a fish, salamander, or frog. There is precedence for the proposal of a microbial infection which causes toxicity in animals in the published preliminary experiments of Moore et al. (48), who found evidence that palytoxin from the coelenterate Palythoa toxica may in fact arise from a vibrio-like bacterium which infects this particular marine animal. Also, Shimizu in a recent review article (47) on saxitoxin has referred to the suggestion made by Jackim and Gentile (49) and subsequently reinforced by the recent studies of Shimizu and Kobayshi (50) that "endosymbiotic bacterium or a transferred plasmoid is actually responsible for the toxin production" in Gonyaulax. Finally, the lack of metabolic incorporation of possible radioactive precursors into tetrodotoxin in Taricha salamanders (47,50) is evidence against the endogenous origin of tetrodotoxin in these animals. One possible explanation proposed for this lack of incorporation was that "the toxin stored by the newts is synthesized by a symbiotic microorganism, but in captivity the conditions are not suitable for the organism or toxin production." Thus Shimizu has explicitly suggested that the biosynthesis of tetrodotoxin may actually be controlled by microorganisms in the salamander.

A final point to be made that may be pertinent to the three possible origins of tetrodotoxin addresses the question: how can the tetrodotoxin-containing animals such as the pufferfish, Atelopid frogs and Taricha salamanders tolerate a concentration of tetrodotoxin in their bodies which would be fatal for all otherwise normal animals? To address this question requires a brief summary of the accepted mechanism of action of tetrodotoxin. It has been established that tetrodotoxin acts by preventing nerve conduction by an extremely specific and reversible blockage of the inward movement of sodium ions through the cell membrane of an activated neuron (52, 54). It does not affect potassium transport. Available evidence suggests that it acts as an easily removable but closely fitting stopper (53,55) which specifically plugs the outer terminus of the voltage-sensitive sodium channel in the nerve membrane; or it acts as a lid which specifically covers the sodium channel (56). A normal nerve, e.g., the giant axon of the squid, has about 95% of its action potential blocked by bathing in a 1×10^{-7} M tetrodotoxin solution. However, the blockage is not complete and a few percent transmission remains even when the concentration is raised 100-fold above the maximum blocking stage. This can be interpreted to mean that normal axons have a few percent of their sodium channels which are not tetrodotoxin sensitive. The action potential of an axon from the tetrodotoxin-containing Taricha salamander or Atelopid frog is unaffected by bathing in a tetrodotoxin solution (45). This resistance to tetrodotoxin clearly does not result from some impermeable cellular barrier. It has been reported that the response of the pectoral fin muscle of some fish closely related to the toxic pufferfish are less resistant than pufferfish themselves to tetrodotoxin but are more resistant than more distantly related species (46). This has been taken as evidence that the resistance of pufferfish to tetrodotoxin is a result of an evolutionary process brought on by tetrodotoxin itself. Such an evolutionary change is certainly compatible with the development of a symbiotic relationship between the host and a tetrodotoxin-producing microorganism.

There appear to be at least two kinds of sodium channels, the common ones which are blocked by tetrodotoxin and alternate ones which are not affected by tetrodotoxin (57,58). It seems reasonable that species with a preponderance of this alternate type of sodium channel might evolve as the predominate form in response to the presence of tetrodotoxin. It is completely reasonable that in the presence of tetrodotoxin, species would be selected in which the nerve membranes had large numbers of functional and tetrodotoxin-insensitive, sodium channels. Lazdunsky and coworkers (58,59) have been studying the effect of culturing cells in the presence of tetrodotoxin in order to obtain more information on this point.

It is self-evident that tetrodotoxin-containing animals are those which have been able to adapt to the presence of tetrodotoxin in their tissues. However, the reason for resistance of certain species to tetrodotoxin does not seem to have much bearing on the question of the origin of tetrodotoxin since these evolutionary arguments appear to be equally compatible with the exogenous, endogenous or symbiotic microorganism theories.

Summary

A critical review of the current information available to us seems to render unlikely the endogenous theory for the origin of tetrodotoxin. The exogenous theory has many points in its favor, but postulates of a common unknown tetrodotoxin source in the food chain which is eaten and concentrated by pufferfish, Taricha salamanders, and Atelopid frogs make the possibility unlikely. It may be, however, that this mechanism does operate for ivory shell and/or trumpet shell. Finally, the theory of infection of these animals by a tetrodotoxin-producing, symbiotic microorganism can rationalize all the findings but suffers from the basic fact that no such organism has as yet been observed.

Conclusion

We hope that these speculations concerning the origins of tetrodotoxin will stimulate experiments to test these theories. Attempts should be made to culture the postulated tetrodotoxin-producing microorganism. Experiments should be undertaken to raise tetrodotoxin-containing animals under sterile and non-sterile conditions.

Acknowledgment

We gratefully acknowledge the continued support of our research on animal toxins by the National Institutes of Health (most recently NIH NS14345-06). The ideas we have developed here would not have evolved without this sustaining aid for our collaborative efforts.

Literature Cited

1. Tani, I. "A Study of the Toxicity of Japanese Fugu"; Teikoku Tosho: Tokyo, 1945.
2. Mosher, H. S.; Fuhrman, F. A.; Buchwald, H. D.; Fischer, H. G. *Science* 1964, 144, 1100-1110.
3. Kao, C. Y. *Pharmacol. Rev.* 1966, 18, 997.
4. Tsuda, K. *Naturwissenschaften* 1966, 53, 171-176.
5. Fuhrman, F. A. *Sci. Am.* 1967, 61, 61.
6. Halstead, B. W. "Poisonous and Venomous Marine Animals of the World"; U. S. Government Printing Office: Washington, D. C., 1967; Vol. I, pp. 83-87. Vol. II, pp. 679-844.
7. Evans, M. H. *Intl. Rev. Neurobiol.* 1972, 15, 83-166.

8. Fuhrman, F. A. In "CRC Handbook of Naturally Occurring Food Toxicants"; Rechcigl, M., Jr., Ed.; CRC Press: Boca Raton, Florida, 1983.
9. Tahara, Y. *J. Pharm. Soc. Japan* 1909, 29, 587.
10. Tahara, Y. *Biochem. Zeit.* 1911, 30, 255.
11. Yokoo, A. *J. Chem. Soc. Japan* 1950, 71, 590.
12. Tsuda, K.; Kawamura, M. *Pharm. Bull. Japan* 1953, 1, 112.
13. Tsuda, K. et al. *Chem. Pharm. Bull. Japan* 1963, 11, 1473; 1964, 12, 634, 642, 1357.
14. Goto, T.; Kishi, Y.; Takahashi, S.; Hirata, Y. *Tetrahedron* 1965, 21, 2059.
15. Woodward, R. B. *Pure and Appl. Chem.* 1964, 9, 49.
16. Furusaki, A.; Tomiie, Y.; Nitta, I. *Bull. Chem. Soc. Japan* 1970, 43, 3332.
17. Noguchi, T.; Hashimoto, Y. *Toxicon* 1973, 11, 305.
18. Elam, K. S.; Fuhrman, F. A.; Kim, Y. H.; Mosher, H. S. *Toxicon* 1977, 14, 45.
19. Wakely, J. F.; Fuhrman, G. J.; Fuhrman, F. A.; Fischer, H. G.; Mosher, H. S. *Toxicon* 1966, 3, 195.
20. Bradley, S. C.; Klika, L. J. *J. Am. Med. Assoc.* 1981, 246, 247.
21. Kim, Y. H.; Brown, G. B.; Mosher, H. S.; Fuhrman, F. A. *Science* 1975, 189, 151.
22. Pavelka, L. A.; Kim, Y. H.; Mosher, H. S. *Toxicon* 1977, 15, 135.
23. Sheumack, D. D.; Howden, M. E. H.; Spence, I.; Quinn, R. J. *Science* 1978, 199, 188.
24. Sutherland, S. K.; Lane, W. R. *Med. J. Aust.* 1969, 1, 893.
25. Jarvis, M. W.; Crone, H. D.; Freeman, S. E.; Turner, R. J. *Toxicon* 1975, 13, 177.
26. Sutherland, S. K.; Broad, A. J.; Lane, W. R. *Toxicon* 1970, 8, 249.
27. Dulhunty, A.; Gage, P. W. *J. Physiol.* 1971, 218, 433.
28. Croft, J. A.; Howden, M. E. H. *Toxicon* 1972, 10, 645.
29. Savage, I. V. E.; Howden, M. E. H. *Toxicon* 1977, 15, 463.
30. Yasumoto, T.; Nakamura, M.; Oshima, Y.; Takahata, J. *Bull. Japan Soc. Sci. Fish.* 1982, 48, 1481.
31. Nakayama, T.; Terakawa, S. *Anal. Biochem.* 1982, 126, 153.
32. Yasumoto, T.; Oshima, Y.; Hosaka, M.; Miyakoshi, S. *Bull. Japan Soc. Sci. Fish.* 1981, 47, 929.
33. Noguchi, T.; Maruyama, J.; Ueda, Y.; Hashimoto, K.; Harada, T. *Bull. Japan Soc. Sci. Fish.* 1981, 47, 909.
34. Kosuge, H. Z.; Ochiai, K.; Masaki, N. *Tetrahedron Lett.* 1972, 2545.
35. Narita, H.; Noguchi, T.; Maruyama, J.; Ueda, Y.; Hashimoto, K.; Watanabe, Y.; Hida, K. *Bull. Japan Soc. Sci. Fish.* 1981, 47, 935.
36. Noguchi, T.; Narita, H.; Maruyama, J.; Hashimoto, K. *Bull. Japan Soc. Sci. Fish.* 1982, 48, 1173.
37. Hashimoto, Y.; Kamiya, H. *Bull. Japan Soc. Sci. Fish.* 1970, 36, 425.
38. Kodama, M.; Noguchi, T.; Maruyama, J.; Ogata, T.; Hashimoto, K. *J. Biochem.* 1983, 93, 243.
39. Koehn, F. E.; Hall, S.; Wichmann, C. F.; Schnoes, H. K.; Reichardt, P. B. *Tetrahedron Lett.* 1982, 23, 2247.
40. Matsui, T.; Hamada, S.; Yamamori, K. *Bull. Japan Soc. Sci. Fish.* 1982, 48, 1179.

41. Suyama, M.; Uno, H. *Bull. Japan Soc. Sci. Fish.* 1957, 23, 438.
42. Matsui, T.; Hamada, S.; Konosu, S. *Bull. Japan Soc. Sci. Fish.* 1981, 47, 535.
43. Matsui, T.; Sato, H.; Hamada, S.; Shimizu, C. *Bull. Japan Soc. Sci. Fish.* 1982, 48, 253.
44. Ishihara, F. *Mittheil. Med. Fak. Tokio Univ.* 1918, 20, 375.
45. Kao, C. Y.; Fuhrman, F. A. *Toxicon* 1967, 5, 25.
46. Kidokoro, Y.; Grinell, A. D.; Eaton, D. C. *J. Comp. Physiol.* 1974, 89, 59.
47. Shimizu, Y. *Pure and Appl. Chem.* 1982, 54, 1973.
48. Moore, R. E.; Helfrich, P.; Patterson, G. M. L. *Oceanus* 1982, 25, 54.
49. Jackim, E.; Gentile, J. *Science* 1968, 162, 915.
50. Shimizu, Y.; Kobayshi, M. *Chem. Pharm. Bull. (Japan)* 1983, 31, 3625.
51. Alan, M.; Shimizu, Y.; Ikawa, M.; Sasner, J. J. *J. Environ. Sci. Health* 1978, A13, 493.
52. Narahashi, T. *Physiol. Rev.* 1974, 54, 813.
53. Takata, M.; Moore, J. W.; Kao, C. Y.; Fuhrman, F. A. *J. Gen. Physiol.* 1966, 49, 977.
54. Kao, C. Y.; Yoek, P. N.; Goldfinger, M. D.; Fuhrman, F. A.; Mosher, H. S. *J. Pharmacol. Exp. Therap.* 1981, 217, 416.
55. Hille, B. *Biophys. J.* 1975, 15, 615.
56. Kao, C. Y.; Walker, S. E. *J. Physiol.* 1982, 323, 619.
57. Redfern, P.; Thesleff, S. *Acta Scand.* 1971, 82, 70.
58. Jacques, Y.; Fosset, M.; Lazdunsky, M. *J. Biol. Chem.* 1978, 253, 7383.
59. Lazdunsky, M., private communication, 1983.

RECEIVED February 6, 1984

29

Tetrodotoxin Determination Methods

YOSHIO ONOUE[1], TAMAO NOGUCHI[2], and KANEHISA HASHIMOTO[2]

[1]Laboratory of Marine Botany and Environmental Science, Faculty of Fisheries, Kagoshima University, Kagoshima, Japan
[2]Laboratory of Marine Biochemistry, Faculty of Agriculture, University of Tokyo, Tokyo, Japan

High-performance liquid chromatography-fluorometry, mass spectrometry and capillary isotachophoresis have been applied successfully to identify or quantitate tetrodotoxin in small volumes of toxin extract from pufferfish. A mouse bioassay method is useful for screening of the toxicity of various organisms from affected areas, although this method may not be sufficient for the identification of the toxin. Recently, distribution of tetrodotoxin in the marine ecosystem has expanded from pufferfishes to some other animals. Rapid and accurate determination of the toxin occurring in those organisms is becoming increasingly important from the public health standpoint.

It is common knowledge that many species of pufferfish are toxic to man. In spite of such recognition, a great number of persons have been intoxicated from ingesting pufferfish in Japan. According to the 1982 food poisoning statistics (Japan), 80% of the victims were associated with this fish.

Tetrodotoxin (abbreviated TTX below, Figure 1) is named after the family Tetraodontidae into which most pufferfishes are classified. TTX has not been detected in any other fishes except for the goby *Gobius criniger*.

During 1935-1945 Professor Tani of Kyushu Imperial University, Japan, surveyed toxicity of various tissues from 19 species of pufferfish inhabiting the surrounding waters of the northern Kyushu Island (Table I) (1, 2). He published a book "Toxicological Studies on Japanese Pufferfishes", which is even recently cited by many researchers dealing with TTX.

As he describes, pufferfishes are mostly toxic, irrespective of the species, tissue and the season of catch. In addition, the toxic potency widely differs even among specimens of the same species by one catch. The muscle of pufferfish from temperate waters is believed to be little or nontoxic, but that from tropical waters to be toxic. Several persons were killed by ingestion of the flesh of the pufferfish caught off Vietnam in 1959 (2). This pufferfish was later identified as *Lagocephalus lunaris lunaris*. It has recently been

0097-6156/84/0262-0345$06.00/0
© 1984 American Chemical Society

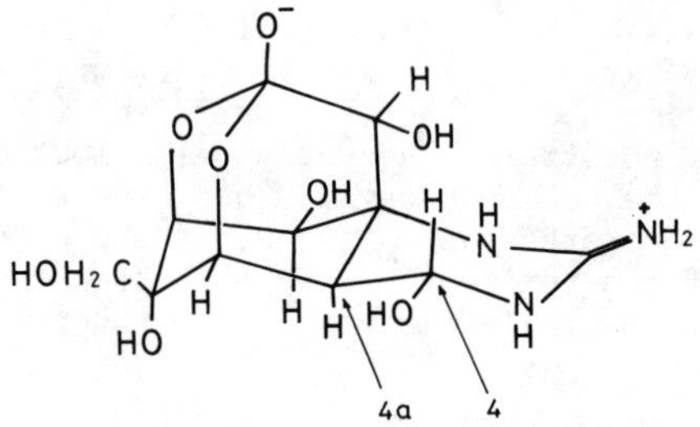

Figure 1. Structure of tetrodotoxin.

Table I. Toxicity of Japanese Pufferfish

Species	Ovary	Testis	Liver	Skin	Intestine	Muscle
Fugu niphobles	A	C	A	B	A	C
F. poecilonotum	A	B	A	B	B	C
F. vermicularis vermicularis	A	D	A	B	B	C
F. pardalis	A	C	A	B	B	D
F. vermicularis porphyreus	A	D	A	B	B	D
F. ocellatus obscurum	A	D	B	B	B	D
F. chrysops	B	D	B	B	C	D
F. rubripes rubripes	B	D	B	D	C	D
F. xanthopterum	B	D	B	B	C	D
F. stictonotum	B	D	B	C	D	D
Lagocephalus laevigatus inermis	D	D	B	D	D	D
L. lunaris spadiceus	D	D	D	D	D	D
Liosaccus cutaneus	D	D	D	D	D	D
Canthigaster rivulata	D	-	C	B	C	D
Diodon holacanthus	D	-	D	D	D	D
Chilomycterus affinis	D	-	D	D	D	D
Ostracion cmmaculatum	D	D	D	D	D	D
Lactoria diaphana	D	D	D	D	D	D
Aracana aculeata	D	D	D	D	D	D

A: Strongly toxic, lethal at less than 10 g.
B: Moderately toxic, not lethal at less than 10 g.
C: Weakly toxic, not lethal at less than 100 g.
D: Negative, not lethal at less than 1000 g.
-: No data available.
Adapted with permission from Ref. 2. Copyright 1979, Japan Scientific Societies Press.

found that the muscle of pufferfishes inhabiting the Sanriku Coast of Japan is often toxified up to a level of several hundred mouse units (MU) per g (3). Since the lethal dose of TTX in human is about 10000 MU (4), the flesh of these pufferfishes may kill a man even when eaten in a small amount less than 100 g.

On the other hand, some pufferfishes which have not been eaten before, are sometimes marketed in Japan. In our mouse bioassay test a high toxicity was detected in the tissues from *Tetraodon alboreticulatus*, one of such pufferfishes; 2870 MU/g ovary and 31 MU/g liver (5). Thus, re-examination seems to be necessary for toxic potency of the pufferfishes landed from Japanese and adjacent waters.

Recent Findings on the Distribution of Tetrodotoxin in Nature

TTX had long been believed to distribute exclusively in pufferfishes. Twenty years ago, a paralytic toxin was found in the ovaries of the California newt *Taricha torosa* and named tarichatoxin (6). Identity of this toxin with TTX was demonstrated later. Tarichatoxin or TTX was also detected in the skin, muscle, blood of the California newt, as well as other species of the genus *Taricha*.

The authors noticed the presence of this paralytic toxin in a goby *Gobius criniger* inhabiting tropical to subtropical seas (7). In the goby as well, TTX was recognized in the tissues such as skin, viscera and muscle. Kim et al. (8) isolated TTX from Costa Rican frogs of the genus *Atelopus*. Sheumack et al. (9) found a paralytic toxin in secretions from the posterior salivary gland of the blue-ringed octopus *Octopus maculosus*, and named maculotoxin. However, this toxin was also identified as TTX.

In 1979, a food poisoning case due to ingestion of the digestive gland of a trumpet shell *Charonia sauliae* occurred in Shimizu, Shizuoka, Japan. Rather unexpectedly, TTX was found to be responsible for this incidence (10). Subsequent surveys showed that most of the digestive glands of trumpet shell specimens collected from the adjacent waters were moderately to highly toxic. The highest toxicity score recorded was 1950 MU/g digestive gland. Since the average weight of digestive gland of this shellfish is about 50 g, ingestion of the single digestive gland may have killed ten persons.

TTX was also found in some other gastropods such as the Japanese ivory shell *Babylonia japonica* (11, 12), and the frog shell *Tutufa lissostoma* (13). In connection with this, some starfishes fed by those gastropods were found to contain TTX at significant levels (14).

In tropical to subtropical waters, there live toxic crabs. They contain paralytic shellfish poisons in most cases. One of those toxic crabs, *Atergatis floridus*,inhabits also along the Pacific Coast of the Japan Proper. Very recently, it was excavated that the crabs collected from Miura Peninsula near Tokyo possess TTX as the major toxin, along with some paralytic shellfish poisons as the minor (15). The distribution of TTX may be expanded further in the future.

Present Status of Determination of TTX

TTX has so far been determined mainly by mouse assay (4). This assay is featured by simplicity, but has some demerits such as low sensitivity, low accuracy, rather high cost of suitable mice, and local difficulty of its procurement. Meanwhile, the distribution of TTX has been expanded from pufferfishes to some animals of various phyla, as described above. More sensitive and accurate assay methods may promote these lines of research.

Under these circumstances, attempts have been made to develop chemical assay methods of TTX. In the near future, the mouse assay method may be replaced by chemical methods.

Chemical Assay Methods

Two example-determinations of TTX using fluorometry coupled with high-performance liquid chromatography (HPLC) are introduced here, along with some other promising assay techniques such as mass spectrometry.

HPLC with Fluorometry Using Alkali (16)

Gonads (10 g) of the pufferfish *Fugu pardalis* are extracted with 25 mL of 0.02 M acetic acid in a boiling water bath for 10 min. After being cooled to room temperature, the extract is filtered. The filtrate is passed through an Amberlite CG-50 column (NH_4^+, 1.2 x 3 cm). The column is developed with 50 mL each of water and 0.5 M acetic acid. A defatting step may be introduced prior to the column treatment when fat-rich tissues like livers are employed. The toxic eluate is concentrated in vacuo and made up to 20 mL with water. A 50- µL portion of the solution is analyzed by HPLC.

The HPLC units include a stainless-steel column (400 x 5 mm I.D.) with a cationic exchanger (Hitachi 3011 C), a loop injector (Kyowa Seimitsu Co.), two constant flow pumps (Seishin Pharmaceutical Co.) for eluant and reagent, reaction coil (Teflon tubing, 30 m x 0.5 mm I.D.) and a fluorometer (Japan Spectroscopic Co., FP-110). The column is eluted with 0.06 M citrate buffer (pH 4.0) at a flow rate of 0.5 mL/min. The eluate is mixed with an equal volume of 4 N NaOH. On alkaline treatment, TTX gives rise to a fluorogenic substance, 2-amino-6-hydroxymethyl-8-hydroxyquinazoline (C_9 base) (17, 18). The excitation and emission wavelengths are set at 357 and 510 nm, respectively. The chromatogram is recorded on a National Pen Recorder VP-6611A.

TTX gives a peak with a retention time of 26 min. The toxicity-fluorescence relation is linear over the range of 0.1-20 MU. The maximum variation of the peak heights is 3%, as tested on varied toxin levels.

HPLC with Fluorometry Using o-Phthalaldehyde (OPA) (19, 20)

Ovaries (100 g) of the pufferfish *Fugu vermicularis porphyreus* are extracted with 1% acetic acid in methanol. The extract is concentrated in vacuo and defatted by shaking with chloroform. The defatted extract is treated with activated charcoal. The toxin adsorbed is eluted with 1% acetic acid in 20% ethanol. The eluate is evaporated in vacuo to dryness. The residue is dissolved in a small amount of water and adjusted to pH 6 with 1 N NaOH. The toxic solution is applied to an Amberlite IRC-50 column (NH_4^+, 2.5 x 45 cm) and developed with 2 L of water, and then 1 L each of 1 and 10% acetic acid. The toxic fractions are freeze-dried, dissolved in 1 mL of water and analyzed by HPLC.

The HPLC system comprises a Hitachi 638-50 analyzer with a 650-10 spectrofluorometer, a 056 recorder operating at 10 mV full scale, a stainless-steel column (150 x 4 mm I.D.) with a Hitachi 3013 C ion-exchanger, and an OPA unit with a reaction coil (10 m x 0.3 mm I.D.).

Ten microlitres of toxic solution are placed on the column equilibrated with 0.005 M acetic acid. A three-step gradient elution with acetic acid is then applied: 0.005-0.015 M, 0-15 min; 0.015-0.15 M, 15-30 min; 0.15-0.50 M, 30-80 min. The flow rate of the elution and the column pressure are 1 mL/min and 110 kg/cm^2, respectively.

The OPA reagent for HPLC is prepared according to the method of Benson and Hare (21). The fluorescence reaction is performed in a 55 °C water bath. OPA reacts with the guanidino group of TTX presumably to form a fluorescent product, 1-alkylthio-2-alkylisoindole (22, 23). TTX is monitored at 453 nm with 332-nm excitation. Peak areas are calculated by a data processing system of the analyzer. The chromatogram of TTX from the column is shown in Figure 2.

Although a multitude of peaks appear on analysis of a crude puffer toxin preparation, only one of them provides a retention time (20.0 min) compatible to that of TTX standard. The identity of TTX and other main contaminants has been confirmed by mouse bioassay, TLC, electrophoresis and amino acid analysis.

The peak area is proportional to the amount of toxin applied. The linearity of this relationship is maintained up to 100 nM TTX or more. The lower limit of OPA detection is 1-2 nM for TTX. Five or more assays indicate that the relative standard deviation for TTX is $\pm$ 2.5%.

Major interfering substances found in the crude toxin preparation are citrulline, ethanolamine, ornithine, lysine and arginine. The presence of more than 10 nM of these amines makes difficult the detection of TTX because of their intense fluorescence. A similar effect is also noted with 100 µM ammonium ions.

Mass Spectrometry (24)

The mass spectrum of TTX can directly be measured by fast atom bombardment- or secondary ion-mass spectrometry. In the former a JEOL JMS DX-300 mass spectrometer equipped with a JEOL JMA-3100 data system is used; xenon provides the primary beam of atoms. Acceleration voltage of the primary ion is 3 kV. Scanning is repeated within a mass range of m/z 100 to 1000.

TTX is dissolved in 0.05 M acetic acid at a concentration of approximately 10 µg/µL. One microlitre each of this TTX solution and glycerol as matrix are placed on the sample stage of the mass spectrometer, mixed well, and introduced into the ion chamber of the spectrometer. Both positive and negative mass spectra of TTX are then measured.

As shown in Figure 3, TTX exhibits $(M+H)^+$ and $(M+H-H_2O)^+$ ion peaks at m/z 320 and 302, respectively, in the positive mass spectrum, and an $(M-H)^-$ peak at m/z 318 in the negative.

In the secondary ion-mass spectrometry, a Hitachi M-80B mass spectrometer equipped with a Hitachi M-0101 data system is applied as above.

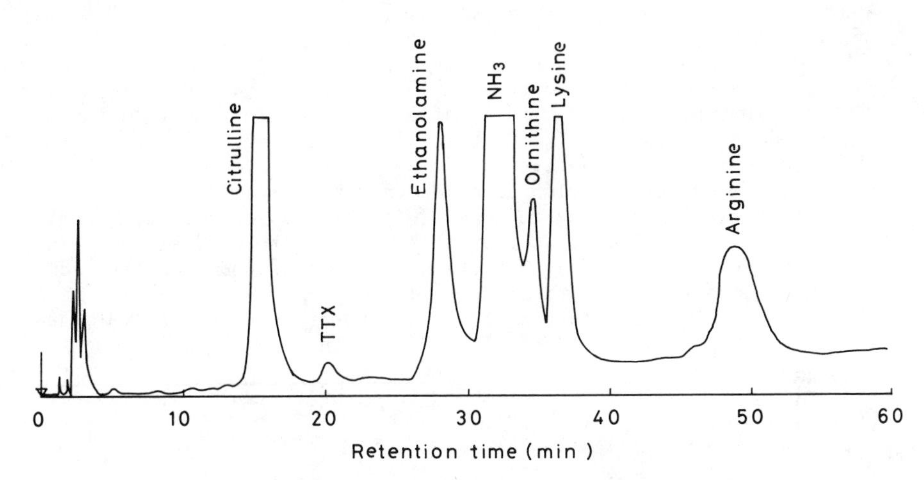

Figure 2. Chromatogram of a crude puffer toxin preparation. Reproduced with permission from Ref. 19. Copyright 1983, Elsevier Scientific Publishing Company.

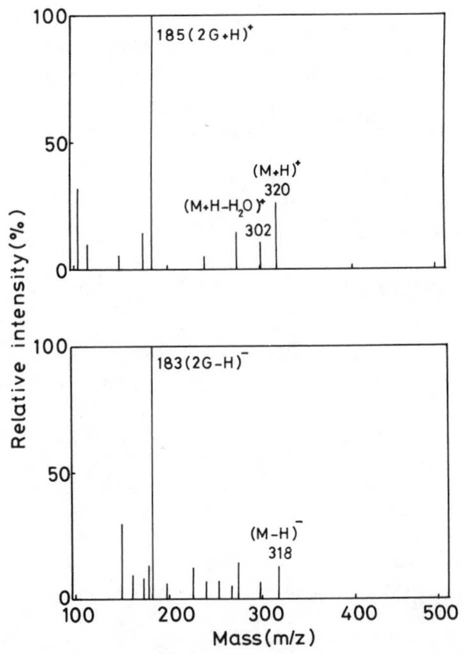

Figure 3. Positive (upper) and negative (lower) mass spectra of TTX, as measured by fast atom bombardment mass spectrometry.

Gas Chromatography-Mass Spectrometry (GC-MS) (10)

A 30-mL flask containing 0.2 mL of TTX solution (250 MU) and 0.5 mL of 1.5 N NaOH is heated at 80-90 °C for 30 min to derive C_9-base from TTX (18, 25). After being cooled to room temperature, the reaction mixture is adjusted to pH 3-5 with 10% HCl and extracted 3 times with 2 mL of n-butanol. The combined extracts are evaporated to dryness under reduced pressure. The C_9-base in the residue is converted to the trimethylsilyl derivative in the presence of N,O-bis(trimethylsilyl)acetamide, trimethylchlorosilane and pyridine (2:1:1). This derivative is then applied to GC-MS (Hitachi 063 gas chromatograph: column (2 m x 3 mm I.D.), Chromosorb W coated with 3% OV-1, temperature 180-250 °C (5 °C/min); Hitachi RMU 6 MG mass spectrometer: ionization voltage 70 eV, carrier gas helium).

An example of gas chromatogram for the trimethylsilyl (TMS) derivative of alkali-hydrolyzed TTX is shown in Figure 4. Although a variety of peaks are seen in this chart, only the peak with a retention time of 7.8 min is associated with C_9-base-$(TMS)_3$, as evidenced by mass spectroscopy (Figure 5).

Sharp fragment ions appear at m/z (%); 407 (41, M^+), 392 (100, M^+-CH_3), 377 (9, M^+-Me_2), 334 (4, M^+-$SiMe_3$), 320 (11, M^+-$NSiMe_3$), 318 (13, M^+-$OSiMe_3$), 304 (8), 282 (6), 230 (9), 207 (21), 203 (14), 174 (9) and 147 (27, TMS-O-$SiMe_2$). M^+ denotes C_9-base-$(TMS)_3$ (MW = 407, $C_9H_6O_2N_3(TMS)_3$.

^{1}H-NMR Spectrometry (10)

TTX (1-5 mg) is dissolved in 0.3 mL of 3% CD_3COOD in D_2O, placed in an NMR tube, and analyzed by a 400-MHz JEOL JNM FX-400 spectrometer with tetramethylsilane (Me_4Si) as the external standard. TTX elicits signals at 1.75 (s, CH_3COOH), 2.09 (d, J = 9.7 Hz), 5.22 (d, J = 9.7 Hz), 3.75 (m) and 3.93 (m) ppm (Figure 6). Coupling (d, J = 9.7 Hz) of the signals on the basis of C_4-H and C_{4a}-H (refer to Figure 1) is recognized at the chemical shifts between 2.09 or 2.15 and 5.22 or 5.31 ppm.

Thin-Layer Chromatography (TLC) (10, 11)

Acetyl ester of TTX (26) is derived as: About 100 MU of TTX is added to a 2 mL mixture of acetic anhydride and pyridine (1:1). The mixture is left overnight at room temperature, and evaporated to dryness under reduced pressure.

TLC is carried out on a 5 x 20 cm silica gel precoated plate (Merck) with a solvent system of pyridine-ethylacetate-acetic acid-water (75:25:15:30), *tert*-butanol-acetic acid-water (2:1:1) or n-butanol-acetic acid-water (60:15:25) (11).

TTX and derivatives are visualized as a pink spot by spraying the Weber reagent, or as a yellow fluorescent spot under a UV lamp (365 nm) after spraying 10% KOH followed by heating at 100 °C for 10 min.

The Rf values of TTX and its derivatives are as follows: 0.70 (TTX) and 0.93 (acetyl TTX) on pyridine-ethylacetate-acetic acid-water (75:25:15:30); 0.21 (TTX) and 0.45 (acetyl TTX) on n-butanol-acetic acid-water (60:15:25); 0.45 (TTX) on *tert*-butanol-acetic acid-water (2:1:1).

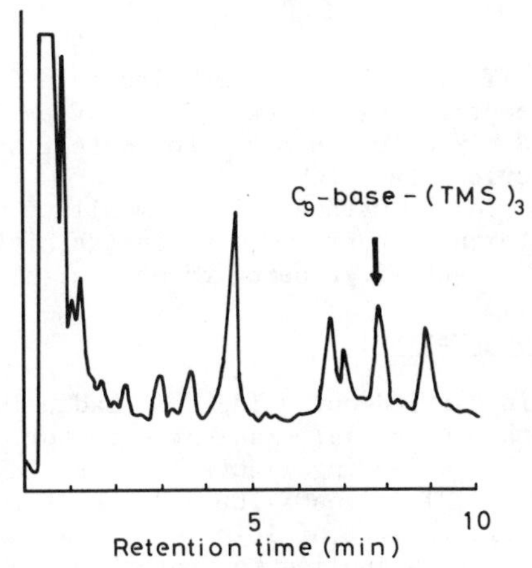

Figure 4. Gas chromatogram of TMS derivative of alkali-hydrolyzed puffer toxin. The arrow shows C_9-base-$(TMS)_3$ derivative.

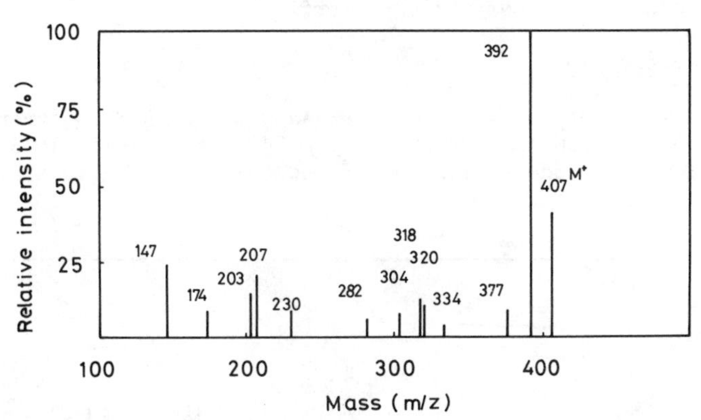

Figure 5. Mass spectrum of C_9-base-$(TMS)_3$ derived from puffer toxin.

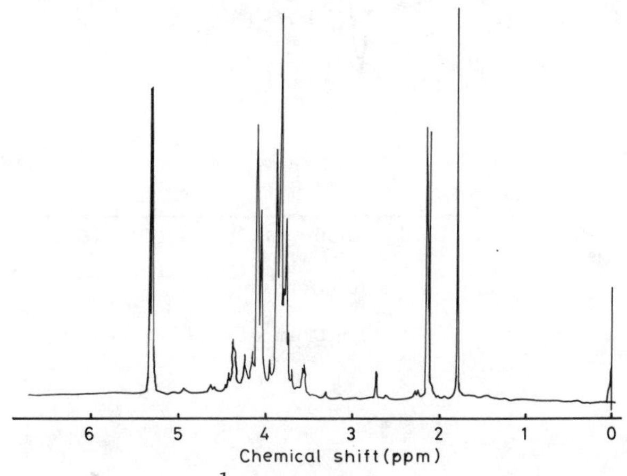

Figure 6. ^{1}H-NMR spectrum of puffer toxin.

Electrophoresis (10)

One microlitre of TTX (10 MU/μL) is applied to a 5 x 18 cm cellulose acetate strip (Chemetron) with 0.08 M Tris buffer (pH 8.7) and electrophoresed at 0.8 mA/cm for 20 min. For detection of TTX the strip is treated as described in "TLC".

TTX moves toward the cathode with a mobility clearly smaller than that of saxitoxin standard (27) which exhibits a blue fluorescent spot under UV light after being treated with alkali.

Capillary Isotachophoresis (28)

Isotachophoresis is carried out using a Shimadzu IR-2A isotachophoretic analyzer with a potential gradient detector. Conditions for the analysis comprise a leading electrolyte of 5 mM potassium acetate (pH 6.0) containing 0.2% Triton X-100 and a half volume of dioxane, and a terminating electrolyte of 10 mM β-alanine adjusted to pH 4.5.

When TTX (2.1 μg) is applied to isotachophoresis, a zone appears at the position with a potential unit value of 0.32. Analytical time is less than 20 min. It is also possible to quantitate the TTX content in crude extracts of toxic pufferfish.

IR Spectrometry (10)

TTX in a KBr pellet is analyzed by a JASCO model IR-S spectrophotometer. The spectrum is complex (3350, 3240, 1670, 1612, 1075 cm^{-1}), but is helpful for identification of the toxin from various sources (Figure 7).

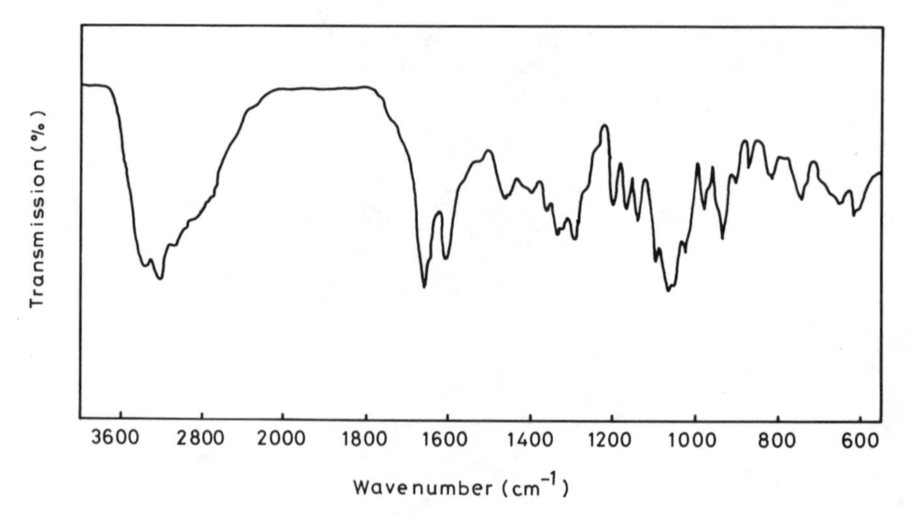

Figure 7. IR spectrum of puffer toxin.

Literature Cited

1. Tani, I. "Toxicological Studies on Japanese Pufferfishes"; Teikoku Tosho: Tokyo, 1945; 103 pp.
2. Hashimoto, Y. "Marine Toxins and Other Bioactive Marine Metabolites"; Japan Scientific Societies Press: Tokyo, 1979; p. 79.
3. Kanoh, S.; Noguchi, T.; Kamimura, S.; Hashimoto, K. J. Food Hyg. Soc. Jpn. 1984, 25, 24-29.
4. Environmental Health Bureau, Ministry of Health and Welfare. "Food Hygiene Inspection Manual"; Food Hygiene Association: Tokyo, 1978; pp. 232-240.
5. Kanoh, S.; Noguchi, T.; Koyama, K.; Hashimoto, K. Bull. Jpn. Soc. Sci. Fish. 1982, 48, 591-592.
6. Brown, M. S.; Mosher, H. S. Science 1963, 140, 295-296.
7. Noguchi, T.; Hashimoto, Y. Toxicon 1973, 11, 305-307.
8. Kim, Y. H.; Brown, G. B.; Mosher, H. S.; Fuhrman, J. Science 1975, 189, 151-152.
9. Sheumack, D. D.; Howden, M. E. H.; Spence, I.; Quinn, R. J. Science 1978, 199, 188-189.
10. Narita, H.; Noguchi, T.; Maruyama, J.; Ueda, Y.; Hashimoto, K.; Watanabe, Y.; Hida, K. Bull. Jpn. Soc. Sci. Fish. 1981, 47,
11. Noguchi, T.; Maruyama, J.; Ueda, Y.; Hashimoto, K.; Harada, T. Bull. Jpn. Soc. Sci. Fish. 1981, 47, 903-913.
12. Yasumoto, T.; Oshima, Y.; Hosaka, N.; Miyakoshi, H. Bull. Jpn. Soc. Sci. Fish. 1981, 47, 929-934.
13. Noguchi, T.; Maruyama, J.; Narita, H.; Hashimoto, K. Toxicon 1984, 22, 219-226.
14. Noguchi, T.; Narita, H.; Maruyama, J.; Hashimoto, K. Bull. Jpn. Soc. Sci. Fish. 1982, 48, 1173-1177.
15. Noguchi, T.; Uzu, A.; Koyama, K.; Maruyama, J.; Nagashima, Y.; Hashimoto, K. Bull. Jpn. Soc. Sci. Fish. 1983, 49, 1887-1892.
16. Yasumoto, T.; Nakamura, M.; Oshima, Y.; Takahata, J. Bull. Jpn. Soc. Sci. Fish. 1982, 48, 1481-1485.
17. Nunez, M. T.; Fischer, S.; Jaimovich, E. Anal. Biochem. 1976, 72, 320-325.
18. Suenaga, K. Jpn. J. Legal Med. 1978, 32, 97-111.
19. Onoue, Y.; Noguchi, T.; Nagashima, Y.; Hashimoto, K.; Kanoh, S.; Ito, M.; Tsukada, K. J. Chromatogr. 1983, 257, 373-379.
20. Onoue, Y.; Noguchi, T.; Nagashima, Y.; Hashimoto, K.; Kanoh, S.; Ito, M.; Tsukada, K. Hitachi Sci. Instr. News 1983, 26, 2220-2223.
21. Benson, J. R.; Hare, P. E. Proc. Nat. Acad. Sci. U. S. 1975 72, 619-622.
22. Roth, M. Anal. Chem. 1971, 43, 880-882.
23. Simons, S. S., Jr.; Johnson, K. F. J. Org. Chem. 1978, 43, 2886-2891.
24. Maruyama, J.; Noguchi, T.; Matsunaga, S.; Hashimoto, K. Agric. Biol. Chem. in press.
25. Suenaga, K.; Kotoku, S. Arch. Toxicol. 1980, 44, 291-297.
26. Knapp, K. R. "Handbook of Analytical Derivatization Reactions"; John Wiley and Sons: New York, 1979; p. 267.
27. Bates, H. A.; Kostriken, R.; Rapoport, H. J. Agric. Food Chem. 1978, 26, 252-254.
28. Shimada, K.; Ohtsuru, M.; Yamaguchi, T.; Nigota, K. J. Food Sci. 1983, 48, 665-667.

RECEIVED June 7, 1984

TOXINS FROM RED TIDE AND CYANOBACTERIA

30

Toxins from Florida's Red Tide Dinoflagellate *Ptychodiscus brevis*

DANIEL G. BADEN[1], THOMAS J. MENDE[1], MARK A. POLI[1], and RONALD E. BLOCK[2]

[1] University of Miami School of Medicine, Department of Biochemistry, Miami, FL 33101
[2] Papanicolaou Cancer Research Institute, Miami, FL 33136

Toxins T17 and T34, isolated from Florida's red tide dinoflagellate Ptychodiscus brevis, cause a dose-dependent response in several in vivo and in vitro biological systems. Potent fractions responsible for in situ fish lethalities, respiratory irritation, and neurotoxic shellfish poisoning have been identified. The identity of potent fractions from three different research groups as well as the effects of chemical derivatization of the fractions on potency has been determined. Potency has been linked to membrane depolarization through endogenous sodium channels and specific binding to excitable membranes has been observed using tritiated toxin.

The marine dinoflagellate Ptychodiscus brevis is the causative organism in Florida's red tides(1). A similar toxic species occurs in Spain(2). The organism is toxic in both the field and in the laboratory(3) and produces both neurotoxic and hemolytic components. Florida red tides have a profound effect on the environment, causing extensive fish kills and destroying other marine life. This is the most visible consequence of exposure to P. brevis toxins. Ichthyolethalities occur as a result of exposure to the neurotoxins produced by the dinoflagellates in the bloom(3), although there is some evidence that hemolytic components may play a part in the observed maladies(4). Hemolytic fractions are not potent in mice(LD_{50}>10 mg/kg (5).

Besides dead fish, which pose enormous economic and sanitation problems, there is an airborne irritant which burns the conjunctivae and mucous membranes and induces persistent non-productive coughing and sneezing(6,7). This irritation is thought to arise from particles of the toxic organism entrapped in seaspray(3).

The neurotoxins may be transmitted to man through bioaccumulation in an intermediate marine host. Toxic bivalves result from the filter-feeding of P. brevis cells during red tides, and if consumed result in neurotoxic shellfish poisoning, or NSP. Human oral intoxication is rarely fatal.

Our laboratory has been interested in the toxinology of the purified toxins and we have recently been able to identify potent frac-

0097-6156/84/0262-0359$06.00/0
© 1984 American Chemical Society

tions responsible for ichthyolethalities and for oral and inhalation intoxication in humans. We have, through collaborative studies, also begun to determine the mechanism of action of these potent compounds in biological systems.

Summary of Previous Work

From our laboratory cultures of *P. brevis*, we are able to purify to homogeneity and crystallinity two toxins, namely T17 and T34(8,9). Both toxins have been subjected to a variety of *in vivo* and *in vitro* test systems in order to ascribe specific actions to each(10-13). The results are summarized in Table I below.

Table I. Comparative Potency of *P. brevis* Toxins T17 and T34

Test System	T17	T34	Reference
In vivo			
Fish(*Gambusia affinis*) (24 hr LC_{50},mole/liter)	1.40×10^{-9}	6.70×10^{-10}	(8)
Mouse(Swiss white)(24 hr LD_{50},mole/Kg)			(9)
intravenous	1.05×10^{-7}	2.20×10^{-7}	
intraperitoneal	1.89×10^{-7}	2.20×10^{-7}	
oral	5.80×10^{-7}	7.40×10^{-6}	
Guinea Pig bronchoconstriction(equivalent to 0.05 mg/Kg acetylcholine) (mole/Kg intravenous)	1.10×10^{-9}	no effect	(11)
In vitro			
Rat Phrenic Nerve Hemidiaphragm, neuromuscular block(IC_{50},mole/liter)			(12)
tetanus	1.10×10^{-10}	1.50×10^{-12}	
twitch	5.00×10^{-9}	4.50×10^{-12}	
Crayfish Giant Axon(ED_{50}, mole/liter)	1.50×10^{-9}	$\sim 10^{-9}$	(14,2)
(maximum depolarization)	30 mV	30 mV	(14)

Both toxins are potent ichthyotoxins. This is not surprising as most investigators use a fish bioassay to indentify potent fractions during purification. Thus, these two toxins, in addition to others which may be present *in situ*, are likely responsible for fish kills during red tides(8). A comparison of potency in mice by three different routes of administration illustrates that only T17 is acutely toxic in an oral sense, indicating it as a likely agent responsible for neurotoxic shellfish poisoning(9). Likewise, only T17 causes a bronchoconstriction in anesthetized guinea pigs, suggesting it as a likely airborne respiratory irritant(11). The differences in potency are currently regarded by us as deriving from their respective lipid solubilities, discussed in the following section. We believe that biological assays should, whenever possible, be conducted using the route of exposure experienced during normal episodes of intoxication.

In the phrenic-nerve hemidiaphragm preparation, both T17 and T34 cause a concentration-dependent increase in resting tension of the muscle, leading to a concentration-related neuromuscular block. The initial contracture observed is inhibited by tetrodotoxin or d-tubocurarine and is enhanced by 4-aminopyridine or cholinesterase inhibitors. Inhibitory concentrations are in the nanomolar to picomolar concentration ranges, with T34 being consistently more potent than T17(12). Using intracellular microelectrode techniques in the rat phrenic-nerve hemidiaphragm neuromuscular junction, T17 was observed to cause a concentration-related increase in miniature endplate potential frequency, accompanied by spontaneous endplate potentials. Persistent sodium channel-mediated nerve depolarization is postulated to be sufficient to inhibit neuromuscular transmission(13), but the increase in spontaneous m.e.p.p. frequency suggested that perhaps neurotransmitter depletion played a role in the development of the neuromuscular block. This possibility was eliminated by electron microscopic examination of blocked neuromuscular junctions. A complete complement of acetylcholine vesicles was observed(12). Radioactive acetylcholine tracer experiments support this indication.

Phrenic-nerve hemidiaphragm preparations were preincubated in modified Krebs solution containing 1.0 x 10^{-6}M [methyl-^{3}H]choline (specific activity= 8 Ci/mmole) and 0.5 mM eserine salicylate for 30 min, followed by a brief rinse in eserine-containing modified Krebs solution but no label. Preparations were then incubated at 37°C in modified Krebs solution containing eserine for 120 min, taking duplicate 0.1 ml samples of the supernatant fluid at 10 min intervals. At 60 min, 1 x 10^{-7}M T17 was added to the bath. Samples taken both before and after T17 addition were subjected to high voltage paper electrophoresis to separate [^{3}H]choline from [^{3}H]acetylcholine. Fractions were visualized with iodine vapor, and were cut out and placed in scintillation vials with 10 ml Aquasol. Radioactivity was estimated using liquid scintillation techniques. Of the total radioactivity present in the preparations following preincubation, T17 caused the release of 15 to 25% of the label. The release was immediate and complete within 10 to 20 min of application. As can be seen from Figure 1, [^{3}H]choline is slowly released from the preparation during the entire experiment, except for the 10 to 20 min following T17 addition. Conversely, [^{3}H]acetylcholine is not released until T17 is added to the bath fluid, and then its release is only transient and does not continue, even in the presence of T17. This pattern of release is consistent with the results we have presented previously, i.e. the transient increase in resting tension of the muscle upon toxin addition. The slowed rate of release of [^{3}H]choline upon T17 addition cannot be explained at this time. However, with 75 to 85% of the [^{3}H]acetylcholine still associated with the preparation at the termination of the experiment, we are certain that neurotransmitter depletion does not take place under these conditions.

The neurotoxic actions of T17 on membrane excitability were examined in squid giant axon initially and in more detail using crayfish giant axon and intracellular microelectrode techniques(14). Detailed studies utilizing T34 are not available due to technical problems associated with its extreme hydrophobicity and resulting difficulties in removing it from Plexiglass testing chambers. The toxins are effective when applied either externally or internally. When applied externally, T17 caused a dose-dependent depolarization and a depression of the amplitudeand rate of rise of the action potential. At 1 x 10^{-6}M, a complete block of excitability occurred. T17-induced depolarization was effectively reversed by 3 x 10^{-7}M tetrodotoxin or 1 x 10^{-3}M external sodium solution. Pretreatment with tetrodotoxin completely protected the axon from T17's effects but depolarization occurred upon removal of tetrodotoxin from solution. Anthopleurin A enhances the effects of T17. These results indicate that T17 activates sodium channels by binding at a site separate from that of tetrodotoxin or Anthopleurin A.

Correlation of Potent Fractions

Several potent fractions isolated from laboratory cultures of *P. brevis* have been prepared by a number of research groups, but due to a lack of standardized nomenclature, the number of toxins produced by the organism is uncertain. The structures for three "brevetoxins" have now been reported(15-17). BTX-B, the major toxin reported by Lin *et al.*(15)(Figure 2a), is thought to be identical to GB-2 isolated earlier by Shimizu(16). Likewise, T34 isolated from our laboratory cultures of the organism(8), is also thought to be identical to GB-2. GB--3, isolated by Chou and Shimizu(17), was identified as Figure 2b and sodium cyanoborohydride reduction of GB-2 yielded a toxin containing an hydroxymethylene function in place of the aldehyde group. ^{1}H-nmr spectra of GB-3 and reduced GB-2 confirmed their identical nature(17). BTX-C, an halogenated brevetoxin(Figure 2c) has also been reported(18). BTX-C is not present in our cultures, nor is it present in the cultures of Shimizu(Y. Shimizu, personal communication). From our cultures of *P. brevis*, we have purified T34 in a yield of 5.6±0.7 mg/10^9cells; T17 in a yield of 1.5±0.3 mg/10^9cells.

Using authentic samples of BTX-B, we subjected it and T34 to chemical reduction akin to that employed by Chou and Shimizu to reduce GB-2 to GB-3. Employing equimolar sodium borohydride in acetonitrile solution at room temperature for 3.5 min, both T34 and BTX-B were reduced to identical fractions which comigrated with T17 in three silica gel thin-layer chromatographic systems, possessed the same ^{1}H-nmr spectra as T17 and GB-3, and were all equipotent in mouse bioassay. Chemical oxidation of BTX-B and T34 using argentic oxide in the presence of sodium cyanide in dry methanol(19) resulted in a single homogeneous but essentially non-toxic fraction upon thin-layer chromatography(Figure 2d)(Table II).

The 100 MHz ^{1}H-nmr spectra of (a)-(d) are virtually identical and (a)-(c) have been reported previously(15,16,18). The ^{1}H-nmr spectra of BTX-B, T34, and GB-2 are identical to one another(2), and the respective spectra of reduced T34, T17 and GB-3 are also identical to one another. Thus, in T17 and T34 spectra, there are seven methyl signals, each corresponding to signals present in GB-3 and GB-2(BTX-B) respectively(16,18). T17 and T34 differ from one another in the lower field region of the ^{1}H-nmr spectrum. In T17(or reduced T34), the aldehyde signal of T34($CDCl_3$) at (δ9.50) is absent and is replaced by a 2 proton singlet(δ4.06). There is a simultaneous shift of the terminal methylene protons in T34(δ6.06, 6.29) to upper fields in T17(δ4.94, 5.10). The remaining olefinic protons in their respec-

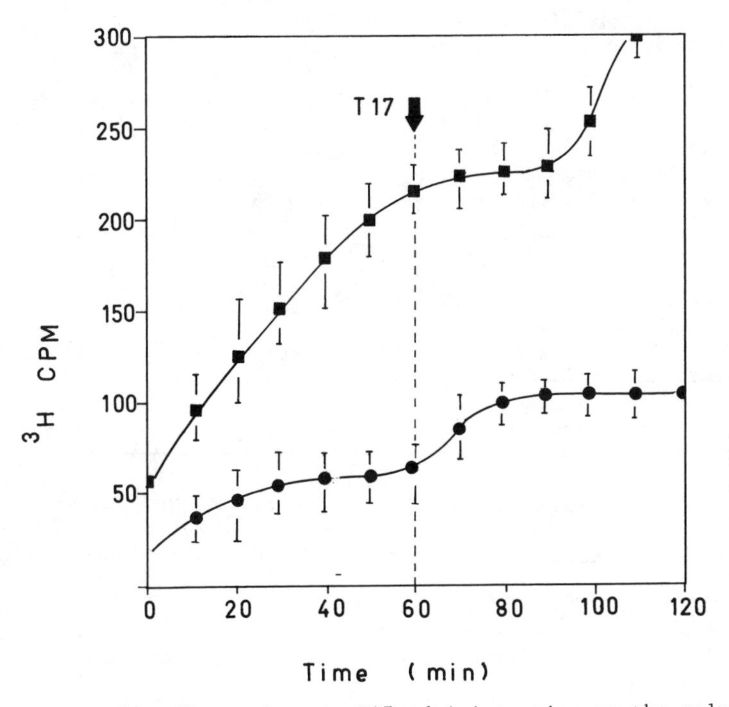

Figure 1. The effect of toxin T17 administration on the release of [^{3}H]choline (■) and [^{3}H]acetylcholine (●) from rat phrenicnerve hemidiaphragm preparations.

Figure 2. The structures of the brevetoxins. Key: a=T34, BTX-B, GB-2; b=T17, GB-3; c=BTX-C; d=oxidized T34.

tive spectra $\delta(CDCl_3)$: 5.71(m,1H), 5.75(m,2H), 6.29(d,1H) do not shift upon reduction of the aldehyde function. The chemical shifts described above reproduce the spectral changes which accompany the reduction of GB-2 to GB-3(18). ^{1}H-nmr spectra of Figure 2d are identical to those of Figure 2a with the exception of the aldehyde proton (δ9.50), which is replaced by an acid proton(broad, δ11.2). In addition, the terminal methylene proton signals in Figure 2a (δ6.08, 6.31) are shifted upfield in Figure 2d (δ4.75,4.93).

Table II.
Silica Gel Thin-Layer Chromatographic Analysis
of
Brevetoxins and Their Derivatives

Compound	System 1	System 2	System 3
T34, BTX-B	0.34	0.79	0.41
T17, reduced T34, reduced BTX-B	0.17	0.66	0.22
Oxidized T34, oxidized BTX-B	0.74	0.82	0.69

system 1= acetone/light petroleum (30/70); system 2= chloroform/methanol/trifluoroacetic acid (100/10/1); system 3= ethyl acetate/light petroleum (70/30).

Therefore, by preparing identical reduction products of BTX-B and T34 and by determining their relative migrations in three TLC systems, we could tentatively conclude that BTX-B and T34 were equivalent compounds. Further, since the ^{1}H-nmr spectrum of reduced T34 is identical to that of GB-3, we could conclude that GB-2 and T34 were identical compounds. The EI mass spectrum of T17 (and reduced T34) at 16 eV also confirms that GB-3 and T17 are identical. Thus, the major fragment ions obtained were: m/z 725(1%), 681(50%), 349 (13%), 291(15%), 109(46%0, and 44(100%). The oxidation of both T34 and BTX-B to the corresponding acid, and determination of comigration in the three TLC systems is a further confirmation of their equivalence.

Potency of Figure 2 a,b,and d in intraperitoneal mouse bioassay illustrates that the character of the chemical substituent at carbon 40 in part determines potency. Thus, T17 (GB-3; LD_{50}=0.17 mg/Kg body weight) is more potent than T34 (BTX-B,GB-2; LD_{50}=0.20 mg/Kg) and both are more potent than oxidized T34 (oxidized BTX-B; LD_{50}=1.50 mg/Kg.

It should be noted that all three are lethal, however, and so the acute toxicity of these compounds is not entirely due to C-40 substituent effects. Potency does follow the oxidation series from alcohol to aldehyde to acid *in vivo*, suggesting that perhaps these substituents influence the degree of accessibility of each lipid-solvent soluble toxin to its membrane site of action. Being that the toxins in their natural forms are so soluble in non-polar solvents, and tend to bind to or solubilize in the lipid components of membrane it is not surprising that the acid derivative is the least potent of the three. The exact mechanism by which these toxins exert their lethal effects is not known with certainty. Based on earlier work of McFarren *et al.*(20), incubation of brevetoxins in dilute alkali totally destroys activity. The potent portion of the molecules may therefore be postulated to reside in the α,β-unsaturated lactone of these toxins and may hence act in a manner similar to digitalis(21). We are continuing chemical modification studies, which will further delineate substituent effects on the potency of this new class of polyether toxins.

The potency of the brevetoxins in biological systems leads us to believe that specific interactions, presumably associated with the voltage-dependent sodium channel, take place. Thus, it may be expected that a specific site or sites on excitable membranes bind(s) the brevetoxins in a concentration-dependent manner. Membrane-binding assays, performed according to the methods of Catterall *et al.* (22), show a specific binding of T17 to rat brain synaptosomes in a dose-dependent manner(Figure 3). The radioactive probe, $[^3H]$T17, was prepared by NaB^3H_4 reduction of T34, and yielded a specific activity of 15-20 Ci/mmole. It was necessary to add an emulsifying agent to the binding medium to prevent micelle formation when T17 reached higher concentrations. The K_D observed (a preliminary observation) with (6.6×10^{-7}M) or without(2.3×10^{-7}M) emulsifier differs by about three-fold, a result we attribute to differences in T17 solubility. Our studies are continuing in this direction to obtain a total Scatchard analysis of brevetoxin binding to syaptosomes and to determine the relationship of its binding to the binding of other sodium channel-perturbing toxins.

Acknowledgments

The authors thank Prof. Koji Nakanishi for his generous gift of BTX-B and to Prof. Yuzuru Shimizu for 500 MHz spectra of GB-2. Work from our laboratory was funded by NIH grant number ES-02299 and ES-02651, DHHS.

Legend of Symbols

LD_{50} lethal dose for 50% of the experimental population.

LC_{50} lethal concentration (fish assay) for 50% of the experimental population.

IC_{50} concentration required to inhibit action (muscle twitch or tetanus strength) to 50% of its control level.

ED_{50} concentration required to effect a response of 50% of the maximal response(squid/crayfish giant axon depolarization).

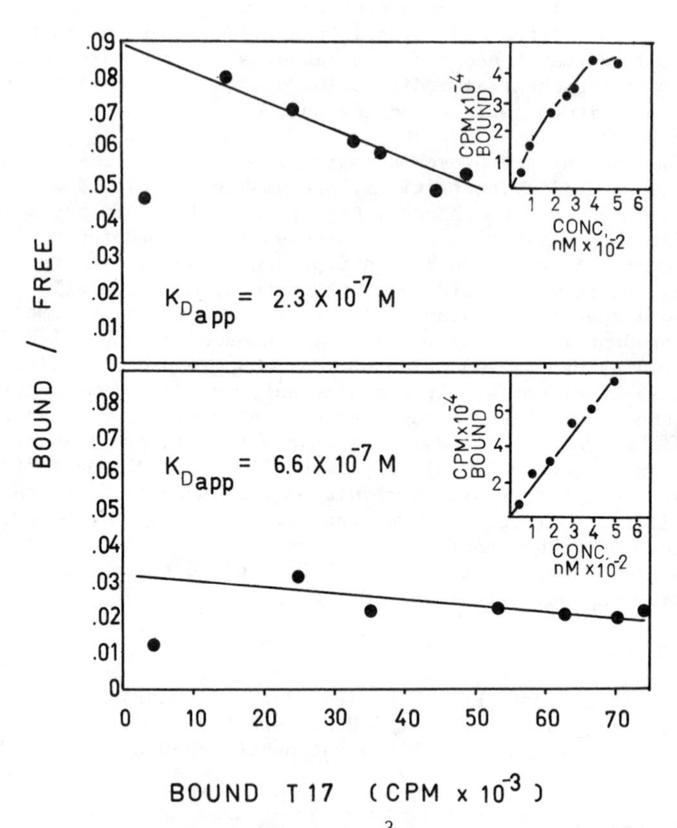

Figure 3. The specific binding of $[^3H]$T17 to rat brain synaptosomes in the absence (top panel) or presence (bottom panel) of emulsifier.

Literature Cited

1. Davis, C.C. Bot. Gaz. 1947, 109, 358.
2. Baden, D.G. Int. Rev. Cytol. 1983, 82, 99.
3. Steidinger, K.A.; Burklew, M.A.; Ingle, R.M. In "Marine Pharmacognosy"; Martin, D.F.; Padilla, G.M., Eds.; Academic Press: New York, 1973; p. 179.
4. Quick, J.A.; Henderson, G.E. Int. Conf. Tox. Dinoflagellate Blooms, 1st, 1975, p. 413.
5. Trieff, N.M.; Ramanujam, V.M.S.; Alam, M.; Ray, S.M. Int. Conf. Tox. Dinoflagellate Blooms, 1st, 1975, p. 309.
6. Music, S.L.; Howell, J.T.; Brumback, C.L. J. Fla. Med. Assoc. 1973, 60, 27.
7. Weech, A.A. J. Fla. Med. Assoc. 1976, 63, 409.
8. Baden, D.G.; Mende, T.J.; Lichter, W.; Wellham, L. Toxicon. 1982, 19, 455.
9. Baden, D.G.; Mende, T.J. Toxicon. 1982, 20, 457.
10. Baden, D.G.; Mende, T.J.; Block, R.E. In "Toxic Dinoflagellate Blooms"; Taylor, D.L.; Seliger, H.H., Eds.; Elsevier:Amsterdam.
11. Baden, D.G.; Mende, T.J.; Bikhazi, G.M.; Leung, I. Toxicon. 1982 20, 929.
12. Baden, D.G.; Bikhazi, G.M.; Decker, S.J.; Foldes, F.F.; Leung, I. Toxicon. in press.
13. Vogel, S.M.; Atchison, W.D.; Narahashi, T. Fed. Proc. Am. Soc. Exp. Biol. 1982, 41, 8487(Abstr.).
14. Huang, J.M.; Wu, C.H.; Baden, D.G. J. Pharm. Exp. Ther. in press.
15. Lin, Y.Y.; Risk, M.; Ray, S.M.; Van Engen, D.; Clardy, J.; Golik, J.; James, J.C.; Nakanishi, K. J. Am. Chem. Soc. 1981, 103, 6773.
16. Shimizu, Y. Pure Appl. Chem. 1982, 54, 1973.
17. Chou, H.N.; Shimizu, Y. Tetrahedron Lett. 1982, 23, 5521.
18. Golik, J.; James, J.C.; Nakanishi, K.; Lin, Y.Y. Tetrahedron Lett. 1982, 23, 2535.
19. Corey, E.J.; Gilman, N.W.; Ganem, B.E. J. Am. Chem. Soc. 1968, 90, 5616.
20. McFarren, E.F.; Tanabe, H.; Silva, F.J.; Wilson, W.B.; Campbell, J.E.; Lewis, K.H. Toxicon 1965, 3, 111.
21. Campbell, S.F.; Danilewiez, J.C. Ann. Rep. Med. Chem. 1978, 13, 92.
22. Catterall, W.A.; Morrow, C.S.; Hartshorne, R.P. J. Biol. Chem. 1979, 254, 11379.

RECEIVED February 6, 1984

Public Health and Toxins from Marine Blue-Green Algae

RICHARD E. MOORE

Department of Chemistry, University of Hawaii, Honolulu, HI 96822

Marine blue-green algae belonging to the Oscillatoriaceae are frequently toxic. Lyngbya majuscula, for example, is the causative agent of a severe contact dermatitis that sometimes affects swimmers and bathers at beaches on the windward side of Oahu, Hawaii during the summer months. In August, 1980 an outbreak of this dermatitis occurred and 86 persons with symptoms were reported to the Hawaii Department of Health. The active principles in the blue-green alga are aplysiatoxin and debromoaplysiatoxin, two highly inflammatory acetogenic substances that were first found in the digestive tract of the sea hare Stylocheilus longicauda. A smaller amount of a third inflammatory substance, lyngbyatoxin A, an indole alkaloid that is structurally related to teleocidin B from Streptomyces mediocidicus, is also present in L. majuscula and is responsible in part for the dermatitis. Aplysiatoxin, debromoaplysiatoxin, and lyngbyatoxin A have been shown to be potent tumor promoters in vivo, comparable in potency and biological effects with 12-O-tetradecanoylphorbol-13-acetate (TPA) from Croton oil. The discovery of the cocarcinogenic properties of these toxins suggests that L. majuscula may be an important public health concern.

Only one marine blue-green alga, Lyngbya majuscula, poses any potential public health concern. This filamentous cyanophyte is the causative agent of a severe contact dermatitis that affects several swimmers and bathers using the windward beaches of the Hawaiian Islands during the summer months (1-3). The active principles have been isolated and identified as two phenolic bis-lactones, aplysiatoxin and debromoaplysiatoxin (4,5), and an indole alkaloid, lyngbyatoxin A (6). All three of these substances have been shown to be potent irritants, producing erythema, blisters and necrosis when applied to the skin (6,7).

0097-6156/84/0262-0369$06.00/0
© 1984 American Chemical Society

The most recent major outbreak of this dermatitis on the windward side of Oahu occurred in August, 1980 at Kailua, Kalama, and Pilapu beaches. A total of 86 cases were reported to the Hawaii State Department of Health. The dermatitis was described as similar to a burn and generally involved the genital and perianal areas. The initial symptoms, which appeared after a few hours, were erythema and a burning sensation, followed by blister formation and deep desquamation which lasted for several days. Aplysiatoxin and debromoaplysiatoxin were identified as the dermatitis-producing agents of this outbreak, since relatively large quantities of both toxins were isolated from specimens of L. majuscula floating in the ocean at the time (8). Debromoaplysiatoxin had already been shown to be present in the L. majuscula that was responsible for an outbreak of seaweed dermatitis at Laie Bay on windward Oahu in 1977 (9).

Lyngbyatoxin A is generally a minor component in dermatitis-producing L. majuscula (10). Lyngbyatoxin A, however, is the major irritant in a variety of L. majuscula growing at Kahala beach near Diamond Head, Oahu. Curiously outbreaks of seaweed dermatitis have not been reported in the Kahala area.

L. majuscula is a common blue-green alga which grows abundantly in many areas of the sub-tropical and tropical Pacific Basin and also in the Caribbean. Outside of the Hawaiian Islands, however, seaweed dermatitis has only been reported in Japan. A large outbreak of skin dermatitis occurred at Gushikawa beach in Okinawa in July, 1968 and affected 242 persons (11,12). Although a blue-green alga had been considered to be one of the possible causes of this mass attack, samples of L. majuscula were not collected and examined at the time of the outbreak and so it is not absolutely certain that L. majuscula was the causative agent. Debromoaplysiatoxin and aplysiatoxin, however, have been shown to be present in L. majuscula growing at Gushikawa beach by Fujiki at the National Cancer Center Research Institute in Tokyo, Japan (private communication).

Structure Determination

Debromoaplysiatoxin and aplysiatoxin were first isolated from the midgut gland of the sea hare Stylocheilus longicauda by Kato and Scheuer in 1974 (4). The two toxins were accompanied by the relatively non-toxic compounds, anhydrodebromoaplysiatoxin and anhydroaplysiatoxin, which Kato and Scheuer found to be the products of mild acid treatment of the corresponding toxins. From elegant chemical and spectral studies, Kato and Scheuer deduced the gross structures for the two toxins and two anhydrotoxins in this four-component mixture without ever isolating any of the pure compounds. Although they strongly suspected that the aplysiatoxins had an algal origin, the dietary source was not determined.

In a search for new antineoplastic agents from blue-green algae, Mynderse isolated a cytotoxic substance that was active against P-388 lymphocytic mouse leukemia from a deep-water variety of L. majuscula growing on the pinnacles in Enewetak Atoll and showed that it was identical with debromoaplysiatoxin (9). Since Kato and Scheuer had not isolated pure debromoaplysiatoxin, comparison of the 20,30-diacetates was made to establish their

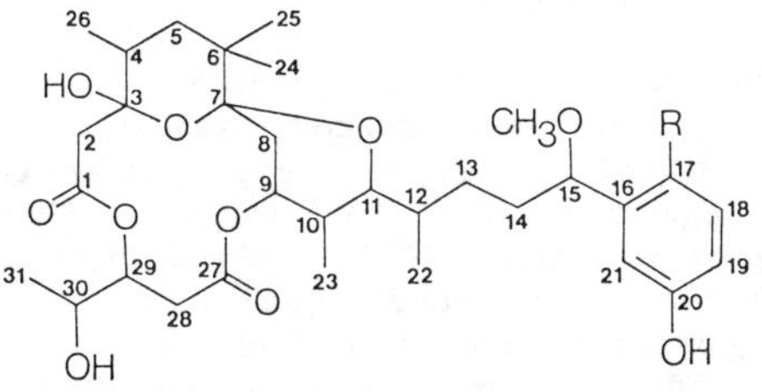

APLYSIATOXIN, R = Br
DEBROMOAPLYSIATOXIN, R = H

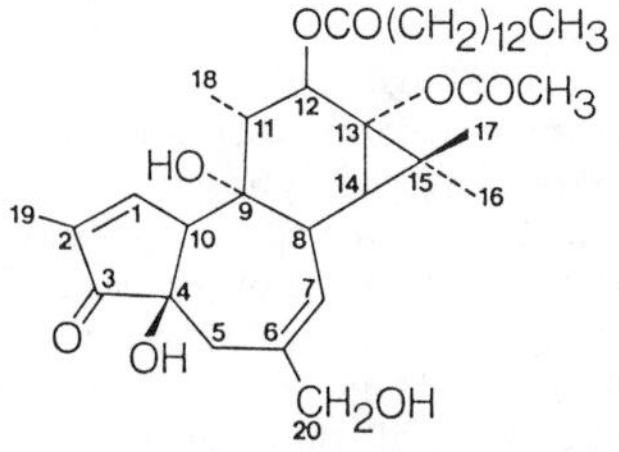

TPA

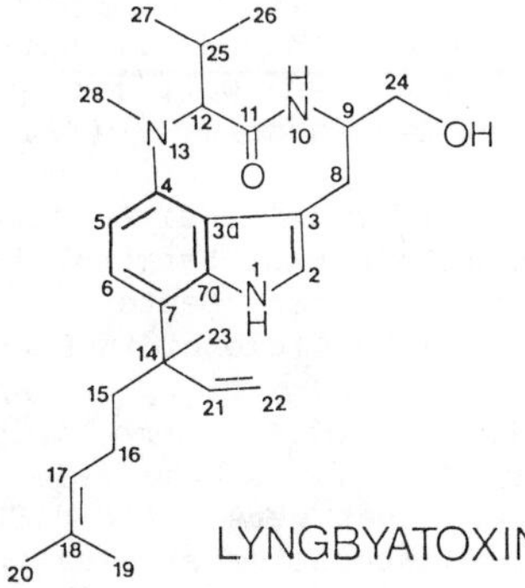

LYNGBYATOXIN A

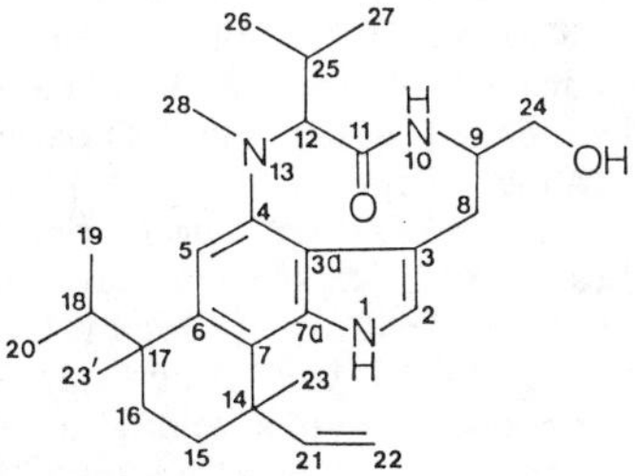

TELEOCIDIN B

identities. Mynderse managed to crystallize debromoaplysiatoxin, but to date it has not been possible to solve its structure by X-ray crystallography. Circular dichroism and difference nOe ^1H NMR studies and chemical degradation, however, indicated that the absolute stereochemistry was probably 3S,4R,7S,9S,10R,11R,12S,15S,-29R,30R (10). Recently 19,21-dibromoaplysiatoxin, formed by bromination of debromoaplysiatoxin in aqueous methanol at pH 6, was crystallized by C. Cheuk in my laboratory and X-ray studies by G. Matsumoto and J. Clardy at Cornell University have now verified the proposed absolute stereochemistry.

Lyngbyatoxin A was first isolated from the variety of *L. majuscula* growing at Kahala beach, Oahu (6). Its structure, which was deduced mainly from spectral data, proved to be similar to that of teleocidin B (13), a highly irritating substance that is produced by certain Streptomyces, e.g. *S. mediocidicus*. These soil fungi had been shown to be responsible for a contact dermatitis affecting workers in the antibiotic industry. Recently Fijiki and Sugimura at the National Cancer Center Research Institute in Japan have found that one of the two components of teleocidin A, another dermatitis-producing agent from these *Streptomyces*, is identical with lyngbyatoxin A. The absolute stereochemistries of the teleocidins and lyngbyatoxin A are unknown at this writing; however, optical studies indicate that the absolute configuration of the nine-membered lactam ring is the same in both the fungal and algal toxins (6).

Tumor-Promoting Properties

In two-stage chemical carcinogenesis, certain chemical agents (tumor promoters), which alone do not cause cancer, amplify the development of tumorous cells from cells that have been initially exposed to a single, subcarcinogenic dose of carcinogen. Skin tumors on mice can be made to appear in about 10-12 weeks following a single sub-carcinogenic application of carcinogen with twice weekly applications of a tumor promoter. Tumors are not formed if the order of treatment is reversed, i.e. months of twice weekly applications of tumor promoter are followed by a onetime application of a subcarcinogenic dose of carcinogen. Unlike carcinogens which act directly on the cellular DNA, tumor promoters exert their effects by binding to receptors associated with the cell membrane (14). These receptors somehow control cell growth and differentiation, for some cells can be induced to proliferate while others are induced to differentiate on treatment with very small quantities of tumor promoter. The most well known tumor promoter is 12-O-tetradecanoylphorbol-13-acetate (TPA), a deterpenoid ester from Croton oil.

About four years ago phorbol esters and related diterpenes were the only class of tumor promoters known to act at hormonal levels. Fujiki and Sugimura at the National Cancer Center Research Institute in Tokyo, Japan then discovered that teleocidin B and its dihydro derivative were powerful tumor promoters (15), acting also at hormonal levels and sharing with TPA many of the same biological effects. This was a very interesting finding as the teleocidins were indole alkaloids and had structures that were quite different from the phorbol esters. Subsequent collaborative studies between our laboratory and the Japanese established that lyngbyatoxin A was also a potent tumor promoter (16) and furthermore that debromoaplysiatoxin and aplysiatoxin belonged to a third class of powerful tumor promoters which behaved like the phorbol esters and teleocidin alkaloids (17).

Lyngbyatoxin A, debromoaplysiatoxin, and aplysiatoxin induce irritancy in mouse skin to the same degree as TPA (16,17). Significant reddening of mouse ear skin is observed after 24 hours when 0.1 nmol of toxin is applied. Each compound exhibits a potency similar to that of TPA in inducing ornithine decarboxylase (ODC) activity in dorsal mouse skin. Increased ODC activity is characteristic of fast-growing neoplasms and uptake of putrescine by tumorous cells is much more rapid compared with normal cells (18). Each toxin shows the same effects as TPA in several cell culture systems, including stimulation of arachidonic acid release, prostaglandin production, and choline turnover (19,20); formation of superoxide anions and hydrogen peroxide (21); induction of Epstein-Barr virus expression and enhancement of Epstein-Barr virus-induced transformation (22); induction of terminal differentiation and adhesion of HL-60 cells (17,23); aggregation of human lymphoblastoid cells (22); inhibition of terminal differentiation of Friend erythroleukemia cells (16); aggregation of NL-3 cells (17); stimulation of 2-deoxyglucose transport; enhancement of transformation by adenovirus; enhanced cloning efficiency of adenovirus-transformed cells; inhibition of melanogenesis in B16 cells and inhibition of myogenesis in human myoblast cultures (24);

stimulation of immune interferon, T cell growth factor (interleukin-2), and lymphotoxin production in human peripheral blood lymphocytes (25); and inhibition of binding of [^{125}I]EGF (epidermal growth factor) to its membrane receptors in mouse and rat embryo cell lines (20). Lyngbyatoxin A, debromoaplysiatoxin, and aplysiatoxin have also been found to inhibit the specific binding of [^{3}H]-phorbol-12,13-dipropionate (26) and [^{3}H]-phorbol-12,13-dibutyrate (20) to membrane-associated cellular receptors at potencies similar to those of TPA. These findings provide evidence that phorbol esters, teleocidin alkaloids, and aplysiatoxin bis-lactones have certain structural features in common which enable them to bind to the same receptors, thereby producing similar biological responses in the cell.

Recent studies suggest that the phorbol ester, teleocidin, and aplysiatoxin tumor promoters operate by activating a phospholipid and calcium ion dependent phosphorylating enzyme, protein kinase C (27-30). The activity of protein kinase C is also stimulated by unsaturated diacylglycerol. Since these tumor promoters induce increased turnover of membrane phospholipids, diacylglycerol may be generated to further increase protein kinase C activity. It has been suggested that the putative endogenous analog of these tumor promoters might actually be a diacylglycerol (31) and that protein kinase C may be a receptor for the tumor promoters or at least a component of the receptor complex (31). When intact cells are treated with the tumor promoter, protein kinase C appears to move from the cytosol to the cell membranes. Whether the tumor promoter is bound to protein or phospholipid in the apparent quaternary complex of phospholipid, protein kinase C, calcium ion, and tumor promoter has not been established unambiguously.

To date the evidence seems to favor the binding of tumor promoter to phospholipid in the cell membrane. Specific binding of [^{3}H]TPA to mouse epidermal particulate matter is susceptible to phospholipases C and A_2, less susceptible to protease, and completely resistant to glycosidase (32). Photoaffinity labelling studies with [20-^{3}H]-phorbol 12-p-azidobenzoate 13-benzoate indicates that the irreversible binding of this photolabile phorbol ester to mouse brain membrane is predominantly to the phospholipid (specifically phosphatidylethanolamine and phosphatidylserine) portion rather than to the protein portion (33).

It is interesting that the digestive tracts of sea hares and fish that feed upon L. majuscula are not affected by the aplysiatoxins and lyngbyatoxin A. Preliminary studies by Fujiki suggest that the cells of sea hares may lack the tumor promoter-binding receptors that are so common in mammalian cells (34).

X-ray structural data are available for TPA (35) and dihydroteleocidin B (36-37), but only the absolute stereochemistry of TPA is known. As mentioned above the structure of aplysiatoxin has been solved by X-ray crystallography and its absolute stereochemistry has been determined from other data (10). These three tumor promoters, which represent three distinct classes of natural products, appear to have common structural features that enable it to bind to the same membrane receptors.

Evidence in the literature (38-40) indicates that the α,β-unsaturated keto group at C-3, the primary allylic hydroxyl at C-20, the tertiary hydroxy group at C-4, and the long-chain ester group at C-12 are needed for the high activity of TPA. The Weinstein group at Columbia University has used computer graphic analysis to compare the three-dimensional structures of TPA and dihydroteleocidin B (41). Their best fit is obtained when the absolute configurations of the two amino acid residues in dihydroteleocidin B are both D. In their model the C-11 carbonyl, N-13, N-1, and the OH on C-24 in dihydroteleocidin occupy very similar positions in space with the C-3 carbonyl, the OH at C-4, the OH at C-9, and the OH on C-24 in TPA, respectively; the monoterpenoidal portion of dihydroteleocidin B and the C_{14} ester group of TPA, which are essential for effective binding of these promoters to the hydrophobic regions of the receptor, are also in similar positions.

Possible Human Intoxication

It is not clear if the toxins associated with *L. majuscula* enter the human food chain. Hashimoto had observed rabbitfish (*Siganus fuscescens*) feeding on sea grasses entangled with *L. majuscula* during his investigation of seaweed dermatitis in Okinawa and wondered if there could be any connection between the toxicity of *L. majuscula* and human intoxication in the Ryukyus Islands resulting from ingestion of the viscera of rabbitfish (11). No follow-up study, however, was made.

There is a recent case report of a local resident of Hawaii who inadvertently attempted to orally ingest *L. majuscula* (42). Upon placing the alga in his mouth, he noted an instant burning sensation and several hours later the mucous membranes of the anterior portion of his mouth appeared as if they had been scalded. The discomfort persisted for three days, but after two weeks all manifestations had completely disappeared. Interestingly, *L. majuscula* grossly resembles the edible, filamentous green alga *Enteromorpha prolifera*, known as limu'ele'ele to the Hawaiians (43).

During our collections we have noticed that *L. majuscula* is frequently entangled with other seaweeds, some of which are edible. One wonders if some of the seaweed that is eaten by the Hawaiians could be contaminated by small amounts of *L. majuscula*. Unlike in Japan where increased urbanization has resulted in a decrease in stomach cancer, the Hawaiian race continues to show the highest incidence of gastrointestinal cancer in the world. The Hawaiians eat at least two seaweeds that contain carcinogenic and mutagenic halogencontaining compounds, viz. *Asparagopsis taxiformis* (Hawaiian name: limu kohu) (44,45) and *Laurencia nidifica* (Hawaiian name: limu mane'ono'o) (46,47). In their diet alone the Hawaiians may be getting all of the necessary agents for the initiation and promotion of gastrointestinal cancer. Epidemiological studies, however, are needed to evaluate the actual role of seaweed diet in the incidence of stomach cancer in Hawaii (48).

To date there is no evidence that the toxins of *L. majuscula* are involved in the development of human cancer. Certainly frequent ingestion of edible seaweeds that are contaminated with *L. majuscula* could increase the probability of gastrointestinal cancer. There is precedence for this. The black and Creole population of the Caribbean island of Curacao suffers from an exceedingly high rate of esophageal cancer which appears to be related to the daily intake of a tea prepared from the leaves of a bush Croton flavens. Analysis of the leaf extract shows the presence of diterpene di- and triesters that are structurally related to TPA (49). The diesters exhibit strong tumor promoting activity in dorsal mouse skin. Each cup of tea contains more tumor promoter than is required to maintain chronic irritation of the human esophagus, a necessary requirement for the promotion of esophageal cancer.

Acknowledgment

Research in the author's laboratory was supported by PHS Grant CA12623, awarded by the National Cancer Institute, Department of Health and Human Services.

Literature Cited

1. Grauer, F. H. Hawaii Med. J. 1959, 19, 32.
2. Banner, A. H.; Scheuer, P. J.; Sasaki, S.; Helfrich, P.; Alender, C. B. Ann. N.Y. Acad. Sci. 1960, 90, 770.
3. Grauer, F. H.; Arnold Jr,. H. L. Arch. Dermatol. 1961, 84, 720.
4. Kato, Y.; Scheuer, P. J. Pure Appl. Chem. 1975, 41, 1.
5. Kato, Y.; Scheuer, P. J. Pure Appl. Chem. 1976, 48, 29.
6. Cardellina II, J. H.; Marner, F-J.; Moore, R. E. Science 1979, 204, 193.
7. Solomon, A. E.; Stoughton, R. B. Arch. Dermatol. 1978, 114, 1333.
8. Serdula, M.; Bartolini, G.; Moore, R. E.; Gooch, J.; Wiebenga, N. Hawaii Med. J. 1982, 41, 200.
9. Mynderse, J. S.; Moore, R. E.; Kashiwagi, M.; Norton, T. R. Science 1977, 196, 538.
10. Moore, R. E. Pure Appl. Chem. 1982, 54, 1919.
11. Hashimoto, Y.; Kamiya, H.; Yamazato, K.; Nozawa, K. In "Animal, Plant, and Microbial Toxins"; Ohsaka, A.; Hayashi, K.; Sawai, Y., Eds.; Plenum: New York, 1976; Vol. 1, pp. 333-338.
12. Hashimoto, Y. In "Marine Toxins and Other Bioactive Marine Metabolites"; Japan Scientific Societies Press, Tokyo, 1979; pp. 210-215.
13. Nakata, H.; Harada, H.; Hirata, Y. Tetrahedron Lett. 1966, 2515.
14. Driedger, P.; Blumberg, P. Proc. Natl. Acad. Sci. USA 1980, 77, 567.
15. Fujiki, H.; Suganuma, M.; Matsukura, N.; Sugimura, T.; Takayama, S. Carcinogenesis 1982, 3, 895.
16. Fujiki, H.; Mori, M.; Nakayasu, M.; Terada, M.; Sugimura, T.; Moore, R. E. Proc. Natl. Acad. Sci. USA 1981, 78, 3872.
17. Fujiki, H.; Suganuma, M.; Nakayasu, M.; Hoshino, H.; Moore, R. E.; Sugimura, T. Gann 1982, 73, 495.
18. Volkow, N.; Flamm, E.; Goldman, S.; Cravioto, H.; Wolf, A.; Brodie, J. Science 1983, 221, 673.
19. Sakamoto, H.; Terada, M.; Fujiki, H.; Mori, M.; Nakayasu, M.; Sugimura, T. Weinstein, I. B. Biochem. Biophys. Res. Commun. 1981, 102, 100.
20. Horowitz, A.; Fujiki, H.; Weinstein, I. B.; Jeffrey A.; Okin, E.; Moore, R. E.; Sugimura, T. Cancer Res. 1983, 43, 1529.
21. Goldstein, B.; Witz, G.; Amoruso, M.; Stone, D.; Troll, W. Cancer Letters, 1981, 11, 257.

22. Eliasson, L.; Kallin, B.; Patarroyo, M.; Klein, G.; Fujiki, H.; Sugimura, T. Int. J. Cancer 1983, 31, 7.
23. Nakayasu, M.; Fujiki, H.; Mori, M.; Sugimura, T.; Moore, R. E. Cancer Letters 1981, 12, 271.
24. Fisher, P.; Miranda, A.; Mufson, A.; Weinstein, L.; Fujiki, H.; Sugimura, T.; Weinstein, I. B. Cancer Res. 1982, 42, 2829.
25. Yip. Y. K.; Kelker, H. C.; Stone-Wolff, D. S.; Pearlstein, K.; Urban, C.; Vilcek, J. Cellular Immunology, in press.
26. Schmidt, R.; Adolf, W.; Marston, A.; Roeser, H.; Sorg, B.; Fujiki, H.; Sugimura, T.; Moore, R. E.; Hecker, E. Carcinogenesis, 1983, 4, 77.
27. Castagna, M.; Takai, Y.; Kaibuchi, K.; Sano, K., Kikkawa, U.; Nishizuka, Y.; J. Biol. Chem. 1982, 257, 7847.
28. Niedel, J.; Kuhn, L.; Vandenbark, G. Proc. Natl. Acad. Sci. USA, 1983, 80, 36.
29. Kraft, A. S.; Anderson, W. B. Nature 1983, 301, 621.
30. Ashendel, C. L.; Staller, J. M.; Boutwell, R. K. Biochem. Biophys. Res. Commun. 1983, 111, 340.
31. Weinstein, I. B. Nature 1983, 302, 750.
32. Esumi, M.; Fujiki, H. Biochem. Biophys. Res. Commun. 1983, 112, 709.
33. Delclos, K. B.; Yeh, E.; Blumberg, P. Proc. Natl. Acad. Sci. USA 1983, 80, 3054.
34. Nagle, D. S.; Jaken, S.; Castagna, M.; Blumberg, P. M. Cancer Res. 1981, 41, 89.
35. Brandl, F.; Rohrl, M.; Zechmeister, K.; Hoppe, W. Acta Cryst. B 1971, 27, 1718.
36. Harada, H.; Sakabe, N.; Hirata, Y. Bull. Chem. Soc. Japan 1966, 39, 1773.
37. Sakabe, N.; Harada, H.; Hirata, Y. Tetrahedron Lett. 1966, 2523.
38. Hecker, E. In "Carcinogenesis: Mechanisms of Tumor Promotion and Cocarcinogenesis"; Slaga, T. J.; Sivak, A.; Boutwell, R. K., Eds.; Raven Press: New York, 1978; Vol. 2, pp. 11-48.
39. Van Duuren, B. L.; Tseng, S. S.; Segal, A.; Smith, A. C.; Melchionne, S.; Seidmann, I. Cancer Res. 1979, 39, 2644.
40. Yamasaki, H.; Weinstein, I. B.; Van Duuren, B. L. Carcinogenesis 1981, 2, 537.
41. Weinstein, I. B.; Horowitz, A. D.; Jeffrey, A. M.; Ivanovic, V. In "Genes and Proteins in Oncogenesis"; Weinstein, I. B.; Vogel, H. J., Eds.; Academic Press: New York, 1983; pp. 99-109..
42. Sims, J.; Zandee Van Rilland, R. Hawaii Med. J. 1981, 40, 243.
43. Abbott, A.; Williamson, E. H. Limu: An Ethnobotanical Study of Some Edible Hawaiian Seaweeds. Pacific Tropical Botanical Garden, Lawai, Kauai, Hawaii, 1974.
44. Burreson, B. J.; Moore, R. E.; Roller, P. P. J. Agric. Food Chem. 1976, 24, 856.
45. Moore, R. E. Accounts Chem. Res. 1977, 10, 40.
46. Waraszkiewicz, S. M.; Sun, H. H.; Erickson, K. L. Tetrahedron Lett. 1976, 3021.
47. Sun, H. H.; Waraszkiewicz, S. M.; Erickson, K. L.; Tetrahedron Lett. 1976, 4227.
48. Kolonel, L. N.; Nomura, A. M. Y.; Hinds, M. W.; Hirohata, T.; Hankin, J. H. Cancer Res. 1983, 43, 2397.
49. Hecker, E. Submitted for publication.

RECEIVED March 22, 1984

32

Toxins from Freshwater Cyanobacteria

WAYNE W. CARMICHAEL and NIK A. MAHMOOD

Department of Biological Sciences and Biomedical Sciences Program, Wright State University, Dayton, OH 45435

Marine and freshwater cyanobacteria produce exotoxins. The main genera responsible for freshwater toxic blooms are Microcystis, Anabaena, Aphanizomenon and Oscillatoria. Toxins produced include: 1. anatoxins, alkaloids and peptides of Anabaena; 2. the peptide microcystin and related peptides of Microcystis; 3. aphantoxins, compounds of Aphanizomenon with properties similar to some paralytic shellfish poisons. Properties of Oscillatoria toxin suggest they are peptides similar to those of Microcystis. Microcystis toxins are peptides (M.W. approx. 1200) which contain three invariant D-amino acids, alanine, erythro-β-methyl aspartic and glutamic acids, two variant L-amino acids, N-methyl dehydroalanine and a β-amino acid. Individual toxic strains have one or more multiples of this peptide toxin. The one anatoxin characterized is a bicylic secondary amine called anatoxin-a (M.W. 165). The aphantoxin isolated in our laboratory contains two main toxic fractions. On TLC and HPLC the fractions have the same characteristics as saxitoxin and neosaxitoxin.

Toxins are produced by several planktonic bloom forming genera of freshwater cyanobacteria. The toxins are primarily a threat to domestic and wild animals which drink from the shores of ponds, lakes and other waters where the surface accumulation of the bloom allows a localized concentration of toxin both in the cyanobacteria cells and in the surrounding water. These surface accumulations of toxins and cells are a particular concern in late spring, summer and fall in the eutrophic water bodies of both southern and northern temperate latitudes. Toxic cyanobacteria were recognized about 100 years ago (1) following death of domestic animals drinking surface blooms containing toxic cells. Since that time research has focused on occurrence and distribution of toxic blooms in relation to animal death. Direct human contact with toxic blooms has been rare and the present threat to humans is through drinking water supplies (2), recreational water (3) and the increasing use

0097-6156/84/0262-0377$06.00/0
© 1984 American Chemical Society

of cyanobacteria as a source of single cell protein (4). Since cyanobacteria toxins have fewer vectors than other phyco or mycotoxins, by which they can enter the human body they have not received the same amount of research effort. While they are not considered to be seafood toxins there are several points of similarity making it useful to consider them along with these other toxins. These include:

1) Both seafood toxins and cyanobacteria toxins are water-based diseases.
2) Both groups of toxins are produced by phytoplankton.
3) While retained within the cells to varying degrees both groups are exotoxins.
4) The toxins produced are fast acting neuro or organ toxins absorbed via the oral route.
5) Certain cyanobacteria toxins have structural/functional similarities to certain paralytic shellfish toxins, especially saxitoxin and neosaxitoxin.

Types of Toxic Cyanobacteria

Toxic cyanobacteria are unicellular or filamentous in morphology. Early researchers reported about nine genera of cyanobacteria considered responsible for toxic water blooms (5,6). The list of confirmed toxic genera, as verified by laboratory culture of toxic isolates is not as large. Toxic strains of *Anabaena flos-aquae*, *Aphanizomenon flos-aquae*, *Microcystis aeruginosa*, *Oscillatoria agardhii* and *Oscillatoria rubescens* are maintained in laboratory culture collections (7,8). These toxic species can occur worldwide with the following countries reporting toxic blooms for the 1979-1983 period: Australia, Bangladesh, Canada, India, Israel, Japan, New Zealand, South Africa, United States, Union of Soviet Socialist Republics and 11 European countries (7,9). Only a few of these countries have correlated and confirmed animal losses, human contact irritations or gastroenteritis with the presence of these toxic water blooms. There is a need for a greater awareness of the occurrence of toxic cyanobacteria and their role in animal and human water based diseases.

Types of Toxins

Peptide toxins. Of all the toxins produced by freshwater cyanobacteria, the peptide toxins of *Microcystis aeruginosa* have received the most attention. All research on these peptide toxins indicates they are small, possibly cyclic, with molecular weights estimated at 1200 to 2600 (10,11). Recent work has become more definitive in the estimation of molecular weight and amino acid composition. In 1978 Elleman et al. (12) reported that they had isolated and characterized the peptide toxin of an Australian strain of *Microcystis* which was a pentapeptide with a minimum molecular weight of 654. It consisted of equimolar amounts of alanine, tyrosine, methionine, glutamic and β-methyl aspartic acid and methylamine.

Two other research groups have come to similar conclusions about the peptide toxins from *Microcystis*. Eloff, Siegelman and

Kycia (13) used freeze dried cells of 13 strains from Canada, Israel, South Africa and the United States. The toxins were purified by solvent extraction, gel filtration and high performance liquid chromatography. They found similar toxins present in the various strains but certain amino acids were different. In all cases the toxin was a pentapeptide containing β-methyl aspartic acid, glutamic acid, and alanine with methylamine attached to the glutamic acid. These three amino acids could be found as repeating monomers, up to seven times in one of the strains, but usually one to three times. If more than three units were present one or two of the toxic units accounted for >90% of the toxic activity. The two additional amino acids present occurred as pairs consisting of: leucine and arginine, leucine and alanine, tyrosine and arginine, methionine and arginine, leucine and tyrosine, alanine and tyrosine or arginine and arginine. The peptide containing leucine and arginine was present in nine of the thirteen toxic isolates.

Botes, et al. (14) isolated and analyzed two toxic Microcystis from South Africa. All toxic monomers were pentapeptides with one residue each of D-glutamic acid, D-alanine and erythro-β-methyl aspartic acid. Methylamine was also present and thought to be a hydrolytic breakdown product of N-methyl dehydroalanine (15). One of the Microcystis isolates cultured in the laboratory contained four toxic monomers labeled BE-2 to BE-5. Configuration assignments of the amino acids present confirmed that the invariant amino acids were all in the D form while the variant amino acids were in the L form. The variant L-amino acids for the monomers were Arg and Leu for BE-2, Arg and Tyr for BE-3, Ala and Leu for BE-4 and Ala and Tyr for BE-5 (15). The Australian toxin (12) had the same invariant amino acids and L-Met and Tyr for the variant amino acids. N-methyldehydroalanine was also found in the BE-2 to BE-5 monomers but was not described for the Australian toxin. An apolar side chain of about 20 C-atoms was present in BE-2 to BE-5 but was not reported in the Australian toxin. The N-methyldehydroalanine and the 20-carbon side chain were postulated as contributing to the high hydrophobicity of the toxin and suggests a possible mechanism of action since it should have a good affinity for cell membranes. In a later publication (16) the BE-4 monomer was analyzed by fast atom bombardment mass spectrometry. The authors concluded that the toxin has a molecular weight of 909 with an estimated molecular weight for the 20-carbon side chain of 313. It was also felt that the toxin was not a linear peptide as reported earlier (14,15) but a monocyclic peptide. The 20-carbon side chain was thought to be a novel β-amino acid.

Other Peptide Toxin Producing Cyanobacteria. No other cyanobacteria genera have been analyzed for the presence of toxic peptides. Based on toxicity signs however it is thought that peptide toxins are present in certain isolates of *Anabaena flos-aquae* (7) and *Oscillatoria agardhii* (9,17).

Alkaloid Toxins. Only one alkaloid toxin has been chemically defined from the cyanobacteria. This is the secondary amine, 2-acetyl-9-azabicyclo (4-2-1) non-2-ene, called anatoxin-a. It is isolated from the filamentous strain *Anabaena flos-aquae* NRC-44-1 (18,19).

Synthesis of antx-a has been done through a ring expansion of cocaine (20), from 1, 5-cyclooctadiene (21) and by intramolecular cyclization between an iminium salt and a nucleophilic carbon atom (22). The structure of antx-a is given in Fig. 1.

Two other suspected alkaloid producing cyanobacteria strains, Anabaena flos-aquae NRC-525-17 and Aphanizomenon flos-aquae NH-5, are now being studied. The toxin of An. flos-aquae NRC-525-17 (anatoxin-a(s)) is thought to have CNS stimulating properties (7) and that of Aph. flos-aquae NH-5 (aphantoxin) is thought to produce the paralytic shellfish poisons saxitoxin and neosaxitoxin (Fig. 1) (7,23). Earlier research had already suggested that certain blooms of Aph. flos-aquae could produce paralytic shellfish poisons. These studies used water blooms collected from Kezar Lake, New Hampshire (25,30). In 1980 Carmichael isolated a neurotoxic strain of Aph. flos-aquae from a small pond in New Hampshire. These strains have also been shown to produce toxins similar to saxitoxin and neosaxitoxin (23) and are the ones used in the studies presented here. In our laboratory crude preparations of aphantoxins and anatoxin-a(s) are extracted similarly except at the final stages of purification (Fig. 2). A Bio-gel P-2 column (2.2 x 80 cm) is used for aphantoxins gel filtration and a Sephadex G-15 (2.6 x 42 cm) column for anatoxin-(s). Both toxins are eluted with 0.1 M acetic acid at 1.5 ml/min. Fractions of aphantoxins from Bio-gel P-2 run are spotted on thin-layer chromatography plates (Silica gel-60, EM reagents) and developed according to Buckley et al. (1976) (31). The Rf values for the aphantoxins, saxitoxin and neosaxitoxin standards (Table 1) indicates that two of the aphantoxins (i.e. I and II) are similar to saxitoxin and neosaxitoxin.

The first toxic peak(s) from the Bio-gel P-2 run (volume reduced to 1 ml) is passed through a preparative mini column (Sep-pak C_{18}) to separate the toxin(s) from yellowish pigments. The eluant is then subjected to HPLC using a semipreparative column (CN bonded phase 9.4 mm x 25 cm). Fig. 3 illustrates the presence of neosaxitoxin (second last peak) and saxitoxin (last peak). A total of 150 mu (mouse units) was loaded. The profile in Fig. 4 shows only neosaxitoxin (500 mu) is present because only a portion of the first toxic peak(s) from the P-2 run was injected. The bottom profile is that of a standard neosaxitoxin (200 mu).

The HPLC elution pattern of anatoxin-s(s) is shown in Fig. 5 and is the last peak in the profile. Determination of LD_{50} gave an approximate value of 40 μg/kg body weight, which is five times more potent than anatoxin-a.

Types of Toxic Effects

Organtoxic. Of all the cyanobacteria toxins, those of the colonial Microcystis aeruginosa are the most widespread. They have caused the death of livestock and wild animals in several countries of the world. While reports on toxicity of Microcystis have some conflicting signs of poisoning the consistent pathological findings include swollen blood-engorged liver with hemorrhagic necrosis and mildly edematous lungs. The variation in other signs, such as survival time, can be attributed to the dosage, animal species, age and condition of the bloom, presence of other toxic or nontoxic cyano-

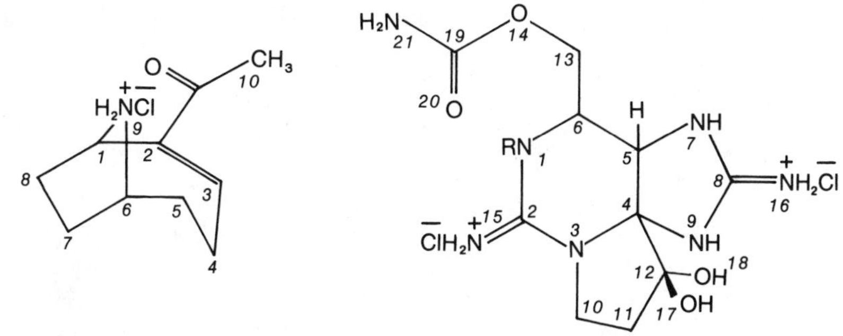

anatoxin - a hydrochloride

R=H; saxitoxin dihydrochloride
R=OH; neosaxitoxin dihydrochloride

Figure 1. Structure of anatoxin-a from *Anabaena flos-aquae* NRC-44-1, saxitoxin and neosaxitoxin. Saxitoxin and neosaxitoxin is produced by certain species of marine algae and by the fresh-water cyanobacteria *Aphanizomenon flos-aquae* NH-5.

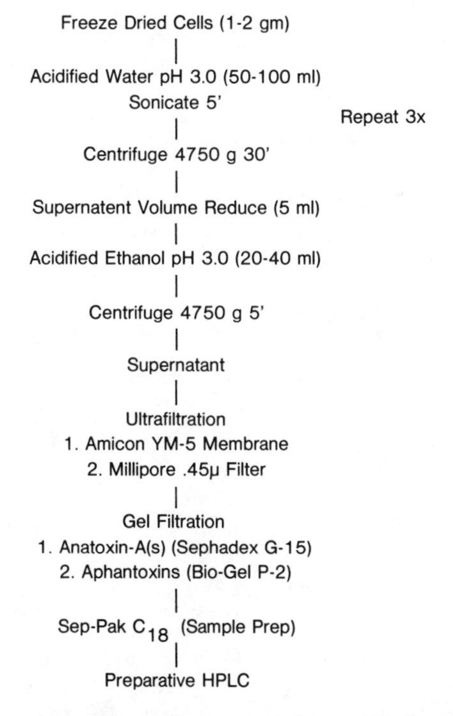

Figure 2. Flow diagram for the extraction of Anatoxin-a(s) and Aphantoxins.

TABLE 1. Thin-layer chromatography of Aphanizomenon flos-aquae NH-5 fractions from Bio-gel P-2 gel filtration

Fractions	Flourescence[+]	RF
Aphantoxin I[a]	Yellow-green	.74
Aphantoxin II[a]	Blue	.69
Aphantoxin III[b]	Blue	.62
Aphantoxin IV[b]	Blue	.77
Saxitoxin[a,c]	Blue	.68
Neosaxitoxin[a,c]	Yellow-green	.73

Note: The solvent was pyridine:ethyl acetate:acetic acid:water (75:25:15:30) (24)

[+]Spots are detected by spraying with 1% Hydrogen peroxide. Heated 110°C 15 min. and observed under 366 nm UV.

[a]LD_{50} i.p. mouse; 10 μg/kg

[c]Provided by Dr. Sherwood Hall, Woods Hole Oceanographic Institute.

[b]Possibly Aphantoxin III is the B_1 toxin of Hall, 1980 (24); and Aphantoxin IV is gonyautoxin II of Alam, 1978 (27).

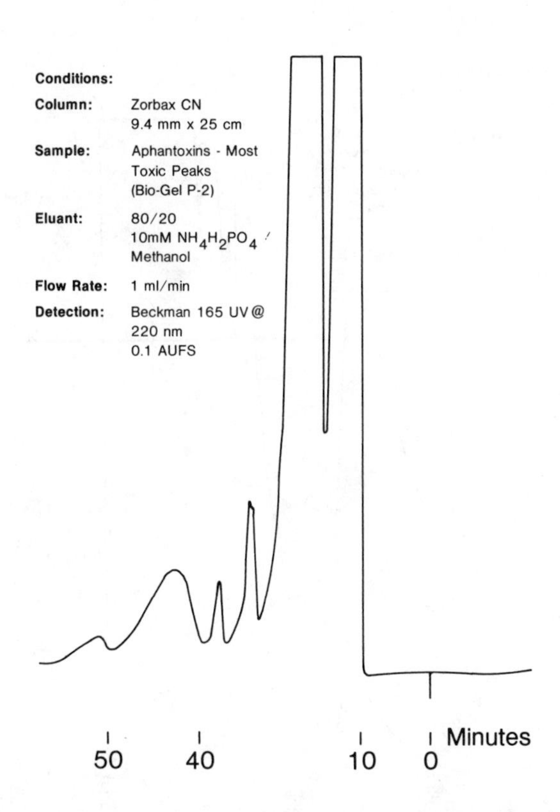

Figure 3. HPLC profile of Aphantoxin most toxic peak (left; corresponds to saxitoxin) and the major toxic component (second from left; corresponds to neosaxitoxin). A total of 150 mouse units were loaded. A mouse unit is equal to the amount of toxin needed to kill a 20 g mouse in 15 min.

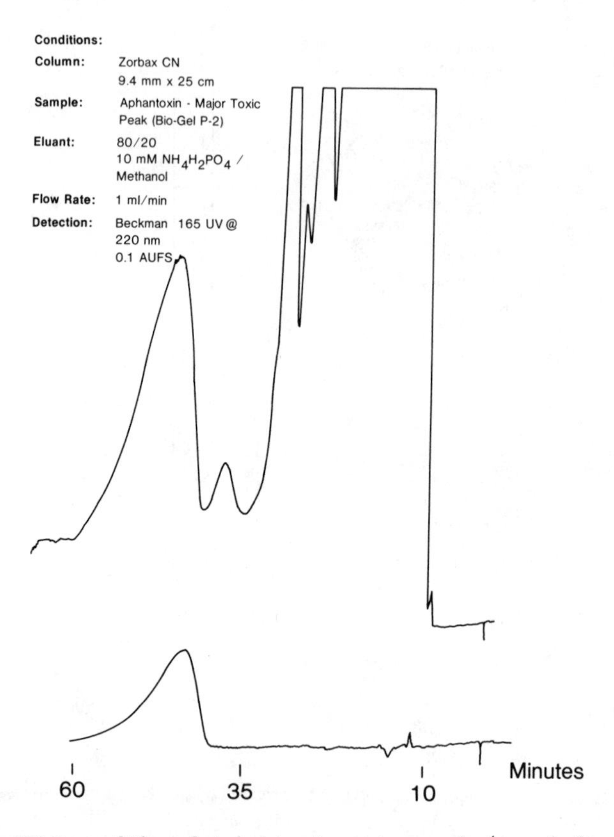

Figure 4. HPLC profile of Aphantoxin major peak (top-left; 500 mouse units) and neosaxatoxin standard (24) (bottom-left; 200 mouse units).

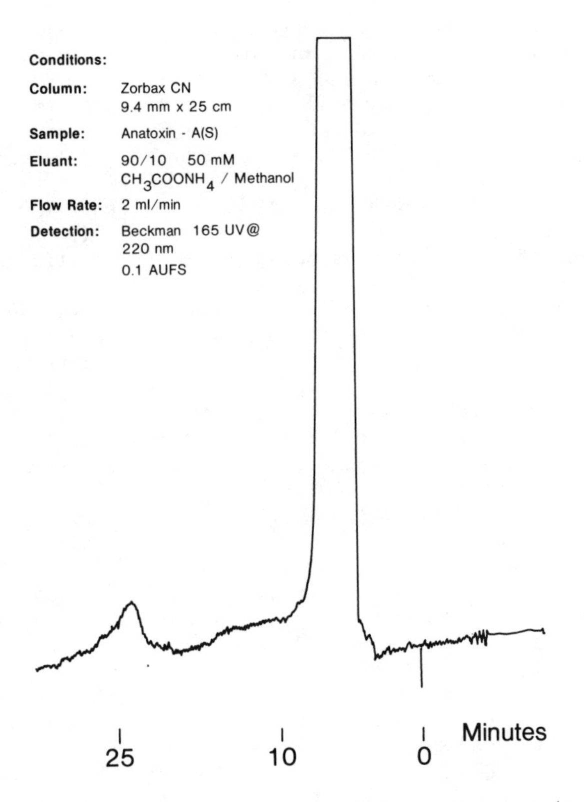

Figure 5. HPLC profile of Anatoxin-a(s); toxic peak (far-left; 200 mouse units). (LD_{50} i.p. mouse equal to 40 ug/kg body weight).

bacteria and amounts of toxin(s) outside the cells. Of special importance is the condition of the cells within the bloom and the bloom concentration since the peptides of Microcystis are retained largely within the cells until cell lysis, either in the water or in the animals, has occurred. This leakage of toxin into the water supply means that toxins can be expected to be present even in finished water when a toxic bloom is present.

In laboratory studies rats and mice injected iv with toxin extracts or ip with toxic cells or extracts die within 1/2 to 3 hours. Death is preceded by pallor and prostration, with terminal episodes of unprovoked leaping and twitching. Upon necropsy, the animals show grossly enlarged livers engorged with blood, with the remainder of the carcass being exsanguinated. Liver weight is increased and at death composes about 9 to 10 percent of body weight of mice as opposed to about 5 percent in controls (32,33). The blood content of livers of mice poisoned by Microcystis increases from 6.6 ml blood/100 g liver in controls to 53.4 ml blood/100 g for mice killed 45 min after toxin injection (33).

Histological examination of the liver reveals extensive centrilobular necrosis with loss of characteristic architecture of the hepatic cords. TEM examination indicated that both hepatocytes and hepatic endothelial cells were destroyed. The only alterations noted prior to cell rupture were slight mitocondrial swelling and slight cell swelling. Damaged cells had extensive fragmentation and vesiculation of the membrane (34).

Gross and histological examination of intestine, heart, spleen, kidneys and stomach have shown no consistent abnormalities; lungs were mildly congested with occasional patches of debris. Platelet thrombi have been reported in the lungs of affected animals (32,33). It has been suggested that these platelet thrombi may be a direct effect of the toxin and may secondarily cause the liver effects by creating sufficient pulmonary-congestion to cause right heart failure which, in turn, could cause blood pooling and congestion in the liver (32). However, in time course studies, Falconer et al. (33) reported that the platelet plugs did not appear in histological preparations taken at 15 and 30 minutes after toxin injection and were present only in later preparations. This evidence, along with other evidence of effects on isolated hepatocytes and the rapid onset of the liver effects *in vivo* have led other researchers to believe that the liver damage is a direct effect of the toxin on the hepatocyte membrane and that the immediate cause of death in acutely dosed animals is hemorrhagic shock (33,35). Occasionally, hemorrhages have been noted in other organs (17). This could possibly be due to coagulation problems associated with the liver damage.

Neurotoxic. Toxins of An. flos-aquae are referred to as anatoxins. Anatoxin-a (LD_{50} i.p. mouse, 200 μg/kg) (36,37) produced by An. flos-aquae NRC-44-1 is a potent nicotinic depolarizing neuromuscular blocking agent. Anatoxin-a(s) (LD_{50} i.p. mouse, 40 μg/kg) produced by An. flos-aquae NRC-525-17 has not been pharmacologically characterized. It does have neuromuscular blocking action producing opisthotonos and salivation in laboratory animals. No effect of the toxin was seen on isolated muscle preparations of frog rectus, chick biventer, rat hemidiaphragm-phrenic nerve or guinea pig ilium. The opistothonos and salivation could be reversed in chicks by prior injection of atropine (7).

The neurotoxins isolated from Aph. flos-aquae were shown to have similar chemical and biological properties to paralytic shellfish poisons (PSP) (25,29,38). Sawyer et al. in 1968 (25) were the first to demonstrate that the crude preparation of aphantoxins behave like saxitoxin, the major paralytic shellfish poison. They showed that the toxins had no effect on the resting membrane potential of frog sartorius muscle; blocked action potential on desheathed frog sciatic nerve and also abolished spontaneous contractions in frog heart. Sasner et al. (1981) (29) using the lab cultured strain reported similar results.

Further confirmation of the similarities in biological activities between aphantoxin and PSP was shown by Adelman et al. (1982) (30). They showed that crude preparations of aphantoxins blocked the Na^+ channel of giant squid axon with equal potency as saxitoxin.

Summary

Cyanobacterial toxins (both marine and freshwater) are functionally and chemically a diverse group of secondary chemicals. They show structure and function similarities to higher plant and algal toxins. Of particular importance to this publication is the production of toxins which appear to be identical with saxitoxin and neosaxitoxin. Since these are the primary toxins involved in cases of paralytic shellfish poisons, these aphantoxins could be a source of PSP standards and the study of their production by Aphanizomenon can provide information on the biosynthesis of PSP's. The cyanobacteria toxins have not received extensive attention since they have fewer vectors by which they come in contact with humans. As freshwater supplies become more eutrophicated and as cyanobacteria are increasingly used as a source of single cell protein toxic cyanobacteria will have increased importance (39). The study of these cyanobacterial toxins can contribute to a better understanding of seafood poisons.

Acknowledgments

N.A.M. would like to thank the Science University of Malaysia for a research fellowship and the Biomedical Sciences Program at Wright State University for travel support.

Literature Cited

1. Francis, G. Nature (London). 1878, 18,11.
2. Sykora, J.L.; Keleti, G. In "The Water Environment: Algal Toxins and Health"; Carmichael, W., Ed.; Environmental Science Research Vol. 20, Plenum Press: New York, 1981; pp. 285-302.
3. Billings, W.H. In "The Water Environment: Algal Toxins and Health"; Carmichael, W., Ed.; Environmental Science Research Vol. 20, Plenum Press: New York, 1981, pp. 243-256.
4. Carmichael, W.W.; Gorham, P.R. In "The Production and Use of Micro-Algae Biomass"; Shelef, G.; Soeder, C.J. Eds.; Elsevier/North Holland: New York, 1980, pp. 437-448.

5. Grant, G.A.; Hughes, E.O. Can. J. Public Health. 1953, 44, 334.
6. Schwimmer, D.; Schwimmer, M. In "Algae and Man"; Jackson, D.F. Ed.; Plenum Press: New York, 1964; pp. 368-412.
7. Carmichael, W.W. S. Af. J. Sci. 1982, 78, 367.
8. Skulberg, O.M. "Culture Collection of Algae, Norwegian Institute for Water Research (NIVA)"; Oslo, Norway, 1983, 22 pp.
9. Skulberg, O.M., Codd, G.A., Carmichael, W.W. Ambio. (In Press).
10. Runnegar, M.T.; Falconer, I.R. In "The Water Environment: Algal Toxins and Health"; Carmichael, W., Ed.; Environmental Science Research Vol. 20, Plenum Press: New York, 1981; pp. 325-342.
11. Watanabe, M.F.; Oishi, S. App. Envir. Micro. 1982, 43, 819.
12. Elleman, T.C.; Falconer, I.R.; Jackson, A.R.B.; Runnegar, M.T. Aust. J. Biol. Sci. 1978, 31, 209.
13. Eloff, J.N.; Siegelman, H.W.; Kycia, H. S. Af. J. Sci. 1982, 78, 377 (Abstr.).
14. Botes, D.P.; Kruger, H.; Viljoen, C.C. Toxicon. 1982, 20, 945.
15. Botes, D.P.; Viljoen, C.C.; Kruger, H.; Wessels, P.L.; Williams, D.H. Toxicon. 1982, 20, 1037.
16. Santikarn, S.; Williams, D.H.; Smith, R.J.; Hammond, S.J.; Botes, D.P.; Tuinman, A.; Wessels, P.L.; Viljoen, C.C.; Kruger, H.J. Chem. Soc., Chem. Commun. 1983, 12, 652.
17. Østensvik, Ø.; Skulberg, O.M.; Søli, N.E. In "The Water Environment: Algal Toxins and Health"; Carmichael, W., Ed.; Environmental Science Research. Vol. 20, Plenum Press: New York, 1981; pp. 315-324.
18. Huber, C.S. Acta Crystallogr. Sect. B. 1972, 28, 25-77.
19. Devlin, J.P.; Edwards, O.E.; Gorham, P.R.; Hunter, N.R.; Pike, R.K.; Stavric, B Can. J. Chem. 1977, 55, 1367.
20. Campbell, H.F.; Edwards, D.E.; Kolt, R. Can. J. Chem. 1977, 55, 1372.
21. Campbell, H.F.; Edwards, D.E.; Elder, J.W.; Kolt, R.S. Pol. J. Chem. 1979, 53,27.
22. Bates, H.A.; Rapoport, H.J. Am. Chem. Soc. 1979, 101, 1259.
23. Ikawa, M.; Wegener, K.; Foxall, T.C.; Sasaer, J.J. Jr. Toxicon. 1982, 20, 747.
24. Hall, S.; Reichardt, P.B.; Neve, R.A. Biochem. and Biophysical Res. Comm. 1980, 97, 649.
25. Sawyer, P.J.; Gentile, J.H.; Sasner, J.J. Jr. Can. J. Microbiol. 1968, 14, 1199.
26. Alam, M.; Ikawa, M.; Sasner, J.J. Jr.; Sawyer, P.J. Toxicon. 1973, 11, 65.
27. Alam, M.; Shimizu, Y.; Ikawa, M.; Sasner, J.J. Jr. J. Environ. Sci. Health, 1978, A13, 493.
28. Alam, M.; Euler, K.L. In "The Water Environment: Algal Toxins and Health"; Carmichael, W. Ed.; Environmental Science Research Vol. 20, Plenum Press: New York, 1981; pp. 405-414.
29. Sasner, J.J.; Ikawa, M.; Foxall, T.L.; Watson, W.H. In "The Water Environment: Algal Toxins and Health"; Carmichael, W. Ed.; Environmental Science Research Vol. 20, Plenum Press: New York, 1981; pp. 389-404.
30. Adelman, W.J. Jr.; Fohlmeister, J.F.; Sasner, J.J. Jr.; Ikawa, M. Toxicon 1982, 20, 513.

31. Buckley, L.J.; Ikawa, M.; Sasner, J.J. Jr. J. Agr. Food Chem. 1976, 24, 107.
32. Slatkin, D.N.; Stoner, R.D.; Adams, W.H.; Kycia, J.H.; Siegelman, H.W. Science 1983, 220, 1383.
33. Falconer, I.R.; Jackson, A.R.B.; Langley, J.; Runnegar, M.T. Aus. J. Biol. Sci. 1981, 34, 179.
34. Foxall, T.L.; Sasner, J.J. Jr. In "The Water Environment: Algal Toxins and Health"; Carmichael, W. Ed.; Environmental Science Research Vol. 20, Plenum Press: New York, 1981; pp. 365-388.
35. Runnegar, M.T.; Falconer, I.R.; Silver, J. Archives of Pharm. 1981, 317, 268.
36. Carmichael, W.W.; Biggs, D.F.; Peterson, M.A. Toxicon 1979, 17, 229.
37. Spivak, C.E.; Witkop, B.; Albuquerque, E.X. Mol. Pharmacol. 1980, 18, 384.
38. Jakim, E.; Gentile, J. Science 1968, 162, 915.
39. Shelef, G.; Soeder, C.J. Eds. "The Production and Use of Micro-Algae Biomass"; Elsevier: New York, 1980; 852 pp.

RECEIVED March 2, 1984

Studies on *Aphanizomenon* and *Microcystis* Toxins

JOHN J. SASNER, JR., MIYOSHI IKAWA, and THOMAS L. FOXALL[1]

Departments of Zoology and Biochemistry, Spaulding Life Science Building, University of New Hampshire, Durham, NH 03824

The aphantoxins from Aphanizomenon flos-aquae (freshwater Cyanobacteria) are similar in chemical characteristics and physiological effects to paralytic shellfish poisons from marine sources. Aphantoxin isolation and characterization was done using molecular weight filtration, solvent separation, column chromatography, electrophoresis, and fluorometry. These methods revealed the presence of the toxic cations neoSTX (90%) and STX (5-10%); and an anionic substance as yet unnamed. The toxins from A. flos-aquae reversibly blocked voltage-dependent Na^+ channels in crayfish and squid giant axons, similar to TTX, STX, and gonyautoxins. Microcystin, from Microcystis aeruginosa, is a family of related peptides which causes hepatomegaly, liver hemorrhaging, destruction of hepatocytes and endothelia, and associated shock in endothermic vertebrates.

Autotrophic microorganisms are the primary producers which comprise the basis of the food chain and energy budget in aquatic ecosystems. They utilize dissolved inorganic nutrients and synthesize new organic materials necessary for the support of animal life at higher trophic levels, i.e., primary, secondary and tertiary consumers (1). These primary producers, which play a vital role in productivity, may also synthesize products (biotoxins) which adversely affect other organisms elsewhere in the food web. Recent concern about aquatic biotoxins, particularly from microorganisms, stems from an increasing dependency on marine and freshwater environments for food and/or potable water, as a potential source for drugs and other chemicals, and modern applications of genetic engineering (2). There is an urgent need, therefore, to (a) identify the sources of biotoxins, (b) detect the

[1]Current address: Human Nutrition Research Center, New England Medical Center, Boston, MA 02111

0097-6156/84/0262-0391$06.00/0
© 1984 American Chemical Society

presence of the toxins in the food chain, (c) develop accurate and rapid methods for the detection, identification, and assay of the toxins, and (d) understand the effects of the toxins on animal systems. These natural products are of more than just academic interest, since they may cause massive animal mortalities as well as environmental, legal, recreational, and health-related problems.

The predominant marine biotoxins are paralytic shellfish poisons (PSP), tetrodotoxin (TTX), and ciguatera toxin. The latter two are ichthyosarcotoxins that pose serious health problems in tropical and subtropical countries. Paralytic shellfish poisons, on the other hand, are commonly associated with marine molluscs which filter-feed on phytoplankton and act as "biological storage depots" for toxin accumulation. Both PSP and ciguatera toxin originate in several species of dinoflagellates -- a group of microorganisms that are second in abundance only to the diatoms as the primary producers in marine environments (3). The predominant freshwater biotoxins are produced by 3 species of cyanobacteria, *Aphanizomenon flos-aquae* (aphantoxins), *Microcystis aeruginosa* (microcystin), and *Anabaena flos-aquae* (anatoxins). Blooms of these organisms have been linked to massive animal kills, including zooplankton, fish, larval amphibians, the poisoning of farm animals, and reduced water quality for drinking and recreational uses. Toxicity problems have been reported from all of the continents of the world (4). *Aphanizomenon flos-aquae* and *Microcystic aeruginosa* occur in New England, intermittently, in both toxic and non-toxic forms.

Occurrence of Aphanizomenon and Microcystis in New Hampshire

Kezar Lake is a typical glacial ice-scour lake in North Sutton, N.H. A brief history of this lake is included here because (a) toxic blooms of *Anabaena*, *Aphanizomenon*, and *Microcystis* have occurred seasonly over the past 20 years, (b) it has provided aphantoxins and microcystin for several research laboratories, (c) it is one of relatively few lakes where extensive physical, chemical, and biological data have been collected over extended time periods, and (d) various attempts at experimental corrective measures for controlling cyanobacterial blooms have been made here with some success. Kezar Lake covers 182 acres, has a maximum depth of 8.0 meters and an average depth of 3.7 meters. Some 50 years ago, Wadleigh State Park was established on its southeastern shore and became a popular recreational area. The town of New London built a sewage treatment plant in the early 1930's and began discharging secondary waste water into Lion Brook, some 5.6 kilometers upstream from the lake. Over time, the secondary effluent from the treatment plant contributed high concentrations of phosphates and nitrates into Lion Brook -- the major tributary into the lake. Approximately 70% of the phosphate load in Kezar Lake was attributed to the effluent from the treatment plant (5). Cyanobacterial blooms (*Anabaena*) were first reported in the early 1960's. Water clarity was reduced, offensive taste and odor occurred, and the first fish mortalities were noted. Copper sulphate treatment of the lake was successfully accomplished to retard blooms of *Anabaena*. Within a few years the blooms became more intense and included another species, Aphanizomenon. Similar control measures using the algicide were not as effective on the latter species and massive fish mortalities resulted -- an estimated 11 tons of fish in 1966. The recreational utility of Kezar Lake diminished, property values decreased, the N.H. State Tax Commission reduced property values by 30%, and legal action was successfully taken by property owners against the town of New London for remuneration of value lost. In 1968, an attempt was made to artifically destratify the lake using on-shore compressors connected to aerators sunk in the deepest part of the lake. This mixing of the water reduced thermal stratification, increased water clarity, and reduced dense accumulations of cyanobacteria in the upper surface waters. The positive effects of destratification were short-lived, however, and were abandoned after a few summers (6). Nutrient stripping and the initiation of tertiary treatment at the sewage treatment plant reduced the nutrient discharge levels. However, blooms of Aphanizomenon and Microcystis continued into the mid-1970's, presumably thriving on residual nutrients. In the summers of the latter 1970's and early 1980's, the Aphanizomenon blooms diminished and were replaced by Microcystis blooms. Thus, the recent history of cyanobacterial blooms in Kezar Lake involved the 3 problem-causing species. In the spring of 1983, approximately 25 acres (14%) of the lake bottom were treated with alum in an attempt to chelate the nutrients which "fuel" the initiation of blooms during the spring turnover. Preliminary observations indicate that the lake improved in quality, blooms were retarded, water clarity doubled, and recreation, instead of fish mortality, dominated the environment for the first time in approximately 20 years.

In the mid-1960's, toxic cyanobacterial blooms in Kezar Lake, Winnisquam Lake, and Skatutakee Lake in New Hampshire caused moderate to heavy fish mortalities at these 3 sites. Samples of the dominant microorganisms were identified by F. Drouet, Botany Department, Academy of Natural Sciences, Philadelphia, PA and C. M. Palmer, Taft Sanitation Center, Cincinnati, OH, as Aphanizomenon flos-aquae (L.) Ralfs. This species is also known as A. holsaticum Richt (preferred by Drouet). However, we chose to use A. flos-aquae because it is more widely noted in the scientific literature. The essentially monospecific blooms of A. flos-aquae (99.5%) occurred as single trichromes (filaments) with 25 to 70 cells/filament. Previous descriptions of this species indicated that filaments were usually bound together in a "rafted" or fasciculate form. Our samples from these lakes were, apparently, atypical for the species. Cell dimensions, however, approximate those reported (7) for A. flos-aquae cells cultured from the Rideau Canal, Ottawa, Canada in 1961 and designated NRC 28, 31, and 32. These dimensions were 3 x 5 um for vegetative cells, 5 x 7 um for heterocysts, and 6 x 20 um for akinetes. We have not found any fasciculate, toxic blooms of A. flos-aquae in New England. However, both "rafted" and "non-rafted" cultures from Klamath Lake, Oregon were found to be toxic (8). The toxicity of Aphanizomenon was firmly established in 1968 (9).

Intermittently, we have sampled almost monospecific blooms of both Aphanizomenon and Microcystis. Samples of M. aeruginosa were collected from several sources. Kezar Lake and Marsh Pond provided both toxic and non-toxic clones which were identified by A. Baker, Botany Department, University of New Hampshire. Cell concentrations generally exceeded 5 x 10^4/ml, and during dense blooms were > 10^6 cells/ml. An effective method of collecting large amounts of A. flos-aquae employed DeLaval separators, at lakeside or in the laboratory, to spin cells from large volumes of water. To collect Microcystis in the field, plastic storage containers were used to skim surface colonies. Samples were placed in large separatory funnels and the floating cell masses were concentrated by flotation and removal of water. Crude materials from both species were stored, either wet or lyophilized, in the frozen state.

In order to verify the specific sources of toxins and ensure large quantities of working material, it was also necessary to maintain mass cultures of A. flos-aquae and M. aeruginosa. Monospecific cultures of the latter species (NRC-1, SS-17 strain) was obtained from other laboratories and maintained in mass culture for several years. Unialgal isolates of A. flos-aquae were prepared from Kezar Lake and maintained in our laboratory from 1968 to 1978. The NH-1 strain of this species was isolated at Wright State University, by W. Carmichael, in 1980, from a bloom in a farm pond near Durham, NH. Laboratory cultures of both species (A. flos-aquae and M. aeruginosa) were grown in ASM-1 medium (10) buffered with Tris (pH 8.2) and containing 1% soil extract. Test tube cultures were expanded to 20 l glass carboys, then to 160 l plastic tanks. In mass culture the soil extract was omitted. All cultures were maintained under constant illumination (ca. 4000 f.c.) at 21 ± 2°C. They were aerated with air passing through sterile cotton filters. Cultures were harvested after approximately 2 weeks, using a continuous flow DeLaval Separator, as with the samples from natural blooms. The cells were stored frozen as wet or lyophilized samples. Periodic mouse bioassays were used to check on the potency of the material. Cyanobacterial samples retained potency under these conditions for several years.

Physiological Effects - Aphantoxins

The effects displayed by fish and mammals challenged with aphantoxins were qualitatively similar to symptoms reported for paralytic shellfish poisons from marine sources. In mice, the characteristic symptoms include coordination loss, irregular ventilation, spastic twitching, gaping mouth, and death by respiratory failure. Plankton samples collected during or shortly after A. flos-aquae blooms were usually devoid of common zooplankters. In the laboratory, Daphnia magna were affected when placed in aerated cultures or media containing extracts of A. flos-aquae. Within a short time, the characteristic movements of the second antennae were erratic or stopped, causing the animals to settle to the bottom. Daphnia removed to clean water recovered while those retained in toxin-containing media perished in from 2 to 24 hours. Other species among the primary consumers are also sensitive to the presence of aphantoxins (8, 11). The freshwater bivalves, Elliptio camplanatus and Corbicula fluminea, stored small amounts of toxin after feeding on A. flos-aquae cultures. Mouse bioassay of entire bivalve meat produced symptoms similar to those described for cell extracts or PSP. However, the bivalves stored approximately only 40 to 60 micrograms of toxin (STX equivalents) per 100 grams of meat before they were affected by the aphantoxins. Earlier work (12) showed that nerve preparations from Elliptio were more sensitive to STX and TTX than nerves from several marine bivalves. We have fed marine mussels (Mytilus edulis) on concentrated cell masses from laboratory cultures of A. flos-aquae and found significantly higher toxin accumulations (200 to 300 micrograms/100 grams meat) than occurred in the freshwater bivalve tissues. The freshwater filter feeders displayed a flaccid paralysis of the foot and mantle after 2-3 days in A. flos-aquae cultures, which was reversed upon return to non-toxic water. Isolated tissue and cell preparations were used to determine the site and mode of action of the toxin. Lyophilized Aphanizomenon cells were suspended in distilled water, sonicated, and centrifuged to remove particulate material. Molecular weight filters were used to "clean up" the crude material. The supernatant was passed through a 10 K dalton Millipore filter, a 500 dalton Amicon filter, and the final filtrate lyophilized. The latter was reconstituted in appropriate Ringers or artificial seawater for testing on giant axons of crayfish (Cambarus sp.) or the squid (Loligo pealiei). The crayfish ventral nerve cords were excised, the medial and lateral giant axons partially desheathed, and the preparations placed in a chamber divided into 3 (1 cm^3) sections. Each section was separated by petroleum jelly. Experiments were performed at room temperature (20-22°C). The crayfish Ringers was composed of (in mM/l) NaCl (146); KCl (4); $CaCl_2$ (8); $MgCl_2$ (4); and Tris-maleate buffer, pH 4.0, (10 mM/l).

In one series of experiments, stimulation of the axons occurred in the first section of the chamber, perfusion in the second, and intracellular recording was from the third section. Aphantoxins (0.4 to 10 ug/ml) reversibly blocked the conduction of action potentials in the crayfish giant axons. Following the addition of toxin, the latency between the stimulus, and action potential increased. When this latency period increased by 1 msec (25 to 30%), then conduction was blocked. The block time, after addition of toxin, and the recovery time, after Ringer wash, was dose-dependent. In other experiments, bathing solutions containing aphantoxin, STX, and TTX were adjusted to give similar and reproducible block times. The subsequent recovery times after Ringer wash was measured (to 90% control). The recovery times for aphantoxin and STX blocked axons were similar and significantly shorter than for TTX, Table I.

Table I. Comparative Reversibility: Aphantoxin versus STX and TTX

Toxin	No. of Preps	Conc.g/ml	Block Time (sec.)	Recovery Time (sec.)
Aphantoxin	5	4.0×10^{-5}	20 ± 2	115 ± 20
STX	5	3.5×10^{-8}	28 ± 8	95 ± 19
TTX	5	2.5×10^{-8}	24 ± 5	257 ± 34

The relative recovery times after Ringer wash suggest that aphantoxins acted more like STX than TTX in crayfish lateral and medial giant axons (11, 13).

In order to examine the effect of aphantoxin on the generation of the action potential and the membrane resistance, intracellular recordings were obtained from desheathed nerve cords. When the toxin (0.8 ug/ml) and recording microelectrode were placed in the same chamber, the spike amplitude gradually decreased to total block in 25 min. The rise time from baseline to peak amplitude increased 3 to 5 fold before complete block occurred, and the slope of the falling or recovery phase remained unchanged. In Figure I, for example, the rise time increased from 0.5 msec to 1.8 msec, while the falling phases were approximately 1.3 msec in all traces. Toxin application did not alter the transmembrane resting potential (-80 mV) or the membrane resistance (11, 13). The Na^+ dependency of the crayfish preparations was verified by challenging them with Na^+-free Ringers (choline chloride substitution). Records similar to Figure I, with toxin were obtained, showing reversible blocking of action potential generation. On the other hand, calcium-free Ringers only slightly affected the waveform and the subsequent addition of toxin decreased the rise rate, just as in the normal Ringers with toxin. The rising or depolarizing phase of the action potential is associated with a transient increase in Na^+ conductance (14). The data shows that aphantoxins blocked the Na^+ dependent, depolarizing phase without affecting the K^+-dependent repolarization phase.

Verification of this interpretation came from voltage-clamp studies using squid giant axons. Aphantoxin samples were prepared by molecular weight separation, using ultrafiltration membranes, as described above for the extracts used on crayfish. Squid axons were voltage clamped to a series of depolarizing potentials and the corresponding membrane currents were recorded. The bathing medium was changed to aphantoxin/seawater solutions and the series of depolarizing potentials was repeated (Figure II). The voltage-clamp data shows aphantoxins are very potent and specific inhibitors of voltage-dependent Na^+ channels in the squid axon. There was no effect on the K^+ conductance, and the blocking of excitability was completely reversible. The dose-response relationship of the transient peak current was plotted as a function of toxin concentration, expressed as STX equivalents in artificial seawater. The data points fell on a curve which was representative of 1:1 stoichiometry (15). The dissociation constant from this relationship indicated that 3.47 nM aphantoxin (as STX equivalents) blocks 50% of the membranes' Na^+ channels in the squid axon (16).

Aphantoxin was found to contain Neo-STX and STX in an approximate ratio of 9:1 (17). These two toxic components were found to be equipotent when tested on rise rate of action potentials and current-voltage relations in amphibian sartorius muscle fibers (18). The equipotency of STX and Neo-STX was also determined in voltage-clamped squid giant axons (19). The specificity of Neo-STX for the Na^+ channel was established, as well as the identification of the 7,8,9 guanidinium moeity as the active group involved in binding/blocking the channels (18).

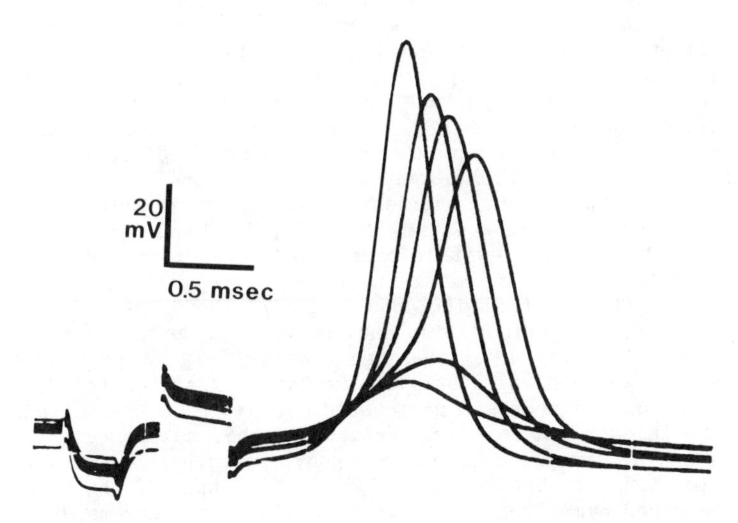

Figure I. Action Potentials from crayfish giant axons. Control = Top Trace. Toxin (0.8 ug/ml) reduced rise rate and amplitude after 5, 10, 15, 20, and 25 min. Reversible. Stim. = 2 V; 0.25 msec pulse width. (from Sasner et al., 1981)

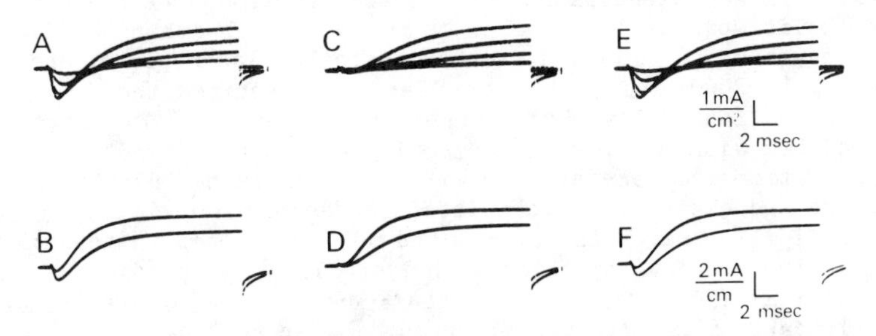

Figure II. Membrane currents in voltage-clamped squid axons. Current traces from holding potentials of -60, -30, -20, -10, and 0 mV in (A), (C), and (E). (A) = control; (C) = 30 nM aphantoxin (STX equiv.); (E) = recovery after wash (10 min). (B), (D), and (F) = 20, and 40 mV. (B) = control; (D) = toxin; (F) = recovery. Temp. = 6.2°C. Reproduced with permission from Ref. 16. Copyright 1982, Pergamon Press, Ltd.

Physiological Effects - Microcystin

Three different toxins have been associated with *Microcystis aeruginosa*. The most common of these is a fast death factor (FDF, 1 hr), which is a family of related peptides collectively called microcystin. A slow death factor (SDF, 24-48 hrs) may also be present in some strains, as well as a diarrheagenic factor (DF). The latter 2 originate in bacteria co-existing with the *Microcystis* (20-22), and are not always present or measurable by bioassay in the presence of microcystin. The complex of peptides (microcystin) are secondary metabolites (23) which mainly affect endothermic vertebrates (birds and mammals) after ingestion of toxin by drinking from waterblooms. Ectothermic vertebrates (fish and amphibians) and several invertebrates were not affected by cell extracts (i.p. injection or immersion) (24). When small mammals were challenged (i.p.) with reconstituted lyophilized cellular material (50 mg/Kg), survival times were aproximately 1 hour. Characteristic symptoms included pallor of the extremities, lethargy, labored ventilation and non-violent death. A histological survey of tissues from the major organ systems of mice revealed drastic changes only in the liver. Hepatocyte degeneration and necrosis progressed outward from the centrilobular regions in poisoned animals. Hepatocytes became progressively swollen over time, and showed a cloudy, eosinophilic cytoplasm. Hepatic sinusoids became distended, cells ruptured, the integrity of the parenchymal cord structure was disrupted, and massive hemorrhaging into the liver was routinely observed (Figure III). The progressive liver damage was examined at the ultrastructural level over the course of 1 hour after i.p. treatment, with a lethal dose of microcystin (50 mg/Kg). After 10 minutes, membrane fragments were noted in the sinusoids. Within 20 minutes, bile canaliculi and microvilli in the space of Disse were distorted, and endothelia were either swollen or fragmented. Within 40 minutes, hepatocyte disruption was seen in centrilobular regions, and the pooling of RBC's and cell organelles characterized the disintegration of liver structure and the massive hemorrhaging associated with microcystin poisoning. Within 60 minutes, cell degeneration and necrosis was extensive in the centrilobular regions and extended outward into the parenchymal cords, (24, 25). Table II illustrates the weight and volume changes in mouse liver at 60 minutes post injection with lyophilized samples of microcystin (50 mg/Kg). Using the F test for significance, (a) no significant difference was observed between control and experimental body weights and (b) a highly significant difference was noted between control and experimental liver weights after 1 hour (99% confidence interval). In adult mice, the total circulating blood volume is estimated at 77.8 to 86.4 cc/Kg body weight (26). For the 20 gram animals used in our study, this calculates to 1.6 to 1.7 cc blood/animal. The increase in liver weight and volume and the observed hemorrhaging over time suggests a 36% decrease in circulating blood volume. The 70% increase in liver volume also must be viewed as a significant decrease in circulating blood volume leading to cardiovascular shock.

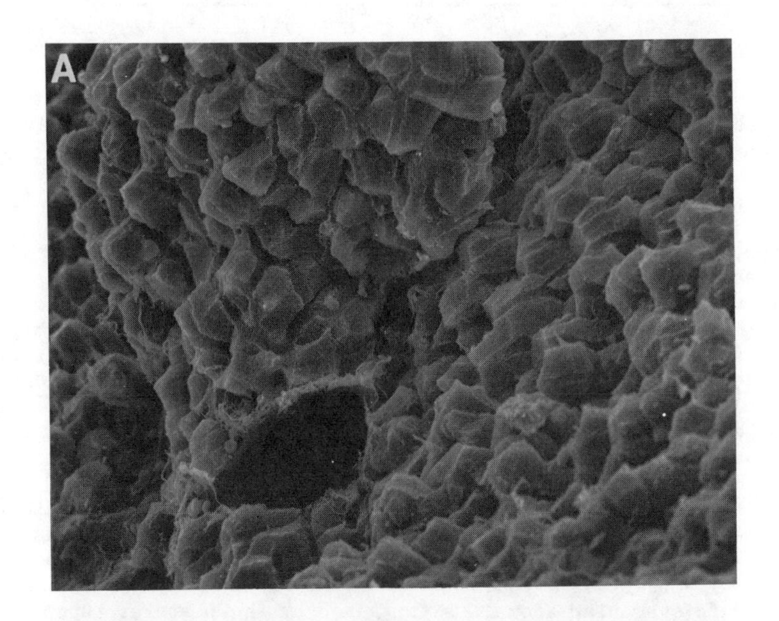

Figure III. Scanning Electron Micrographs of Mouse Liver. A = Control (440X); B = 30 min. after microcystin (790X). Note RBC's and damaged hepatocytes.

Table II. Mouse Liver Weight and Volume Changes after Microcystin

	N	Body Wt. (g)	Liver Wt. (g)	% Liver/ Body Wt.	Liver Vol. (cc)
Controls (X)	15	20.66	1.05	5.02	0.86
Experimentals (X)	15	19.28	1.51	7.77	1.47
Percent Change	--	--	+44%	+55%	+71%

Other workers (27) measured a 48% increase in hepatic blood volume within 45 minutes post injection with microcystin. This is consistent with our estimates of the extensive liver damage. Blood pressure data from rats treated with lethal doses of microcystin (50 to 200 mg/Kg body weight) (28) fell to less than half the initial or control pressure within 45 minutes. In addition, concentrations of liver enzymes increased markedly in the blood, thereby lending chemical support to the anatomical observations (29). The current histological, ultrastructural, blood pressure, and blood chemistry information supports the view that extensive and rapid liver damage with massive hemorrhaging into this organ is symptomatic of microcystin poisoning. Other recent work (30) does not agree with this interpretation and suggests that the toxin causes pulmonary platelet thrombi formation. This would lead to hypoxemia, heart failure and shock, and the liver damage described above is viewed as a secondary effect after a cascade of pulmonary and cardiovascular anomalies triggered by the platelet thrombi. To resolve this controversy, carefully controlled experiments must be done to establish the specific timing of events after poisoning. It is important to resolve whether primary structural damage in the liver causes subsequent lung and circulatory effects or whether primary pulmonary blockage is the cause of the circulatory effects and liver damage.

Chemistry - Aphantoxins

The physiological effects of aphantoxins from toxic *Aphanizomenon*, noted above, were similar to those described for paralytic shellfish poisons (PSP) from the marine dinoflagellate, *Gonyaulax tamarensis*. In addition, certain chemical similarities were also found early, i.e., IR spectra and color reactions, using stored materials from Kezar Lake (31). Subsequent chemical data suggested the presence of 4 toxic components from *A. flos-aquae* -- one of which was saxitoxin (STX) (32). Using stored cell samples collected from natural blooms and laboratory cultures, we have studied the chemical properties, methods of detection and quantitation, and stability of aphantoxins. This work was a "spin off" from our initial goal of developing chromatographic and fluorometric procedures for the detection, quantitation, and isolation of STX-related toxins from crude extracts of marine bivalve tissues (33-36).

Weak cation exchange resins (carboxylic acid type) have been widely employed for the isolation and detection of marine PSP's. However, the less basic gonyautoxins (GTX's) tended to elute with the solvent front with these resins -- especially when very crude extracts were used. We currently use strong cation exchange resins (sulfonic acid type) for the separation of the toxic components in PSP and Aphanizomenon (17, 37). The fluorescence generated on oxidation of the STX group of toxins (STX, GTX2, and GTX3) was used for their detection. Unfortunately, the neosaxitoxin (neo-STX) group (neo-STX, GTX1, and GTX4) did not fluoresce, under the conditions of the tube assay, and was routinely estimated using the standard mouse bioassay.

Figure (IV) shows the fluorescence and toxicity profiles of A. flos-aquae extracts from Kezar lake. The profiles show a close correlation between toxicity and fluorescence. The early eluting peak is of unknown nature while the later peak was coincident with STX standards. The early peak suggested the possible presence of gonyautoxins which were routinely found in marine shellfish samples contaminated with Gonyaulax tamarensis (Figure V B). However, hot acid treatment (0.02 M HCl) of the A. flos-aquae extract caused this early peak to disappear. Similar acid treatment of the gonyautoxins from Mytilus showed them to be more stable in acid than the Kezar aphantoxin. The results showed that the early-eluting peak from Kezar lake samples was very labile, but not very potent and, therefore, different from gonyautoxins II and III (38). It is possible that this early eluting peak could be a cryptic form of PSP such as B_1 (39), which is relatively non-toxic but highly fluorescing, and which yields STX on mild acid treatment. However, no particular increase in STX was observed upon acid treatment. The toxin profile was different for the Kezar Lake and NH-1 strains or clones of A. flos-aquae (Figures IV and V A). When compared with elution profiles from Mytilus, the NH-1 strain showed only one fluorescent toxic peak (fractions 70 to 80) which corresponded to the Mytilus peak for STX, and a non-fluorescent but toxic peak (fractions 60 to 70) which corresponded to neo-STX (Figure V A & B).

Samples of the NH-1 strain of A. flos-aquae were purified on a Bio-Gel P-2 column (Bio Rad Laboratories). The cell extracts were pre-treated with ammonium sulphate and/or molecular filtration (10 K dalton, Millipore Corp.) to remove interfering proteins and pigments. The mobile phase of the P-2 column was changed from dist. H_2O to 0.1 M acetic acid prior to elution of the second peak. The toxicity profile (Figure VI) from the P-2 column procedure, showed an early, toxic, water eluting peak, and a larger (4X) later peak that eluted with the acid fractions. Fractions representing the larger (acid) peak were pooled and chromatographed on a Bio-Rex 70 (H+) column (0.8 x 60 cm) using a linear acetic acid gradient. The gradient was 0.0-0.3 M acetic acid to tube 40, and 0.3-1.3 M acetic acid to tube 80. The presence of gonyautoxins in A. flos-aquae NH-1 was not detected, since no measurable toxicity or fluorescence was seen at low acetic acid concentrations (Figure VII). At higher acetic acid concentrations, a toxic but non-fluorescent peak immediately preceeded a toxic fluorescent peak. This profile was similar to the neo-STX - STX pair reported for scallop extracts (40). The early, toxic, water-eluted peak from the P-2 column (Figure VI) showed the presence of two new toxic components in the A. flos-aquae, NH-1 strain. These compounds behaved as anionic substances on electrophoresis at pH 8.6 (41).

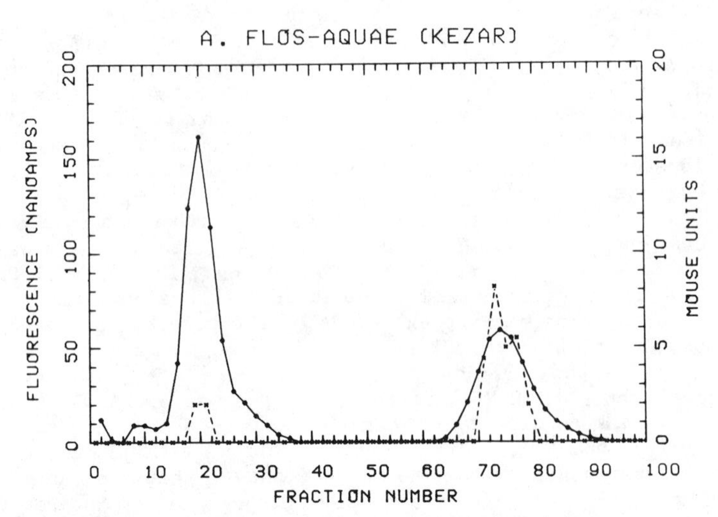

Figure IV. Fluorescence and Toxicity Profile of Kezar Aphanizomenon. Solid line = fluorescence; excitation at 330 nM and emission at 380 nM. Dash line = MU from mouse bioassay.

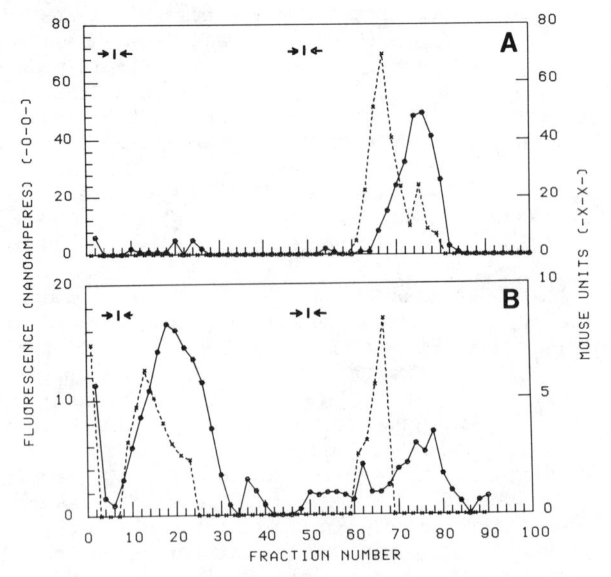

Figure V. Fluorescence and Toxicity Profiles of Aphanizomenon, NH-I, (A) and Mytilus edulis Extracts (B). Solid line = fluorescence; Dash line = MU from mouse bioassay. Note difference in Kezar and NH-I profile. Fluorescence peak in A and 2nd fluor. in B coincide with STX standard. Reproduced with permission from Ref. 17. Copyright 1982, Pergamon Press, Ltd.

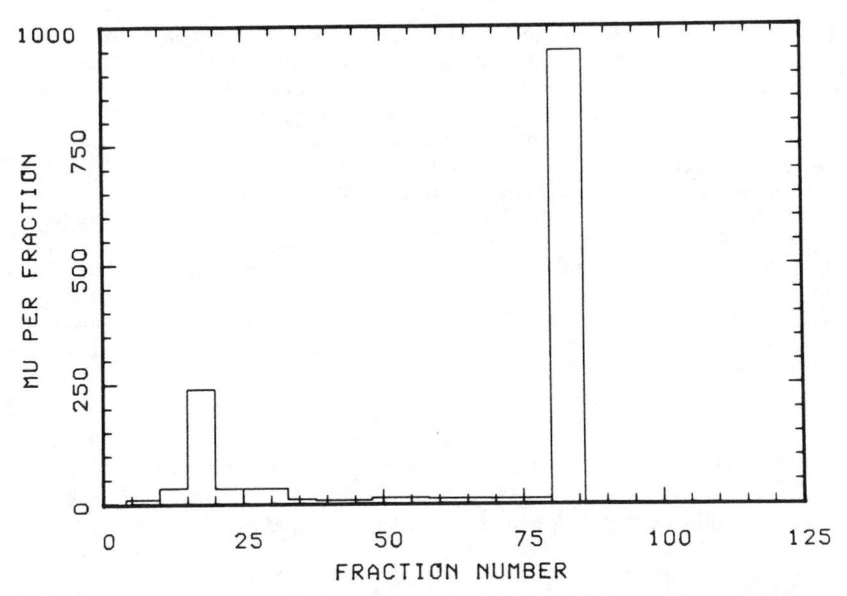

Figure VI. Toxicity Profile from Bio-Gel P-2 Column. 1st Peak mobile phase = water; 2nd peak mobile phase = 0.1 M acetic acid. 1st peak = 2 anionic, unnamed toxins. Fractions from 2nd peak pooled and re-chromatographed in Fig. VII, (neo-STX and STX). (from Auger, 1983)

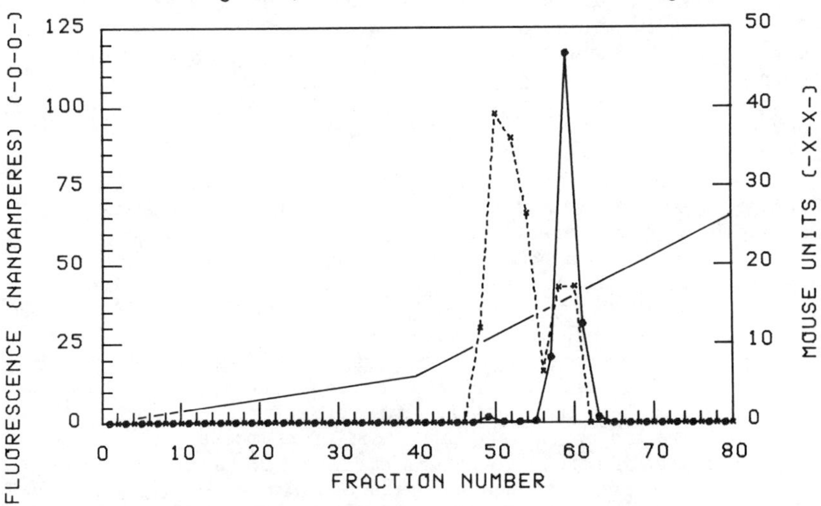

Figure VII. Bio-Rex 70 (H+) Chromatography of Partially Purified <u>A. flos-aquae</u>, NH-1 Toxin. Solid line = fluorescence; Dash line = MU. Column (0.8 X 60 cm) developed with 0.0-0.3 M linear acetic acid gradient to tube 40, and 0.3-1.3 M linear gradient to tube 80. 5 ml fractions were collected. Reproduced with permission from Ref. 17. Copyright 1982, Pergamon Press, Ltd.

Preliminary testing on whole animal (bioassay) and isolated tissue preparations (voltage-clamped squid giant axons) showed effects similar to those measured for STX and neo-STX. Again the presence of anionic cryptic forms of the toxins was indicated.

The inability of the fluorometric method to directly assess the presence of the neo-STX group (neo-STX, GTX1, and GTX4), due to the presence of an OH group at position N_1, presents problems in the evaluation of shellfish samples by public health agencies that monitor PSP. We have recently found that neo-STX forms a blue color with Folin-Ciocalteu phenol reagent and that it can be detected in crude extracts of *A. flos-aquae* and of toxic shellfish when it is present in appreciable amounts. It is not as sensitive as the fluorometric detection of STX (42).

Summary and Conclusions

Aquatic biotoxins attract attention from researchers because of their potency, specificity of action, and in some cases, their reversibility. The increasing demands on aquatic environments makes it necessary to accurately detect, chemically identify, and continuously monitor potentially harmful natural products. The chemistry and toxicology of active materials from aquatic microorganisms has focussed attention on marine species because of "red tides", and associated PSP in molluscs, and TTX and ciguatera toxin in fish. At least 2 of these originate in marine dinoflagellates (PSP and ciguatera), which along with the diatoms, comprise the bulk of the phytoplankton. The freshwater cyanobacteria (blue-green algae) also may produce potent toxins which affect organisms higher in the food chain. The presence of STX, neo-STX, and related compounds in *Aphanizomenon* has only recently been described for procaryotes (17) and may offer interesting alternatives for studying the biosynthesis of natural products. The chemical nature of aphantoxins and their site and mode of action in biological systems is understood, at least when compared to similar toxins (PSP) from marine sources (16). However, since the sensitivity of freshwater animals to aphantoxins is, apparently, different from their marine counterparts, more work is needed to understand the potential impact of massive blooms on resident animal populations, i.e. fish and bivalves. Toxic waterblooms, where *Aphanizomenon flos-aquae* is the dominant species, may be rare, when compared to the more cosmopolitan *Microcystis aeruginosa*. The latter has caused extensive mortality of farm animals after ingestion of bloom materials. Recent work which characterized microcystin is important for developing methods of detection, isolation, and purification of these peptides (43-45). While the hepatotoxic nature of microcystin has been known for some time, and was recently described in detail (46), recent evidence was presented which implicates the presence of pulmonary platelet thrombi (30). The precise sequence of physiological anomalies which kills endothermic vertebrates must be studied, preferably using pure toxic materials. The identification and purification of microcystin for such studies may come from HPLC methods developed by Siegelman and presented elsewhere in these proceedings.

Acknowledgments

This work was supported by NSF Grant No. DAR-8003523. Early work on the toxins was supported by the New Hampshire Water Resources Research Center of the University of New Hampshire, Grant No. AO-47NH, from the Office of Water Research and Technology, United States Department of the Interior as authorized under the Water Research and Development Act of 1978, Public Law 95-467. We also thank Dr. William J. Adelman, Jr., NINCDS, Woods Hole, for performing the voltage-clamp experiments; Kurt Auger for superior work in biochemistry; and Toshinori Hoshi for studies on the crayfish axons.

Literature Cited

1. Greeson, P. E. Water Resources 1969, 5, 16-30.
2. Colwell, R. R. Science 1983, 222, 19-24.
3. Ryther, J. H. In "Luminescence of Biological Systems"; Johnson, F. H., Ed.; Amer. Assn. Adv. Sci.: Washington, D. C., 1955; pp. 387-414.
4. Carmichael, W. W. "The Water Environment"; Plenum Press: New York and London, 1981.
5. "A Historical Outlook on Kezar Lake", New Hampshire Water Supply and Pollution Control Commission, 1981, No. 124.
6. Haynes, R. Ph.D. Thesis, University of New Hampshire, Durham, 1971.
7. McLachlan, J.; Hammer, V. T.; Gorham, P. R. Phycologia 1963, 2, 157-68.
8. Gentile, J. H. In "Microbial Toxins"; Aji, S. J.; Ciegler, A.; Kadis, S.; Monte, T. C.; Weinbaum, G., Eds; Academic Press: New York and London, 1971; Chap. 2.
9. Sawyer, P. J.; Gentile, J. H.; Sasner, J. J. Can. J. Microbiol. 1968, 14, 1199-1204.
10. Carmichael, W. W.; Gorham, P. R. J. Phycol. 1974, 10, 238-40.
11. Sasner, J. J., Jr.; Ikawa, M.; Foxall, T. L.; Watson, W. H. In "The Water Environment"; Carmichael, W. W., Ed.; Plenum Press: New York & London, 1981; pp. 389-403.
12. Twarog, B. M.; Yamaguchi, H. In "The 1st Internat'l. Conf. on Toxic Dinoflag. Blooms"; LoCicero, V. R., Ed.; Massachusetts Sci. and Tech. Foundation: Wakefield, Massachusetts, 1975, pp. 381-93.
13. Hoshi, T., unpublished data.
14. Adelman, W. J.; French, R. I. Oceanus 1976, 19, 6-16.
15. Cuervo, L. A.; Adelman, W. J., Jr. J. Gen. Physiol. 1970, 55, 309.
16. Adelman, W. J., Jr.; Fohlmeister, J. F.; Sasner, J. J., Jr.; Ikawa, M. Toxicon 1982, 20, 513-16.
17. Ikawa, M.; Wegener, K.; Foxall, T. L.; Sasner, J. J., Jr. Toxicon 1982, 20, 747-51.
18. Kao, C. Y.; Walker, S. E. J. Physiol. 1982, 323, 619-37.
19. Adelman, W. J., Jr.; Sasner, J. J., Jr. Unpublished data.
20. Aziz, K. Science 1974, 183, 1206-7.
21. Hughes, E. O.; Gorham, P. R.; Zehnder, A. Phycological News Bull. 1955, 8(5), 5.

22. Thompson, W. K.; Laing, A. C.; Grant, G. A. In "Toxic Algae"; 1957, IV, Report No. 51, Defense Research Board, Canada.
23. Runnegar, M. T. C.; Falconer, I. R.; Jackson, A. R. B.; McInnes, A. Toxicon 1983, Suppl. 3, pp. 337-80.
24. Foxall, T. L.; Sasner, J. J., Jr. In "The Water Environment"; Carmichael, W. W., Ed.; Plenum Press: New York & London, 1981; pp. 365-387.
25. Runnegar, M. T.; Falconer, I. R.; Silver, J. Arch. Pharmacol. 1981, 317, 268-72.
26. Crispens, C. G., Jr. "Handbook on the Laboratory Mouse"; Charles C. Thomas Publisher, 1975; p. 95.
27. Falconer, L. R.; Runnegar, M. T. C.; Jackson, A. R. B.; McInnes, A. Toxicon 1983, Suppl. 3, pp. 119-21.
28. Ostensvik, O.; Skulberg, O. M.; Soli, N.E. In "The Water Environment"; Carmichael, W. W., Ed.; Plenum Press: New York & London, 1981; pp. 315-324.
29. Falconer, I. R.; Beresford, A. M.; Runnegar, M. T. C. Med. J. Aust. 1983, 1, 511-4.
30. Slatkin, D. N.; Stoner, R. D.; Adams, W. H.; Kycia, J. H.; Siegelman, H. W. Science 1983, 220, 1383-5.
31. Jackim, E.; Gentile, J. Science 1968, 162, 915-6.
32. Alam, M.; Shimizu, Y.; Ikawa, M.; Sasner, J. J., Jr. J. Environ. Sci. Health 1978, A13, 493-9.
33. Buckley, L. J.; Ikawa, M.; Sasner, J. J., Jr. J. Agric. Food Chem. 1976, 24, 107-11.
34. Buckley, L. Ph.D. Thesis, University of New Hampshire, Durham, 1976.
35. Shoptaugh, N. H.; Buckley, L. J.; Ikawa, M.; Sasner, J. J., Jr. Toxicon 1978, 16, 509-13.
36. Shoptaugh, N. H.; Carter, P. W.; Foxall, T. L.; Sasner, J. J., Jr.; Ikawa, M. J. Agric. Food Chem. 1981, 29, 198-200.
37. Ikawa, M.; Wegener, K.; Foxall, T. L.; Sasner, J. J., Jr.; Noguchi, T.; Hashimoto, K. J. Agric. Food Chem. 1982, 30, 526-8.
38. Ikawa, M.; Wegener, K.; Foxall, T. L.; Sasner, J. J., Jr.; Carter, P. W.; Shoptaugh, N. H. In "The Water Environment"; Carmichael, W. W., Ed.; Plenum Press: New York & London, 1981; pp. 415-25.
39. Hall, S.; Reichardt, P. B.; Neve, R. A. Biochem. Biophys. Res. Communs. 1980, 97, 649-53.
40. Shimizu, Y.; Yoshioka, M. Science 1981, 212, 547.
41. Auger, K., unpublished data.
42. Mosley, S., unpublished data.
43. Botes, B. P.; Kruger, H.; Viljoen, C. C. Toxicon 1982, 20, 945-54.
44. Botes, D. P.; Viljoen, C. C.; Kruger, H.; Wessels, P. L.; Williams, D. H. Toxicon 1982, 20, 1037-42.
45. Santikarn, S.; Williams, D. H.; Smith, R. J.; Hammond, S. J.; Botes, D. P.; Tuinman, A.; Wessels, P. L.; Viljoen, C. C.; Kruger, H. J. Chem. Soc., Chem. Commun. 1983, 652-4.
46. Foxall, T. L. Ph.D. Thesis, University of New Hampshire, Durham, 1980.

RECEIVED March 2, 1984

34

Toxins of *Microcystis aeruginosa* and Their Hematological and Histopathological Effects

HAROLD W. SIEGELMAN, WILLIAM H. ADAMS, RICHARD D. STONER, and DANIEL N. SLATKIN

Biology and Medical Departments, Brookhaven National Laboratory, Upton, NY 11973

The peptide toxins of the cyanobacterium Microcystis aeruginosa extracted from small samples of cells (about 20 mg dry weight) are readily detectable by high performance liquid chromatography. A mouse injected with a median lethal dose of purified toxin (about 0.06 μg per gram of body weight) shows no evidence of toxicity for about 30 minutes. The mouse then develops multiple signs of toxicity and may die within 30 to 60 minutes. The toxicity is associated with marked reduction in the number of blood platelets, multiple microscopic clots in the lung, and engorgement of the liver with blood. The toxicity is not inhibited by prior administration of several anticoagulants to the mouse.

The cosmopolitan cyanobacterium Microcystis aeruginosa is frequently the major component of freshwater cyanobacterial blooms. These blooms can cause serious water management problems and are occasionally associated with animal poisoning. The M. aeruginosa toxins are potent lethal peptides which contain three invariant D-amino acids (Ala, erythro-β-methyl Asp, and Glu), two variant L-amino acids, N-methyl dehydroalanine, and a β-amino acid (1-3). Multiple toxins have been purified from clonal isolates (1,4). The toxic peptide described in this chapter is denoted toxin-LR using the standard one-letter abbreviations for its two variant amino acids, leucine and arginine.

Extracts of toxic isolates of M. aeruginosa or a purified peptide toxin injected into the bloodstream or the peritoneal cavity of mice elicit hepatotoxic effects (5-7,8). The fresh weight of the liver is greatly increased, and sinusoidal congestion, hemorrhage, and necrosis of the liver are observed. A lethal dose of purified toxin also induces multiple thrombi in the lungs and causes a marked reduction in the number of circulating blood platelets (6,7).

In this chapter, a procedure for initial evaluation of the M. aeruginosa toxins from laboratory cultures or field collections using high performance liquid chromatography (HPLC) is described.

0097-6156/84/0262-0407$06.00/0
© 1984 American Chemical Society

The hematological and histopathological effects of toxin-LR in mice are reviewed.

Toxin Detection by HPLC

A simplified procedure for detecting the *M. aeruginosa* toxins was devised. A 20.0 mg sample of lyophilized cells was extracted for 1 hr with 1 ml of 38% ethanol, 5% n-butanol, 50 mM ammonium acetate pH 6.0, shaking the suspension manually for a few seconds several times. The suspension was then centrifuged (5 min at 12,000 Xg). One ml of n-butanol and then 1 ml of water were added to the supernatant solution with vortexing. The sample was centrifuged (10 min at 500 Xg), and the upper n-butanol layer was collected and dried by rotary evaporation. The residue was extracted with 1 ml of 26% acetonitrile, 500 mM ammonium acetate pH 6.0, stored for 16 hr at -10°C, and then centrifuged (10 min at 12,000 Xg). All operations were conducted at room temperature except as noted. The partially purified extract was stored at -10°C and was apparently stable for several months. Aliquots of 2.5 to 25 µl of extract were examined by HPLC under the following conditions:

Precolumn	pellicular octadecyl beads (SynChropak RSC, SynChrom, Linden, IN) (0.4 x 5.0 cm)
Column	3 µm octadecyl silica (Hypersil, Shandon) (0.46 x 5.0 cm)
Solvent	26% acetonitrile, 500 mM ammonium acetate, pH 6.0
Flow	1.0 ml/min
Detection	238 nm at 0.04 AUFS (Absorbance Units Full Scale)

Isolate NRC-1 and the SS-17 clone of NRC-1 of *M. aeruginosa* were cultured in BG-11 medium (9). They were kindly provided by Dr. W. W. Carmichael, Wright State University, Dayton, Ohio. A lyophilized sample of a toxic bloom of *M. aeruginosa* collected from Kezar Lake, New Hampshire on October 10, 1978 was generously provided by Dr. J. J. Sasner, Jr., University of New Hampshire, Durham, New Hampshire.

Partially purified extracts of the two isolates and the Kezar Lake toxic bloom sample of *M. aeruginosa* were chromatographed. The HPLC elution profiles of these are shown in Figure 1. All components of the elution profiles of isolates NRC-1 and SS-17 are closely similar, and they are each characterized by one major toxin, toxin-LR. The elution profile of the Kezar Lake bloom extract also consists primarily of toxin-LR. Moreover, it does not contain some of the minor components observed in the elution patterns of NRC-1 and SS-17. The toxin-LR content of the Kezar Lake bloom was higher than that of the isolates.

The detection of *M. aeruginosa* toxins is readily achieved by HPLC and ultraviolet absorbance. Provisionally, components with identical retention times are considered identical. Confirmation is made by preparative isolation, amino acid analysis, and toxicity testing. The frequent, but not invariable, presence of toxin-LR as a principal toxin provides a useful HPLC marker. The wavelength used for detection of the toxins, 238 nm, is the principal absorbance maximum of toxin-LR. The 238 nm absorbance is probably

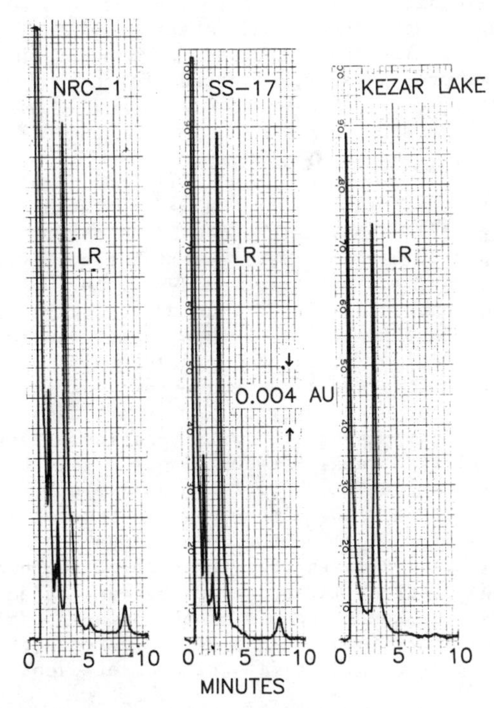

Figure 1. HPLC elution patterns of extracts of M. aeruginosa isolate NRC-1, clone SS-17 of isolate NRC-1, and a toxic bloom sample from Kezar Lake, New Hampshire. Toxin-LR eluted at about 2.7 minutes. Smaple volumes injected were: NRC-1 (17 µl), SS-17 (20 µl), and Kezar Lake (4 ul).

associated with the N-methyl-dehydroalanine content of the toxin (1). Relative amounts of toxin are indicated by peak heights. Exhaustive extractions and authentic standards can provide quantitative estimates of toxin content. The detection of 40 ng of toxin-LR, about 2% of a lethal dose in a mouse, is readily achieved by the HPLC system described above. Five µm-octadecyl or octyl silica columns (0.46 x 15 cm) can also be used for the HPLC of the toxins. However they have longer retention times and lower peak heights. More sensitive ultraviolet detectors and microbore columns may reduce sample requirements and detection limits by an order of magnitude or more. The characteristic components observed in HPLC elution profiles provide a useful method for characterizing a specific isolate.

Hematology and Histopathology

Injections of toxin-LR into mice at a median lethal dose (LD_{50}) of about 0.06 µg per gram body weight caused no observable behavioral changes for about 30 min (6,7,9). The mice then developed hunched posture, immobility, and piloerection during increasingly frequent and longer time intervals. Their reduced activity was interrupted by apparently unprovoked leaps. Such behavior was followed by lassitude, piloerection, tachypnea, and subcostal retraction. The ears and digits of affected mice became pale but not cyanotic. If death ensued, it was preceded by pallor of the eyes and tail, syncope, and then coma associated with occasional respiratory gasps. The usual time range between marginally lethal doses (~ LD_{50}) of toxin and death, 1 to 2 hours, was not altered when supralethal doses (~ 4 x LD_{50}) were given. Doses that were more than or less than ~ 30% of the LD_{50} were lethal or non lethal, respectively. This indicates that the dose-effect curve is quite steep. From the distinct latent period between injection and onset of toxicity and from the steepness of the dose-effect curve, we suggest that there may be one class of toxin receptors which when activated initiates an unidentified cascade of events leading to death.

Necropsies for pathological study were performed promptly on moribund mice which were killed within 2 hours of injection of toxin-LR. Mouse livers were dark red, markedly enlarged, and engorged with blood. The observed 50 percent increase in fresh weight of the liver is attributed to the accumulation of about one-half of the total circulating blood in and around hepatic blood channels. The cerebral cortex was slightly swollen and pale, and a thin film of pink ascitic fluid was sometimes seen. No other abnormalities were observed.

The vital organs were also examined histologically. Mice dying from toxin-LR injections have multiple pulmonary thrombi such as those shown in Figure 2. These thrombi are not attached to vascular endothelium, and nearby endothelial cells are not swollen. The thrombus shown in Figure 2 is comprised of irregular clusters (< 15 µm diameter) of granules. Individual granules are about 1 µm in diameter. Such thrombi were readily detected by their purple color after staining with phosphotungstic acid-hematoxylin. No fibrin strands or erythrocytes were observed microscopically in the thrombi. There was an exclusive correlation between pulmonary thrombi and rapid death of mice from toxin-LR.

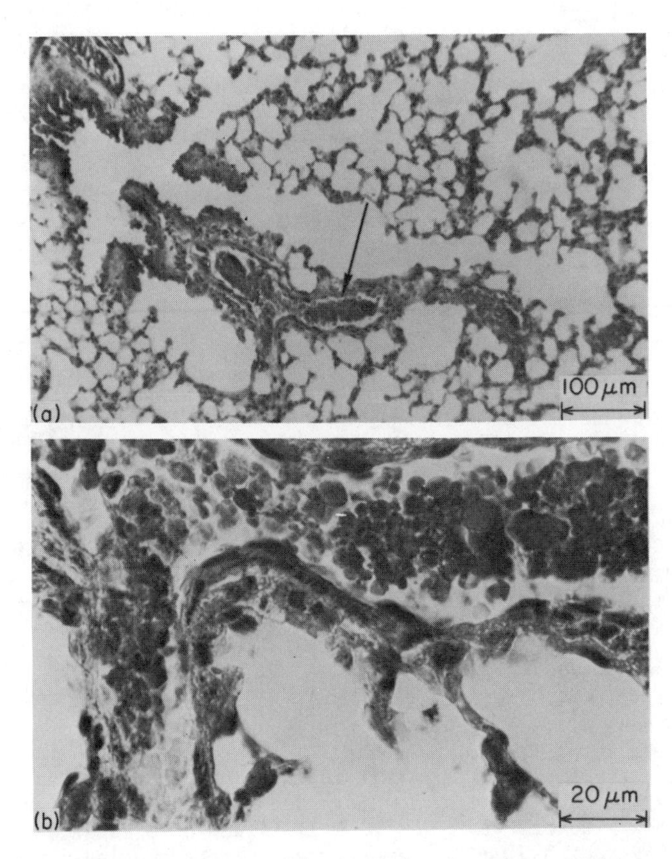

Figure 2. Photomicrographs (a) and (b) show low and high power views, respectively of a phosphotungstic acid hematoxylin-stained 5 μm thick section of lung from a mouse given a lethal dose of toxin-LR, and then killed when moribund. The arrow in (a) points to a thrombus in a pulmonary artery. The same thrombus is shown at higher magnification in (b). The thrombus consists of irregular clusters of small granules.

Livers from toxin-injected mice showed severe congestion whether or not the mice survived the initial critical 2-hour post injection period. Occasionally, a mouse injected with a nominally lethal dose of toxin-LR survived the critical two hours but remained listless until it died several hours or several days later. The reasons for the delayed death are unknown. Such mice developed focal fatty degeneration of the liver and active regeneration of liver cells.

The consistent occurrence of pulmonary thrombi after toxin-LR injection suggested that anticoagulant therapy might reverse the toxin action. Several different types of anticoagulants including heparin, acetylsalicylic acid, Malayan pit viper venom, streptokinase, and warfarin were tested (7). In preliminary studies, the time required for each anticoagulant to become effective was determined at two dose levels. Groups of 10 mice were injected at each dose level of anticoagulant. After the time required for the anticoagulant to become effective, a lethal injection of toxin-LR was given to the mice. The lethality of toxin-LR and the occurrence of pulmonary thrombi was not affected by any of the anticoagulant treatments. These results suggest that toxin-LR may initiate an unknown thrombogenic cascade.

The number of circulating blood platelet is markedly reduced by injection of a lethal dose of toxin-LR into mice (7). The time course of the decrease of the blood platelets is closely paralleled by the increase in fresh weight of the liver. The sharp rise in Spearman's rank correlation between platelet count and liver weight 30 minutes after injection indicates that thromocytopenia and hepatomegaly were almost concurrent.

The fatal complications caused by toxin-LR in mice are unusual. Toxin-LR is a potent thrombogenic agent. Generally, such agents bring about symptoms of illness within a few minutes of injection of a lethal dose. However, toxin-LR produces no overt signs of illness until about one-half hour has elapsed. Moreover, pretreatment of mice with known antithrombotic agents does not affect the lethality of the toxin.

At present, we suspect that toxin-LR causes heart failure in mice, perhaps due to suddenly increased resistance to pulmonary blood flow. Heart failure in mammals is known to cause engorgement of the liver with blood. Pulmonary vascular occlusion may also cause secondary hypoxemia and shock. However, biochemical pathways that are initiated by toxin-LR and that lead to the onset of discernable signs of illness after 30 min are unidentified. The 30 min asymptomatic period following toxin injection may be associated with a toxin-initiated cascade of biochemical events which lead to overt signs of illness.

Literature Cited

1. Botes, D. P.; Kruger, H.; Viljoen, C. C. Toxicon 1982, 20, 945-954.
2. Botes, D. P.; Viljoen, C. C.; Kruger, H.; Wessels, P. L.; Williams, D. H. Toxicon 1982, 20, 1037-1042.
3. Santikarn, S.; Williams, D. H.; Smith, R. J.; Hammond, S. J.; Botes, D. P.; Tuinman, A.; Wessels, P. L.; Viljoen, C. C.; Kruger, H. J. Chem. Soc. Commun. 1983, 652-654.
4. Eloff, J. N.; Siegelman, H. W.; Kycia, J. H. S. Afr. J. Sci. 1982, 78, 377 (Abstr.).
5. Elleman, T. C.; Falconer, I. R.; Jackson, A. R. B.; Runnegar, M. T. Aust. J. Biol. Sci. 1978, 31, 209-218.
6. Falconer, I. R.; Jackson, A. R. B.; Langley, J.; Runnegar, M. T. Aust. J. Biol. Sci. 1981, 34, 179-187.
7. Slatkin, D. N.; Stoner, R. D.; Adams, W. A.; Kycia, J. H.; Siegelman, H. W. Science 1983, 220, 1383-1385.
8. Eloff, J. N.; Steinitz, Y.; Shilo, M. Appl. Environm. Microbiol. 1976, 31, 119-126.
9. Leeuwangh, P.; Kappers, F. I.; Dekker, M.; Korselman, W. Aquatic Toxicology 1983, 4, 63-72.

RECEIVED March 2, 1984

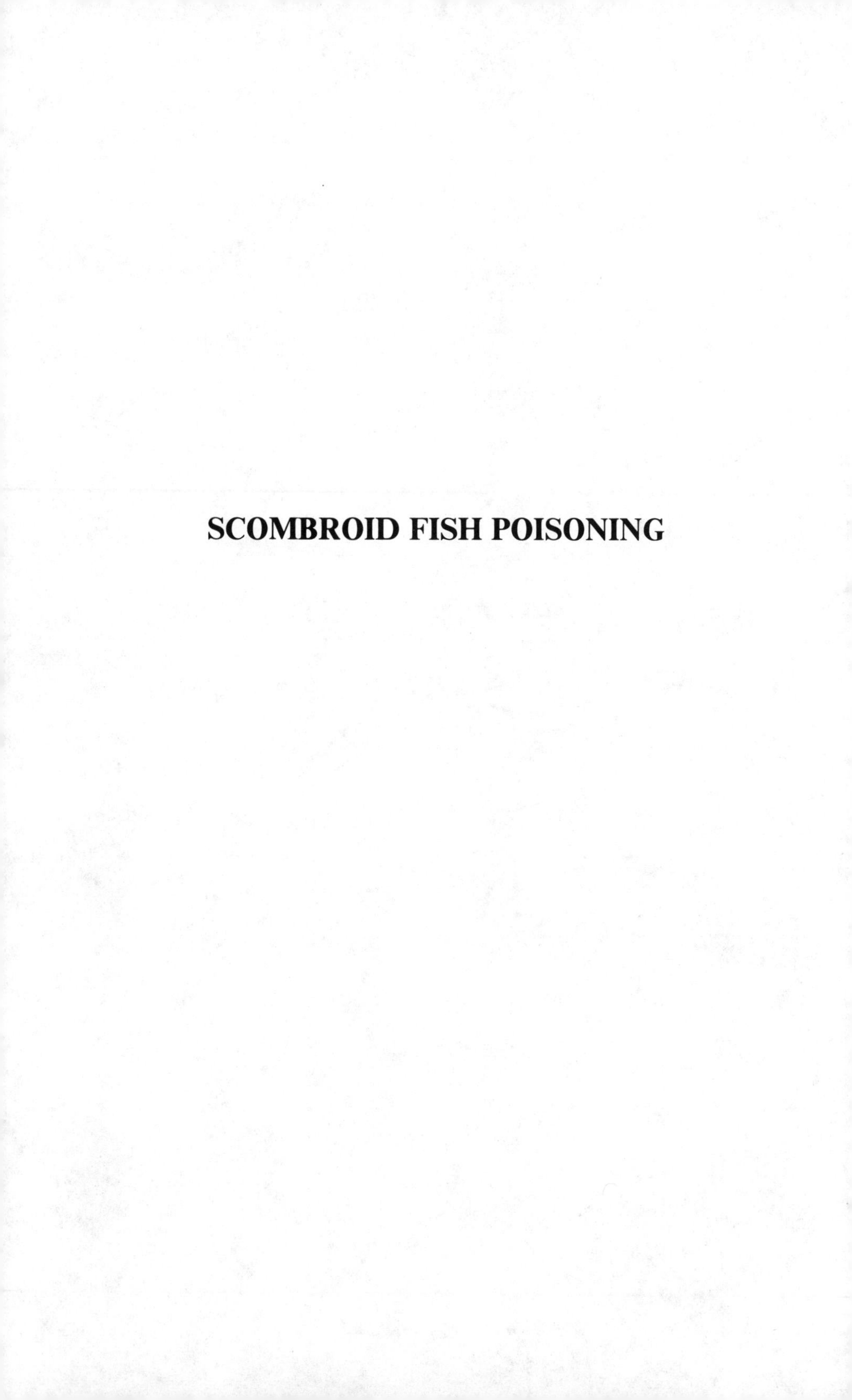

SCOMBROID FISH POISONING

35

Toxicology of Scombroid Poisoning

STEVE L. TAYLOR, JULIA Y. HUI, and DAVID E. LYONS[1]

Food Research Institute, University of Wisconsin, Madison, WI 53706

Scombroid poisoning is caused by ingestion of foods containing unusually high levels of histamine. Scombroid poisoning occurs worldwide in all countries where fish is consumed. The evidence supporting the role of histamine as the causative agent in scombroid poisoning is compelling. However, histamine ingested with spoiled fish is much more toxic than histamine ingested in an aqueous solution. This paradox may be explained by the presence of potentiators of histamine toxicity in spoiled fish. Several substances, including cadaverine and putrescine, have been identified as possible potentiators of histamine toxicity that would be expected to be present in spoiled fish. The mechanism of action of these potentiators has not been completely elucidated, but they appear to act by inhibition of intestinal histamine-metabolizing enzymes. This enzyme inhibition increases the intestinal uptake of unmetabolized histamine.

Scombroid poisoning is a chemical intoxication occurring after the ingestion of foods that contain unusually high levels of histamine. The incubation period for this foodborne disease is short ranging from several minutes to several hours following ingestion. The duration of the illness is typically only a few hours, but symptoms lasting up to several days have been reported (1,2).

A variety of symptoms can occur in cases of scombroid poisoning. The symptoms can be cutaneous (rash, urticaria, edema, localized inflammation), gastrointestinal (nausea, vomiting, diarrhea), hemodynamic (hypotension), and neurological (headache, palpitations, tingling, flushing or burning, itching). Most individuals suffering from scombroid poisoning will experience only a few of these symptoms. Even in group outbreaks, it may not be possible to observe all of these symptoms.

[1]Current address: Cooper Biomedical/Therapeutics Division, San Jose, CA 95119

0097-6156/84/0262-0417$06.00/0
© 1984 American Chemical Society

The symptoms vary in frequency of occurrence. Merson et al. (3), in an investigation of a large outbreak, listed nausea, cramps, and an oral burning sensation as the most common symptoms. Murray et al. (4), summarizing a series of outbreaks in England, reported rash, flushing with sweating, and an oral burning sensation as the most common symptoms.

Because the symptoms of scombroid poisoning mimic those observed in allergic reactions, the illness is sometimes misdiagnosed as an allergy. Scombroid poisoning can be easily distinguished from an allergic reaction (1). In group outbreaks, the attack rate often approaches 100%, while fish allergies are relatively rare. The patients involved in outbreaks of scombroid poisoning will often indicate that they have previously eaten the fish on many occasions without ill effects. Also, analysis of the remaining fish from an outbreak will reveal high levels of histamine in the fish.

Epidemiology of Scombroid Poisoning

Complete reporting of scombroid poisoning certainly does not occur on a worldwide basis. Many countries do not keep statistical data on the incidence of foodborne diseases. However, even in countries with good record-keeping, scombroid poisoning is incompletely reported for several reasons: (a) many patients do not seek medical attention for this relatively mild, short-lived malady, (b) physicians often misdiagnose scombroid poisoning as food allergy or some other foodborne disease, and (c) in most countries, scombroid poisoning is not a notifiable disease.

Worldwide Incidence. Table 1 lists the number and size of outbreaks of scombroid poisoning occurring in the United States, England, Japan, Canada, New Zealand, and Denmark during the period of 1971 through 1980. In recent years, England has experienced the largest number of outbreaks although most have been small. Japan has had the largest outbreaks and the greatest total number of cases of scombroid poisoning. Japanese outbreaks of scombroid poisoning have declined dramatically since 1980 following the adoption of strict control measures aimed at insuring that fish are held at 5°C or below after catching. Although less complete statistics have been kept, outbreaks have also occurred in West Germany, East Germany, France, Norway, Finland, Sweden, Portugal, Poland, Czechoslovakia, Sri Lanka, Indonesia, South Africa, Egypt, Turkey, Australia, South Korea, Hong Kong, People's Republic of China, and Republic of China (Taiwan).

Types of Fish. The fish most commonly implicated in these outbreaks are the so-called scombroid fish belonging to the families Scomberesocidae and Scombridae. These fish would include the many varieties of tuna, skipjack, bonito, albacore, mackerel, Spanish mackerel, bluefish, saury, butterfly kingfish, and seerfish. Tuna, skipjack, and mackerel are the most commonly involved scombroid fish.

Several types of non-scombroid fish can also be incriminated in outbreaks of scombroid poisoning. Thus, scombroid poisoning is a

TABLE I. Number and Size of Outbreaks of Histamine Poisoning Occurring in Certain Countries from 1971-1980

Year	Country	Number of Outbreaks	Number of Cases
1971	Japan	2	70
1972	Japan	4	137
	United States	6	?
1973	Japan	3	2702
	New Zealand	3	10
	United States	12	326
1974	Japan	1	33
	New Zealand	1	1
	United States	10	26
1975	Canada	1	1
	Japan	7	396
	United States	6	16
1976	Canada	1	2
	Great Britain	3	9
	Japan	4	31
	United States	3	43
1977	Great Britain	2	3
	Japan	3	69
	United States	13	71
1978	Canada	1	5
	Denmark	6	?
	Great Britain	3	9
	Japan	2	32
	United States	7	30
1979	Denmark	3	?
	Great Britain	43	179
	Japan	7	321
	United States	12	132
1980	Canada	2	3
	Denmark	8	?
	Great Britain	28	77
	Japan	4	153
	United States	NA	NA

misnomer. A more appropriate name for this foodborne disease would be histamine poisoning.

Among the non-scombroid fish, mahi-mahi is the most common fish implicated in histamine poisoning in the U.S. Other non-scombroid fish that have been involved in outbreaks of histamine poisoning are sardines, pilchards, anchovies, herring, black marlin, and kahawai.

Other Foods. Cheese has been implicated in several outbreaks of histamine poisoning in the U.S., Canada, France, and the Netherlands (5-8). Swiss cheese has been involved in all of the U.S. incidents and the French outbreak, while cheddar and Gouda cheese were involved in the Canadian and Dutch outbreaks, respectively. Ham has allegedly been involved in an outbreak in France; sauerkraut was implicated in one case in Germany (9); chicken was associated with an outbreak in Japan.

Evidence that Histamine is the Causative Agent

The evidence supporting the role of histamine as the causative agent of scombroid poisoning is rather compelling (1,10). Many incidents of scombroid poisoning have been investigated over the years and, in most cases where samples were available for analysis, the levels of histamine were high. The symptoms of scombroid poisoning are consistent with the involvement of histamine; they mimic the symptoms observed in allergic reactions or following intravenous histamine administration. Finally, the efficacy of antihistamines in the treatment of scombroid poisoning provides perhaps the strongest evidence for the involvement of histamine in this disease.

The results of animal and human challenge studies also support the role of histamine in scombroid poisoning to some extent. Oral administration of histamine to pigs and dogs elicits an emetic reaction (11,12). Intraduodenal injection of histamine in rats and cats produced transient hypotension (13). Intraduodenal injection of a histamine containing yeast extract into cats produced a variety of histamine like effects including increased volume and acidity of gastric secretion, increased hematocrit and limb volume, and enhanced electromyographic activity (13). Motil and Scrimshaw (14) observed some mild and transient symptoms reminiscent of scombroid poisoning in human challenge studies following oral administration of 100-180 mg of histamine in combination with wholesome tuna.

While the evidence implicating histamine as the causative agent of scombroid poisoning is compelling, Japanese investigators at one time isolated a histamine like substance called saurine that was possibly involved in scombroid poisoning (15). Saurine has since been identified as the phosphate salt of histamine (16).

The Paradox

Given the compelling evidence that histamine is the causative agent in scombroid poisoning, orally administered histamine is remarkably non-toxic to humans (14,17,18). Weiss et al. (17) first demonstrated the lack of toxicity of orally administered histamine to humans in 1932. They showed that 180 mg of histamine was without noticeable effect when administered orally while 7 μg of histamine would cause vasodilation and increases in the heart rate when given intravenously. Granerus (18) challenged human subjects orally with up to 67.5 mg of histamine and observed a similar lack of toxic response. Motil and Scrimshaw (14) administered 100-180 mg of histamine in combination with tuna and noted some transient toxic effects as mentioned earlier. However, the doses necessary to elicit these mild symptoms were several times higher than the doses producing more severe symptoms when consumed with spoiled scombroid fish.

This paradox between the lack of toxicity of pure histamine and the apparent toxicity of equivalent doses of histamine in spoiled fish could be explained by the existence of potentiators of histamine toxicity in spoiled fish. These potentiators would serve to lower the threshold dose of histamine necessary to elicit scombroid poisoning symptoms in humans.

Evidence for the Involvement of Potentiators of Histamine Toxicity

Several possible potentiators of histamine toxicity have been suggested by various *in vivo* and *in vitro* experiments, although none of these substances has been clearly implicated in scombroid poisoning.

Miyaki and Hayashi (19) isolated an unidentified synergistic factor from dried seasoned saury. Subsequently, Hayashi (20) reported that trimethylamine, trimethylamine oxide, agmatine, and choline could potentiate the contractile effect of histamine on guinea pig uterus. This report was later disputed by Kawabata et al. (21) who found that trimethylamine and trimethylamine oxide were ineffective in promoting the action of histamine on guinea pig uterus. Mongar (22) observed that short-chain aliphatic diamines, such as cadaverine and putrescine, could competitively inhibit diamine oxidase (DAO), an enzyme active in detoxification of histamine. These diamines could potentiate histamine-induced contractions of guinea pig ileum (22). Inhibitors of DAO, such as cadaverine, were shown to potentiate the contraction of guinea pig ileum, trachea, and uterus (23).

Several *in vivo* experiments in animals and humans suggest the existence of potentiators of histamine toxicity in spoiled fish. Parrot and Nicot (24) demonstrated that putrescine enhanced the lethality of orally administered histamine in guinea pigs. Cadaverine has been shown to have a similar effect on the oral toxicity of histamine in guinea pigs (25). Weiss et al. (17) showed that 180 mg of histamine was without effect when administered orally to humans, while Motil and Scrimshaw (14) did observe some toxic symptoms after oral administration of an equivalent dose of histamine with "wholesome" tuna.

Further experiments will be necessary with fish incriminated in histamine poisoning outbreaks to confirm the presence and identity of any potentiators of histamine toxicity.

Histamine Metabolism and its Inhibition

The principal pathways for the biogenesis and metabolism of histamine are well known. Histamine is formed by decarboxylation of the amino acid, L-histidine, a reaction catalyzed by the enzyme, histidine decarboxylase. This decarboxylase is found in both mammalian and non-mammalian species. The mammalian enzyme requires pyridoxal phosphate as a cofactor. The bacterial enzyme has a different pH optimum and utilizes pyruvate as a cofactor (26,27).

Two major histamine-metabolizing enzymes, DAO (histaminase) and histamine-N-methyltransferase (HMT) are found in mammalian tissues. The product of HMT, N^{τ}-methyl-histamine (N^{τ}-MH), is mostly converted by monoamine oxidase (MAO) to N^{τ}-methylimidazole acetic acid (N^{τ}-MIAA). In the other pathway, the product of DAO, imidazole acetic acid (IAA), can be conjugated to form imidazole acetic acid riboside (IAA-riboside). Very little histamine is excreted unchanged. The relative importance of the two enzymatic routes varies among species (28). In man, the pathways for parenterally administered [^{14}C]-histamine have been mapped (29). Analysis of urine samples showed that N^{τ}-MIAA was the principal metabolite (42-47%), with smaller quantities of IAA-riboside (16-23%), IAA (9-11%), N^{τ}-MH (4-8%), and unchanged histamine (2-3%). When [^{14}C]-histamine was given orally instead of intravenously, much less radioactivity could be recovered in the urine (18). Sjaastad and Sjaastad (30) explained this low recovery by a finding that part of the radioactive metabolites were excreted in the feces. They also found that some of the radioactivity was exhaled as $^{14}CO_2$, due to metabolism of the imidazole ring in the gut. A portion of the orally administered histamine is also converted by intestinal bacteria to N-acetylhistamine. HMT and DAO activities are quite high in the small intestinal mucosa (31), and it is thought that the majority of an oral dose of histamine is metabolized as it traverses the intestinal mucosa or as it circulates through the liver (32).

HMT Inhibitors. HMT is very selective toward histamine. In the HMT reaction, S-adenosylmethionine serves as the methyl donor. HMT is inhibited by S-adenosylmethionine analogues, such as S-adenosylhomocysteine (33). Other inhibitors of HMT include antimalarial drugs, e.g. chloroquin and amodiaquin, and antihistaminic agents (34-36). Together with DAO, this enzyme is subject to substrate inhibition by high concentrations of histamine (37). HMT is also inhibited by the histamine H_2 receptor agonists, dimaprit and impromidine (38,39).

DAO Inhibitors. Unlike HMT, DAO is not selective towards histamine. It oxidizes other diamines, e.g. putrescine. Many inhibitors of DAO have been identified; the most well-known example is aminoguanidine. Many antihistaminic agents and related drugs are good inhibitors of DAO (40). Other DAO inhibitors include bases, such as amidines and guanidines, carbonyl agents, substituted hydrazines, and chelating agents (41,42). Many of these known DAO inhibitors belong to a class of drugs called monoamine oxidase inhibitors (43).

MAO Inhibitors. MAO is an enzyme which oxidizes a variety of monoamines. Among the substrates of this enzyme are tyramine, tryptamine, 5-hydroxytryptamine, histamine, and short chain aliphatic monoamines (44). Oxidation of histamine to imidazoleacetaldehyde can be carried out by DAO as well as MAO. MAO is also responsible for the conversion of N^{τ}-MH, the product of HMT, to N^{τ}-MIAA. Many MAO inhibitors have been identified; they are conventionally divided into hydrazides, hydrazines and amines (44). Some MAO inhibitors, e.g. the hydrazines, are non-selective; they also inhibit DAO.

Potentiation via Enzyme Inhibition

Since the majority of an oral dose of histamine is excreted as various histamine metabolites, the metabolism of histamine probably serves as a detoxification mechanism. As noted earlier, the small intestine and liver are particularly active in histamine metabolism and would be expected to protect against the toxicity of orally administered histamine.

The hypothesis that histamine toxicity could be potentiated by inhibition of histamine-metabolizing enzymes was supported by early experiments that demonstrated a potentiation of histamine-induced contractions of smooth muscle by inhibitors of DAO (22,23). However, these experiments were not directed at the oral toxicity of histamine. Taylor and Lieber (45) showed that rat intestinal HMT and DAO could be inhibited *in vitro* by certain amines, including some putrefactive amines that are known to occur in spoiled fish (46,47). Many of these amines inhibited only one of the two histamine-metabolizing enzymes, but several including cadaverine and aminoguanidine were effective inhibitors of both HMT and DAO (45). Mixtures of the inhibitors were not tested, but would be predicted to be quite effective in inhibiting intestinal histamine metabolism.

The *in vitro* experiments of Lyons et al. (48) provide the best evidence that potentiation operates through the inhibition of the histamine-metabolizing enzymes. In these experiments, rat ileal segments were perfused *in vitro* with either histamine alone or histamine plus an equimolar amount of cadaverine, aminoguanidine, or anserine. A tracer quantity of ^{3}H-histamine was added to the perfusate. The ileal segments were bathed in an oxygenated, nutrient medium throughout the experiment. The serosal fluid was sampled periodically during a 2-h incubation, counted for total tritium, and monitored for histamine and its major metabolites after separation by thin-layer chromatography. The transport of radioactivity from the lumen to the serosal fluid was not affected by the presence of cadaverine, aminoguanidine, or anserine. However, as shown in Table II, the ratio of histamine to its metabolites was altered substantially by the presence of cadaverine and aminoguanidine. In the presence of these substances, more unmetabolized histamine reached the serosal fluid indicating that inhibition of intestinal histamine metabolism had been effective in potentiating the uptake of histamine. Anserine was ineffective.

Table II. Percent of Serosal Radioactivity Represented by Histamine after 2-h Intestinal Perfusions in the Presence or Absence of Potentiators(48)

Additions to Luminal Perfusate[a]	% Histamine[b]
Histamine	22.9 ± 5.9
Histamine + Cadaverine	67.0 ± 11.9
Histamine + Aminoguanidine	60.4 ± 5.1
Histamine + Anserine	30.6 ± 6.4

[a]Initial concentration of each amine in the luminal perfusate was 5.4 mM.

[b]Average ± SEM of 4 trials.

Further research will be necessary to demonstrate conclusively that inhibition of histamine metabolism is responsible for the potentiation of histamine toxicity that is apparently observed in scombroid poisoning. In vivo experiments will be necessary to show that hepatic histamine metabolism is also compromised by the ingestion of suspected potentiators. Also, the effectiveness of cadaverine and other possible potentiators must be demonstrated under conditions where the histamine level exceeds the potentiator concentration by a factor of approximately 10. This concentration ratio would parallel that found in spoiled tuna more closely than the levels used in the experiments of Lyons et al. (48).

Additional circumstantial evidence for the validity of the inhibition hypothesis of potentiator action may come from several recent outbreaks of histamine poisoning. Isoniazid, an antituberculosis agent and known inhibitor of DAO, has played a role in several outbreaks of histamine poisoning (49-52). These outbreaks all involved patients on isoniazid therapy who developed histamine poisoning symptoms after ingestion of fish or cheddar cheese containing marginally toxic amounts of histamine. In these cases, inhibition of DAO by isoniazid may have enhanced the toxicity of these marginally toxic levels of histamine.

Other Possible Mechanisms of Potentiation

Another possible mechanism of potentiator action, the barrier disruption hypothesis, has received considerable attention. This theory of potentiator action, first proposed by Parrot and Nicot (24), suggests that the potentiators may interfere with the protective actions of intestinal mucin. Mucin is known to bind histamine in vitro (53), and Parrot and Nicot (24) suggested that this binding was essential to prevent passage of histamine across the intestinal wall. Potentiators such as putrescine and cadaverine may bind competitively and preferentially with mucin facilitating the movement of histamine into the circulation (24,54).

The results of the intestinal perfusion studies of Lyons et al. (48) cannot be explained on the basis of the barrier disruption hypothesis. The barrier disruption hypothesis would predict an increase in the overall transport of radioactivity in the presence of potentiators; this was not observed by Lyons et al. (48). The changes in the ratio of histamine to its metabolites obtained in the experiments of Lyons et al. (48) would not be predicted by the barrier disruption hypothesis either.

In support of the barrier disruption hypothesis, Jung and Bjeldanes (54) demonstrated that intestinal transport of radioactivity was potentiated by incubation of equimolar amounts of cadaverine and ^{14}C-histamine in non-everted gut sacs over a 2-h incubation period. Additionally, the ratio of histamine to its metabolites was not altered in the presence of cadaverine. These results are in direct conflict with those obtained by Lyons et al. (48). However, the experiments of Jung and Bjeldanes (54) appear to have been flawed. Plattner et al. (55) showed that everted gut sacs are useful for only 30 min in oxygenated, nutrient medium before considerable cell membrane damage occurs. Non-everted gut sacs, as used by Jung and Bjeldanes (54) would be predicted to be less viable due to the greater lack of oxygen and nutrient exposure of the epithelial cells of the villi. Jung and Bjeldanes (54) used 3-h incubations of the non-everted gut sacs and provided no histological evidence of tissue integrity. Lyons et al. (48) demonstrated extensive histological damage in non-everted gut sacs of rats after 30 min of incubation.

Later, Chu and Bjeldanes (53) showed that the binding histamine to mucin could be inhibited *in vitro* by spermine, spermidine, putrescine, cadaverine, and a basic extract of tuna. The inhibition required relatively high amine concentrations, and the concentrated tuna extract exerted only a 23% inhibition of binding (53). Each mole of intestinal mucin can bind 2.5 moles of histamine (53). Given the vast molar excess of mucin to histamine in the gastrointestinal tract, it is difficult to envision that such inhibition of binding would play more than a secondary role in the potentiation of histamine toxicity.

Pharmacological-Toxicological Actions of Histamine

H_1 and H_2 Receptors. Histamine exerts its actions by binding to receptors on cell membranes. Two types of histamine receptors, the H_1 and H_2 receptors, are known; specific agonists and antagonists exist for each of these receptors. Black et al. (55) differentiated H_1 and H_2 receptors with the compounds, 2-methylhistamine and 4-methylhistamine. 2-Methylhistamine is active on tissues with H_1 receptors; 4-methylhistamine is active on tissues with H_2 receptors. Classical antihistaminic drugs were developed in the 1930's; these compounds block H_1 but not H_2 receptors. Among the clinically used H_1-blockers are derivatives of ethanolamine, ethylenediamine, alkylamine, piperazine and phenothiazine (32). These agents are valuable in the treatment of various allergic responses, common cold, and motion sickness. Many of these agents are also used as hypnotics, and as local anesthetics to relieve pain and itching. The classical antihistaminics block many but not all of the manifestations of allergic reactions.

This lack of complete effectiveness led to the hypothesis that a second type of histamine receptor existed. In 1972, Black et al. (55) discovered a new series of antagonists which they called H_2 receptor blockers. Burimamide was the first highly effective H_2-blocker, but it was poorly absorbed orally. The modified compound, metiamide, had better absorption but was found to cause granulocytopenia (57,58). Finally, cimetidine was tested and found to be a potent and relatively non-toxic antagonist (59). Cimetidine is now widely used clinically to treat duodenal ulcers, Zollinger-Ellison Syndrome and other gastric hypersecretory diseases (32).

Mast Cells and Basophils. The chief sites of histamine storage are mast cells in the tissues and basophils in blood. These cells synthesize histamine and store it in secretory granules along with a heparin-protein complex. In response to specific antigens, mast cells or basophils are sensitized. Histamine is then secreted from the storage granules. Besides the histamine stores in mast cells and basophils, there is evidence of non-mast cell histamine in some tissues, particularly gastric and intestinal mucosa (60).

Pharmacological-Toxicological Responses to Histamine. The principal pharmacological actions of histamine were outlined by Dale and Laidlaw in the 1910's (61,62). They noted that the predominant effects of histamine varied between species. Many of these effects are similar to the symptoms seen in histamine poisoning outbreaks (3,10). In man, the effects of histamine on the cardiovascular system are very important. The actions on this system involve both H_1 and H_2 receptors. Histamine causes dilatation of small blood vessels and capillaries, and constriction of larger blood vessels (63). The symptoms of flushing and headache experienced by many of the victims are caused by the dilating action of histamine on small blood vessels, capillaries, and venules. Another classical effect of histamine on small blood vessels is the increased permeability (32). This is responsible for the urticaria (hives) seen in many outbreak cases. In a small percentage of victims, cardiac palpitation was reported (9), which results from the direct effects of histamine on the heart. Histamine increases both the contractility and the pacemaker rate of the heart (32). In very severe cases, histamine shock can occur (10) due to a profound fall in blood pressure, but this requires large doses of histamine (32).

Histamine also acts on extravascular smooth muscles to cause contraction or relaxation. Most often, contraction is due to activation of H_1 receptors and relaxation to activation of H_2 receptors (32). In man, histamine causes contraction of bronchial and intestinal smooth muscles. Histamine-induced contraction of guinea pig ileum is a standard bioassay for histamine. Its effects on smooth muscle of the eye and genitourinary tract are important in some species but not in human (64). In scombroid poisoning cases, most of the victims experienced gastrointestinal symptoms, i.e. abdominal cramps, diarrhea, nausea, and vomiting (34). It is known that contraction of intestinal smooth muscle leads to cramps and diarrhea (64). However, the mechanism that leads to nausea and vomiting is not clearly understood.

Histamine also evokes a copious secretion of highly acidic gastric juice from the gastric glands at doses below those that influence blood pressure (32). This effect of histamine is mediated through H_2 receptors on the parietal cells. The importance of this effect in scombroid poisoning is not known. Histamine also has some stimulant actions on salivary, pancreatic, intestinal, bronchial, and lacrimal secretions (32), but these effects are relatively unimportant.

Histamine is a powerful stimulant of nerve endings, both motor and sensory nerves (32). Its stimulation is important in producing pain and itching, which are important components of the urticarial response and reactions to insect stings (64). This effect is mediated through H_1 receptors. Such nerve stimulation might conceivably be important in the initiation of the emetic response.

Threshold Toxic Dose for Histamine in Foods and Regulatory Limits

The threshold toxic dose for histamine in foods is not precisely known. Estimates are difficult to acquire from outbreaks of histamine poisoning because of the variability in histamine content in the fish (1,65). Simidu and Hibiki (66) estimated the threshold toxic dose for histamine in fish to be approximately 60 mg/100 g, but their methods were not terribly precise. Based on experience acquired in the investigation of hundreds of scombroid poisoning incidents, the U.S. Food and Drug Administration recently established 50 mg/100 g as the hazard action level for histamine in tuna. They have not yet established regulatory limits for histamine in other fish or cheese.

The existence of potentiators could dramatically influence the threshold toxic dose for histamine in foods. Since different fish might be expected to vary in the type and amount of the various potentiators, the threshold toxic dose for histamine would be expected to vary from fish to fish also. The differences in type and amount of the potentiators would be predicted from expected differences in the types of microflora, the metabolic capabilities of the microflora, and conditions of spoilage. Even greater differences would be expected in the comparisons of different species of fish or in comparisons of fish and cheese. Consequently, although the health hazard associated with ingestion of tuna containing 50 mg histamine per 100 g is established, the hazard associated with 50 mg histamine/100 g in other fish and cheese remains to be determined.

Until the role of the potentiators can be more completely elucidated, it may be premature to establish regulatory limits for histamine in fish or other foods on the basis of health hazards.

Literature Cited

1. Lerke, P. A.; Werner, S. B.; Taylor, S. L.; Guthertz, L. S. West. J. Med. 1978, 129, 381-6.
2. Todd, E.; Lavallee, J.; Martin, R. S.; Greene, V. C.; Gilgan, M. W.; Landry, G.; Murphy, S. S. Canada Diseases Wkly. Rpt. 1981, 7, 77-8.
3. Merson, M. H.; Baine, W. B.; Gangarosa, E. J.; Swanson, R. C. J. Am. Med. Assoc. 1974, 228, 1268-9.
4. Murray, C. K.; Hobbs, G.; Gilbert, R. G. J. Hyg. Camb. 1982, 88, 215-20.
5. Doeglas, H. M. G.; Huisman, J.; Nater, J. P. Lancet 1967, ii, 1361-2.
6. Chambers, T. L.; Staruszkiewicz, W. F., Jr. J. Assoc. Off. Anal. Chem. 1978, 61,1092-7.
7. Taylor, S. L.; Keefe, T. J.; Windham, E. S.; Howell, J. F. J. Food Prot. 1982, 45, 455-7.
8. Kahana, L. M.; Todd, E. Canada Diseases Wkly. Rpt. 1981, 7, 79-80.
9. Mayer, K.; Pause, G. Lebensm.-Wiss. Technol. 1972, 5, 108-9.
10. Arnold, S. H.; Brown, W. D. Adv. Food Res. 1978, 24, 113-54.
11. Blonz, E. R.; Olcott, H. S. Comp. Biochem. Physiol. 1978, 61C, 161-3.
12. Henry, S.; Sobotka, T.; Staruszkiewicz, W.; Olivito, V.; Michel, T. Abst. 19th Ann. Mtg. Soc. Toxicol. 1980, A27.
13. Blackwell, B.; Marley, E. Br. J. Pharmacol. 1966, 26, 142-61.
14. Motil, K. J.; Scrimshaw, N. S. Toxicol. Lett. 1979, 3, 219-23.
15. Kawabata, T.; Ishizaka, K.; Miura, T. Jpn. J. Med. Sci. Biol. 1955, 8, 521-8.
16. Foo, L. Y. J. Sci. Fd. Agric. 1976, 27, 807-10.
17. Weiss, S.; Robb, G. P.; Ellis, L. B. Arch. Intern. Med. 1932, 49, 360-96.
18. Granerus, G. Scand. J. Lab. Clin. Invest. 1968, 22 (Suppl. 104), 49-58.
19. Miyaki, K.; Hayashi, M. J. Pharm. Soc. Japan 1954, 74, 145-48.
20. Hiyashi, M. J. Pharm. Soc. Japan 1954, 74, 1148-51.
21. Kawabata, T.; Ishizaka, K.; Miura, T. Bull. Jpn. Soc. Sci. Fish. 1956, 21, 1177-80.
22. Mongar, J.-L. Br. J. Pharmacol. 1957, 12, 140-148.
23. Arunlakshana, O.; Mongar, J. L.; Schild, H. O. J. Physiol., Lond. 1954, 123, 32-.
24. Parrot, J.-L.; Nicot, G. In "Handbook of Experimental Pharmacology"; Eichler, O.; Farah, A., Eds.; Springer-Verlag: New York, 1966; Vol. 18, Pt. 1, pp. 148-61.
25. Bjeldanes, L. F.; Schutz, D. E.; Morris, M. M. Food Cosmet. Toxicol. 1978, 16, 157-9.
26. Beavan, M. A. In "Histamine: Its Role in Physiological and Pathological Processes"; Dukor, P.; Kallos, P.; Truka, Z.; Walesman, B. H.; deWeck, A. L., Eds.; MONOGRAPHS IN ALLERGY Vol. 13, S. Karger: Basel, Switzerland, 1978; pp. 4-30.
27. Recsei, P. A.; Moore, W. M.; Snell, E. E. J. Biol. Chem. 1983, 258, 439-44.
28. Schayer, R. W. In "Handbook of Experimental Pharmacology"; Rocha e Silva, Ed.; Springer: Berlin, 1978; Vol. 18, Pt. 2, pp. 131-50.

29. Schayer, R. W.; Cooper, J. A. D. J. Appl. Physiol. 1956, 9, 481-3.
30. Sjaastad, O.; Sjaastad, O. V. Acta Pharmacol. Toxicol. 1974, 34, 33-45.
31. Kim, K. S.; Backus, B.; Harris, M.; Rourke, P. Comp. Biochem. Physiol. 1969, 31, 137-45.
32. Douglas, W. W. In "The Pharmacological Basis of Therapeutics"; Gilman, A. G.; Goodman, L. S.; Gilman, A., Eds.; Macmillan Publ. Co., Inc.: New York, 1980, 6th ed.; pp. 609-46.
33. Baudry, M.; Chast, F. ; Schwartz, J. C. J. Neurochem. 1973, 20, 13-21.
34. Cohn, V. H. Biochem. Pharmacol. 1965, 14, 1686-8.
35. Taylor, K. M.; Snyder, S. H. Mol. Pharmacol. 1972, 8, 300-10.
36. Taylor, K. M. Biochem. Pharmacol. 1973, 22, 2775-6.
37. Taylor, S. L.; Lieber, E. R. Comp. Biochem. Physiol. 1979, 63C, 21-6.
38. Shaff, R. E.; Beaven, M. A. Biochem. Pharmacol. 1977, 26, 2075-8.
39. Beaven, M. A.; Roderick, N. B. Biochem. Pharmacol. 1980, 29, 2897-900.
40. Finazzi-Agro, A.; Floris, G.; Fadda, M. B.; Crifo, C. Agents and Actions 1979, 9, 244-7.
41. Buffoni, F. Pharmacol. Review 1966, 18, 1163-99.
42. Zeller, E. A. In "Diamine oxidases"; Boyer, P. D.; Lardy, H.; Myrback, K., Eds.; THE ENZYMES 2nd Ed., Academic Press: New York, 1963; p. 313-35.
43. Crabbe, M. J. C.; Bardsley, W. G. Biochem. Pharmacol. 1974, 23, 2983-90.
44. Marley, E.; Blackwell, B. Adv. Pharmacol. Chemother. 1970, 8, 185-249.
45. Taylor, S. L.; Lieber, E. R. Food Cosmet. Toxicol. 1979, 17, 237-40.
46. Lukton, A.; Olcott, H. S. Food Res. 1958, 23, 611-8.
47. Mietz, J. L.; Karmas, E. J. Food Sci. 1977, 42, 155-8.
48. Lyons, D. E.; Beery, J. T.; Lyons, S. A.; Taylor, S. L. Toxicol. Appl. Pharmacol. 1983, 70, 445-58.
49. Kahana, L. M.; Todd, E. Canada Diseases Wkly. Rpt. 1981, 7, 117-8.
50. Uragoda, C. G.; Kottegoda, S. R. Tubercle 1977, 58, 83-9.
51. Senanayake, N.; Vyravanathan, S. Toxicon 1981, 19, 184-5.
52. Senanayake, N.; Vyravanathan, S.; Kanagasuriyam, S. Br. Med. J. 1978, 2, 1127-8.
53. Chu, C.-H.; Bjeldanes, L. F. J. Food Sci. 1981, 47, 79-80, 88.
54. Jung, H.-Y. P.; Bjeldanes, L. F. Food Cosmet. Toxicol. 1979, 17, 629-32.
55. Plattner, H.; Klima, J.; Mehnert, A.; Berger, H. Virchows Arch. Abt. B Zellopath. 1970, 6, 337-49.
56. Black, J. W.; Duncan, W. A. M.; Durant, C. J.; Ganellin, C. R.; Parsons, E. M. Nature (Lond.) 1972, 236, 385-90.
57. Black, J. W.; Durant, G. J.; Emmett, J. C.; Ganellin, C. R. Nature (Lond.) 1974, 248, 65-7.
58. Wood, C. J.; Simkins, M. A. Int. Symp. on Histamine H_2-Receptor Antagonists, 1973, Smith Kline and French Lab. Ltd., Welwyn, Garden City.

59. Brimblecoombe, R. W.; Duncan, W. A. M.; Durant, G. J.; Emmett, J. C.; Ganellin, C. R.; Parsons, M. E. J. Int. Med. Res. 1975, 3, 86-92.
60. Beavan, M. A.; Horakova, Z.; Severs, W. B.; Brodie, B. B. J. Pharmacol. Exp. Ther. 1968, 161, 320-8.
61. Dale, H. H.; Laidlaw, P. P. J. Physiol. (Lond.) 1910, 41, 318-44.
62. Dale, H. H.; Laidlaw, P. P. J. Physiol. (Lond.) 1911, 43, 182-95.
63. Beaven, M. A. "Histamine: Its Role in Physiological and Pathological Processes"; Karger, S.: Basel (Switzerland), 1978; p. 36-50.
64. Burkhalter, A.; Frick, O. L. In "Histamine, Serotonin, and the Ergot Alkaloids"; Katzung, B. G., Ed.; BASIC AND CLINICAL PHARMACOLOGY, Lange Medical Publications: Los Altos, Calif., 1982; p. 169-85.
65. Frank, H. A.; Yoshinaga, D. H.; Nip, W.-K. Mar. Fish. Rev. 1981, 43(10), 9-14.
66. Simidu, W.; Hibiki, S. Bull. Japan, Soc. Sci. Fish. 1955, 21, 365-7.

RECEIVED February 6, 1984

36

Enzymatic Mechanisms for Amine Formation in Fish

RONALD R. EITENMILLER and SELWYN C. DE SOUZA

Department of Food Science, University of Georgia, Athens, GA 30602

Amine build-up in fish muscle usually results from decarboxylation of amino acids in the muscle by enzymes of bacterial origin. This review will present information on the activity of bacterial decarboxylases and the formation of amines in fish. Mechanisms of decarboxylase action and production of bacterial decarboxylases in fish muscle are discussed. Emphasis is placed upon studies dealing with formation of histidine decarboxylase and histamine. Histamine, because of its involvement in Scombroid food poisoning, has been extensively studied with regard to its formation in fish and fishery products.

Fish muscle has the ability to support formation of a wide variety of amine compounds that result from the direct enzymatic decarboxylation of amino acids. Because decarboxylation requires the presence of a free amino acid, the substrate for the enzyme, amine build-up normally occurs during a decomposition or spoilage process involving formation of free amino acids through proteolysis together with bacterial production and action of an amino acid decarboxylase. Interest in the enzymatic formation of amines in fish stems from the capacity of specific amines to produce organoleptic changes associated with decomposition, the potential to utilize such amines as quality indicators and the ability of some amines to produce a physiological response in the consumer when ingested.

Examples of amines that frequently occur in fish muscle include cadaverine from lysine, putrescine from ornithine and histamine from histidine. Histamine, because of its involvement in Scombroid food poisoning (histamine intoxication), has been extensively studied with respect to factors influencing its formation in many different fish species. Tuna and other fish from the families Scomberesocidae and Scombridae and a non-scombroid fish, mahi-mahi (dolphin fish) have been most

0097-6156/84/0262-0431$06.00/0
© 1984 American Chemical Society

commonly noted to contain elevated levels of histamine. These species have been associated with incidents of histamine intoxication (1). Muscle of the mahi-mahi, like the Scombroid fish, contains large amounts of histidine and is readily susceptible to formation of toxic levels of histamine. Histamine in such species poses significant detection problems due to the lack of rapid detection methods, the potential for histamine to be produced to toxic levels without formation of organoleptic spoilage indicators and unequal distribution of the toxic agent in the fish (1).

Because amine formation in fish muscle and other foods usually results from bacterial growth with concomitant production of a bacterial decarboxylase, this paper will concentrate on the mechanisms of bacterial decarboxylation and factors influencing the production and activity of the enzymes. Also, because of the overall scope of the subject, the availability of excellent reviews on bacterial decarboxylation (2, 3) and the public health importance of histamine in fish and fishery products, this paper will primarily be limited to a discussion of histidine decarboxylase (EC 4.1.1.22) and the formation of histamine in fish muscle.

History of Research on Bacterial Decarboxylases

Table I presents some of the historically important events in research completed to date on bacterial decarboxylation. This area of research has been important to the understanding of microbial metabolism and mechanisms of enzyme action. As early as 1928, Koessler *et al* (4) postulated that bacterial amine formation may be a protective mechanism against an acid environment. This phenomenon became known as the protective action theory, whereby bacteria will synthesize a decarboxylase(s) in an attempt to produce amines and increase the pH of the environment. During the period 1940-1946, E. F. Gale and coworkers (5-7) established the existence of six inducible amino acid decarboxylases most of which required an unidentified cofactor. In 1945, Bellamy and Gunsalus (8) and Baddiley and Gale (6) showed that decarboxylases for tyrosine, lysine, arginine and ornithine required a phosphorylated derivative of pyridoxal as the cofactor. Further, Baddiley and Gale (6) concluded that the activity of glutamate and histidine decarboxylase did not depend upon the phosphorylated pyridoxal derivative. It was not until the early 1950's that the phosphorylated pyridoxal derivative was identified as pyridoxal -5'- phosphate (9). Further, in 1960 Shukuya and Schwert (10) showed that the glutamate decarboxylase did depend upon the presence of a tightly bound pyridoxal phosphate moiety.

Considerable work has been completed on the histidine decarboxylase from *Lactobacillus* 30a, a strain of

Table I. Historical Perspective on Bacterial Amino Acid Decarboxylases

Date	Findings	Reference
1928	Theory that bacterial amine formation may be a protective mechanism against acid environment postulated.	4
1940-46	Presence of cofactor-dependent inducible amino acid decarboxylases in bacteria established.	5, 6, 7
1945	Dependence of tyrosine, lysine, arginine and ornithine decarboxylase on a phosphorylated pyridoxal derivative shown.	6, 8
	Non-dependence of glutamate and histidine decarboxylases on a phosphorylated pyridoxal derivative suggested.	6
1951	Phosphorylated pyridoxal derivative proved to be pyridoxal 5'-phosphate.	9
1953	Non-dependence of histidine decarboxylase from Lactobacillus 30a on pyridoxal phosphate suggested.	11
1954	Mechanism of action of amino acid decarboxylation proposed.	18, 19
1960	Glutamate decarboxylase found to be dependent on pyridoxal phsophate.	10
1965	Histidine decarboxylase from Lactobacillus 30a isolated, free of pyridoxal phosphate.	12
1968	Pyruvoyl residue on histidine decarboxylase from Lactobacillus 30a identified. Subunit structure proposed.	13, 14
1970	Participation of a pyruvoyl moiety in Schiff base formation shown.	14
1981	Subunit structure of histidine decarboxylase from Lactobacillus 30a found to be related to that of Micrococcus sp.	15

Lactobacillus delbrueckii. In 1953, Rodwell suggested that the histidine decarboxylase of *Lactobacillus* 30a was not dependent upon pyridoxal phosphate (11). Rodwell based his suggestion upon the fact that the organism lost its ability to decarboxylate ornithine but retained high histidine decarboxylase activity when grown in media deficient in pyridoxine. It was not until 1965 that E. E. Snell and coworkers (12) isolated the enzyme and showed that it was, indeed, free of pyridoxal phosphate. Further advances in characterization of the enzyme were made by Riley and Snell (13) and Recsei and Snell (14) who demonstrated the existence of a pyruvoyl residue and the participation of the pyruvoyl residue in histidine catalysis by forming a Schiff base intermediate in a manner similar to pyridoxal phosphate dependent enzymes. Recent studies by Hackert et al. (15) established the subunit structure of the enzyme which is similar to the subunit structure of a pyruvoyl decarboxylase of a *Micrococcus* species (16).

Amino Acid Decarboxylase Mechanisms

Amino acid decarboxylation takes place by the removal of the α-carboxyl group to give the corresponding amine. Two mechanisms of action have been identified which include a pyridoxal phosphate dependent reaction and a non-pyridoxal phosphate dependent reaction.

Pyridoxal Phosphate Dependent. A diagram of the pyridoxal phosphate reaction is given in Figure 1. Pyridoxal phosphate joined in a Schiff base linkage to the amino group of a lysyl residue forms the active site of the enzyme (17). Pyridoxal phosphate by itself can catalyze many amino acid reactions that are usually brought about by pyridoxal phosphate dependent enzymes and thus can be considered to be the portion of the enzyme that actually takes part in the reaction. The carbonyl group of pyridoxal phosphate reacts readily with amino acids to form Schiff base intermediates, which are then decarboxylated with the elimination of water to yield the corresponding amines and the original pyridoxal phosphate moiety. A mechanism for this reaction was first proposed by Werle and Koch (18) and was later modified by Mandeles et al (19), who after studies using D_2O, proposed that the hydrogen atom on the amino carbon remained fixed during the reaction, rather than being removed before decarboxylation, as suggested by Werle and Koch. The reaction takes place in the following stages: 1) When the enzyme comes in contact with an amino acid, transaldimation occurs and an adduct is formed between the amino group on the amino acid and the aldehyde group on pyridoxal -5'- phosphate to form a pyridoxylidene amino acid (Schiff base), which has two resonance forms, 2) The bond to the carboxyl group of the amino acid is cleaved, liberating CO_2 and yielding a quinonoid intermediate, 3) With the addition of a water molecule, the original site of decarboxylation is protonated and the Schiff base splits up into pyridoxal phosphate and an amine corresponding to the amino acid.

Non-pyridoxal Phosphate Dependent. Figure 2 depicts the postulated mechanism for a non-pyridoxal phosphate catalyzed decarboxylation of histidine to histamine involving a pyruvoyl residue instead of pyridoxal -5'- phosphate (20). Histidine decarboxylases from *Lactobacillus* 30a and a *Micrococcus* sp. have been shown to contain a covalently bound pyruvoyl residue on the active site. The pyruvoyl group is covalently bound to the amino group of a phenylalanine residue on the enzyme, and is derived from a serine residue (21) of an inactive proenzyme (22). The pyruvoyl residue acts in a manner similar to pyridoxal phosphate in the decarboxylation reaction.

The reaction proceeds as follows: 1) An imine (Schiff base) is formed between a carboxyl group on the bound pyruvate and the amino group of histidine, with the elimination of a water molecule, 2) The bond to the carboxyl group of histidine is cleaved, liberating CO_2, 3) The second imine intermediate so formed is hydrolyzed to the free enzyme and histamine.

Factors that Influence Bacterial Decarboxylase Production

Gale (7) identified several factors that influence the production of amino acid decarboxylases by bacteria. These factors include the following: (a) The organism must have the genetic potential to synthesize the enzyme, (b) The specific amino acid substrate needs to be supplied in the growth medium, (c) Pyridoxine and/or nicotinic acid must be supplied in the media, (d) Greatest decarboxylase activity usually occurs at acid pH levels, (e) Temperatures less than 30°C are usually optimal for decarboxylase production and (6) Enzyme activity is usually greatest at the time cell division ceases. Researchers have found instances where other factors such as oxygen tension, carbon dioxide tension, metal ions, various amino acids and other variables have affected enzyme yields and specific activities (23). Bellamy and Gunsalus (8) showed that conditions inducive to maximal growth are not always the same as the ones that give maximal enzyme activity. The following discussion will briefly review parameters that pertain to histidine decarboxylase production and its relationship to histamine formation in fish.

Bacteria that Produce Histidine Decarboxylase. Early studies demonstrated that histidine decarboxylase is widely distributed in the genera *Escherichia*, *Salmonella*, *Clostridium*, *Bacillus* and *Lactobacillus* (24). Evidence for the enzyme in *E. coli* included the finding by Gale (7) that 14 of 155 cultures possessed histidine decarboxylase. A study by Hanke and Koessler (25) found that six of 20 coliform cultures synthesized the enzyme. Havelka (26) found that 71.4% of all *Enterobacteriaceae* isolated from imported marine fish produced histamine. Eggerth (27) found 40 strains of bacteria of intestinal origin capable of decarboxylating histidine. Rodwell (11) found various

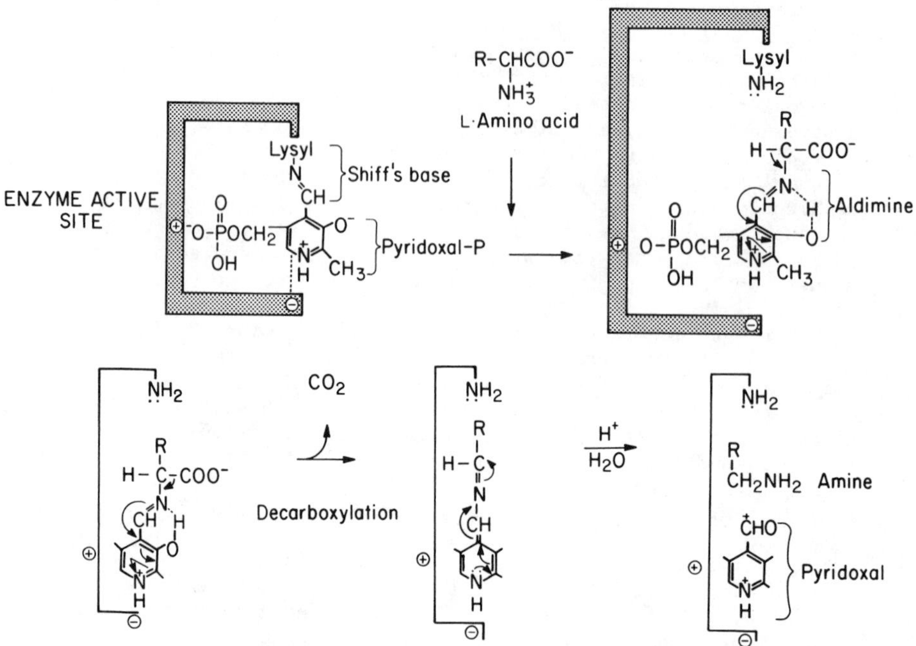

Figure 1. Pyridoxal Phosphate Dependent Decarboxylase Mechanism

Reproduced with permission from Ref. 17. Copyright 1981, John Wiley and Sons, Inc., New York.

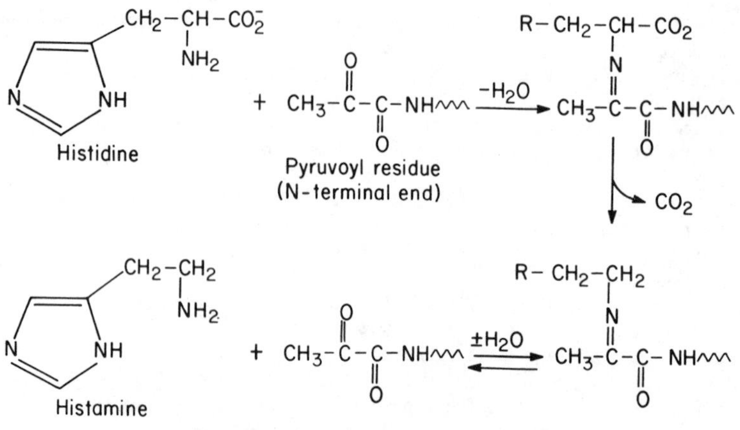

Histidine

Pyruvoyl residue (N-terminal end)

$-H_2O$

CO_2

Histamine

$\pm H_2O$

Figure 2. Pyruvoyl Dependent Decarboxylase Mechanism

Reproduced with permission from Ref. 20. Copyright 1981, Springer-Verlag, New York.

lactobacilli including Lactobacillus 30a with histidine decarboxylase activity, but the number of dairy-type lactobacilli found to contain the enzyme has been small. Horakova et al (28) isolated a Lactobacillus thought to be related to Lactobacillus acidophilus that decarboxylated histidine, and DeKoning (29) reported that a Lactobacillus was responsible for formation of toxic amounts of histamine in a Gouda cheese. Voigt and Eitenmiller (30) reported that only two of 38 dairy-related bacteria produced histidine decarboxylase activity. The two cultures that produced the enzyme were E. coli and C. perfringens that were included in the study as reference organisms with known histidine decarboxylating capability.

It appears that bacteria important to the production of toxic levels of histamine in fish are limited in number. Only species of Proteus morganii, Klebsiella pneumoniae and Hafnia alvei have been isolated from fish or fishery products involved in Scombroid poisoning incidents (26, 31, 32, 33). Of these bacteria, only P. morganii appears to universally possess the enzyme. Also, Aiso et al (34) reported that 84 of 84 P. morganii strains produced histidine decarboxylase. Likewise, Eitenmiller et al (35) showed 22 of 22 P. morganii cultures to possess the enzyme. Taylor et al (36) in a comprehensive study of 112 food-borne bacterial species found that 13 of 15 P. morganii strains produced histamine in trypticase -soy broth- histidine media to levels greater than 50 nanomoles/ml; whereas, only 3 of 12 Hafnia alvei strains produced histamine to similar levels. Taylor et al (33) in a later study isolated a histamine producing strain of Klebsiella pneumoniae from a sample of tuna sashimi implicated in an outbreak of scombroid poisoning. This strain of K. pneumoniae produced 19,900 nanomoles of histamine in 7 h of incubation in tuna fish infusion broth compared to production of less than 50 nanomoles/ml for other bacteria isolated from the sashimi. Examination of 50 additional K. pneumoniae strains revealed that 12 of the strains produced histamine in amounts approximately equivalent to that formed by the isolated strain. Recent research by Yoshinaga and Frank (37) indicated that C. perfringens and other intestinal organisms that can be liberated into the body cavity through degeneration of the intestinal tract may play an important role in histamine formation in tuna and other fish species. The significance of such anaerobic bacteria to histamine formation in fish muscle needs to be thoroughly studied.

Relationship of Bacterial Histidine Decarboxylase Production to Histamine Formation. Many studies have been completed with the objective of understanding factors such as storage time and temperature that influence production of histamine in fish. The majority of the investigations have considered only the histamine content of the product ,and, consequently, only limited information is available concerning the relationship of histidine decarboxylase formation by the microflora to histamine build-up.

Edmunds and Eitenmiller (38) in a study of the effect of storage time and temperature on histamine content and histidine decarboxylase activity of several fresh water and marine species demonstrated that the microflora developing on species such as speckled trout (*Cynoscion nebulosus*) and channel catfish (*Ictalurus punctatus*) under ambient temperature conditions will synthesize measurable histidine decarboxylase activities even though the free histidine content of the muscle is low. In Spanish mackerel (*Scomberomorus maculatus*) muscle, higher enzyme levels were observed which probably resulted from enzyme induction resulting from the relatively large amount of free histidine characteristically present in Scombroid species. Low levels of histidine decarboxylase were noted during 4°C storage trials in all species indicating that the psychrophilic microflora developing on the species included in the study would not readily lead to elevated histamine levels. Studies by Behling and Taylor (39) and Arnold et al (40) showed that histamine could reach significant levels when large numbers of viable cells of *Proteus morganii* or *Klebsiella pneumoniae* were inoculated into tuna fish infusion broth at temperatures as low as 7°C. This low temperature does not necessarily indicate the ability of such bacteria to reproduce at refrigeration temperatures but does indicate that, if an active decarboxylase is present, refrigeration will only slow the reaction and not stop histamine formation.

Eitenmiller et al (35) studied production of histidine decarboxylase and histamine by *Proteus morganii* in tuna infusion broth and in tuna fillets. The study showed that maximal enzyme activity (nanomoles CO_2/mg cells/h) was obtained in the tuna infusion broth at pH levels that tended to inhibit growth, suggesting that the protective action theory as postulated by Koessler et al (4) pertains to *P. morganii*. *P. morganii* also produced significant amounts of enzyme from pH 5.5-7.5, indicating that the normal pH of fish muscle would be suitable for production of high levels of decarboxylase. Also, maximal levels of enzyme were produced at temperatures in the ambient range. Temperature above 30°C led to lower decarboxylase levels. Omura et al (41) also indicated that *P. Morganii* produced significant levels of histidine decarboxylase at 30°C. As first observed by Bellamy and Gunsalus (8) for other bacteria, the conditions that maximized enzyme synthesis were different from those conducive to maximal cell growth. The pH and temperature optima for the decarboxylation reaction were pH 6.5 and 37°C, respectively.

Data on the effects of storage temperature on formation of histidine decarboxylase and histamine in tuna fillets are presented in Table II and Table III, respectively. The study included fillets with the normal microflora and fillets inoculated with *P. morganii* cells. Storage at 15°C resulted in only a slight increase in measurable histidine decarboxylase activity in the control fillets containing the normal microflora (Table 2). Fillets inoculated with *P. morganii* and stored at 15°C developed higher levels of histidine decarboxylase after 24 h of storage (61.6 nanomoles of CO_2/g/h). The histidine decarboxylase activity decreased slightly after 48 h of storage. At 15°C, histamine increased from 12.1 mg/100 g in the control to 21.4 mg/100 g and 43.7 mg/100 g in the uninoculated and

Table II. Histidine Decarboxylase Activity in Tuna Fish Fillets Stored at 15, 24 and 30C[a,b]

	Time (h)					
	0	6	12	18	24	48
15°C						
Uninoculated	0.7[c]	1.8	2.2	-	2.8	3.5
Inoculated	3.6	18.1	46.4		61.6	54.3
24°C						
Uninoculated	-	3.6	11.5	13.3	16.6	-
Inoculated	-	30.7	170.2	109.8	76.8	-
30°C						
Uninoculated	-	10.6	20.1	24.0	31.3	-
Inoculated	-	102.1	250.8	214.9	176.5	-

[a] Reproduced with permission from Ref. 35. Copyright 1981, International Association of Milk, Food and Environmental Sanitarians.
[b] Values represent averages of duplicate storage trials.
[c] Nanomoles of CO_2/g of muscle/h.

Table III. Histamine (mg/100 g) in Tuna Fish Fillets Stored at 15, 24 and 30C[a,b]

	Time (h)					
	0	6	12	18	24	48
15°C						
Uninoculated	12.1 (1236)[c]	13.0	14.7	-	14.9	21.4 (989)
Inoculated	-	13.3	19.6	-	31.0	43.7 (1030)
24°C						
Uninoculated	-	13.1	17.2	28.5	29.4 (1076)[c]	-
Inoculated	-	13.3	37.5	164.1	520.0 (279)	-
30°C						
Uninoculated	-	13.0	19.8	27.5	37.1 (1169)	-
Inoculated	-	13.5	52.8	280.0	608.3 (127)	-

[a] Reproduced with permission from Ref. 35. Copyright 1981, International Association of Milk, Food and Environmental Sanitarians.
[b] Values represent averages of duplicate storage trials.
[c] Histidine (mg/100 g).

inoculated tuna, respectively, after 48 h of storage (Table 3). During the same storage period, free histidine decreased from an initial level of 1236 mg/100 g to 989 mg/100 g in the uninoculated fillets and 1030 mg/100 g in the inoculated fillets. The slight decreases in free histidine in the control are most likely due to microbial utilization and loss of the amino acid in small amounts of drip that occurred unavoidably during storage. The increase in histamine concentration at 15°C was not large enough to account for the lowering of free histidine in the fillets. However, the observation that the histamine concentration reached 43.7 mg/100 g in the fillets inoculated with P. morganii and stored at 15°C indicates that histamine could be formed to appreciable and even toxic levels at storage temperatures not optimal for bacterial growth or histidine decarboxylase activity.

Rapid enzyme and histamine formation occurred in the inoculated fillets stored at 24 and 30°C. (Table 2 and Table 3). Maximal histidine decarboxylase activity occurred after 12 h of storage at both temperatures and decreased threafter. Histamine levels reached 520 mg/100 g and 608 mg/100 g at 24 and 30°C after 24 h of storage. At 24°C , the free histidine content decreased to 279 mg/100 g. In the uninoculated fillets, histidine decarboxylase activity was somewhat higher than noted in fillets stored at 15°C; however, histamine content remained low after 24 h at 24°C and 30°C.

This study along with several other investigations (38, 39, 40, 42), demonstrates that storage temperature is a critical factor influencing formation of histamine in fish muscle. The requirement for rapid and uninterrupted refrigeration after catch cannot be over-emphasized. As shown for P. morganii (35), normal muscle pH conditions are close to pH levels required for the optimal activity of histidine decarboxylase (pH 6.5) from P. morganii and are low enough to permit rapid enzyme synthesis. The free histidine in the muscle of mahi-mahi and Scombroid fish can act both as an enzyme inducer and a substrate. If the microflora on the fish includes a strong histidine decarboxylase organism such as P. morganii, accumulation of toxic levels of histamine can rapidly occur.

Literature Cited

1. Arnold, S. H.; Brown, W. D. In "Advances in Food Research"; Chichester, C. O;, Mrak, E. M.; Stewart, G. F., Eds.; Academic: New York, 1978; Vol. 24, p. 113.
2. Boeker, E. A.; Snell, E. E. In "The Enzymes"; Boyer, P.D., Ed.; Academic: New York, 1972; Vol. VI, p. 217.
3. Snell, E. E.; Recsei, P. A. In "Structural and Functional Aspects of Enzyme Catalysis", 1981, p. 177.
4. Koessler, K. K.; Hanke, M. T.; Sheppard, M. S. J. Infect. Dis. 1928, 43, 363.
5. Gale, E. F.; Epps, H. M. R. Biochem. J. 1944, 38, 232.
6. Baddiley, J.; Gale, E. F. Nature 1945, 155, 727.
7. Gale, E. F. In "Advances in Enzymology"; Nord, F. F., Ed.; Interscience: New York, 1946; Vol. VI, p. 1.

8. Bellamy, W. D.; Gunsalus, I. C. J. Bacteriol. 1944, 48, 191.
9. Heyl, D; Luz, E.; Harris, S. A.; Folkers, K. J. Am. Chem. Soc. 1951, 73, 3436.
10. Shukuya, R.; Schwert, G. W. J. Biol. Chem. 1960, 235, 1653.
11. Rodwell, A. W. J. Gen. Microbiol. 1953, 8, 233.
12. Rosenthaler, J.; Guirard, B. M.; Chang, G. W.; Snell, E. E. Proc. Natl. Acad. Sci. U.S., 1965, 54, 152.
13. Riley, W. D.; Snell, E. E. Biochemistry 1968, 7, 3520.
14. Recsei, P.; Snell, E. E. Biochemistry 1970, 9, 1492. 15.
15. Hackert, M. L.; Meador, W. E.; Oliver, R. M.; Salmon, J. B; Recsei, P. A.; Snell, E. E. J. Biol. Chem. 1981, 256, 687.
16. Gonchar, N. A.; Katsnel'son, A. A.; L'Vov, Y. M.; Semina, L. A.; Feigin, L. A. Biofizika 1977, 5, 801.
17. Conn, E. E.; Stumpf, P. K. "Outlines of Biochemistry"; John Wiley and Sons: New York,
18. Werle, E.; Koch, W. Biochem. Z. 1949, 319, 305.
19. Mandeles, S.; Koppelman, R.; Hanke, M. E. J. Biol. Chem. 1954, 209, 327.
20. Snell, E. E.; Recsei, P. A.; Misono, H. In "Metabolic Interconversion of Enzymes 1975"; Shaltiel, S., Ed.; Springer-Verlag: New York, 1976, p. 213.
21. Recsei, P. A.; Snell, E. E. Biochemistry 1973, 12, 365.
22. Recsei, P. A.; Snell, E. E. In "Metabolic Interconversion of Enzymes 1980"; Holzer, H., Ed.; Springer-Verlag: New York, 1981; p. 335.
23. Voigt, M. N. M.S. Thesis, University of Georgia, Athens, 1973.
24. Ienistea, C. In "The Microbiological Safety of Food"; Hobbs, B. C.; Christian, J. B., Eds.; Academic: London, 1973 p. 327.
25. Hanke, M. T.; Koessler, K. K. J. Biol. Chem. 1924, 59, 855.
26. Havelka, B. Cesk. Hyg. 1967, 12, 343.
27. Eggerth, A. H. J. Bacteriol. 1939, 37, 205.
28. Horakova, Z.; Zierdt, C. H.; Beaven, M. A. Eur. J. Pharmacol. 1971, 16, 67.
29. DeKoning, P. J. Neth. Milk Dairy J. 1968, 22, 153.
30. Voigt, M. N.; Eitenmiller, R. R. J. Food Protect. 1977, 40, 241.
31. Kawabata, T; Ishizaka, T.; Miura, T.; Sasaki, T. Bull. Jpn. Soc. Sci. Fish. 1956, 22, 41.
32. Sakabe, Y. J. Nara Med. Assn. 1973, 24, 248.
33. Taylor, S. L.; Guthertz, L. S.; Leatherwood, M.; Lieber, E. R. Appl. Environ. Microbiol. 1979, 37, 274.
34. Aiso, K.; Iida, H.; Nakayama, J.; Nakano, K. Chiba Daigaka Fuhai Kenkyusho Hokoku 1958, 11, 1.
35. Eitenmiller, R. R.; Wallis, J. W.; Orr, J. H.; Phillips, R. D. J. Food Protect. 1981, 44, 815.
36. Taylor, S. L.; Guthertz, L. A.; Leatherwood, M.; Tillman, F.; Lieber, E. R. J. Food Safety 1978, 1, 173.
37. Yoshinaga, D. H.; Frank, H. A. Appl. Environ. Microbiol. 1982, 44, 447.

38. Edmunds, W. J.; Eitenmiller, R. R. J. Food Sci. 1975, 40, 516.
39. Behling, A. R.; Taylor, S. L. J. Food Sci. 1982, 47, 1311.
40. Arnold, S. H.; Price, R. J.; Brown, W. D. Bull. Jpn. Soc. Sci. Fish. 1980, 46, 991.
41. Omura, Y; Price, R. J; Olcott, H. S. J. Food Sci. 1978, 43, 1779.
42. Frank, H. A.; Yoshinaga, D. H.; Nip, W. Mar. Fish. Rev. 1981, 43, 9.

RECEIVED February 6, 1984

Histamine Formation in Tuna

HILMER A. FRANK and DERRICK H. YOSHINAGA

Department of Food Science and Human Nutrition, University of Hawaii, Honolulu, HI 96822

Spoilage of skipjack tuna was studied under controlled incubation conditions, using histamine formation as an indicator of decomposition. The bacteria in skipjack loin tissue were enumerated during incubation at 38°C, the optimum decomposition temperature, and representative spoilage organisms were identified. Histamine-forming isolates included four Gram-negative, facultatively anaerobic bacteria (*Enterobacter aerogenes*, *Klebsiella pneumoniae*, *Proteus mirabilis*, and *Vibrio alginolyticus*) and a single anaerobic sporeformer, *Clostridium perfringens*. Several isolates were used to determine if histamine-forming bacteria could grow and if whole cell suspensions could decarboxylate histidine at low temperatures. At 4°C *Klebsiella pneumoniae* grew in skipjack infusion broth and its resting cells produced histamine from histidine. Although *Clostridium perfringens* strains did not grow at low temperatures, cells grown at 38°C were able to produce histamine at 4°C.

Decomposition of tuna and other scombroid marine fish (albacore, bonito, mackerel, sardines, etc.) is usually characterized by the formation of high levels of histamine (1-3), sometimes in excess of 5 mg per g of loin tissue. Since fresh tuna has essentially no free histamine (4-8), the presence of histamine is considered to be an indication of decomposition (8-10). Histamine is particularly suitable as an indicator of prior spoilage in canned tuna (2, 4, 11-13) because its heat stability allows histamine to survive commercial heat processing.

Histamine is produced by bacteria that decarboxylate histidine (1-3, 10, 14-17), an amino acid that is abundant in the muscle of scombroid fish and other marine fish such as mahimahi, the dolphinfish (7, 18-25). Histidine decarboxylating bacteria have been isolated from spoiled marine fish (1, 2, 17, 25-27) and from fish involved in food poisoning incidents (1-3, 10, 16, 28-31). With the exception of *Clostridium perfringens* and *Vibrio alginolyticus* found

0097-6156/84/0262-0443$06.00/0
© 1984 American Chemical Society

in skipjack tuna (25), histamine-forming bacteria isolated from decomposed fish belong to the family Enterobacteriaceae. Histamine-forming bacteria obtained from fish implicated in scombroid poisoning also were Enterobacteriaceae, and these included Proteus morganii (1, 28), Klebsiella pneumoniae (30, 31), and Hafnia alvei (24, 29).

The results given below were collected in a study of decomposition in skipjack tuna (Katsuwonus pelamis) under controlled incubation conditions and were described in part previously (8, 25). Skipjack was chosen as the test species for this investigation because:

(1) Skipjack contains the highest histidine levels reported among various tunas (20, 24); consequently, histamine formation in decomposed skipjack should be relatively high.

(2) An adequate supply of live skipjack tuna was available locally. By incubating the fish immediately after expiration, we avoided undesired postmortem spoilage changes caused by improper handling before incubation.

Experimental Procedures

Material and methods were described in detail previously (8, 25). Live skipjack tuna were allowed to expire, placed separately in plastic bags containing seawater, and incubated for the times desired at various temperatures. After incubation, samples of tissue were removed for various bacteriological analyses and histamine estimations.

Results and Discussion

Fresh skipjack tuna contained about 0.1 mg of histamine per 100 g (8).

Incubation Temperature. The effect of temperature, the most important external factor affecting histamine formation, has been studied in a number of scombroid fish and bacterial cultures (1-3, 6-8, 27, 32-34). However, variations in types of fish, handling procedures, and test organisms have contributed to wide differences in the results reported.

Figure 1 shows the extent of histamine formation in whole skipjack tuna at temperatures over the range of 15.6°C to 48.9°C. After incubation for 24 h, samples were taken from the anterior section of loin in each fish and assayed for histamine content. The optimum temperature for histamine formation in skipjack tuna was 38°C, where 643 mg of histamine were present per 100 g of tuna (8).

Figure 1 also shows that histamine formation is negligible at about 50°C and below 20°C. However, when longer incubations were used, histamine was produced at low temperatures. For example, we found 16.9 mg per 100 g after 12 days at 10°C, 7.3 mg per 100 g after 24 days at 4.4°C, and 2 mg of histamine per 100 g after 42 days at -1.1°C (35).

Histamine Distribution in Decomposed Tuna. The level of histamine produced is not uniform throughout decomposed fish (2, 30, 36). In spoiled skipjack tuna, we found a histamine gradient that was highest in the anterior section and decreased gradually in sections approaching the tail (8, 25).

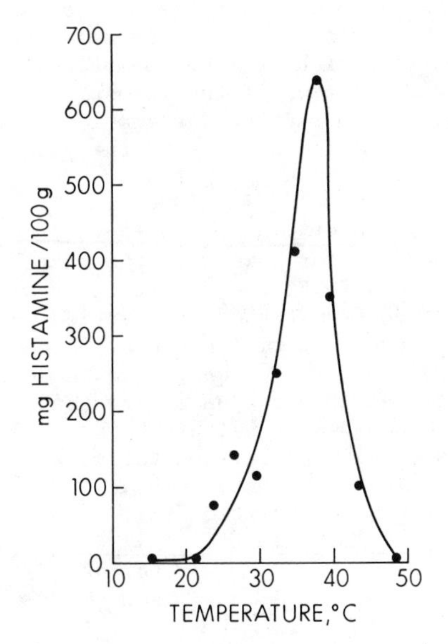

Figure 1. Effect of incubation temperature on histamine formation in skipjack tuna.

Figure 2 shows the amounts of histamine in four sections of skipjack tuna incubated at 38°C, the optimum temperature for decomposition (Figure 1). Histamine was found first in the anterior section, exceeding 100 mg per 100 g at 12 h, and reached 680 mg by 24 h. Histamine production was slower in the remaining sections and did not reach significant levels until 18 h of incubation had elapsed. A histamine gradient was evident at 24 h, with the highest level in the anterior section.

The intestine is believed to be a major source of bacteria responsible for histamine formation in skipjack tuna (25). The higher level and rate of histamine formation in the anterior section are related to the location of the intestinal tract in the forward end of the fish. Postmortem disintegration of the intestine liberates its microbial contents into the visceral cavity and anterior muscle tissue.

Bacterial Growth and Histamine Formation. The quantitative relationship between histamine formation and the microbial flora in skipjack tuna at 38°C is shown in Figure 3. At intervals during incubation samples were removed from the second section and assayed for histamine content and bacterial numbers. After 24 h the anaerobic bacterial count was 3.5×10^5 per g, and the histamine content was 297 mg per 100 g. Anaerobic counts were used to measure the microbial population because over 92% of the bacteria found in decomposed skipjack tuna were obligate or facultative anaerobes (Table I).

Table I. Bacteria Isolated From Decomposed Skipjack Tuna

Organisms	No. of isolates	No. of histidine decarboxylase-positive isolates
Obligately aerobic		
Micrococcus luteus	6	0
Planococcus citreus	4	0
Obligately anaerobic		
Clostridium bifermentans	1	0
C. botulinum type C	2	0
C. ghoni	7	0
C. mangenotii	11	0
C. novyi type B	1	0
C. perfringens	9	9
C. sardiniensis	1	0
Facultatively anaerobic		
Aeromonas spp.	30	0
Enterobacter aerogenes	1	1
Klebsiella pneumoniae	2	2
Pediococcus halophilus	2	0
Proteus mirabilis	3	3
Vibrio alginolyticus	23	3
V. anguillarum	31	0

Reproduced with permission from Ref. 25.
Copyright 1982, American Society for Microbiology.

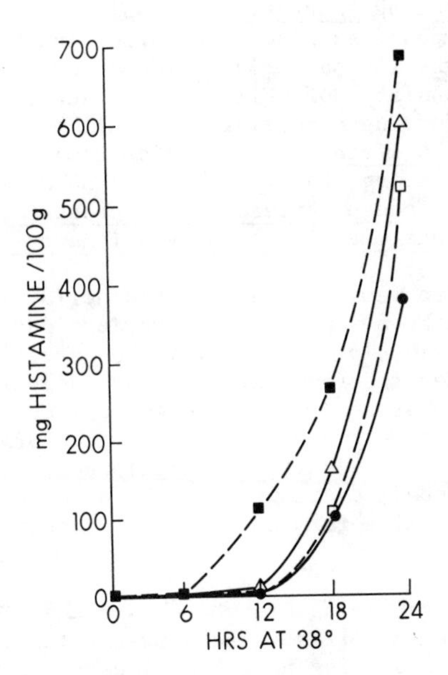

Figure 2. Histamine distribution in sections of skipjack tuna at 38 °C. Section legends: ■ = anterior; △ = second; □ = third; ● = posterior.

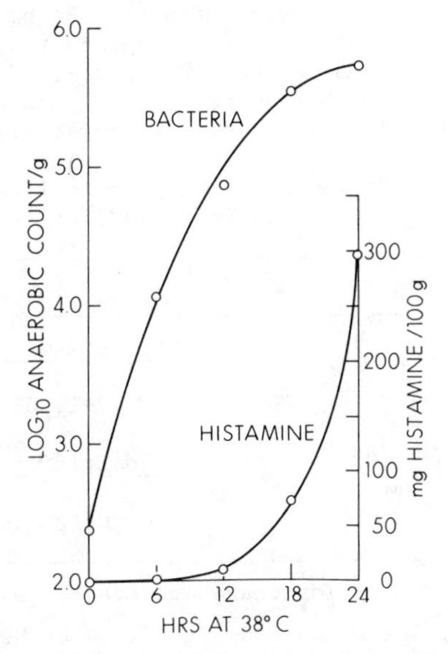

Figure 3. Bacterial growth and histamine formation in skipjack tuna at 38 °C. Reproduced with permission from Ref. 25. Copyright 1982, American Society for Microbiology.

Identification of Spoilage Microflora. Table I lists the bacteria recovered from decomposed skipjack tuna after incubation for 24 h at 38°C. Eighteen of the 134 bacteria isolated were histamine formers, and these strains consisted of obligate and facultative anaerobes. The nine histamine-forming clostridia were all members of a single species, *Clostridium perfringens*. The nine remaining histamine formers were Gram-negative, facultative anaerobes belonging to the families *Enterobacteriaceae* (*Enterobacter aerogenes*, *Klebsiella pneumoniae*, and *Proteus mirabilis*) and *Vibrionaceae* (*Vibrio alginolyticus*).

Many bacteria can decarboxylate histidine (3, 15-17, 27), and a number of these have been isolated from various decomposed and toxic marine fish (1-3, 8, 10, 16, 17, 25-31). To date, all histamine forming bacteria recovered from marine fish, including *Clostridium perfringens* (37), are present in the intestinal contents of man and other animals. Kimata estimated that histamine-forming bacteria constituted about 1% of the surface microflora of live marine fish (1). However, few detailed studies have been conducted to determine the incidence of histamine-forming bacteria among marine fish and environments (3).

Low Temperatures. After being brought aboard the fishing vessel, tuna usually are kept in refrigerated seawater (about 0°C) until they can be frozen (33). Depending upon the times and temperatures employed, considerable histamine may form in the tuna during refrigerated storage.

A number of studies have dealt with histamine formation at low temperatures in several fish and in bacterial cultures (3, 6, 7, 27, 32, 33, 38, 39), but with varying results. We have observed histamine formation in skipjack tuna under controlled incubation conditions at temperatures between 10°C and -1.1°C (35). We also studied growth of several histamine-forming bacteria at low temperatures (Table II). Histamine-forming strains were inoculated

Table II. Growth of Histamine-forming Bacteria at 4°C in Tuna Fish Infusion Broth (25)

Organisms	
Growth	No Growth
Klebsiella pneumoniae	*Proteus mirabilis*
UH-1[a], UH-2, no. T2[b]	UH-1, UH-2
Proteus morganii no. 180[c]	*Clostridium perfringens*
	UH-1, UH-2, UH-3

[a]UH = strains isolated by Yoshinaga and Frank (25).

[b]From S. L. Taylor, Letterman Army Institute of Research, San Francisco.

[c]From D. A. Corlett, Jr., Del Monte Research Center, Walnut Creek, Calif.

into tuna fish infusion broth (25) and incubated for 9 days at 4°C. Growth was determined by the appearance of turbidity in the cultures and by microscopic examination. Table II shows that K. pneumoniae strains and P. morganii grew at 4°C, but that none of the P. mirabilis or C. perfringens grew.

Table III. Histamine Formation at 4°C by Whole Cell Suspensions of Selected Bacteria

Organisms	Histamine formed, µg per 10^7 cells
Proteus morganii no. 180[a]	0.01
Klebsiella pneumoniae	
UH-1[b]	0.54
UH-2	0.88
no. T2[c]	0.36
Clostridium perfringens	
UH-1	9.0
UH-3	13.7

[a]From D. A. Corlett, Jr., Del Monte Research Center, Walnut Creek, Calif.

[b]UH = strains isolated by Yoshinaga and Frank (25).

[c]From S. L. Taylor, Letterman Army Institute of Research, San Francisco.

Histamine formation at 4°C was studied with resting cell suspensions of several histidine decarboxylating bacteria (Table III). Cells were harvested from cultures grown at 38°C in Trypticase soy broth (BBL, Cockeysville, MD.), suspended in 0.2 M phosphate buffer (pH 6.0) containing 0.1% histidine, and incubated anaerobically at 4°C for 21 days. After incubation, the cells were removed by filtration, and histamine was measured in the supernatant liquid. Of the two species capable of growing at low temperatures (Table II), histamine formation at 4°C was negligible with P. morganii cells and moderate with K. pneumoniae. However, cells of C. perfringens, which did not grow at low temperatures (Table II), were active producers of histamine at 4°C.

These results suggest that histamine in refrigerated tuna could be formed by:

(1) bacteria such as Klebsiella pneumoniae which can grow as well as decarboxylate histidine at low temperatures; or

(2) bacteria such as Clostridium perfringens which grow at moderate temperatures but produce enzymes that decarboxylate histidine at low temperatures.

Conceivably, histamine might be produced during a fishing voyage when fish are stored under inadequate refrigeration in the storage wells of tuna seiners.

Acknowledgments

Journal series no. 2819 of the Hawaii Institute of Tropical Agriculture and Human Resources. This investigation was supported by contract no. 03-6-208-35369 from the National Marine Fisheries Service, NOAA, U. S. Department of Commerce.

Literature Cited

1. Kimata, M. In "Fish as Food. Vol. 1. Production, Biochemistry, and Microbiology"; Borgstrom, G., Ed.; Academic Press: New York, 1961, p. 329-352.
2. Ienistea, C. In "The Microbiological Safety of Food"; Hobbs, B. C.; Christian, J. H. B., Ed.; Academic Press: London, 1973; p. 327-343.
3. Arnold, S. H.; Brown, W. D. Adv. Food Res. 1978, 24, 113-154.
4. Geiger, E. Food Res. 1944, 9, 293-297.
5. Geiger, E.; Courtney, G.; Schnakenberg, G. Arch. Biochem. 1944, 3, 311-319.
6. Hardy, R.; Smith, J. G. M. J. Sci. Food Agric. 1976, 27, 595-599.
7. Fernandez-Salguero, J.; Mackie, I. M. J. Food Technol. 1979, 14, 131-139.
8. Frank, H. A.; Yoshinaga, D. H.; Nip, W.-K. Marine Fisheries Rev. 1981, 43(10), 9-14.
9. Ferencik, M.; Krcmery, V.; Kriska, J. J. Hyg., Epidemiol., Microbiol., Immunol. 1961, 5, 341-348.
10. Hobbs, G.; Hodgkiss, W. Devel. Food Microbiol. 1982, 1, 71-117.
11. Williams, D. W. J. Assoc. Offic. Agric. Chemists 1954, 37, 567-572.
12. Hillig, F. J. Assoc. Offic. Agric. Chemists 1954, 37, 927-931.
13. Lieber, E. R.; Taylor, S. L. J. Chromatog. 1978, 153, 143-152.
14. Geiger, E. Arch. Biochem. 1948, 17, 391-395.
15. Mossel, D. A. A. In "The Safety of Foods"; Graham, H. D., Ed.; Avi Publishing Co.: Westport, Conn., 1968; p. 168-182.
16. Taylor, S. L.; Guthertz, L. S.; Leatherwood, M.; Tillman, F.; Lieber, E. R. J. Food Safety 1978, 1, 173-187.
17. Niven, C. F., Jr.; Jeffrey, M. B.; Corlett, D. A., Jr. Appl. Env. Microbiol. 1981, 41, 321-322.
18. Shewan, J. M. J. Sci. Food Agric. 1955, 6, 99-104.
19. Simidu, W.; Hibiki, S. Bull. Japan. Soc. Sci. Fisheries 1955, 21, 361-364.
20. Lukton, A.; Olcott, H. S. Food Research 1958, 23, 611-618.
21. Hughes, R. B. J. Sci. Food Agric. 1959, 10, 558-564.
22. Hibiki, S.; Simidu, W. Bull. Japan. Soc. Sci. Fisheries 1959, 24, 916-919.
23. Takagi, M.; Iida, A.; Murayama, H.; Soma, S. Bull. Fac. Fisheries Hokkaido Univ. 1969, 20, 227-234.
24. Ferencik, M. J. Hyg., Epidemiol., Microbiol., Immunol. 1970, 14, 52-60.
25. Yoshinaga, D. H.; Frank, H. A. Appl. Env. Microbiol. 1982, 44, 447-452.
26. Omura, Y.; Price, R. J.; Olcott, H. S. J. Food Sci. 1978, 43, 1779-1781.

27. Arnold, S. H.; Price, R. J.; Brown, W. D. Bull. Japan. Soc. Sci. Fisheries 1980, 46, 991-995.
28. Kawabata, T.; Ishizaka, K.; Miura, T.; Sasaki, T. Bull. Japan. Soc. Sci. Fisheries 1956, 22, 41-47.
29. Havelka, B. Cesk. Hyg. 1967, 12, 343-352. Chem. Abstr. 1968, 21019g.
30. Lerke, P. A.; Werner, S. B.; Taylor, S. L.; Guthertz, L. S. West. J. Med. 1978, 129, 381-386.
31. Taylor, S. L.; Guthertz, L. S.; Leatherwood, M.; Lieber, E. R. Appl. Env. Microbiol. 1979, 37, 274-278.
32. Edmunds, W. J.; Eitenmiller, R. R. J. Food Sci. 1975, 40, 516-519.
33. Behling, A. R.; Taylor, S. L. J. Food Sci. 1982, 1311-1314, 1317.
34. Eitenmiller, R. R.; Orr, J. H.; Wallis, W. W. In "Chemistry & Biochemistry of Marine Food Products"; Martin, R. E.; Flick, G. J.; Hebard, C. E.; Ward, D. R., Ed.; Avi Publishing Co.: Westport, Conn., 1982; p. 39-50.
35. Frank, H. A.; Yoshinaga, D. H., unpublished data.
36. Hillig, F. J. Assoc. Offic. Agric. Chemists 1956, 39, 773-800.
37. Smith, L. DS.; Holdeman, L. V. "The Pathogenic Anaerobic Bacteria"; Charles C Thomas: Springfield, Ill., 1968; p. 203.
38. Kimata, M.; Kawai, A. Mem. Res. Inst. Food Sci., Kyoto Univ. 1953, 5, 25-54.
39. Shewan, J. M.; Liston, J. J. Appl. Bacteriol. 1955, 18, 522-534.

RECEIVED March 2, 1984

Author Index

Subject Index

Production by Anne Riesberg
Indexing by Karen McCeney
Jacket design by Pamela Lewis

Elements typeset by Hot Type Ltd., Washington, D.C.
Printed and bound by Maple Press Co., York, Pa.

NOV 05 1984